Identification of Onset, End, Magnitude
(Spi and Spei) and Composite Index

DROUGHT EARLY WARNING SYSTEM

Identification of Onset, End, Magnitude
(Spi and Spei) and Composite Index

DROUGHT EARLY WARNING SYSTEM

MUNI RATNAM PANTULA

Notion Press

Old No. 38, New No. 6
McNichols Road, Chetpet
Chennai - 600 031

First Published by Notion Press 2016
Copyright © Muni Rathnam Pantula 2016
All Rights Reserved.

ISBN 978-1-945497-42-1

This Book is dedicated to the Memory of my Father and Mother
Sri P.Subbanna & Smt. Subbamma & My Wife P.Prema Kumari

CONTENTS

Dr. N. Janardhan Raju, M.Sc., Ph.D.
Professor-Alexander von Humboldt Fellow
School of Environmental Sciences
Jawaharlal Nehru University,
New Delhi, INDIA-110067

Phone: 09910629336
Email: rajunj7@gmail.com

FOREWORD

I am very glad to review the Book authored by Sri P. Muniratnam a Civil Engineer and Hydrologist. He introduced and institutionalized a new comprehensive approach to drought management system which is based on the technological advances and new innovations in drought onset, duration, end along with severity, relief and mitigation in his book.

Drought is among the most climatic phenomena, affecting society and environment. The approach to drought is generally reactive and tends to focus on crisis management in India. India has become more drought prone, intense and frequent as a result of climate change over the past quarter-century. What is required at present is monitoring and early warning systems to deliver timely information to decision makers. It is imperative to shift from managing crisis to preparing for droughts.

In the past, drought management strategies were worked out generally during and after the onset of drought which lacked preventive interventions. They did not integrate new technologies for early warning nor did it emphasize mitigation as an essential element of drought management.

The Present book encompasses the characterisation of meteorological drought by newly invented index called "SPI-Standardised Precipitation index" approved by World Meteorological Organization (WMO) during June 2011. It is a simple index with Precipitation is the only Parameter and can be computed for different scales (1-3-6-12-24 months) and compared across regions of different climatic zones. It identifies meteorological drought at short term scales 2 to 3 months and hydrological drought at longer scales 12 and 24 months. In the context of global warming on drought conditions characterisation has also been done with a new variation of Standardized Precipitation Index (SPI) called "Standardized Precipitation Evapotranspiration Index (SPEI). It is also multi-scalar index.

The author has depicted graphs with regard to trends, onset, end, magnitude with dates of drought occurrence for 102 years of rainfall and temperature with the concept of SPI & SPEI, for Anantapur District of Andhra Pradesh, India.

To characterise the agricultural drought climatological water balance has been carried out in this book for 30 years for Anantapur district, A.P. This is adopted as per international seminar on agriculture drought at Genava June 2010 conveyed by WMO.

Eight drought Parameters were computed from the water balance study, on the basis of book keeping procedure of Thornth Waite and Mathur 1955. Analysis of drought and their intensities are made from the percentage departure of aridity index (Ia) from the median value. Categorization of droughts, Decennial frequency of droughts, climate shifts for planning short term operations with regard to agricultural development. Based on the percentage of moisture values the suitability of crops that can be grown in absence of supplemental irrigation has been dealt in this book.

The space Technology for agricultural drought monitoring has been discussed in this book about Normalized Difference Vegetation Index (NDVI) deviation from historic NDVI and VCI (Vegetation Condition index).

Hydrological drought indices based on Precipitation, stream flow, low flows, run off, ground water levels and water balance are discussed in this book as a separate chapter. Remote sensing data such as NDVI, WSVI along with SPI and SWI has been used to identify the hydrological drought. The conclusions for such a decision has been discussed in this book.

A Separate chapter on Socio-economic drought also included in the book which relates the supply and demand of goods and services with the above three mentioned types of droughts.

Regarding to meteorological drought characterising a consensus has been reached and recommended to use SPI as per Lincoln declaration recommendation on drought indices and early warning system of drought during December 2009 and recommended for creating a new "Composite Index" for characterizing agricultural & hydrological droughts since various indices presently used suffer from various lucunas and consensus was not arrived among 22 countries including India.

So it is imperative to have a combined index along with some remote sensing indices to characterize the respective droughts just like U.S. drought monitor which incorporates multiple indices and indicators of drought including impacts into assessment process fully.

In this book a "Composite Index" to be called as "Indian drought monitor" with ten indicators and indices has been developed for releasing drought information weekly considering and incorporating review from a group of climatologists, extension agents and others across the nation. This will lead the country economically forward. Another chapter regarding mitigation of drought is dealt in this book. This book is useful for University students and Researchers who are working in the fields of hydrology, hydrogeology, climatology, agricultural engineering and authorities to serve the people of this nation.

As an immediate measure, it is suggested that "Standardized Precipitation Evapotranspiration Index" (SPEI) may be developed based on the computer programme given in this book as input parameters precipitation, temperature and latitude of site for each village having 30 years of monthly data from the month required (backward) and drought occurrence depicted on the graph drawn between SPEI values and years. SPEI maps for each district with village data as one station for 1-3-6-12-24 monthly scales, different maps may be generated by using Grid analysis and display systems (GrADS). The discrete station SPEI data are interpolated using Crossman objective analysis. The grid resolution is 0.4 degrees. The preparedness plan for each village depending on SPEI values is made and depending on the severity values of drought, priority list of villages for taking up mitigation is prepared and action taken for redrassal.

This process may be continued till all the required data required for indicators and indices for development of "Composite index" as suggested in the book is collected and additional rainfall gauge and climatological stations setup for new villages.

This is a welcome suggestion put forth by the author in this book. Hence I earnestly hope that IMD and state Governments will take appropriate action in this regard in the greater well-being of the people of this nation.

Dr. N. Janardhan Raju, M.Sc., Ph.D.
Professor-Alexander von Humboldt Fellow
School of Environmental Sciences
Jawaharlal Nehru University,
New Delhi, INDIA-110067

1

PREFACE AND INTRODUCTION TO FIRST EDITION

CHAPTER OUTLINE

- Drought identified after onset at present in India. New Technologies for early warning (i.e.) SPI, SPEI are introduced for this purpose. They work for all places, times and for all types of droughts.

"In Many Parts of the world, the approach to droughts is generally reactive and tends to focus on crisis management. But to the national and regional scale, responses are known to be often untimely, poorly coordinated and lacking the necessary integration. As a result, the economic Social and environmental impacts of droughts have increased significantly, in many regions of the world. We simply cannot afford to continue in a piecemeal mode, driven by crisis rather than prevention. We have the knowledge, we have the experience and we can reduce the impact of droughts. Without coordinated national drought Policies, nations will continue to respond to drought in a reactive way. **What we need are monitoring and early warning systems to deliver timely information to decision makers. We must have effective impact assessment Procedures, Proactive risk management measures, preparedness plans to increase coping capabilities and effective emergency response programmes to reduce the impact of drought".**

(During the opening Session of the of the high-level meeting on National drought Policy in March 2013 – the secretary - General of the world meteorological organization, Michel Jarraud – Statement).

"Over The Past Quarter — Century, the world has become more drought – Prone and droughts are projected to become more wide spread, intense and frequent as a result of climate change. The long-term impacts of prolonged drought on eco-systems are profound, accelerating, land degradation and desertification. The consequences include impoverishment and the risk of local conflict over water resources and productive land.

Droughts are hard to avert, but their effects can be mitigated. Because they rarely observe national borders, they demand a collective response.

The price of preparedness is minimal compared to the cost of disaster relief. Let us therefore Shift from managing crises to preparing for droughts and building resilience by fully implementing the out comes of the High-level meeting on National drought Policy held in Geneva last march".

(In 2013, the Secretary – General of the United Nations. Ban Ki-moon-Statement).

Drought is an insidious (treacherous) hazard of nature. It is a normal, recurrent feature of climate. It occurs almost every where, although its features vary from region to region. Drought is usual dryness of soil resulting in crop failure and shortage of water for other uses, caused by significantly lower rainfall than average over a prolonged period.

Hot dry winds, shortage of water, high temperature and consequent evaporation of moisture from the ground can contribute to conditions of drought. Scientists still to know why weather patterns change. However, this can be caused by changes in the oceans. Every few years, a change in sea temperatures in eastern pacific, called EL Nino affects Weather around the world and causes drought.

Droughts are disastrous for people, animals and plants. A shortage of rain means, crops can not grow properly and herds of animals cannot get enough drinking water. So People face food and water shortages. Dried – out grass and trees can easily catch fire and loose dust can blow up into blinding dust storms – Droughts can also cause wars, where people are forced to leave their lands and flock into other areas.

EL Nino is a warming of surface of ocean waters in the eastern pacific that can lead to flooding and drought around the world. Throughout its history earth has warmed up and cooled down. Experts think that today's warming is down to humans and its happenings faster than normal. Carbon dioxide and methane gases are released into the air as pollution. They are known as green house gasses and can stop the sun's heat escaping from the atmosphere.

A Global Warming happens, when green house gases collect in the earth's atmosphere. They let the heat from the sun through, but as it bounces back it gets trapped close to the earth, making the planet heat up.

Rising global temperatures, appear to be a major factor, provoking more frequent and intense droughts in sub-tropical areas of Asia and Africa. The climate models predict a global worming of about 1.4°C–5.8°C between 1990 & 2100 and the sea level to rise by 9 to 88 cms. In this book a chapter on weather has been dealt.

Drought in India has resulted, in tens of millions of deaths over the course of the 18th or 19th and 20th centuries. Indian agriculture is heavily dependent on the climate of India. A favourable south west summer is critical in securing water for irrigating Indian crops.

In some parts of India the failure of the monsoons result in water shortages resulting in below–average crop yields. This in particularly true of major drought prone regions such as southern and eastern Maharashtra, northern Karnataka, Andhra Pradesh, Odissa, Gujarat and Rajasthan. In the past droughts have periodically led to major Indian famines including the Bengal famine of 1770, in which upon one third of population in affected areas died. The 1876–1877 famine in which, over five million people died and the 1899 famine in which, over 4.5 million died. (From Wikipedia, the free encyclopedia).

All such episodes of severe droughts correlate with Elnino–Southern oscillation (ENSO) events. Elnino–Related droughts have been implicated in periodic declines in Indian agricultural output. Never the less, ENSO events that have coincided with abnormally high sea surface temperatures in the Indian ocean in one instance during 1997 and 1998 by up to 3°C (5°F) have resulted in increased oceanic evaporation resulting in unusually wet weather across India.

Such anomalies have occurred during a sustained Warm spell that began in the 1990s. A Contrasting phenomenon is that instead of the usual high pressure air mass over the Southern Indian Ocean, an ENSO related oceanic low pressure convergence center forms, it then continually pull dry air from central Asia, desiccating India during what should have been the humid summer monsoon season.

This reversed air flow causes India's droughts. The extent that an ENSO event raises sea surface temperatures in the Central Pacific Ocean influences the degree of drought. Around 43% of Elnino events are followed by drought in India.

Generally, rainfall is related to be a mount (determined by air temperature) of water vapour carried by regional atmosphere phone, combined with the upward forcing of the air mass containing that water vapour.

If these combined factors do not support precipitation, volumes sufficient to reach the surface, the result is a drought. Drought has many causes. It can be caused by not receiving rain or snow over a period of time. Weather that changes in the wind patterns that move clouds and moisture through the atmosphere can cause a place not to receive its normal amount of rain or snow over a long period of time.

The fourth assessment report of the inter Governmental Panel on climate change (IPCC) stated that the world has been more drought prone

during the past 25 years and that climate projections indicate an increased frequency in the future. This carries significant implications for the agriculture sector especially in the developing countries like India.

Today the world is facing a great water crises than ever. Droughts of lesser magnitude are resulting in greater impacts. Even in years with normal precipitation water shortage has become wide spread in both developing and developed nations in humid as well as arid climates. When faced with severe drought Governments become eager to act. Unfortunately this eagerness usually wanes when precipitation returns to normal.

The economy of India is greatly dependent on water resources as well as rainfall. The erratic nature of monsoon rainfall give rise to low rainfall in some years (leading to drought) and normal to excess rainfall in others India gets nearly 80% of its rainfall during the south west monsoon season (June to September) Delayed onset of the monsoon, prolonged breaks in the monsoon during the normally most active months of July and August, early with drawl of the monsoon and erratic distribution of rainfall during monsoon make India, especially the low rainfall belts, vulnerable to droughts.

For proper monitoring and assessment of droughts, different drought indices are used. Indian Meteorological Department (IMD) monitors meteorological and agricultural droughts based on **"Percentage of rain fall departure"** and **"Aridity anomaly index"** respectively, where as national remote sensing centre (NRSE) Hyderabad monitors agricultural drought using remote sensing technique.

Although rainfall deviation from long-term mean continues to be a widely adopted indicator for meteorological drought intensity assessment because of its simplicity, its application is strongly limited by its inherent nature of dependence on mean rainfall deviations, cannot be applied uniformly to different areas having different amounts of mean rainfall since a higher and a low rainfall area can have the same rainfall deviation for two different amounts of actual rainfall. Therefore rainfall deviations across space and time need to be interpreted with utmost care (Naresh Kumar et al., 2009).

"Percentage of average rainfall" is not often used to estimate droughts. It can be misleading because the median rainfall is different than the mean. For instance, a few years of extremely heavy rain can significantly increase the mean rainfall, meaning that most years will be less than "100 percent" of the mean, even though they won't signify droughts. **The standardized precipitation index (spi) is often used instead.** It is essentially the number of standard deviations (with some other complicated math because rainfall is not a normal distribution) that Precipitation is from the median value with negative values signifying below normal and positive signifying above. This index can be calculated over any signifying above. This index can be calculated over any period of time from days to years and therefore it can measure drought severity over any time period of interest.

The standardized precipitation index (SPI) expresses the actual rainfall as standardized departure from rainfall probability distribution function and hence this index has gained importance in recent years as potential drought indicator permitting comparisons across space and time (Naresh Kumar *et al.*, 2009).

The international workshop on indices and early warning systems for droughts held at Lincoln Nebraska USA on 8–11 December 2009 recommended to use spi to characterise meteorological droughts and the recommendation to use the spi was approved by WMO in June 2011. The comprehensive user manual for the spi which was developed by WMO is enclosed in this book for future guidance. **As per the procedure indicated in the manual, spi has been worked out for Anantapur district of A.P. India for a historical rainfall record for 102 years.**

A brief introduction to standardized precipitation index (SPI) written by WU Man–chi September 2013 is as follows.

("Deficit of rainfall over a period of time at a certain location, could lead to various degrees of drought conditions, affecting water resources, agriculture and socio-economic activities").

Since rainfall varies significantly among different regions, the concept of drought may differ from places to places.

As such from more effective assessment of the drought phenomena, the World Meteorological Organization (WMO) recommends adopting the standardized precipitation index (SPI) to monitor the severity of drought events.

In simple terms, SPI is a normalized index representing the probability of occurrence of an observed rainfall amount when compared with rainfall climatology at a certain geographical location over a long–term reference period.

Negative SPI values represent rainfall deficit whereas positive SPI values indicate rainfall surplus.

The advantage of SPI Over the difference from average map however is that while the later will only indicate the numerical magnitude of variation (i.e. 40 mm less, 10 mm more etc.) without stating how much less than what. That is (40 mm less than 80 mm or 40 mm less than 200 mm). The spi shows the statistical magnitude of deviation from the average and therefore better portrays the seriousness of the shortage.

This book consists of SPI and SPEI, detailed graphs drawn for Ananthapur district A.P. India for the period 1901 to 2002 years. The severity of drought event at scales, one month, 3 months, 6 months, 9 months and 12 months have been calculated and exhibited as spi and SPEI. **As an example for the years 2000, 2001, 2002, the detailed independent scales wise graphs have been depicted, duly indicating the date of onset, end of drought with duration and severity.**

For remaining years of 1901 to 2002 the output has been enclosed. Those who are interested can develop the graphs with the aid of out put enclosed to know onset, end, duration along severity of drought for each year.

In this connection it is to mention that during 2002 year neither IMD nor any other abroad models could recognize the drought event (during their forecast) which was a major drought for India. But it could be seen that by adopting SPI or SPEI Procedures (as per programme developed by Microsoft) the drought event has been revealed. So this method can be adopted to characterize meteorological drought by IMD instead relying on percentage departure method. This index has been developed by Mc. Kee et. al in 1993. Simple index, precipitation is the only parameter (probability of observed precipitation transformed into an index) with multiple time scales.

The values of SPI can be derived by comparing the total cumulative precipitation for a particular station or region over a specific time interval (for example the last month, the last 3 months the last six months) with the average cumulative precipitation for that same time interval over the entire length of the record.

For example the total precipitation in May of any given year for the region could be compared to average total precipitation for that region for all May's in the historical record (May 1900–2002).

The standardized Precipitation index (SPI) was designed to enhance the detection of onset and monitoring of drought (Me Kee *et al.*, 1993) the key feature of the spi is the flexibility to measure the drought at different time scales, because droughts very greatly in duration, it is important to detect and monitor them at a variety of time scales.

A drought event is defined where the spi is continuously negative and reaches a value of (–) 1.0 or less and Continues until the spi becomes positive.

Drought duration is defined by the interval between the beginning and end of that period.

The magnitude of the drought event is measured by the sum of the SPI values for the months of the drought. The following are the characteristics of standardized precipitation index (spi).

1. Developed by Mc Kee *et al.*, in 1993.
2. Simple index–Precipitation is the only Parameter, (Probability of observed precipitation transformed into an index).
3. Being used in research or operational mode in over 60 countries.
4. Multiple time scales allow for temporal flexibility in evaluation of precipitation conditions and water supply.

It is not simply the difference of Precipitation from the mean divided by the standard deviation. Precipitation is normalized using a probability distribution so that values of spi are actually seen as standard deviations from the median. Normalized distribution allows for estimating both dry and wet Periods. **Accumulated values can be used to analyze drought severity (magnitude) need 30 years of continuous monthly precipitation data** (the longer data the better). The spi can be computed using a weekly time step as well. Spi time scales intervals longer than 24 months may be unreliable. It is spatially invariant in its interpretation. Probability based (probability of observed precipitation transformed into an index) nature is well suited to risk management. **Many drought planners appreciate the spi's versatility. It can be calculated with missing data in the input. It can provide early warning of drought and help assess drought severity.**

The spi values found for the Anantapur district AP at different time scales are briefly interpreted as follows: graphs for each year 1900 to 2002 are drawn with spi values worked out and enclosed. The probability of occurrence of drought for different time scales also exhibited. For example graph using values of the spi for 2001 to 2002 year has been drawn and the onset of drought, end of drought, with magnitude for different time scales, have been exhibited. If any interested to know the details of individual year as stated earlier, based on spi values enclosed in the book. The graphs can be developed and onset, end, magnitude of drought can be obtained.

The one month spi is a short – term value and during the growing season it can be important for correlation of soil moisture and crop stress. The 3-Months spi reflects short and medium term moisture conditions and provides a seasonal estimation of precipitation, the 6th month spi compares the precipitation for that period with the same 6-months period over the historical record (1900–2002). 'The information from a 6-month spi may also begin to be associated with anomalous stream flows and reservoir levels. The 9-Months spi provides an indication of precipitation patterns over a medium time scale. Spi values below (–) 1.5 for these time scales are usually of good indication that fairly significant impacts are occurring in agriculture and may be showing up in other sectors as well. The spi of 12 month reflect long term precipitation patterns. Spi's of these time scales are probably tied to stream flows, reservoir levels and even ground water levels, at longer time scales.

The spi has key limitations as follows: As a measure of water supply, the spi does not account for evapo-transpiration and this limits its ability to capture the effect of increased temperatures (Associated with climate change) on moisture demand and availability.

1. Sensitive to the Quality and reliability of data, used to fit the distribution, 30-50 years recommended.

2. Does not consider the intensity of Precipitation and its potential impacts on run off, stream flow and water availability within the system of interest.

A short coming of SPI as noted by Trenbert *et al.*, "The spi is based on precipitation alone and provide a measure only for water supply. They are very useful as a measure of precipitation deficits or meteorological drought, but are limited because they do not deal the evapo-transpiration side of the issue".

In response to the above a new variation of the spi index by **Vicente-Serrano** *et al.*, **S.M.Begueria S. and Lopez – Moreno J.I. (2010) attempted to address this issue by including a temperature component through the calculation of their new spi index called the standardized precipitation evapo-transpiration index (SPEI) a multi scalar drought index, Sensitive to Global Warming. The inputs required are precipitation, mean temperature and latitude of the site.**

Spi for different time scales for analysis of droughts at Anantapur District, A.P. along with onset termination and magnitude can be seen below. Out of 102 years period (1900–2002) number of times the wet and dry events occurred have been exhibited below.

Table 1.1   Probability of Recurrence: SPI – 1ˢᵗ Month 1900–2002

Description	Category	No. of Times in 102 years	Severity of Event
Extremely wet	2+	22	I in 5 years
Very wet	1.5–1.99	68	I in 2 years
Moderate wet	1.00–1.49	105	I in 1 year
Near normal	(–) 0.99 to 0.9	880	0
Moderate dry	(–) 1 to (–) 1.49	76	I in 1 year
Severely dry	(–)1.5 to (–)1.0	32	I in 3 years
Extremely dry	Less than (–)2	26	I in 4 years

Note

1.   Every alternate year is a moderate drought. Every 3ʳᵈ year is severe drought

2.   For every fourth year is extreme drought. Every fifth year is a wet year

Table 1.2 SPI – 3rd Month – 1900–2002

Description	Category	No. of Times in 102 Years	Severity of Event
Extremely wet	2+	20	I in 5 years
Very wet	1.5 – 1.99	62	I in 2 years
Moderate wet	1.00 –1.49	115	I in 1 years
Near normal	(–) 0.99 to 0.9	802	0
Moderate dry	(–) 1 to (–) 1.49	76	I in 1 year
Severely dry	(–)1.5 to (–)1.0	56	I in 2 year
Extremely dry	Less than (–)2	23	I in 4 year

Note

1. Every alternate year is a moderate drought Every 2nd year is severely drought
2. Every 4th year is extremely drought. Every fifth year is wet year

Table 1.3 SPI – 6th Month – 1900–2002

Description	Category	No. of Times in 102 Years	Severity of Event
Extremely wet	2+	10	I in 10 years
Very wet	1.5–1.99	77	I in 1 year
Moderate wet	1.00–1.49	122	I in 1 year
Near normal	(–) 0.99 to 0.9	796	0
Moderate dry	(–) 1 to (–) 1.49	111	I in 1 year
Severely dry	(–)1.5 to (–)1.0	73	I in 1 year
Extremely dry	Less than (–)2	26	I in 4 years

Note

1. Every alternate year is modern drought
2. Every alternate year also in severely drought
3. Every 4th year is extremely drought

Table 1.4   SPI – 9th Month – 1900 - 2002

Description	Category	No. of Times in 102 Years	Severity of Event
Extremely wet	2+	9	I in 11 years
Very wet	1.5 – 1.99	65	I in 2 years
Moderate wet	1.00–1.49	137	I in 1 years
Near normal	(–) 0.99 to 0.9	791	0
Moderate dry	(–) 1 to (–) 1.49	111	I in 1 year
Severely dry	(–)1.5 to (–)1.0	65	I in 2 years
Extremely dry	Less than (–)2	34	I in 3 years

Note

1. Every alternate year is moderate drought
2. Every 2nd year is severe drought
3. Every 3rd year is extremely drought

Table 1.5   SPI – 12th Month – 1900 - 2002

Description	Category	No of Times in 102 Years	Severity of Event
Extremely wet	2+	15	I in 7 years
Very wet	1.5–1.99	61	I in 2 years
Moderate wet	1.00–1.49	126	I in 1 year
Near normal	(–) 0.99 to 0.9	801	0
Moderate dry	(–) 1 to (–) 1.49	105	I in 1 year
Severely dry	(–)1.5 to (–)1.0	81	I in 1 year
Extremely dry	Less than (–)2	29	I in 4 years

Note

1. Every year moderate drought
2. Every year severe drought
3. Every 4th year extreme drought

SPEI has the advantage of combining a multi-scalar character with the capacity to include the effects of temperature variability on drought assessment. Mathematically, the SPEI is similar to the standardized precipitation index (SPI) but includes the role of temperature. Abramopoulos *et al.* (1988) used a general circulation model experiment to show that evaporation and transpiration can consume upto 80% of rainfall.

There has been a general temperature increase (0.5–2°C) during the last past 150 years (Jones and Moberg 2003) and climate change models predict a marked increase during the 21st century (IPCC, 2007). It is expected that this will have dramatic consequences for drought conditions, with an increase in water demand due to evapo-transpiration (Sheffield and Wood, 2008) Dubrousky *et al.*(2008) recently showed that the drought effects of warming predicted by global climate models can be seen in the (PDSI) whereas the spi (which is based only on precipitation data) does not reflect expected changes in drought conditions.

Therefore, the use of drought indices which include temperature data in their formulation (such as the PDSI) is preferable, especially for applications involving future climate scenarios. **However, the PDSI lacks the multi-scalar character essential for both assessing drought in relation to different hydrological systems, and differentiating among different drought types. Therefore a new index was formulated called as (the standardized precipitation evapo-transpiration index) based on precipitation and PET.** The SPEI combines the sensitivity of PDSI to changes in evaporation demand (caused by temperature fluctuations and trends) with the simplicity of calculation and the multi-temporal nature of the SPI. **The new index in particularly suited to detecting, monitoring and exploring the consequences of global warming on drought conditions.**

The SPEI uses the monthly (or weekly) difference between precipitation and PET. This represents a simple climatic water balance (Thornth Waite, 1948) which is calculated at different time scales to obtain the SPEI. Using the simplest approach to calculate PET (Thornth Waite, 1948) which has the advantage of only requiring data on monthly mean temperature.

The procedure for calculating the SPEI is similar to that for the SPI. However the SPEI uses "climatic water balance" the difference between precipitation and reference evapo-transpiration (P-ETO) rather than precipitation (P) as the input.

The climatic water balance compares the available water (P) with the atmospheric evapo-transpirative demand (ETO) and therefore provides a more reliable measure of drought severity than only considering precipitation.

The climatic water balance is calculated at various time scales (i.e. over one month, two months, 3 months etc..) and the resulting values are fit to a log – logistic probability distribution to transform the original values to standardized units that are comparable in space and time and at different SPEI time scales.

Although the SPEI was only recently developed, it has been used in diverse studies that have analysed drought variability such as (i.e.,) drought reconstruction, drought atmospheric mechanisms, climate change and identification of drought impacts on hydrological, agricultural, ecological systems.

The log logistic distribution was used to calculate the SPEI as per Pearson III and gamma distribution, which used to calculate the SPI. Only when the index was computed at short time scales for some few very low precipitation, PET values (mainly arid locations with a highly variable climatology) there were many problems experienced.

These were minor and already known for SPI calculations when the two parameter gamma distribution is used (Wu *et al.*, 2007). **However the use of three parameter distribution to calculate the SPEI reduced this problem noticeably.**

In summary, the SPEI fulfils the requirements of a drought index as indicated by Nkemdirim and Weber (1999) since its multi scalar character enables it be used by different scientific disciplines to detect, monitor and analyse drought.

The SPEI allows comparison of drought severity through time and space, since it can be calculated over wide range of climates, as can the SPI. Moreover Keyantash and Dracup (2002) indicated that drought indices must be statistically robust and easily calculated, and have a clear and comprehensible calculation procedure. All these requirements are met by the SPEI.

However, a crucial advantage of the SPEI over the most widely used drought indices that consider the effect of PET on drought severity is, that its multi-scalar characteristics enable indentification of different drought types and impacts in the context of global warming.

Software has been created to automatically calculate the SPEI over a wide range of time scales. The soft were is available in the web respiratory of the Spanish national research council.

As an example, the SPEI has been worked out to Anantapur district of A.P. India for the period 1900–2002. The data obtained from IMD was rainfall and temperature for the above period. The output for 102 years has been enclosed in this book. For those who are interested, can draw the graph for each year on computer and the time of onset, end and magnitude of each year can be looked in. As a first hand example for the year 2000 & 2001 for time scales, 1st month, 3rd month, 6th month, 9th month and 12th month the onset, end and magnitude has been shown on the drawn graphs in this book.

The global performance of SPI and SPEI was compared by Sergio M. Vicente Serrano Jesus Julio Camarero and others to predict the changes in soil moisture, stream flow 'Crop yield and forest growth and found a superior capability of SPEI to identify drought impacts as compared with the SPI one. **They detected small differences in comparative performance of SPI and SPEI indices, but the SPEI was the drought index that best captured the responses of the assessed variables to drought in summer, the season in which more drought – related impacts are recorded and in which drought is critical. Hence SPEI index shows improved capability to identify drought impacts as compared with the spi one.**

Statement showing early Warning System (Onset, End, Duration, Magnitude) of (period 1900–2002) SPI and SPEI for the year 2001 (Anantapur District) Andhra Pradesh, India.

Time Scales	Standardized Precipitation Index (SPI)			Standardized Precipitation Evapo-Transpiration Index (SPEI)		
January – March – June – September – December	Onset of Drought	End of Drought	Magnitude	Onset of Drought	End of Drought	Magnitude
1st Month of 2001 year	1st Week of May 2001	3rd week of September 2001	(–) 1.65	2nd week of October 2001	Continued to next year	(–) 2.085
3rd Month of 2001 year	2nd week of May 2001	2nd week of October 2001	(–) 2.70	1st week of September 2001	Continued to next year	(–) 3.416
6th month of 2001 year	1st week of April 2001	2nd week of May 2002	(–) 4.16	1st week of July 2001	Continued to next year	(–) 2.307
9th month of 2001 year	2nd week of May 2001	2nd week of May 2002	(–) 4.27	2nd week of May 2001	1st week of November 2001	(–) 0.741
12th month of 2001 year	Last week of July 2001	1st week of June 2002	(–) 2.77	3rd week of May 2001	1st Week of November 2001	(–) 0.695

Statement showing the probability of drought event in comparison between SPI and SPEI for Anantapur district A.P. India.

Designation	Category	Standardized Precipitation Index (SPI)		Standardized Precipitation Evapo Transpiration Index (SPEI)	
		Number of times in 102 years (average of 5time scales)	Probability of Event	Number of times in 102 Years (average of 5time scales)	Probability of Event
Extremely wet	2+	15	1 in 7 years	16	1 in 6 years
Very wet	1.5 – 1.99	67	1 in 2 years	79	1 in 1 year
Moderate wet	1.00–1.49	121	1 in 1 year	117	1 in 1year
Near normal	(–) 0.99–0.90	814	0	752	0
Moderate dry	(–)1.00 – (–) 1.49	96	1 in 1 year	111	1 in 1 year
Severely dry	(–) 1.50 – (–) 1.00	61	1 in 2 years	81	1 in 1 year
Extremely dry	Less Than (–) 2.00	28	1 in 4 years	8	1 in 13 years (only in 1st month)

Note: Probability of occurrence of events in SPEI is occurring every year since temperature parameter was considered.

The results of comparison to provide a global assessment of the performance of six drought indices prominently in use around the world for monitoring drought impacts on a several hydrological, agricultural and ecological response variables have been enclosed in this book. The work done was by Sergio M. Vicente – Serrano, San I ago Bengueria and others. The six drought indices are (PDSI) Palmer Drought severity index, (WPLM), modified palmer drought severity index (PHDI) palmer hydrological drought index, z–index, palmer z-index (which is derived from the palmer model and it is much more responsive to short-term droughts) spi (standard precipitation index) SPEI (Standardized Precipitation Evapotranspiration Index). The SPEI index shows improved capability to identify drought as compared with spi one. The spi was superior to all the other remaining indices.

Regarding Characterisation of Agricultural Drought

Scenario in India   About two thirds of the geographic area in India receives low rainfall (less than 1000 mm) which is also characterized by uneven and erratic distributions. Out of net sown area of 140 million hectares about 68% is reported to be vulnerable to drought conditions and about 50% of such vulnerable area is classified as "severe" where frequency of drought is almost regular. Abnormally low rainfall in 1979 in India reported to have reduced the overall food grain by as much as 20%. The 1987 drought in India damaged 58.6 million hectares of cropped area affecting over 285 million people. The 2002 drought has reduced the sown area to 112 million hectares from 124 million hectares and the food grains production to 174 million tons from 212 million tons.

The total food grain production in India has to be stepped up from 212 million metric tons to 300 million metric tons by 2020 to meet the food demands of growing population. Therefore, there is a need for effective monitoring of agricultural drought, its onset, progression and impact on crops to minimize the damages.

The Indian Meteorological Department monitors drought using water balance technique which addresses agricultural drought. The aridity index is calculated using water deficit and water need. The departure of aridity index from normal percentage terms is used to define the various drought severity. Anomaly upto 25% is attributed to mild drought, 26–50% to moderate drought and > 50% to severe drought. A review of past agricultural droughts in country reveal the lack of unique relationship between incident ground measured rainfall (only a part of which replenishes soil moisture and thus available to vegetation) and the resulting vegetation development within and between seasons as well as across space. The "rainfall use efficiency" varying over both time and space and the vegetation species dependence, limits the use of rainfall as a sole or major agricultural drought indicator.

Aridity anomaly data currently available is only representation of large areas such as meteorological sub-divisions. The aridity anomaly suffers from the same limitations as that of rainfall.

Conventionally agricultural drought conditions are characterized by ground observations on meteorological parameters such as rainfall aridity and agricultural parameters such as crop sown area and crop condition, crop yield. Though the ground observations of agricultural conditions made by the state departments of Agriculture and Revenue are exhaustive, such a system involves a significant amount of subjective judgement at various stages. The periodicity and extent of ground observations also vary significantly between different states. The nature of sparse ground observations also make it difficult to assess, in near real – time, average drought conditions over the district. **Thus ground monitoring of both causative factors as well as impact of drought assessment suffer from various limitations such as sparse observations subjective data etc.**

There is a need for building up the capabilities by using innovative technologies and management measures for effective management of agricultural droughts in the country. A system for national / regional and subregional assessment and monitoring of agricultural drought conditions through the cropping season to provide periodic information on the prevalence, severity level and persistence of agricultural drought is the utmost need of the hour. Such information provided in time, effective manner, will help the resource managers in optimally allocating scarce financial and other resources to where and when they are most needed.

One of the sectors where the immediate impact is felt is agriculture. Drought impact is also felt due to (1) Deficit ground water recharge (2) Non availability of quality seeds (3) Reduced, power for agricultural operations, and due to distress sale of cattle (4) Land degradation (5) Fall in investment capacity of farmers, rise in prices, reduced grain trade and power supply (6) Shift in agricultural practices (low to moderate water demand crops to high crops) (7) Crop damage due to rain and snow pest. (8) Location of high water consuming milestones at semi arid / arid regions.

The international seminar on agricultural drought index and monitoring on national scale, held at Geneva / Murcia Spain 8 June 2010 conveyed by WMO, attended by 19 Nos. scientists from all regions recommend to characterize agricultural drought by water balance indices (agro-meteorological drought indices) such as (1) climate water balance model (2) crop water balance model (3) required irrigation water apart from four remote sensing based drought indices namely (1) perpendicular drought index (PDI) (2) modified PDI (MPDI) (3) enhanced vegetation index (EVI) and (4) vegetation condition index (VCI).

To characterize the agricultural drought, in this book calculation of climatological water balance for Anantapur Dist. A.P. India has been made for a period of 30 years (1946–1975). This district has been selected due to its continuous drought prone character and this district is largely dependent on rainfed agriculture. This district receives rainfall from South West monsoon,

North East monsoon, i.e., June to Sept. and Oct. to Dec.) Annual mean rainfall from 1911 to 2004 year (94 yrs) is 568.5 mm. More than one half of years the actual rainfall is below annual mean rainfall of 568.5 mm.

The following parameters are found to identify the droughts in the district. Index of aridity (Ia), index of humidity (Ih), Index of moisture (Im), Index of moisture adequacy (Ima), AE actual evapo-transpiration and PET potential evapo-transpiration. W.S. water surplus, W.D. water deficit. The parameters have been computed on the basis of book keeping procedure of Thornthwaite and Mather (1955) which is a simple model, the one proposed by them is the most used around the world for agricultural purpose. The PET is calculated from the mean monthly temperature and rainfall data collected from IMD. The soils are classified as red alfisols accounting for 78% while the black soils are found in 20% of total geographical area (Swaminathan Foundation, Chennai Report on Anantapur Dist.). So the water holding capacity is considered as 200 mm as per Thornthwaite and Mather Tables 1957).

The aridity index maximum 75.98% and minimum 61.1% shows the severity of drought in the district. The average moisture index Im% being 41.847%. So the area lies in semi arid climate as per revised scheme of Thornthwaite and Mather 1955). The negative values of Im% are (–) 45.5% maximum and (–) 36.66% minimum. The index of moisture values are being negative indicating the high values of Index of aridity.

The index of moisture adequacy Ima% average annual value being 37.122% indicates poor moisture availability. The agricultural activity thus restricted only to the rainy season, unless otherwise raised as irrigated crop, as the water deficit is existing in the entire district and there is no water surplus. The index of humidity (Ih) helps to determine the climate of the region and the district enjoys a dry climate. **The climatic water balance on an annual basis is made use of to determine the drought years and their severities in the district. Analysis of droughts and their intensities are made from the percentage departure of aridity index (Ia) from the median value (vide table).**

Categorization of droughts is made by following the scheme of standard deviation technique employed by Subrahmanyam et al. (1965). During the study period, the district experienced droughts as follows moderate – 5 Nos. large – 4 Nos. severe – 6 Nos. out of 30 years period from 1946-1975).

Decennial frequency of droughts has been made to evolve a better drought management system in the district. While knowledging from the point of view of large scale planning, the study of droughts is very important for planning short term operations especially in connection with the agricultural development.

The water balance of any region is not stable over a period of time. The study of climate shifts due to changes in moisture regime of climate and their periodicities is another important of drought climatology. **Hence it is necessary to study the variation in the index of moisture to understand the shifts in the climate to wetter or drier condition. This aspect is of great practical value in long**

term economic planning. To illustrate the climatic shifts, the (Im) values of Anantapur Dist. are plotted for individual years from 1946–1975, the extreme dryness and wetness are identified (vide fig.) **The study has revealed that the climate of the district is semi-arid to dry semi humid.**

Though climatic shifts are temporary such study of climate which may result in severe floods and droughts depending upon the intensity and duration of water surplus or deficiency.

The suitability of a region for agricultural development can be clearly assessed with index of moisture adequacy Ima% based on the percentage of moisture values, the suitability of crops that can be grown. Successfully in the absence of supplemental irrigation can be decided. The index of adequacy values of Anantapur District vary from about 27.00% to 52.00% in 1965 and 1956 respectively. **On an average the area can be classified (as per given by Subrahmanyan *et al.*, 1963) as below 40% range where in drought resistant crops such as Jowar, Ragi, Barja etc., to be grown. Most of the Ima values of the district are also less than 40% (20 years out of thirty years) and remaining ten years also falls in the category of 40–60 range which indicates that the area is also favourable for cultivation of millets.**

The frequency analysis of rainfall for the district is made for the period 1946–1975 (30 years) to find the 75% and 50% confident limits of rainfall. A graph drawn is presented in the book between annual rainfall and corresponding exceedence probability, which indicates 56.66% of timeout of 30 years is below normal rainfall (i.e. period of drought). Alternatively every alternative year is almost a drought year.

Space technology for agricultural drought monitoring has been adopted in India. Although there is many vegetation indices and derived indices, vegetation Condition Index (VCI) and Normalized Difference Vegetation Index (NDVI) deviation from historic NDVI are being widely used as operational basis to assess drought situation. Further in this book about the universal drought indices description and their uses and limitations, are also dealt.

Hydrological drought may be the result of long term meteorological droughts that results in the drying up of reservoirs, lakes streams – rivers and a decline in ground water levels (Rathore, 2004). Hydrological drought indices based on precipitation stream flow, low flows, run off, ground water levels and water balance are discussed in this book as a separate chapter.

Remote sensing data such as NDVI and Water Supply Vegetation Index (WSNI) along with spi and SWI (Standardized water level index) has to be used to identify the hydrological drought. The conclusions for such a decision has been discussed in this book.

The Lincoln declaration on drought indices and early warming system of drought during December 2009 at the break out session (participants from

22 countries including India) recommended for creating a new "composite index" for charactering agricultural and hydrological droughts. Regarding Meteorological drought consensus has been reached and recommended to use spi. So it is imperative to have a combined index along with some remote sensing indices to characterize the respective droughts just like U.S. drought monitor, which incorporates multiple indices and indicators of drought including impacts into the assessment process fully.

Although many countries do not have the range of data available to replicate this process fully any approach that incorporates information beyond precipitation and perhaps temperature data is going to provide a more accurate picture of drought severity.

A separate chapter on socio-economic drought is also included in this book which relates the supply and demand of goods and services with the three above mentioned types of drought. When the supply of some goods and services such as water and electricity are weather dependent, then drought may cause shortages in supply of these economic goods.

The expert group meeting on early warning systems for drought preparedness, sponsored by WMO and others recently examined the status, short comings and needs of (DEWS) Drought early warning systems and made recommendations on how these systems can help in achieving a greater level of drought preparedness (Wilhite *et al.*, 2000). The proceedings of the meeting documented recent efforts in DEWs in countries such as Brazil, China, Hungry India, Nigeria, South Africa and U.S.A. and but also noted the activities of regional drought monitoring centers in Eastern and Southern Africa and efforts in West Asia and Southern Africa. **The short comings of current drought early warning systems (DEWS) were noted in the following areas** (1) Data networks (2) Data sharing (3) Early warming system products (4) Drought forecasts (5) Drought monitoring tools (6) Integrated drought climate monitoring (7) Drought impact assessment methodology (8) Delivery systems (9) Global drought early warning system. These items have been dealt in detailed in this book.

With the above short comings duly rectifying and creating a "composite Index" in India will be a greater help to farmers and food security and enhancement of drinking water. The early onset and end along with qualifying severity of drought will be an imperative task laying on the authorities.

Developing a composite index and releasing drought information weekly duly consulting and incorporating review from a group of climatologists, extension agents and others across the nation, will lead the country economically forward. The following indicators and indices are suggested in this book to be a part of composite index "to be called as "Indian drought monitor" (1) SPI (2) SPEI (3) AAI (4) SWI (5) NDVI (6) NDWI (7) VCI (8) Water balance indices (Agro meteorological drought indices) (9) SWSI.

Another chapter regarding Mitigation of drought is brought out in this book. Drought can be mitigated by two kinds of measures either by adopting preventive measures or by developing a preparedness plan.

While drought mitigation programmes have been in existence for a long time they have not been effective due to many reasons, such as allocations being as thinly scattered over large areas. The peoples involvement is very merge. The programmes are fragmented in their implementation. The accountability mechanisms are very weak and all programmes experience serious lacunae.

The following topics have been dealt to mitigate drought in this book (1) soil and water conservation measures and farm practices (2) Rain water harvesting and conservation (a) Artificial recharge of ground water (b) Traditional methods (c) Long term irrigation management (4) Afforestation (5) Community participation in drought mitigation (6) Climate variability and adoption (7) Soil and water conservation measures for different rainfall regions especially the soil and water conservation measures for different rainfall regions (Arid semi arid and sub-humid regions) are dealt in the book. Success stories regarding roof water harvesting Rejuvenation of Swarnamukhi river basin in Chittoor District, AP (Construction of series of 9 Nos. of sub-surface dams). Improvement of quality and quantity of drinking water used by colonies of department of space from Kalangi river conservation of base flow by construction of subsurface dams – voices from the farmers and end users. Subsurface dams to harvest rain water – A case study — Managed aquifer recharge by construction of sub-surface dam in semi-arid regions – A case study of the Kalangi river basin – AP – Drought – try capturing the rain briefing paper for members of Parliament and state legislatures.

Drought preparedness Planning will also provide substantial benefit in preparing for potential changes in climate historically more emphasis has been given to flood management than drought management. With increasing pressure on water and other natural resources because of increasing and shifting populations (i.e. regional and rural to urban) it in imperative for all parts of India to improve their capacity to manage water supplies during water short years.

Drought risk is a product of region's exposure to the natural hazard and its vulnerability to extended periods of water shortage (D.A Wilier, 2000). If nation is to make progress in reducing the serious consequences of drought they must improve their understanding of the hazard and the factors that influence vulnerability. It is critical for drought prone regions to better understand their drought climatology (i.e. the Probability of drought at different levels of intensity and duration) and establish comprehensive and integrated drought information–system that incorporate climate soil and water supply factors such as precipitation, temperature, soil moisture, snow pack reservoir and lake levels, ground water levels and stream flow.

India being drought prone should develop national drought policies and preparedness plans that place emphasis on risk management rather than following the traditional approach of crisis management, where the

emphasis is on reactive emergency response measures. Crisis management decreases self reliance and increased dependence on government and donors.

ACKNOWLEDGEMENTS

I am grateful to all the authors whose works I have consulted.

The names of the various authors are given in the appropriate articles and a credit lines are also given at the end of the articles wherever proper. It is my bounden duty to acknowledge with a profound sense of gratitude to various publishers from whose relevant books, periodicals I have reproduced adopted some portions in order to do a minimum Justice to the subject.

I would like to express my sincere gratitude to Sri N. Janardhana Raju Professor, Jawaharlal Nehru University, New Delhi, India for kindly writing the foreward for this Book.

I thank my son-in-law Mr. G. Rajesh and his wife Mrs. P. Kalpana, Software Engineer, Microsoft Company at Redmond U.S.A. for helping me in the analysis of spi and SPEI while manuscript was under preparation.

I would like to thank my grand son Mr. G. Abhi deep for providing me certain information from computer and collection of various maps while manuscript was under preparation.

Finally I thank my son Mr. P.G. Kiran Kumar and his wife of Mrs. P. Nutan and my eldest daughter Mr. P. Kavitha and her husband Mr. G.S. Ramamohan, Editor, ABN Andhra Jyothi T.V Channel for their cooperation and encouragement and for their tolerance and understanding during the last one and half years while the manuscript was under preparation.

Any suggestions which would improve the standard of this book and enhance its usefulness will be thankfully received.

MUNIRATNAM PANTULA

2

EXECUTIVE SUMMARY

🗩 CHAPTER OUTLINE

- To identify all types of droughts early, use of SPI, SPEI, composite index is illustrated with example. Mitigation – preparedness are dealt. Success stories included.

Drought is an insidious hazard of nature. It is often referred to as a "Creeping phenomenon" and its impacts vary from region to region. It is a normal recurrent feature of climate. It is a protracted period of deficient precipitation.

Drought sets off a vicious cycle of socio–economic impacts beginning with crop yield failure, unemployment, erosion of assets, decrease in income worsening of socioeconomic conditions, poor nutrition and subsequently decreased risk absorptive capacity and thus increasing vulnerability of the poor to another drought and other shocks.

India gets 70 to 80% of its annual rainfall during the south west monsoon season (June to September). Delayed onset of the monsoon, prolonged breaks in the monsoon during the normally most active months of July and August, early withdrawal of the monsoon and erratic distribution of rainfall during monsoon make India especially the low rain fall belts vulnerable to droughts. Also there seems to be a clear association between El Nino and Lanina events and weak monsoons.

The drought management system that has been practiced in India since its independence in largely a continuation of the systems and schemes instituted during the colonial period. It emphasizes a relief–based approach and provides certain other small concessions, which do little to alleviate the distress caused by widespread crop failure. It functions on the basis of a conclusive evidence of drought as derived from the crop production in a particular year, which takes a lot of time as well as prevents early and timely help to farmers. In the past drought management strategies were worked out generally during or after the onset of drought which lacked preventive interventions. It did not integrate new technologies for early warning nor did it emphasize mitigation as an essential element of drought management. Thus, there is a strong need to introduce and institutionalize a new comprehensive approach to drought management system, which is based on the technological advances and new innovations in drought onset, duration, end along with severity, relief and mitigation.

Drought is among the most climatic phenomena affecting society and the environment (Withite J., 1993) the root of this complexity is related to the difficulty of quantifying drought severity since we identify a drought by its effects or impacts on different types of systems (agriculture, water, resources, ecology, forestry, economy etc.,). But there is not a physical variable with which we can measure to quantify droughts.

Thus droughts are difficult to pinpoint in time and space since it in very complex to identify the moment when a drought starts and ends and also to quantify its duration, magnitude and spatial extent (Burton *et al.*, 1978 Wilhite, 2000).

These characteristics explain the vast scientific efforts devoted to develop tools providing an objective and quantitative evaluation of drought severity. The quantification of drought impacts is commonly done by using the so-called drought indices, which are proxies based on climatic information and assumed to adequately quantify the degree of drought hazard exerted on sensitive systems. Recent works have reviewed the development of drought indices and compared their advantages and drawbacks (Heim 2002, Keyatash and Dracup 2002, Muhra and Singh 2010, Siva Kumar *et al.*, 2010).

However a few studies have performed robust statistical assessments by comparing different drought indices which may allow recommending the preferential use of one of them based on objective criteria (Guttmann 1998, Keyantash and Dracup 2002, Stinemann 2003, Poulo and Pereira 2006, Quiring 2009, Vicente – Serrano *et al.*, 2010, Barua *et al.*, 2011 Anderson *et al.*, 2011).

In addition, few researchers have compared the relative performance of different drought indices to identify drought impacts on several systems: In the case of drought impacts on hydrological systems (Vasiliades *et al.*, 2011) compared five drought indices in Greece. Lorenzo – Lacruz *et al.* (2010) compared the performance of two drought indices to identify hydrological droughts in river discharges and reservoir storages in central Spain, and zhai *et al.* (2010) compared the relationship between the standardized precipitation index (SPI) and the palmer drought severity index

(PDSI) and stream flow data in ten regions of China. Sims *et al.* (2002) compared the PDSI and SPI to assess soil moisture variations in north Carolina USA. In relation to vegetation activity and crop productivity Potop (2011) compared different indices to assess drought impacts on corn yields in Moldava and Marromatis (2007) and Quiring and Papa Kryia Koru (2003) followed a similar approach by quantifying wheat production in Greece and the Canadian prairies respectively. Quiring and Ganesh (2010) compared drought indices to assess the responses in vegetation activity to drought severity in Texas USA. Kemps *et al.* (2008) assessed tree-ring growth response to different drought indices in the south western USA. Recently, Droby Shev *et al.* (2012) analyzed the correlation between drought indices and fire frequency in Sweden. The results of these studies are diverse, since the best drought index for detecting impacts, changes, as a function of the analyzed system and the performance of the drought indices varied spatially.

As a result, at present there is high uncertainty among scientists, managers and end users of drought information, when they aim to select one drought index for a specific purpose.

To best of our knowledge, at present there is no global study analyzing and comparing to which degree, the most widely used drought indices are able to identify drought impacts on vulnerable systems.

This task is necessary in order to have solid and objective criteria for selecting a drought index to be used for specific tasks. In this study authors provide the first global assessment of the performance of different drought indices for monitoring drought impacts on stream flows, soil moisture, forest growth and crop yield (enclosed).

In the study of performance of drought indices for ecological, agricultural and hydrological applications, the authors provide a global assessment of the performance of different drought indices, for monitoring drought impacts on several hydrological, agricultural and ecological response variables. For this purpose the authors compare the performance of several drought indices (the standardized precipitation index (SPI) four versions of the palmer drought severity index (PDSI) and the standardized precipitation evapo-transpiration index (SPEI) to predict changes in stream flow, soil moisture, forest growth and crop yields. The authors found a superior capability of the SPEI and the spi drought indices, which are calculated on different time – scales, than the Palmer indices to capture the drought impacts on the afore-mentioned hydrological, agricultural and ecological variables. The authors detected small differences in the comparative performance of the spi and the SPEI indices, but the SPEI was the drought index that best captured the responses of the assessed variables to drought in summer, the season in which more drought – related impacts are recorded and in which drought monitoring is critical.

The international workshop on indices and early warming systems for drought held at Lincoln Nebraska USA on 8–11 December 2009 recommended to use spi (standardized precipitation index) to characterise meteorological drought and recommendation to use spi was approved by WMO (World Meteorological Organization) congress in June 2011. The comprehensive use manual for the spi which was developed by WMO is enclosed in this book. Spi expresses the actual rainfall as standardized departure from rainfall probability distribution function and hence, this index has gained importance in recent years as a potential drought indicator permitting comparisons across space and time (Naresh Kumar *et al.*, 2009).

The weakness of spi is the use of precipitation only for identifying meteorological drought onset, end and duration, severity. No soil water balance component, thus no ET/PET can be calculated. Lack of PET makes application on climate changes studies unadvised.

So a new variation of the spi index by Vicente – Serrano *et al.* (2010) attempted to address this issue by including a temperature component through the calculation of their new spi index called the standardized precipitation evapo-transpiration index (SPEI). The inputs required are precipitation, mean temperation and latitude of the site (s) to run the program on. More details about SPEI are available at http:\\digital csic.es/handile/10261/10002.

Hence SPEI (Standardized precipitation evapo–transpiration index) shows improved capability to identify drought impacts as compared with the SPI. In conclusion it seems reasonable to recommend the use of the SPEI if the responses of variables of interest to drought are not known apriori.

In India for proper monitoring and assessment of drought, different drought indices are being used. Indian Meteorological Department (IMD) monitors meteorological and agricultural droughts based on "Percentage of rainfall departure" and "aridity anomaly index" respectively, whereas the National Remote Sensing Centre (NRSC) Hyderabad, monitors agricultural drought using remote sensing technique. The spi has also been found to be effective tool in monitoring meteorological drought.

Although rainfall deviation from the long term mean continues to be a widely adopted indicator, for drought intensify assessment because of its simplicity, its application is strongly limited by its inherent nature of dependence on mean rainfall deviations, can not be applied uniformly to different areas having different amounts of mean rainfall since a high and a low rainfall area can have the same rainfall deviation for two different amounts of actual rainfall. Therefore rainfall deviations across space and time need to be interpreted with at most care (Naresh Kumar *et al.*, 2009).

This book consists of SPI and SPEI detailed graphs drawn for Anantapur Dist., AP India for the period 1900 to 2002 years. The severity of drought event at scales of one month, 3 months, 6 months, 9 months and 12 months has been calculated and exhibited for spi and SPEI. As an example for the years 2000, 2001, 2002 the detailed independent scale wise graphs have been depicted, duly indicating the date of onset, end of drought with duration and severity for remaining years of 1901 to 2000 (102 years). The output has been enclosed. Those who are interested can get developed graphs with the aid of output arrived to know onset, end, duration and severity of drought each year.

In this connection it is to mention that during 2002 year neither IMD nor any other abroad models could recognize the drought event which was

a major drought for India. But it could be seen, that by adopting SPI or SPEI, procedures the drought event has been revealed. So this method can be adopted to characterize meteorological drought by IMD and to detect the onset and end of drought along with it's severity.

To characterize the agricultural drought in this book, calculation of climatological water balance for Anantapur District AP India has been made for the period of 30 years (1946–1975) and the following parameters are found to identify the droughts in the district.

Index of aridity (Ia) Index of humidity (Ih) index of moisture (Im) index of moisture adequacy (Ima) actual evapo-transpiration (AE) and (PET) potential evapo-transpiration (WS) Water surplus, (WD) – Water deficit. The parameters have been computed on the basis of book keeping procedure of Thonthwaite and Mather (1955). The PET is calculated from the mean monthly temperature data collected from IMD and similarly for monthly rainfall.

Water balance models in current use for agriculture purpose   Several water balance models are available in the literature ranging from very simple models based only on rainfall and EPT balance for a given volume of soil, to very complex ones, based in detailed weather, crop and soil data and considering the soil as a multi-layer compartment where the water moves up and down depends on the different vertical and horizontal water inputs and outputs.

Among the simple models, the one proposed by Thornthwaite and Mather (1955) is the most used around the world for agricultural purposes. Meteorological services of countries like Argentina, Brazil, China, India, the USA and Uruguary use Thornthwaite and Mathur water balance for monitoring regional soil water conditions for agricultural crops as well as for monitoring drought conditions.

The normal rainfall for the period under consideration (1946–1975)–30 years) is 537.88 mm for Anantapur Dist. The potential evapo-transpiration being highest during May is 29.19 cm and low in the month of December as 7.84 cm. The soils are classified as red alfisols accounting for 78% while the black soils are found in 20% of total geographical area (Swaminathan Foundation, Chennai report on Anantapur District), So the water holding capacity is considered as 200 mm as per Thornthwaite and Mathur Table 1957.

The aridity Index (Ia) has been computed, reveals that a maximum of 75.98% and minimum being 61.1% in the years 1965 and 1954 respectively, indicating the severity of drought in the district. The average aridity index being 69.524% shows intensity of drought in the district for the 30 years period from (1946-1975).

The average moisture index, (Im%) arrived being (–) 41.847% and so the area lies in semi arid climate as per "Moisture index and climatic types" revised scheme of Thornthwaite and Mather 1955). The index of moisture values are being negative indicating the high values of index of aridity. The negative values of Ima% are maximum (–) 45.59 minimum of (–) 36.66 for years 1965, 1954 respectively.

The index of moisture adequacy (Ima%) average annual value being 37.122% and minimum being 27.18% and maximum 52.93%, during 1965 and 1956 respectively, indicating poor moisture availability. Thus the agricultural activity is normally restricted only to the rainy season, unless otherwise raised as irrigated crop; as the water deficit is existing in the entire district and there is no water surplus.

The index of humidity (Ih) helps to determine the climate of the region as 'moist or dry'. The computed figures have shown for the district as zero which indicates that the district enjoys a dry climate.

The climatic water balance on an annual basis is made use of to determine the drought years and their severities in the district. Analysis of droughts and their intensities are made from the percentage departure of aridity index (Ia) from the median value (vide Table) categorization of droughts is made by following the scheme of standard deviation technique employed by Subrahmanyam et. al (1965). During the study period, the district experienced droughts as follows, Moderate – 5 Nos. large – 4 Nos. severe – 6 Nos. out of 30 years period (1946–1975). The type of analysis of droughts i.e. Decennial frequency of droughts is helpful to evolve a better drought management system in the district. While knowledge from the point of view of large scale planning, study of droughts is very important for planning short term operations especially in connection with the agricultural development.

The water balance of any region is not stable over a period of time. The study of climatic shifts due to changes in moisture regime of climate and their periodicity is another important aspect of drought climatology. Hence it is necessary to study the variation in the index of moisture, to understand the shifts in climate to Wetter or drier condition. This aspect is of great practical value in long term economic planning. To illustrate the climatic shifts, the Im values of Anantapur district are plotted for individual years from 1946 to 1975 and the years of extreme dryness and wetness are identified (fig.). The study has revealed that the climate of the district as semi arid to dry semi humid. Though climatic shifts are temporary such study helps to understand the extreme conditions of climate which may result in severe floods and droughts depending upon the intensity and duration of water surplus or deficiency.

The suitability of a region for agricultural development can be clearly assessed with index of moisture adequacy (Ima%). Based on the percentage of moisture values the suitability of crops that can be grown successfully in the absence of supplemental irrigation can be decided. Index of, moisture adequacy values of Anantapur district vary from 27.18% to 52.93% in 1965 and 1956 respectively. On an average the area can be classified (as given by Subrahmanyan et al. (1963)) as below 40% range where in, drought resistant crops such as jowar, ragi, bajra etc., to be grown. Most of the Ima values of the district are less than 40% (20 years out of thirty years) and remaining ten years also falls in the category of 40–60 range which indicates that the area is also favourable for cultivation of millets.

Kharif is the major cropping season in the district. The gross cropped area being 9.75 lakh hectares prior to 1960 more than two thirds of gross cropped area was cultivated with food grains, predominantly millets with some amount of pulses and paddy (G.O. A.P. 2006). Among non-food crops grown are cotton and groundnut were important.

The area under groundnut, has increased four fold from slightly less than 2 lakhs hectares in the early 1960s to eight lakh hectares by 2005–2006. Minor millets have more or less disappeared for cultivation while the area under major millets has reduced by 90 percent.

As groundnut is largely cultivated with pulses as an intercrop, the importance of pulses has remained more or less stable over the years.

At present in the district, groundnut is essentially a Kharif crop with 98% of the total groundnut (rain fed) area being cultivated during Kharif season July to December. Sunflower crop is followed with groundnut in Rabi season (January to August). The south west monsoon (in Anantapur district) period of rainfall is from 1st June to 30th September. During this season the quantity of rainfall is lower than potential evapo-transpiration thus having serious implications for the crop growth. During the period (1946–1975) south west monsoon rainfall is 308.94 mm against PET 782.3 mm average deficient being 81.32%, 67.20%, 61.00% and 22% during June, July, August, September respectively.

Due to this the soil moisture stress condition under different stages of crop growth would result in inadequate plant population, high percentage of flower drop, poor seed scatting etc., and there by implications for crop yield.

The coefficient of variation of rainfall for the period 1946–75 being 18.35% indicating area being in zone of scanty rainfall.

The frequency analysis of rainfall in Anantapur district has been made for the period 1946–1975 – 30 years shows the following results. The normal rainfall being 538 mm 75% dependability is 430 mm for 24 years and 510 mm at 50% dependability for 15 years out of 30 years period. A graph is drawn between annual rainfall and corresponding exceedence probability. For 17 years out of 30 years rainfall is below normal rainfall i.e., 538 mm. Remaining 13 years surplus years (i.e.,) 56.66% of the time is drought period. Alternatively every alternate year is almost a drought year.

Space Technology for agricultural drought monitoring has been adopted in India. Among the various vegetation indices that are now available, normalized difference vegetation index (NDVI) is an universally acceptable index for operational drought assessment because of its simplicity in calculation, easy to interpret and its ability to partially compensate for the effects of atmosphere, illumination geometry etc. NDVI is transformation of reflected radiation in the visible and near infra red bands of NOAA, AVHRR (Polar Orbiting Satellites collecting course resolution imagery world wide with twice daily coverage and synoptic view) and is a function of green leaf area and biomass.

The severity of drought situation is assessed by the extent of NDVI deviation from its long term mean maps produced using relative greenness (i.e. ratio of current NDVI to the historic mean NDVI for the same period) are quite useful to assess drought situation and hence this indicator is being used widely.

Kogan (1990) developed vegetation condition index (VCI) using the range of NDVI and temperature condition index (TCI) with brightness temperature data, are being followed in India.

Although there are many vegetation indices and derived indices, VCI and NDVI deviation from historic NDVI are being widely used as operational basis to assess drought situation.

Andhra Pradesh is basically a agricultural state, with 34% of its GDP contributed by agriculture. Agriculture is the major source of employment to the people. Droughts during the season affect the agricultural production and agricultural droughts of high severity cripple the economy of the state. **To assess the agricultural drought, it is necessary to measure the extent to which rainfall and soil moisture is falling short of the water requirement of crops during the cropping season.** Moisture adequacy index (MAI) which has been provided by the institute of ICAR (Indian Council of Agricultural Research) is a better measure for assessing the degree of adequacy of rainfall and soil moisture to meet the potential water requirement of crops. Hence, to identify the mandals that can be vulnerable to agricultural drought, weekly MAI was worked out for the mandals having more than 20 years of rainfall data.

Methodology for calculating MAI   MAI (Moisture adequacy index) is the ratio of actual evapo-transpiration (AET) to the potential evapo-transpiration (PET). AET can be obtained as an output parameter from water balance calculations. Thonthwaite and Mather (1955) weekly water balance model was used for estimating water balance of mandals. Agricultural droughts during different seasons (years) were classified into four groups based on average MAI during the season.

Table 2.1 Drought Classification Based on MAI

Drought Severity by	MAI
No drought	MAI > 0.75
Mild drought	MAI < 0.75 and > 0.50
Moderate drought	MAI < 0.50 > 0.25
Severe drought	MAI < 0.25

MAI Values need to be applied in conjunction with other indicators such as rainfall figures, area under sowing and NDVI Values.

The central arid zone research institute (CAZRI) Jodhpur monitors agricultural drought in Indian arid regions by using moisture adequacy index (MAI). Drought is specified crop wise on a real time basis. Moisture adequacy index (MAI) which is based on calculation of weekly water balance is equal to the ratio (expressed as a percentage) of actual evapo-transpiration (AET) to the potential evapo-transpiration following a soil – water balancing approach by using the following equation MAI = AET/PET during different phonological stages of a crop.

Index based on aridity anomaly Thornthwaite (1948) water balance technique is generally used to compute the aridity anomaly index at a location. This index is one of the tools to monitor agricultural drought. To represent crop moisture stress, aridity index (Ia) is computed. For monitoring and mapping agricultural drought on a short time interval say a week is generally considered. The difference between the actual (Ia) and its normal value for that week (i.e. $\overline{Ia}$) furnishes an anomaly that is expressed as percentage.

Aridity anomaly $= \dfrac{Ia - \overline{Ia}}{\overline{Ia}} \times 100$. Based on the aridity index weekly / F.N aridity anomaly maps / reports for the south-west monsoon for five meteorological subdivisions (Coastal A.P. Rayalaseema, South interior Karnataka, Tamilnadu and Pondichery and Kerala) are prepared and sent to various user communities on a real time basis for their use in agricultural planning and research purpose.

At the start of rainy season, seed germination and initial crop growth depend on the amount and distribution of rainfall. The beginning and end of growing period is identified based on index of moisture adequacy (IMA) values. The growing season begins when the IMA is above 50% consecutively for at least three weeks. The end of season is identified when the IMA falls below 25% for four consecutive weeks.

The length of growing period or the moisture availability period is an important parameter to assess the suitability of climate for agricultural production. It is a key parameter which helps in assessing climatic suitability to go for particular cropping system. For example mono cropping with short duration pulses can be adopted for areas with 75 days of length of growing period (LGP) and mono cropping with short to medium duration crops for areas with 75–140 days of LGP and double cropping can be adopted for areas with more than 180 days LGP.

WMO/UNISDR united nations international strategy for disaster reduction expert meeting on agriculture drought indices was organized and held on June 2–4, 2010/2011 at Murcia Spain. The meeting reviewed several drought indices currently used around the world for agricultural drought and assessed the capability of these indices to accurately characterize the severity of drought and its impact on agriculture.

The meeting recommended that given the enhanced availability of and access to data, tools and guidance, materials, countries around the world should move beyond the use of Just rainfall data in the computation of indices for the description of agricultural droughts and their impacts.

This issue becomes very relevant, especially in the context of climate change, water scarcity and food security and hence it is important to use more comprehensive data on rainfall, temperature and soils information in computing drought indices.

Recognising the diverse data and information required for the use of a composite approach, such as the U.S. drought monitor, the meeting recommends that all the countries examine this option.

Given the urgency to address drought monitoring and early warning in a comprehensive manner, there is a need to increase the efficiency in maintaining and enhancing weather data collection networks.

There is a strong need for better soils information and establishment of soil moisture monitoring net works where they do not currently exist.

Closer cooperation and applications between meteorological, agricultural, hydrological and remote sensing agencies and institutions is required, for improved drought monitoring and impact assessment.

The systematic collection and archiving of drought impacts on agriculture, is imperative and more efforts should be made in this area.

There is universal interest in understanding and reducing drought risk and impacts on agriculture. In this context the effective communication of drought information to policy makers, managers, the user community and the media is essential.

Deliverables such as maps, reports and press advisories, need to be produced at regular intervals and disseminated in a timely manner.

Realizing the need for early exchange of data, coming from different sources and institutions, enhanced access to a wide range of weather and soil data for drought monitoring is recommended.

Taking into account the increased, importance of applications of "GIS" there is a need to explore existing capabilities of such systems and enhance the inter – operability between different data platforms particularly at the regional level.

In order to encourage the use of common agricultural drought indices around the world, there is an urgent need to develop common frame works for drought monitoring early warning systems.

In order to achieve this goal an inventory of operational capabilities in the areas of data networks, deliverables and indices used/calculated and disseminated, along with an assessment of user needs should be prepared.

To this end, the meeting recommended that the WMO conduct a survey to compile and assess the capacities and future needs of national meteorological and hydrological services, around the world in building such common frame work for national agricultural drought early warning systems.

The goal of the meeting on the selection of appropriate drought indices or indicators to characterize agricultural drought, was to reach consensus on a single index to accomplish on this task. This is a formidable task given the complexities of agricultural drought and the variable institutional capacity of drought prone nations. At best we should strive to identify a series of alternative approaches to characterize agricultural drought in various settings depending on available data and local capacities (i.e., incomplete multiple indices and indicators) to characterize agricultural drought.

Drought is a creeping phenomenon with no universal definition. Definitions of drought must be region and application or impact specific. Many indices and indicators are available to assist, in the quantitative assessment of drought severity and these should be evaluated carefully for their application to each region or location and sector.

To best characterize drought, it is critically important to use a combination of indices and indicators since no single one, can capture the full severity of a particular drought event. This is an especially different assignment for agricultural and hydrological drought.

Early warning systems are the foundation of effective drought mitigation and preparedness plans. In efforts to build as comprehensive and flexible, a drought early warning system (DEWS) as possible, it is important to monitor drought across the many sectors mentioned earlier.

The use of single index will rarely work for all places at all times and for all types of droughts. Most coordinated monitoring efforts, at the national level are going to need, to track all types of drought. In case such as those, it is important to utilize and incorporate a consolidation of indices and indicators into one comprehensive "Composite index". A composite index approach allows for the most robust way of detecting and determining the magnitude (duration + intensity) of droughts as they occur.

Through a convergence of evidence approach, one can best determine (for a particular state, country or region for a particular time of the year) which indices and indicators do the best job, of depicting and tracking all types of droughts.

The users can then determine, which indicator to use and how much weight to give for each indicator/indices in a "blended approach" that incorporates a multiple parameter and weighting scheme. Such approaches have been used in U.S drought monitor (USDM) and North American drought monitor (NADM) as described below and as part of a series of objective blend products which are produced, for the USDM. U.S drought monitor created in 1999, the weekly U.S. Drought Monitor (USDM) was the first to use a composite index/indicator approach (Svoboda *et al.*, 2002). The product is not an index in and of itself, but rather a combination of indicators and indices that are combined using a simple D_0–D_4 scheme and a percentile ranking methodology (Table) to look at, addressing both short and long term drought across the U.S.

The key indicators/indices revolve around monitoring precipitation, Temperature, stream flow, soil moisture, snow pack and snow water equivalent various indices, such us the SPI and PDSI which are incorporated and in–grated with remotely sensed vegetation indices to come up with a "blended convergence of evidence" approach in dealing with drought severity.

The ranking percentile approach, allows the user to compare and contrast indicators originally having different periods of record and units into one comprehensive indicator that addresses the customized needs of only given user.

The approach also allows for flexibility and adaptation to latest indices, indicators and data that become available over time. It is a blending of objective science and art through the integration of impacts and reports from local experts at the field level. The impacts covered and labelled on the map are (A) for agriculture and (H) for hydrological drought. Some 275 local experts from across the country are allowed to view the draft maps and provide their input data and impacts to either support or refute the initial depiction. An interactive process works through all the indicators / indices data and field input until a compromise is found for the week.

The process then repeals itself the next week and so on. In addition, a set of objective blends are used to help guide the process. This method combines a different set of indicators to produce separate short and long term blend maps that take various indices with variable weightings (depending on region and type of drought) to produce a composite classification scheme and the objective blends can be found at http:// drought unl edu/dm.

Table 2.2 The U.S. Drought Monitor Classification and Ranking Percentile Scheme

Category	Description	Ranking Percentile
D0	Abnormally dry	30% tile
D1	Moderate	20% tile
D2	Severe	10% tile
D3	Extreme	5% tile
D4	Exceptional	2% tile

Source: National Drought Mitigation Center USDA NOAA.

Each week, the author revises the previous map based on rain, snow and other events, observers reports of how drought is effecting crops, wild life and other indicators. Authors balance conflicting data and reports to come up with a new map every Wednesday afternoon. It is released the following Thursday morning.

Drought indicators in USA   (1) North American drought monitor (2) U.S. Drought monitor (3) Crop moisture index (4) HPRCC ACIS maps (5) Palmer drought severity index (6) Soil moisture (7) Standardized precipitation index (8) Surface water supply index (9) Hydrological monitoring (10) Local state and Regional (11) Palco climate data (12) Remote sensing (13) Water quality (14) Wildlife.

Experimental objective blends of drought indicators   The drought blend method has been used for U.S. drought monitoring http://www.drought.unl.ed/dm/monitor.html.

Double blend indicators are divided into short term and long term blend. The short term blends include PDSI, Z, SPI 1st, 3rd months and soil moisture. The long term blends induce PHD, SPI 06, 12, 24 and 60 month and soil moisture.

In the short term blends method, the indicators are weighted to the Precipitation and soil moisture which use to identify the impacts of no irrigated agriculture, wild fire damagers, top soil moisture and pasture conditions.

The long term blend index, indicates the impacts of hydrological drought such as reservoir and well levels and irrigated agriculture. The drought indicator used in drought monitor, provides the most widely used map for drought conditions across United States (and is suitable for Indiana).

Source: Drought indices Michael J. Hayes National drought mitigation center.

The following are the important approaches to drought assessment among such several approaches.

1. Single indicator or index
2. Multiple indicator or index
3. A "Composite or 'Hybrid' Approach

Satellite remote sensing technology is widely used for monitoring crops and agricultural drought assessment in India. India's national agricultural drought assessment and monitoring system (NADAMS) project stands as an example for operational use of both moderate resolution and coarse, resolution.

Although there are many vegetation indices and derived indices, (VCI) Vegetation condition index, developed by Kogan 1990 using the range of NDVI is

$$VCI = \left[\frac{NCVI - NDVI\ \min}{NDVI\ \max - NDVI\ \min} \right] \times 100$$

and NDVI (Normalize differences vegetation Index) deviation from historic NDVI are being widely used on operational basis to assess agricultural drought situation.

The WMO defines a drought index as "An index which is related to some of the cumulative effects of a prolonged and abnormal moisture deficiency" Index is a method of deriving "Value added" information related to drought by comparing current conditions to historical information based upon statistical calculations. Indicator is a measure of a meteorological, hydrological agricultural and socio - economic variables that provides an indication of potential drought related stress or deficiency. Indices are indicators as well.

Examples of indicators: Precipitation amounts, river and stream flow levels, soil moisture information, evapo-transpiration information, reservoir storage, impact information, crop status/yield estimate reports–temperature, vegetation health/ stress, short and long term seasonal forecasts, Ground water, snow pack.

Examples of indices: SPI (Standardized Precipitation Index, (PDSI) Palmer Drought Severity Index (SWSI) surface water supply index, percent of normal / departure from normal precipitation, Deciles, SPEI (Standardized Precipitation Evapo-Transpiration index, E.D.I effective drought index and many others. In India the agricultural drought monitoring mechanism is as follows:

1. IMD identifies drought in all the meteorological sub divisions (aridity anomaly index).

2. Monitoring of water level in reservoirs (CWC).

3. National agricultural drought assessment and monitoring system (NADAMS).

4. Inter – ministerial crop weather watch group (CWWG) provides the trigger for activating drought response system.

Drought triggers specific values of an indicator that initiate and/or terminate each level of a drought plan and associated management responses.

HYDROLOGICAL DROUGHT

Hydrological drought may be the result of long term meteorological droughts that results in the drying up of reservoirs, lakes, streams rivers and a decline in ground water levels (Rather, 2004).

Hydrological Drought

(a) Effects of periods of, rain shortfall on surface and subsurface water supply.

(b) They lag behind meteorological and agro-meteorological droughts.

(c) Ground water drought, is out lined, by lower than average annual recharge, for more than one year.

(d) Ground water levels are good indicators in aquifer area.

Hydrological drought indices based on precipitation, streamflow, low flows, run off, Ground water level and water balance.

The following are the details pertaining to the above indices.

1. **Indices based on precipitation**

    (a) **Standardized Precipitation Index (SPI)**

        Used in South Asia to quantify the precipitation deficit in the monsoon and non-monsoon periods.

    (b) **Effective Drought Index (EDI)**

        It calculates daily drought severity. Rapid detection and precise measurement of short term drought.

2. **Indices based on stream flow**

    (a) **Stream flow Drought Index (SDI).**

        It requires stream flow volume values.

(b) **Surface Water Supply Index (SWSI)**

This index integrate, reservoir storage, stream flow and two precipitation types (snow and rain) at high elevations, into a single index. In winter months SWSI is computed using snow peck, precipitation and reservoir storage. In summer, stream flow, precipitation and reservoir storage data are used.

3. **Indices based on low flows**

WMO 1974 defines low flow, as a flow of water in a stream during prolonged dry weather.

4. **Index based on run off**

Run off data should be fit in to follow normal distribution or other type of distribution.

Normalizing run off would convert the probability density function of Pearson type III distribution into the standard normal distribution as function of Z.

Based on the Z value calculated for normal distribution, drought / flood categories could be defined.

5. **Indices based on ground water levels**

(a) **Standardized Water level Index (SWI)**

Standardized water level index has been developed by, to scale ground water recharge deficit. Seasonal water level and long term seasonal mean and standard deviation are the parameters to calculate (SWI). Positive anomalies correspond to drought and negative anomalies correspond to no-drought or normal condition.

(b) **Ground water Resource Index (GRI)**

A ground water resource index has been developed by Mendicino *et al.* (2008) to quantity ground water detention for the assessment of drought condition.

6. **Indices based on Water Balance**

(a) Palmer Drought Severity Index (PDSI)

The major problem associated with using PDSi is that its computation is complex and requires substantial input of meteorological data. It's application in Asia, where observational networks are scarce is therefore limited.

For hydrological drought analysis, standardized water level index (SWI) was used at Aravali regions of Rajasthan India. The data used were of ground water levels of 541 wells of the region. SWI was calculated using the mean seasonal water levels of 20 years (1984–2003). SWI values of the wells were interpolated using the spline interpolation technique in a GIS environment to generate SWI maps of the region.

Applications of hydrological drought indices in Asia are also indicated in this book.

(a) Aravali region of Rajasthan in India using SWI indices.	(b) Drought assessment in Seoul Korea using EDI Index.
(c) Runoff derived drought index for the arid area of Hexi corridor Northwest China.	(d) Application of SPI and using stochastic models and neural network for drought forecasting for kansabati river basin West Bengal, India.

Note: The neural network models were useful for forecasting of drought which could help local administration and water resource planners to take precautions considering severity of drought known in advance. Identification of drought venerable areas using remote sensing data (i.e.,) for South Rajastan, India. Remote sensing data used for drought assessment using NDVI and water supply vegetation index (WSVI) along with (SPI) for south Rajasthan India is enclosed in this book.

Conclusions for Hydrological Drought

1. SPI (Standardized Precipitation Index) is most commonly used index for hydrological drought assessment in conjunction with other indices.
2. Other hydrological drought indices based on stream flow are hindered by the data availability.
3. Indices based on ground water levels in conjunction with other index are used for hydrological drought assessment for non–monsoon periods.
4. Lack of coordination between data monitoring agencies.

Note: Remote sensing data such as NDVI and water supply vegetation index (WSVI) along with Spi and Swi (Standardized water level index) has to be used to identify the hydrological drought.

The Lincoln declaration on drought indices and early warning systems of drought at Lincoln Nebraska U.S.A – international workshop during 8–11 December 2009, at break out session, the following results were expressed on hydrological drought, by participants numbering fifty four from 22 countries including India. The following indices were identified during exercise I: reservoir levels, percent of normal precipitation, standardized precipitation index (SPI) surface water supply index (SWSI) aggregate dryness index (ADI) normalized ADI and the low flow index. The participants also discussed the potential for the utility of the stream flow drought index (SDI) and for using artificial neural networks to monitor and predict global hydrological drought conditions. **Rather than producing a consensus, exercise–II lead participants to recommend, creating a new "composite index" that would ideally be based on stream flow, precipitation, reservoir levels, snow pack and ground water levels.**

Similarly for characterizing meteorological drought, participants reached a consensus that the WMO should recommend the (spi) standardized precipitation index, for wide spread use in countries wanting to track meteorological drought.

The participants discussed during exercise–I a list of 17 indices for identifying agricultural drought. This list was then refined and limited to include, soil moisture index, percentage normal precipitation, normalized difference vegetation index (NDVI), water balance and heat stress.

Similar to the hydrological drought, session exercise–II did not produce a consensus. Instead, participants recommended holding a follow up discussion or workshop to specifically address agricultural indices. As Per recommendation of international workshop held daring 8–11 December 2009 at Lincoln Nebraska USA for holding a follow up discussion or workshop to specifically address agriculture drought indices WMO/UNISDR expert meeting on agriculture drought indices was organized and held on June 24, 2010 in Murcia Spain.

The meeting reviewed several drought indices, currently used around the world for agricultural drought and assessed the capability of these indices to accurately characterize, the severity of drought and its impact on agriculture. This chapter summarizes the agricultural drought indices that were discussed at the meeting in seven distinct categories as follows:

1. Precipitation – Based indices.
2. Temperature – Based indices.
3. Precipitation and temperature based indices.
4. Indices based on precipitation, temperature, relative humidity, solar radiation, wind speed.
5. Soil moisture and soil characteristics.
6. Indices based on remote sensing.
7. Indices based on a composite approach.
   (Multiple indicators/indices).

A brief review of each of these indices is presented earlier in this book. The meeting recommended, that given the enhanced availability of and access to data, tools and guidance, materials, countries around the world should move beyond the use of Just rainfall data in the computation of indices for the description of agricultural droughts and their impacts.

This issue becomes very relevant, especially in the context of climate change, water scarcity and food security and hence it is important to use rainfall, Temperature and soils information.

In the light of all the discussions and conclusions made during several workshops to build an comprehensive and flexible a drought early warning system (DEWS) as possible it is important to monitor drought across the many sectors mentioned earlier.

The use of a single index will rarely work for all places at all times and for all types of droughts. In cases much as those, it is important to consolidation of indices and indicators into one comprehensive "Composite index" A composite index approach allows for the most robust way of detecting and determining the magnitude (duration + intensity) of droughts as they occur.

Considering the complexity of drought and the many indices and indicators necessary to assess it's severity and likely impacts the most successful approach to date (drought.und.edu/dm) is the U.S drought monitor. It incorporates multiple indices and indicators of drought including impacts into the assessment process fully.

Although many countries do not have the range of data available to replicate this process fully, any approach that incorporates information beyond precipitation and perhaps temperature data is going to provide a more accurate picture of drought severity.

Drought monitoring and early warning   Effective drought early warning systems (DEWS) are an integral part of efforts worldwide to improve drought preparedness, timely and reliable data and information must be the corner stone of effective drought policies and plans. Monitoring drought presents some unique challenges because of drought distinctive characteristics.

An expert group meeting on early warning systems for drought preparedness, sponsored by the world meteorological organization (WMO) and others, recently examined the status, short comings and needs of (DEWS). Drought early warning systems and made recommendations on how, these systems can help in achieving a greater level of drought preparedness (Wilhite *et al.*, 2000b).

This meeting was organized as part of WMOs contribution to the UNCCD. The proceedings of this meeting documented recent efforts in drought early warning systems (DEWS) in countries such as Brazil, china, Hungary, India, Nigeria south Africa and United States, but also noted the activities of regional drought monitoring centers in eastern and Southern Africa and efforts in West Asia and North Africa.

The short comings of current Drought early warning systems (DEWS) were noted in the following areas:

1. **Data networks:** Inadequate density and data quality of meteorological and hydrological networks on all major climate and water supply parameters.

2. **Data sharing:** Inadequate data sharing between Government agencies and high cost of data, limit the application of data in drought preparedness mitigation and response.

3. **Early warning system products:** Data and information products are often not user friendly and users are often not trained in the application of this information to decision making.

4. **Drought forecasts:** Unreliable seasonal forecasts and lack of specificity of information provided by forecasts, limit the use of this information by farmers and others.

5. **Drought monitoring tool:** Inadequate indices for detecting the early onset and end of drought, although the standardized precipitation index (SPI) was cited as an important new monitoring tool to detect the early emergence of drought.

6. **Integrated drought/Climate monitoring:** Drought, monitoring systems should be integrated and based on multiple indicators to fully understand drought magnitude, spatial extent and impacts.

7. **Drought impact assessment methodology:** Lack of impact assessment methodology hinders impact estimates and the activation of mitigation and response programs.

8. **Delivery systems:** Data and information on emerging drought conditions, seasonal forecasts and other products are often not delivered to users in a timely manner.

9. **Global drought early warning system:** No historical droughts data base exists and there in no global drought assessment product that is based on one or two key indicators, which could be helpful to international organizations. **Non-Governmental organizations (NGOs) and others.**

With the above short comings duly rectifying and creating a "Composite index" in India will be a greater help to farmers and food security and enhancement of drinking water. The early onset, duration and end along with quantifying severity of drought will be an imperative task lying on the authorities.

Developing a composite index and releasing drought information weakly duly consulting and incorporating review from a group of climatologists, extension agents and others across the nation will lead the country economically forward.

The following indicators and indices are suggested to be a part of "composite index" to be called as "Indian drought monitor".

1. **Standardized Precipitation Index (SPI):** To characterize meteorological drought with onset and end of droughts and severity duration. Long term scales over 6 months are considered for hydrological drought indicators and hydrological drought using stochastic models and neural network for drought forecasting and end of drought & severity, duration.

2. **Standardized Precipitation Evapo-Transpiration index (SPEI):** The new index particularly suited for detecting, monitoring and exploring the consequences of global warming on drought conditions.

Example   Carried for Kanasabati river basin West Bengal India also in south Asia to quantify precipitation deficit in the monsoon and non-monsoon.

3. **Aridity Anomaly Index (AAI):** To monitor agricultural drought during the rainy kharif and post rainy (Rabi) seasons, based on Thornth Waite (1948) water balance technique on weakly or fortnightly basis.

4. **Moisture adequacy/Availability Index (MAI):** It is obtained from weakly water balance developed by central arid zone research institute (CAZRI) Jodhpur for monitoring in Indian arid regions.

5. **Standardized Water level Index (SWI):** Developed to scale ground water recharge deficit, for hydrological drought analysis used to analysis drought dynamics in aravali region of Rajasthan India. Used mean seasonal water levels for 20 years from 541 wells (1984–2003). SWI values of the wells were interpolated using spline interpolation technique in a GIS environment to generate SWI maps of the region (for Non-Monsoon Period hydrological drought).

6. **Normalized Difference Vegetation Index (NDVI):** It is based on the concept that vegetation vigour is an indication of water availability or lack thereof (To characterize agriculture drought along with MAI).

7. **Normalized Difference Wetness Index (NDWI):** It is based on the use of shortwave infra red (SWIR) band which is sensitive to moisture available in soil as well as in the crop canopy. In the beginning of the cropping season, soil back ground is dominant, which makes SWIR sensitive to soil moisture in top 1–2 em. As the crop progresses SWIR becomes sensitive to leaf moisture content NDWI using SWIR can complement NDVI for drought assessment particularly in the beginning of the cropping season. Higher values of NDWI signify more surface wetness. Used to assessment of agricultural drought.

Notes

(a)   The values of NDVI will be received to states through (NADAMS) National agricultural drought assessment and monitoring system instituted by national Remote Sensing Centre (NRSC) issues a bi-weekly drought bulletin and monthly reports on detailed crop and seasonal condition during the kharif season.

(b)   A negative number (NDVI) or a number close to zero means no Vegetation and a number close to +1 (0.8 to 0.7) represents luxurious vegetation. The values obtained from a given (NDVI) always ranges from (−) 1 to (+) 1. It is necessary that the states declare drought only when the deviation of NDVI value from the normal is 0.4 or less. However the NDVI values needs to be applied in conjunction with other indicators and values.

8. **Vegetation Condition Index (VCI):** This index was first suggested by Kogan 1995 and 1997. It shows effectively how close the current months NDVI is to the minimum NDVI. Calculated from the long term record of remote sensing images. This is also used to identify agricultural drought.

9. **Water Balance Indices:** (Agro meteorological drought indices) (For agricultural drought).

Source: FAO irrigation and drainage paper 25 – 1978.

1. Required irrigation water $I_i = Re_i - ET_{ci}$

    I_i = Required irrigation water in month (i) mm

    Re_i = Effective precipitation during month (i) mm that calculated according to USDA.

    ET_{ci} – is the monthly crop evapo-transpiration mm

Note: IMD collects its rainfall data using a network of 2800 rain gauge stations distributed across 36 meteorological subdivisions of the country. In categorizes rainfall in each subdivision as excess, normal, deficient or scanty. On average each meteorological subdivision has more than a dozen districts. It is very likely that within a meteorological subdivision certain districts are effected by drought while others are not. Therefore it is very essential to have a drought early warming system for each village having one rain gauge station. If on set, end, duration and severity of drought, along with magnitude at village level is known preparedness for drought and mitigation can be of ease for farmers and authorities. So the SPI and SPEI and water balance parameters arrived along with remote sensing indices are to be worked out and informed to villagers and farmers to be given to mitigate drought conditions in India.

10. **Surface Water Supply Index (SWSI) (Hydrological drought):** The index integrate reservoir storage, stream flow and two precipitation types (snow and rain) at high elevations into a single index. It represents water supply conditions unique to each basin or water management requirement of each basin.

$$SWSI = \frac{ap_{snow} + bp_{prec} + cp_{stream} + do_{reservoir} - 50}{12}$$

Where *a, b, c, d* = weights of snow, rain, stream flow and reservoir storage respectively ($a + b + c + d = 1$) and *p* = the probability (%) of non-exceedence for each of these four water balance components. The estimation is carried out with a monthly time step. In winter months, SWSI is computed using precipitation, snow peck and reservoir storage. In summer, stream flow, precipitation and reservoir storage data are used.

The range of SWSI is similar to palmer drought severity index (PDSI) as (–) 4.2 to (+) 4.2.

Remote sensing data were used to Hydrological drought assessment using NDVI and water supplying vegetation index (WSVI) along with spi for southern Rajasthan India.

Mitigation of drought Drought can be mitigated by two kinds of measures either by adopting preventive measures or by developing a preparedness plan.

While drought mitigation programmes have been in existence for a long time, they have not been effective due to many reasons, such as allocations being as thinly scattered over large areas, the peoples involvement is very meagre, the programmes are fragmented in their implementation. The accountability mechanisms are very weak and all programmes experience serious lacunae. The following topics have been dealt to mitigate drought in this book.

(a) Soil and water conservation measures and farm practices.

(b) Rain water harvesting and conservation.

(1) Artificial recharge of ground water (2) Traditional methods (3) Long term irrigation management (4) Afforestation (5) Community participation in drought mitigation (6) Climate variability and adoption (7) Soil and water conservation measures for different rainfall regions.

(c) Water saving technologies and resource conservation through laser levelling.

Drought can be mitigated by two kinds of measures either by adopting preventive measures or by developing a preparedness plan. The aspects on the above two methods are discussed in this chapter. The soil and water conservation measures for different rainfall regions (Arid, Semi arid and sub humid regions) are enclosed.

Success stories   Success stories regarding roof water harvesting, Rejuvenation of Swarnamukhi river basin in Chittoor Dist., A.P. India (construction of series of 9 Nos. of sub surface dams). Improvement of quality and quantity of drinking water used for colonies of Department of space from Kalangi river. Conservation of base flow by construction of sub surface dams – voices from farmers and end users of Swarnamukhi river basin in Chittoor Dist. A.P., India.

Managed aquifer recharge by construction of subsurface dams in semi arid regions – A Case Study of Kalangi river basin Nellore Dist. – A.P. India. Drought – try capturing the rain, briefing paper for members of parliament and state legislators, and interlinking of rivers in India.

3

DROUGHT

CHAPTER OUTLINE

- About types of droughts – ELNINO – climate change – causes of drought in India – U.S. drought monitor are explained.

Defining Drought

Drought is a normal, recurrent feature of climate. It occurs almost every where, although its features vary from region to region and defining, it can be difficult. Drought is unusual dryness of soil resulting in crop failure and shortage of water for other uses, caused by significantly lower rainfall than average over a prolonged period. Hot dry winds, shortage of water, high temperature and consequent evaporation of moisture from the ground can contribute to conditions of drought.

Understanding Droughts

Drought differs from other natural hazards such as cyclones, floods, earthquakes, volcanic eruptions and tsunamis that is:

(a) No universal definition exists.

(b) Being of slow – onset, it is difficult to determine the beginning and end of the event till (spi) standardized precipitation index was designed to enhance the detection of onset and monitoring and termination of drought by Mc Kee *et al.*, in 1993, 1995. Further concerns have been raised about the applicability of the SPI as a measure of changes in drought associated with climate change as it does not deal with changes in evapo-transpiration.

Alternative indices that deal with evapo-transpiration have been proposed. It is a new variation of the spi index by Vicente-Serrano *et al.*, in 2010 attempts to address this issue by including a temperature component through the calculation of their new spi index called the Standardized Precipitation Evapotranspiration Index (SPEI). The inputs required are precipitation, mean temperature and latitude of the site(s) to run the program on. The details about SPEI are available in this book.

The SPEI is a multi-scalar drought index based on climate data. It can be used for determining the onset, duration and magnitude of drought conditions with respect to normal conditions in a variety of natural and managed systems such as crops, ecosystems, rivers, water resources etc.

(c) No single indicator or index can identify precisely the onset and severity of the event and its potential impacts; multiple indicators are more effective.

(d) Spatial extent is usually much greater than that for other natural hazards making assessment and response actions difficult, since impacts are spread over longer geographical areas.

(e) Impacts are generally non-structural and difficult to quantify.

(f) Impacts are cumulative and effects magnify when events continue from one season or year to the next year.

(g) Another factor that distinguishes drought from other natural hazards is the absence of a precise and universally accepted definition for it.

Temperature, wind and relative humidity are also important factors to include in characterizing drought from one location to another. Definitions also need to be application specific because drought impacts will vary between sectors.

Drought means something different to a water manager, agricultural producer, hydroelectric power plant operator and wildlife biologist. Even within sectors, there are many different perspectives of drought because impacts may differ markedly. For example, the impacts of drought on crop yield may differ greatly for maize, wheat, soy beans and sorghum because they are planted at different times during the growing season and have different water requirements and different sensitivities at various growth stages to water and temperature stress.

Space and Time Characteristics of Drought

Droughts differ from one another in three essential characteristics: Intensity, duration, and spatial coverage. Intensity refers to the degree of the precipitation shortfall and or the severity of impacts associated with the shortfall. **It is generally measured by the departure of some climatic parameter (eg. Precipitation) indicator (eg. Reservoir levels) or index (eg. Standardized precipitation index) from normal and is closely linked to duration in the determination of impact.**

Many indices and indicators exist and are widely used for monitoring drought.

Another distinguishing feature of drought is its duration. Droughts usually require a minimum of two to three months to become established but can continue for months or years.

The magnitude of drought impacts is closely related to the timing of the onset of the precipitation shortage, its intensity and the duration of the event.

Droughts also differ in terms of their spatial characteristics. The areas affected by severe drought evolve gradually and regions of maximum intensity (eg. epicentre) shift from season to season and from year to year. In larger countries such as Brazil, China, India, the United States, or Australia, drought would rarely, if ever, affect the entire country. During the sever drought of the 1930s in the United States, for example, the affected by severe and extreme drought reached 65% of the country in 1934. This is the maximum spatial extent of drought in the period from 1895 to 2010.

The climate diversity and size of countries such as the United State suggests that drought is likely to occur somewhere in the country each year. In fact, on average 14% of the country is affected by severe to extreme drought annually. From a planning perspective, the spatial characteristics of drought have serious implications for agriculture, energy, transportation, health, recreating and tourism and other sectors.

A conventional attitude to drought, as a phenomenon of arid and semi–arid areas, is changing in India too. Now even regions with high rainfall, often face severe water scarcities.

Cherrapunji in Meghalaya, one of the world's highest rainfall area, with over 11,000 mm of rainfall, now faces drought for almost nine months of the year. On the other hand, the western part of Jaisalmer district of Rajasthan, one of the driest parts of the country is recording around 9 cm of rainfall in a year. Total rainfall increases generally eastwards and with height. Increase in precipitation is high at an elevation of around 1500 meters in the Himalaya Mountains. With average annual rainfall ranging between 20 cm to over 1000 cm, the primary challenge is to store this precious water for the dry season that may follow.

The droughts in Odisha state, which has an average rainfall of 1100 mm, remain a matter for continuing concern. Conditions of water scarcity in Himalayan region are also not uncommon.

Thus, drought is just not the scarcity of lack of rainfall, but an issue related to water resource management. An earlier analysis of incidence of droughts, over the last two centuries in India, does not show any increase in the frequency of drought in the recent years. However the severity appears to have increased.

Table 3.1   Drought Intensity – A Historical Perspective in India

Period	Drought Years	No. of Years
1801–25	1801, 04, 06, 12, 19, 25	6
1826–50	1832 33, 37	3
1851–75	1853, 60, 62 66 68 73	6
1876–1900	$1877^x + 91\ 99^x +$	3
1901–25	$1901^x\ 04\ 05^x\ 07\ 11\ 13\ 15\ 18^x + 20\ 25$	10
1926–50	$1939\ 41^x$	2
1951–75	$1951\ 65^x\ 66\ 68\ 72^x + 74$	6
1976–09	$1979^x\ 82\ 85\ 87 + 2002^x\ 2009^x$	6

x   Severe drought years = 10 (> 39.5% area affected)
+   Phenomenal drought years = 5 (> 49.5% area)
Source: Drought Research Unit (DRU) Indian Meteorological Department (IMD), Pune.

World Situation

Drought conditions have been widespread in North Africa, the mid-East, west Asian countries, India, China are also known to occur and also happen in North Central and South America. The increased frequency and intensity of extreme weather conditions such as droughts, floods, heat / cold waves, cyclones delayed or yearly onset of rains, long dry spells, early withdrawal, during the last two decades have been attributed to global warming.

What is Drought?

Drought is an insidious hazard of nature. It is often referred to as a "creeping phenomenon" and its impacts vary from region to region. Drought can therefore be difficult for people to understand. It is equally difficult to define, because what may be considered a drought in, say, Bali (six days without rain) would certainly not be considered a drought in Libya (annual rainfall less than 180 mm). In the most general sense, drought originates from a deficiency of precipitation over an extended period of time-usually a season or more-resulting in a water shortage for some activity, group, or environmental sector. Often associated with other climatic factors (i.e. temperature, high winds and low relative humidity) that can aggravate the severity of the drought event. Its impacts result from the interplay between the natural event (less precipitation than expected) and the demand people place on water supply, and human activities can exacerbate the impacts of drought. Because drought cannot be viewed solely as a physical phenomenon, it is usually defined both conceptually and operationally.

Conceptual Definitions

Conceptual definitions, formulated in general terms, help people understand the concept of drought. For example:

Drought is a protracted period of deficient precipitation resulting in extensive damage to crops, resulting in loss of yield.

Conceptual definitions may also be important in establishing drought policy. For example, Australian drought policy incorporates an understanding of normal climate variability into its definition of drought. The country provides financial assistance to farmers only under "exceptional drought circumstances," when drought conditions are beyond those that could be considered part of normal risk management. Declarations of exceptional drought are based on science-driven assessments. Previously, when drought was less well defined from a policy standpoint and less well understood by farmers, some farmers in the semiarid Australian climate claimed drought assistance every few years.

Operational Definitions

Operational definitions help define the onset, severity, and end of droughts. No single operational definition of drought works in all circumstances, and this is a big part of why policy makers, resource planners, and others, have more trouble recognizing and planning for drought, than they do for other natural disasters. In fact, most drought planners now rely on mathematic indices to decide when to start implementing water conservation or drought response measures.

To determine the beginning of drought, operational definitions specify the degree of departure from the average of precipitation or some other climatic variable over some time period. This is usually done by comparing the current situation to the historical average, often based on a 30-year period of record. The threshold identified as the beginning of a drought (e.g., 75% of average precipitation over a specified time period) is usually established somewhat arbitrarily, rather than on the basis of its precise relationship to specific impacts.

An operational definition for agriculture might compare daily precipitation values to evapo-transpiration rates to determine the rate of soil moisture depletion, then express these relationships in terms of drought effects on plant behaviour (i.e., growth and yield) at various stages of crop development. A definition such as this one could be used in an operational assessment of drought severity and impacts by tracking meteorological variables, soil moisture, and crop conditions during the growing season, continually re-evaluating the potential impact of these conditions on final yield.

Operational definitions can also be used to analyze drought frequency, severity, and duration for a given historical period. Such definitions, however, require weather data on hourly, daily, monthly, or other time scales and, possibly, impact data (e.g., crop yield), depending on the nature of the definition being applied. Developing a climatology of drought for a region provides a greater understanding of its characteristics and the probability of recurrence at various levels of severity. Information of this type is extremely beneficial in the development of response and mitigation strategies and preparedness plans.

Types of Drought

Research in the early 1980s uncovered more than 150 published definitions of drought. The definitions reflect differences in regions, needs, and disciplinary approaches.

Wilhite and Glantz[1] categorized the definitions in terms of four basic approaches to measuring drought: meteorological, hydrological, agricultural, and socioeconomic. The first three approaches deal with ways to measure drought as a physical phenomenon. The last deals with drought in terms of supply and demand, tracking the effects of water shortfall as it ripples through socioeconomic systems.

Meteorological Drought

Meteorological drought is defined usually on the basis of the degree of dryness (in comparison to some "normal" or average amount) and the duration of the dry period. Definitions of meteorological drought must be considered as region specific since the atmospheric conditions that result in deficiencies of precipitation are highly variable from region to region.

For example, some definitions of meteorological drought identify periods of drought on the basis of the number of days with precipitation less than some specified threshold. This measure is only appropriate for regions characterized by a year-round precipitation regime such as a tropical rainforest, humid subtropical climate, or humid mid-latitude climate. Locations such as Manaus, Brazil; New Orleans, Louisiana (U.S.A.); and London, England, are examples. Other climatic regimes are characterized by a seasonal rainfall pattern, such as the central United States, northeast Brazil, West Africa, and northern Australia. Extended periods without rainfall are common in Omaha, Nebraska (U.S.A.); Fortaleza, Ceará (Brazil); and Darwin, Northwest Territory (Australia), and a definition based on the number of days with precipitation less than some specified threshold is unrealistic in these cases. Other definitions may relate actual precipitation departures to average amounts on monthly, seasonal, or annual time scales.

---

[1] Wilhite, D.A.; and M.H. Glantz. 1985. Understanding the Drought Phenomenon: the Role of Definitions. Water International 10(3): 111–120.

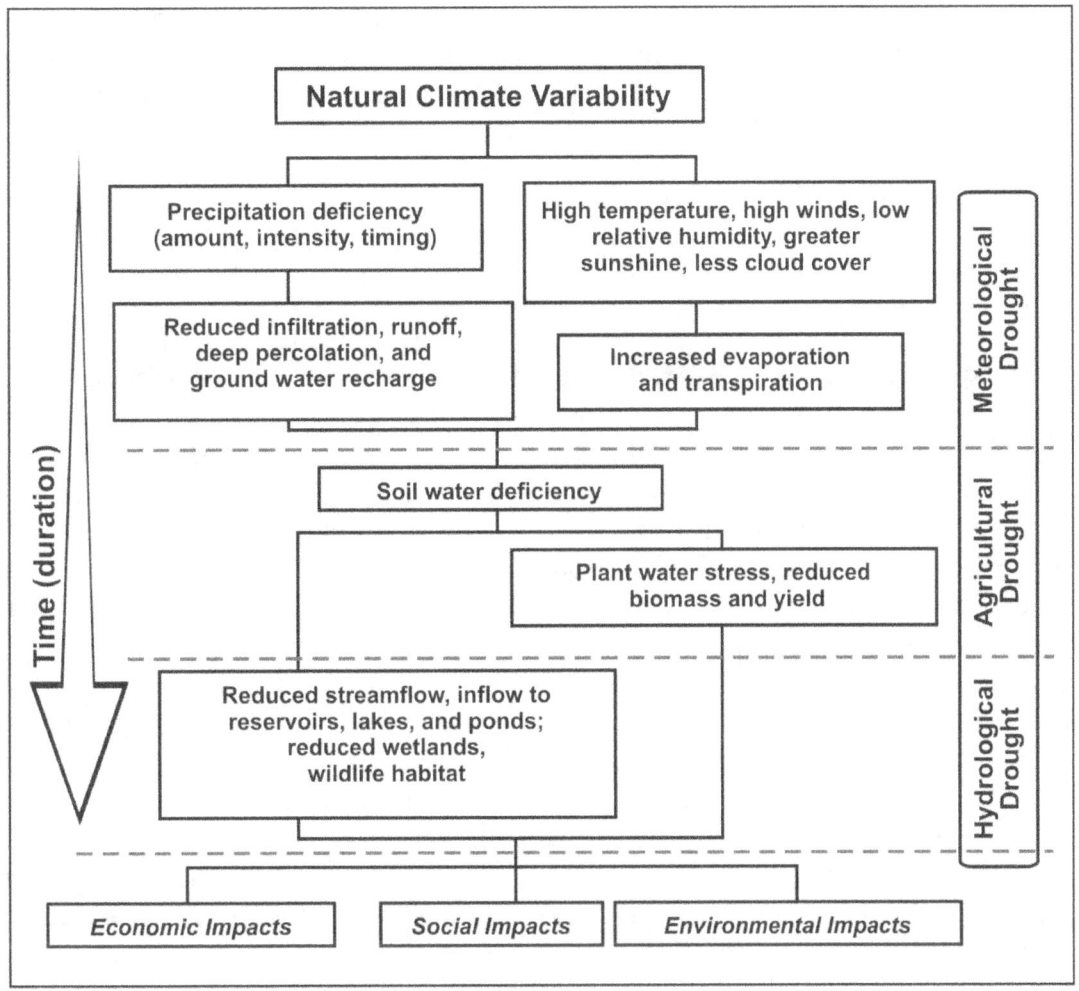

Sequence of drought occurrence and impacts for commonly accepted drought types. All droughts originate from a deficiency of precipitation or meteorological drought but other types of drought and impacts cascade from this deficiency.

Source: National Drought Mitigation Center, University of Nebraska-Lincoln, U.S.A.

Agricultural Drought

Agricultural drought links, various characteristics of meteorological (or hydrological) drought to agricultural impacts, focusing on precipitation shortages, differences between actual and potential evapo-transpiration, soil water deficits, reduced groundwater or reservoir levels, and so forth. Plant water demand depends on prevailing weather conditions, biological characteristics of the specific plant, its stage of growth, and the physical and biological properties of the soil. A good definition of agricultural drought should be able to account for the variable susceptibility of crops during different stages of crop development, from emergence to maturity. Deficient topsoil moisture at planting may hinder germination, leading to low plant populations per hectare and a reduction of final yield. However, if topsoil moisture is sufficient for early growth requirements, deficiencies in subsoil moisture at this early stage may not affect final yield, if subsoil moisture is replenished as the growing season progresses or if rainfall meets plant water needs.

Hydrological Drought

Hydrological drought is associated with the effects of periods of precipitation (including snowfall) shortfalls on surface or subsurface water supply (i.e., stream flow, reservoir and lake levels, groundwater). **The frequency and severity of hydrological drought is often defined on a watershed or river basin scale.** Although all droughts originate with a deficiency of precipitation, hydrologists are more concerned with how this deficiency plays out through the hydrologic system. Hydrological droughts are usually out of phase with or lag the occurrence of meteorological and agricultural droughts. It takes longer for precipitation deficiencies to show up in components of the hydrological system such as soil moisture, stream flow, and groundwater and reservoir levels. As a result, these impacts are out of phase with impacts in other economic sectors. For example, a precipitation deficiency may result in a rapid depletion of soil moisture that is almost immediately discernible to agriculturalists, but the impact of this deficiency on reservoir levels may not affect hydroelectric power production or recreational uses for many months. Also, water in hydrologic storage systems (e.g., reservoirs, rivers) is often used for multiple and competing purposes (e.g., flood control, irrigation, recreation, navigation, hydropower, wildlife habitat), further complicating the sequence and quantification of impacts. Competition for water in these storage systems escalates during drought and conflicts between water users increase significantly.

Socioeconomic Drought

Socioeconomic definitions of drought associate the supply and demand of some economic goods with elements of meteorological, hydrological, and agricultural drought. It differs from the aforementioned types of drought because its occurrence depends on the time and space processes of supply and demand to identify or classify droughts. The supply of many economic goods, such as water, forage, food grains, fish, and hydroelectric power, depends on weather. Because of the natural variability of climate, water supply is ample in some years but unable to meet human and environmental needs in other years. Socioeconomic drought occurs when the demand for an economic good exceeds supply as a result of a weather-related shortfall in water supply. For example, in Uruguay in 1988–89, drought resulted in significantly reduced hydroelectric power production because power plants were dependent on stream flow rather than storage for power generation. Reducing hydroelectric power production required, the government to convert to more expensive (imported) petroleum and implement stringent energy conservation measures to meet the nation's power needs.

In most instances, the demand for economic goods is increasing as a result of increasing population and per capita consumption. Supply may also increase because of improved production efficiency, technology, or the construction of reservoirs that increase surface water storage capacity. If both supply and demand are increasing, the critical factor is the relative rate of change. Is demand increasing more rapidly than supply? If so, vulnerability and the incidence of drought may increase in the future as supply and demand trends converge.

Factors Responsible for Drought Initiation

The factors attributed to drought initiation are as follows:

1. Ocean – atmosphere system, sea surface temperature anomalies, high temperature of the soil during drought and the increase in fine particles in the air, the high albedo in dry areas.

2. Solar – weather relationship

3. Monsoon mechanism and impairment of this mechanism (Radier and Beran, 1979) Sahelain droughts are caused by

   (a) Southward shift of Saharan Trough leading to weak pressure gradient and a weak inter tropical convergence (ITC).

   (b) Progressive degradation of vegetation and increase in albedo.

   (c) Extra tropical factors such as westerlies etc. Precursors to the subsidence have to be understood in terms of atmospheric general circulation, which in turn are conditioned by abnormalities at the land and sea surface.

There are many possible large scale features of the general circulation, which are responsible for a given drought varying according to geographical circumstances.

In tropical regions such as India where rain is due to convergence, the migration or weakening of the inter-tropical convergence zone brings drought.

These departures are in turn explained by anomalous behaviour in large scale features of the circulation such as the St. Helena (Sahel and Brazilian modest) or the Mascarene (India) highs.

In higher altitudes, an upper level high-pressure system can block low-pressure systems. The 1976 drought in Europe has been ascribed to such a blocking high.

Drought in India is correlated to irregular monsoon rainfall in space and time. It is attributed to weakening of Mascarene cell in the southern hemisphere so that ITCZ is weakened to penetrate the Indian peninsula.

The pressure at South Georgic Island (54°s and 36° w) correlates well with drought in India.

In addition to high albedo the local reduction of thermal gradient that sustains the early jet contributes to drought events. The various factors that are responsible for the drought events are shown in figure.

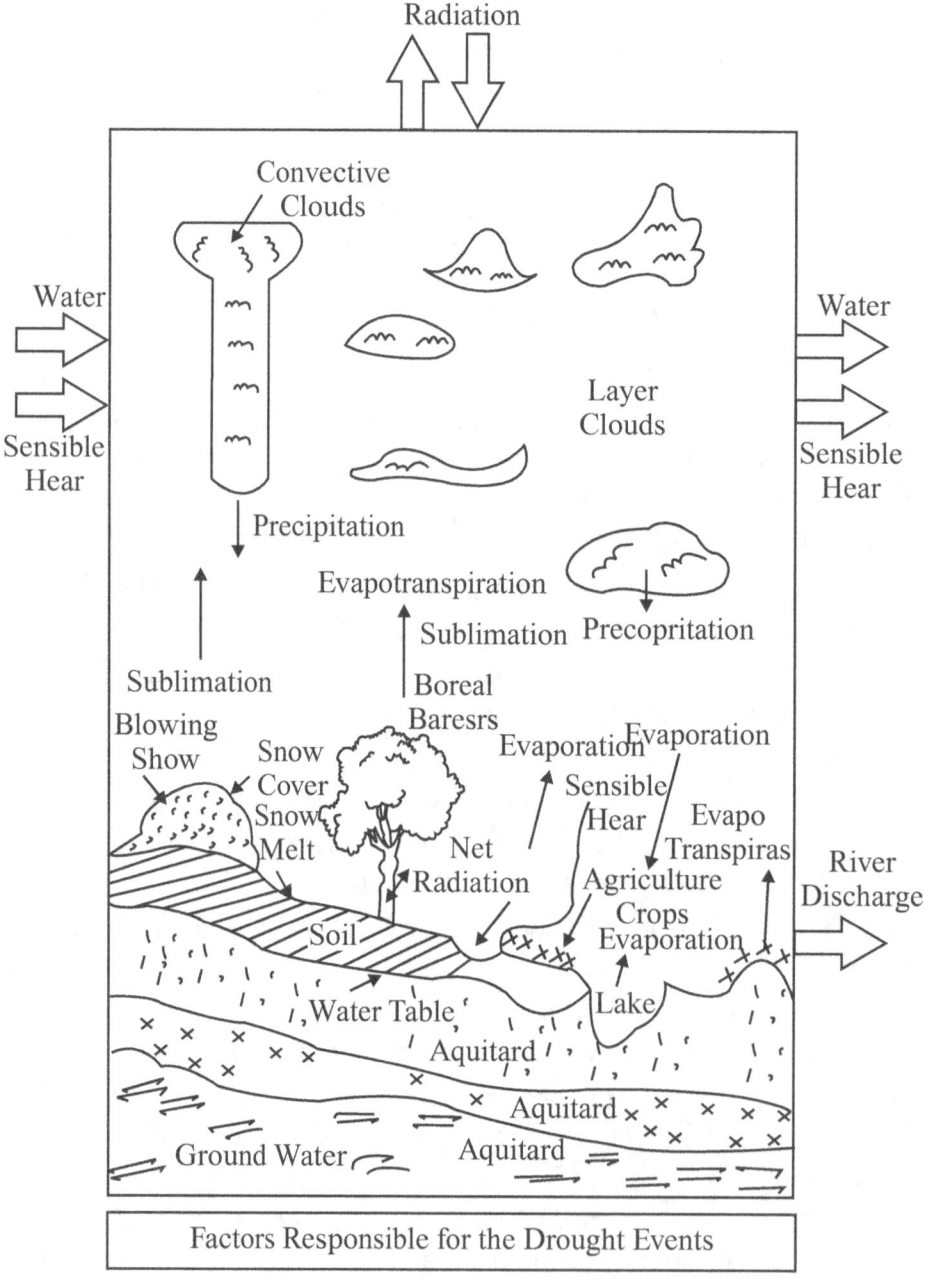

Factors Responsible for the Drought Events

Common causes for Drought in India

1. **Meteorology**

   (a)   Inadequate monsoon rainfall

   (b)   High temperature evaporation, wind speed

   (c)   Unseasonal rains and fog snowfall

2. **Water resources**

   (a)   Inadequate water availability

(b)   High water loss in storage and distribution, utilities

(c)   Over exploitation of surface and ground water.

3. **Agriculture – crop yield**

(a)   Shift in agricultural practices (low to moderate) water demand crops to high crops)

(b)   Crop damage due to rain and snow pest

4. **Population**

(a)   High greater rate of human and animals

(b)   Location of high water consuming milestones at semi arid / arid regions.

Predicting Drought

Empirical studies conducted over the past century have shown that meteorological drought is never the result of a single cause. It is the result of many causes, often synergistic in nature.

Global Weather Patterns

A great deal of research has been conducted in recent years on the role of interacting systems, or teleconnections, in explaining regional and even global patterns of climatic variability. These patterns tend to recur periodically with enough frequency and with similar characteristics over a sufficient length of time that they offer opportunities to improve our ability for long-range climate prediction, particularly in the tropics. One such teleconnection is the El Nino/Southern Oscillation (ENSO).

High Pressure

The immediate cause of drought is the predominant sinking motion of air (subsidence) that results in compressional warming or high pressure, which inhibits cloud formation and results in lower relative humidity and less precipitation. Regions under the influence of semi permanent high pressure during all or a major portion of the year are usually deserts, such as the Sahara and Kalahari deserts of Africa and the Gobi Desert of Asia. Most climatic regions experience varying degrees of dominance by high pressure, often depending on the season. Prolonged droughts occur when large-scale anomalies in atmospheric circulation patterns persist for months or seasons (or longer). The extreme drought that affected the United States and Canada during 1988 resulted from the persistence of a large-scale atmospheric circulation anomaly.

Too Many Variables

Scientists don't know how to predict drought a month or more in advance for most locations. Predicting drought depends on the ability to forecast two fundamental meteorological surface parameters, precipitation and temperature. From the historical record we know that climate is inherently variable. We also know that anomalies of precipitation and temperature may last from several months to several decades. How long they last, depends on air-sea interactions, soil moisture and land surface processes, topography, internal dynamics, and the accumulated influence of dynamically unstable synoptic weather systems at the global scale.

The potential for improved drought predictions in the near future differs by region, season, and climatic regime.

The Tropical Outlook

In the tropics, for example, meteorologists have made significant advances in understanding the climate system. Specifically, it is now known that a major portion of the atmospheric variability that occurs on time scales of months to several years is associated with variations in tropical sea surface temperatures. The Tropical Ocean Global Atmosphere (TOGA) project has produced results that suggest, that it may now be possible, to predict certain climatic conditions associated with ENSO events more than a year in advance. For those regions whose climate is greatly influenced by ENSO-events, TOGA project results may help produce more reliable meteorological forecasts that can reduce risks in those economic sectors most sensitive to climate variability and, particularly, extreme events such as drought.

The Temperate Zone Outlook

In the extratropical regions, current long-range forecasts are of very limited reliability. The ability that does exist is primarily the result of empirical and statistical relationships. In the tropics, empirical relationships have been demonstrated to exist between precipitation and ENSO events, but few such relationships have been confirmed above 30 north latitude. Meteorologists do not believe that reliable forecasts are attainable for all regions a season or more in advance.

Drought Impacts

Why track drought impacts Drought is a slow – moving, natural hazard that affects millions of people world wide each year. The meteorological phenomenon triggers a cascade of impacts across agricultural, hydrological, economic, environmental and social systems. Understanding of these impacts is crucial for drought planning, mitigation and response. It also helps decision makers identify and reduce vulnerability to drought.

Drought produces a complex web of impacts that spans many sectors of the economy and reaches well beyond the area experiencing physical drought. This complexity exists because water is integral to society's ability to produce services.

Common types of drought impacts are as follows:

1. **Economic category:** (a) Agriculture, (b) Industry, (c) Tourism and recreation, (d) Energy, (e) Financial, (f) Transportation.

2. **Environmental category:** (a) Animal / plant, (b) Wet land, (c) Water quality

3. **Social category:** (a) Stress and health, (b) Nutrition, (c) Recreation, (d) Public safety, (e) Cultural values, (f) Aesthetic values.

Drought produces wide ranging impacts that span many sectors of the national economy. Drought produces both direct and indirect impacts. Examples of Direct drought are reduced agricultural production, depleted water levels, higher live stock and wild life mortality rates, increased plant diseases, increased wind erosion, increased insect infestation.

When direct impacts have multiplier effects through the economy and society they are referred to as indirect impacts. These include a reduction in agricultural production that may result in reduced income for farmers and agribusiness, increased prices for food and timber, unemployment, reduced purchasing capacity and demand for consumption, default on agricultural loans, rural unrest and reduction in agricultural employment leading to migration and drought relief programmes.

Economic impacts refer to production losses in agriculture and related sectors especially forestry and fisheries because these sectors rely on surface and subsurface water supplies. It causes a loss of income and purchasing power, particularly among farmers and rural population dependent upon the primary sector for their raw materials would suffer losses due to reduced supply or increased prices. Drought thus has a multiplier effect throughout the economy, which has a dampening impact on employment, flow of credit and tax collections. If the drought is country wide, macro economic indicators at the national level are adversely impacted.

Environmental impacts such as lower water levels in the reservoirs, lakes and ponds as well as reduced flows from springs and streams would reduce the availability of food and drinking water and adversely affect fish and wild life habitat. It may also cause loss of forest cover, migration of wild life and their greater mortality due to increased contact with agricultural producers as animals seek food from farms and and producers are less tolerant of the intrusion.

A prolonged drought may also result in increased stress among endangered species and cause loss of biodiversity.

Reduced stream flow and loss of wet lands may cause changes in the levels of salinity, increased ground water depletion, land subsidence and reduced recharge may damage aquifers and adversity effect the quality of water (i.e. salt concentration, increased water temperature, acidity, dissolved oxygen, turbidity). The degradation of land scape quality including increased soil erosion may lead to a more permanent loss of biological productivity of the landscape.

Social impacts arise from lack of income, causing out migration of the population from the drought affected areas.

People in India seek to cope with drought in several ways which affect their sense of well-being. They withdraw their children from schools, postpone daughters marriages, and sell their assets such as land and cattle.

In addition to economic hardships, it causes a loss of social status and dignity, which people find hard to accept. In adequate food intake may lead to malnutrition and in some extreme cases, cause starvation. Access and use of scarce water resources generate situations of conflict, which could be socially very disruptive. In equalities in the distribution of drought impacts and relief may in exacerbate these social tensions further.

Meteorological history of droughts in India   During 1871–2002, there were 22 major droughts (i.e. anomaly below – 10 per cent) – They are during 1873, 1877, 1899, 1901, 1904, 1905, 1911, 1981, 1920, 1941, 1951, 1965, 1966, 1968, 1972, 1974, 1979, 1982, 1985, 1986, 1987 and 2002. The frequency of drought has varied over the decades. From 1899 to 1920 there were severe drought years.

The incidence of drought came down between 1941 and 1965 when the country witnessed just three drought years.

Again, during 1965–87 of the 21 years 10 were drought years and increased frequency was attributed to the Elnino Southern Oscillation (ENSO). Among the drought years the 1987 drought was one of the worst droughts of the country with an over all rainfed deficiency of 19%. It affected 59–60% of the crop area and a population of 285 million.

In 2002 too, the over all rainfall deficiency for the country as a whole was 19% over 300 million people spread over 18 states were affected by drought in varying degrees. Around 150 million cattle were affected due to lack of fadder and water. Food grains production registered the steepest fall of 29 million tones. No other drought in the past has caused reduction in food grain production in this extent.

Source: Sarma, 2004 (Manual for Drought Management 2009).

Table 3.2   The Administrative Districts Frequently Affected by the Drought in India

Name of the State	Involved Districts
Andhra Pradesh	Anantapur, Chittoor, Cuddapah, Hyderabad, Kurnool, Mahaboobnagar, Nalgonda, Prakasam
Bihar	Munger, Nawdah, Palaman, Roptas, Bhojpur, Aurangabad, Gaya
Gujarat	Ahmedabad, Amrely, Banaskanta, Bhavanagar, Bharuch, Jam Nagar, Kheda, Kutch, Meshana, Panchmahal, Rajkot, Surendranagar.

Drought Impact Reporter

The drought impact reporter is an interactive web-based mapping tool designed to compile and display impact information across the United States of America in near real-time from a variety of sources such as media, Government agencies and the public. Launched in July 2005, this tool is the only nation wide, multi-source archive of drought impact information.

Sponsored by U.S. Department of Agriculture's Risk management agency, the national oceanic and Atmospheric Administration's TRACS Program and Sectoral Applications Research Program.

Information within the drought impact reporter is collected from a variety of sources including the media, Government agencies and reports and citizen observers. Each of these source provides different types of information at different spatial and temporal scales.

The drought impact reporter allows anyone to submit a drought impact. All public contributions are moderated before appearing on the web-site and moderators request additional information as needed.

The information provided by the Drought Impact Reporter (DIR) will help U.S. policy makers and resource managers to identify what types of impacts are occurring and where. It can also help individual agricultural producers a mass evidence of drought conditions.

The U.S. Department of Agriculture's Office of the Chief Economist Publishers Weekly updates showing how drought affects leading crops and cattle.

The U.S. Army Crops of Engineers Missouri river basin water management news releases regularly in press about details of the effects of drought on reservoir storage, releases of water for navigation and hydropower production.

The national inter agency fire center publishes regularly updated statistics and situation reports. Not all wildfires are due to drought, but drought can certainly be one of the underlying factors that makes fires more intense or more widespread. Farmers submit comments on how their crops are doing to farm Journals Agriculture web. Many of their comments are about the effects of drought.

As of June 2008, the nation's 122 weather forecast offices begun producing drought information statements including information about impacts, whenever any part of their service area was in severe drought (D_2) on the U.S. Drought monitor.

The National Agricultural statistics services publishes the weekly weather and crop bulletin and crop progress reports. The USDA's NASS publishes a monthly report on crop production including estimates of upcoming harvest totals for various crops.

NOAA's National climatic data center publishes monthly state of the climate reports including one on drought, which incorporates information about impacts.

U.S. Drought Monitor

This is created in 1999, the weekly U.S. Drought monitor (USDM) was the first to use a composite index / indicator approach (Svoboda et al., 2002). The product is not an index in and of itself but rather a combination of indicators and indices that are combined using a simple $D_0 - D_4$ scheme and a percentile ranking methodology (Table) to look at addressing both short and long term drought across the United States.

The key indicators / indices revolve around monitoring precipitation, temperature, stream flow, soil moisture, snow pack and snow water equivalent. Various indices, such as the SPI and PDSI are incorporated and integrated with remotely sensed vegetation indices to come up with a "blended convergence of evidence" approach in dealing with drought severity.

The ranking percentile approach allows the user to compare and contrast indicators originally having different periods of record and units into one comprehensive indicator that addresses the customized needs of any given user.

The approach also allows for flexibility and adaptation to latest indices indicators and data that become available over time. It is a blending of objective science and art through the integration of impacts and reports from local experts at the field level.

The impacts covered and labelled on the map are (A) for agriculture and (H) for hydrological drought. Some 275 local experts from across the country are allowed to view the draft maps and provide their input data and impacts to either support or refuse the initial depiction.

An interactive process works through all the indicators/indices data and field input until a compromise is found for the week. The process then repeats itself the next week and so on. In addition, a set of objective blends are used to help guide the process. The method combines a different set of indicators to produce separate short – and – log term blend maps that take various indices with variable weightings (depending on region and type of drought) to produce a composite classification scheme and the objective blends can be found at http://drought.unl.u/dm

Table 3.3 The U.S. Drought Monitor Classification and Ranking Percentile Scheme

Category	Description	Ranking Percentile
D_0	Abnormally dry	30
D_1	Moderate	20
D_2	Severe	10
D_3	Extreme	5
D_4	Exceptional	2

Source: National Drought Mitigation Centre, USDA, NOAA

Composite Approach (Multiple Indicators / Indices)

In efforts to build as comprehensive and flexible a drought early warning system (DEWS) as possible, it is important to monitor drought across the many sectors mentioned earlier.

The use of a single index, will rarely work for all places at all times and for all types of drought. Most coordinated monitoring efforts at the national level are going to need to track all types of droughts.

In cases such as those it is important to utilize and incorporate a consolidation of indices and indicators into one comprehensive composite index. A composite index approach, allows for the most robust way of detecting and determining the magnitude (duration + intensity) of droughts as they occur.

Through a convergence of evidence approach one can best determine (for a particular state, country or region for a particular time of the year) which indices and indicators do the best job of depicting and tracking all types of droughts.

The users can then determine which indicators to use and how much weight to give each indicator / indices in a "blended approach" that incorporates a multiple parameter and weighting scheme. Such approaches have been used in the U.S. drought monitor (U.S. DM) and North American drought monitor (NADM) as described below and as part of a series objective blend products which are produced for the USDM.

Tips for using the U.S. Drought Monitor Website

The U.S. drought monitor map is released each Thursday morning along with statistics reflecting the percent area in each category of drought for the entire country and Puerto Rico, for the 48 contiguous states, for each climate region and for individual states.

There are many ways to get the statistical information: To see the percent area statistics for a region or a state, drill down by double – clicking on the main U.S. drought monitor map. For a quick statistical summary for the whole country (as for the 48 contiguous states) click on statistics comparison below the main map.

The weekly comparison link under the maps and data tab in the main navigation bar takes you to the U.S. drought monitor weekly comparison. Where you can choose two maps for a visual comparison of differences in the expanse of drought. Percent area statistics also appear below the two maps. These statistics are for the contiguous United States. You can select a region or a state from the tan drop down box near the top of the archive page.

The tabular data archive page lists all U.S. Drought monitor statistics.

Clicking on the chart icon in the middle of the brown bar below the maps on the archive page gives you a visual display of the area in each category of drought from 2000 through the present. Statistics on the percent area of the United States in each category of drought conditions are automatically updated each week when the new map is released. The current week's status is listed at in the row labelled "current".

To see the percent area statistics for a region or a state, click on the main U.S. drought monitor map.

The archive tables accessible from the archive allow you to scan the U.S. drought monitor values over time, and to select data for the entire country, the contiguous 48 states, a region on an individual state.

What is Climatology?

"Climate is what you expect. Weather is what you get."

Weather is the condition of the atmosphere over a brief period of time. For example, we speak of today's weather or the weather this week. Climate represents the composite of day-to-day weather over a longer period of time.

A climatologist attempts to discover and explain the impacts of climate so that society can plan its activities, design its buildings and infrastructure, and anticipate the effects of adverse conditions. Although climate is not weather, it is defined by the same terms, such as temperature, precipitation, wind, and solar radiation.

People in Minneapolis–St. Paul expect a white Christmas, and people in New Orleans expect very warm, humid summers. And a traveler going to Orlando, Florida, in March will not pack the same kind of clothing as a traveler going to Vail, Colorado, in March. These examples show how climate influences our daily lives. Additionally:

○ Our houses are designed based on the climate where we live.

○ Farmers make plans based on the length of the growing season from the last killing freeze in the spring to the first freeze in the fall.

○ Utility companies base power supplies on what they expect to be the maximum need for heating in the winter and the maximum need for cooling in the summer.

What is "Normal"?

Climate is usually defined by what is expected or "normal", which climatologists traditionally interpret as the 30-year average. By itself, "normal" can be misleading unless we also understand the concept of variability. For example, many people consider sunny, idyllic days "normal" in southern California. History and climatology tell us that this is not the full story. Although sunny weather is frequently associated with southern California, severe floods have had a significant impact there, including major floods in 1862 and 1868, shortly after California became a state. When you also factor in severe droughts, most recently those of 1987–94, a more correct statement would be that precipitation in southern California is highly variable, and that rain is most likely between October and April.

The misconception that weather is usually normal becomes a serious problem when you consider that weather, in one form or another, is the source of water for irrigation, drinking, power supply, industry, wildlife habitat, and other uses. To ensure that our water supply, livelihoods, and lives are secure, it is essential that planners anticipate variation in weather, and that they recognize that drought and flood are both inevitable parts of the normal range of weather.

Weather and Climate

If you look outside right now, do you see sunshine, or mostly clouds? Is it windy? Is it hot or cold? Whatever your answers to those questions are, that is your weather for today. Weather can be defined as the conditions experienced in a place over brief period of time, like day-to-day or even over a week.

When you hear or read a weather forecast, you may hear the word normal. The weather forecaster may say, "today will be warmer than normal" or "we are having below normal precipitation". Weather that we expect or know to be "normal" for a place is one way to define climate.

So, in other words, climate is a place's weather over a longer period of time, like months, seasons, and years. What is normal for a place depends on its elevation (how high above sea level it is), how close it is to the oceans or other large bodies of water, and its latitude (how close or far it is from the equator).

A place's climate can change slightly from year to year or decade to decade. This is what we call natural climate variability. Because these changes or variations can occur, we consider drought to be a normal part of climate just like floods, hurricanes, blizzards, and tornadoes.

What causes these variations? Well, let's start by looking at the water cycle.

Water Cycle

Water covers more than 80% of the earth's surface. It is found in oceans, lakes, rivers and even ice caps and glaciers. Water is also found underground in aquifers. The water that exists today is the same water that existed billions of years ago. This is because water is, what we call a limited renewable resource.

Water is a renewable resource because it travels through the oceans, rivers, ground and atmosphere, it is always moving. It falls from the sky as rain or snow into our oceans, lakes and rivers and onto land. Precipitation that falls on the land enters the ground water through percolation or travels to streams, rivers and lakes as run off. Water in streams and rivers is carried to the oceans, where it evaporates and forms clouds – where the cycle starts all over again.

Water is also a limited resource. We will always have the same amount of water on the earth, but we can not always use as much as we need. One reason is that 97% of the earths water is salt water and salt water is not good for people and wild life and many plants. Of the remaining 3% which is fresh water, nearly 75% is frozen in glaciers, making it unavailable to us. Another reason that water is a limited resource is that pollution, increased human demand of water, and changes in precipitation patterns can decrease both the quality and amount of water available to people, plants and wild life.

Water and Weather on the Move

Water evaporating from the oceans, lakes, rivers and streams moves to the atmosphere. Air carries the moisture up, and if conditions are favourable it forms clouds. Wind moves these clouds around the globe. The wind that caries the clouds that bring rain is called the jet stream. The jet stream changes its pattern with each season. In other words, the jet stream will carry weather patterns, such as precipitation and temperature, in different directions or over different routes during each season. Think about four different routes you could take to a friends house, and during each season you would take a different route – that is what the jet stream does.

If the jet stream changes its pattern or is blocked by 'ridges' or 'troughs' of air in the atmosphere, the normal weather for a place can be much different for a period of time. It is almost as if the jet stream has hit a road block or take a detour. When the jet stream hits a road block or takes a certain detour and is not bringing the clouds that produce rain, a <u>drought can occur</u>. These patterns in the jet stream could change for many reasons. Scientists are still uncovering the answers, but many think that influences such as differences in the amount of snow and ice cover, the amount of vegetation (trees or grasses) covering the land, the moisture in the soil and ocean surface temperature and currents can cause these patterns to change.

ENSO and Drought Forecasting

- o El Niño defined
- o Frequency
- o El Niño and the southern oscillation
- o Tele-connections
- o ENSO and U.S. Droughts
- o ENSO and drought around the World
- o Can we predict ENSO?
- o References

El Niño defined Every two to seven years off the western coast of South America, ocean currents and winds shift, causing water temperatures to warm and displacing the nutrient-rich cold water that normally wells up from deep in the ocean. The invasion of warm water disrupts both the marine food chain and the economies of coastal communities that are based on fishing and related industries. Because the phenomenon peaks around the Christmas season, the fishermen who first observed it named it El Niño ("the Christ Child"). In recent decades, scientists have recognized that El Niño is linked with other shifts in global weather patterns. (For a colorful look at El Niño with real-time graphics, check out NOAA's El Niño Theme Page.)

Frequency The return period of the El Niño event varies from 2 to 7 years. The intensity and duration of the event are also varied and hard to predict. Typically, it lasts anywhere from 14 to 22 months, but it can be much longer or shorter. El Niño often begins early in the year and peaks between the following November and January, but no two events behave in the same way. El Niño is by no means a new phenomenon, and researchers are still working to determine whether global warming would intensify or otherwise affect El Niño. Evidence of the events goes back hundreds of years. This "proxy" or indirect, climatic data can be found in the form of tree ring analysis, sediment or ice cores, coral reef samples, and even historical accounts from early settlers.

El Niño and the southern oscillation   The Southern Oscillation, a "See saw of atmospheric pressure between the eastern equatorial Pacific and Indo–Australian areas" (Glantz *et al.*, 1991), is closely linked with El Niño. During an El Niño–Southern Oscillation (ENSO) event, the Southern Oscillation is reversed. Generally, when pressure is high over the Pacific Ocean, it tends to be low in the eastern Indian Ocean, and vice versa (Maunder, 1992). It is measured by gauging sea-level pressure in the east (at Tahiti) and west (at Darwin, Australia) and calculating the difference. This is then put into an index called the Southern Oscillation Index (SOI) or Tahiti–Darwin Index. High negative values of the SOI represent an El Niño, or "warm event". ENSO events are those in which a Southern Oscillation extreme and an El Niño occur together. El Niño and Southern Oscillation often occur together, but also happen separately. The Climate Prediction Center shows a plot of the SOI.

High positive values of the SOI indicate a La Niña, or "cold event". La Niña is the counterpart of El Niño and represents the other extreme of the ENSO cycle. In this event, the sea surface temperatures in the equatorial Pacific drop well below normal levels and advent to the west while the trade winds are unusually intense rather than weak. La Niña years often (but not always) follow El Niño years. A listing of El Niño and La Niña years since the turn of the century is found here.

ENSO is perhaps a better term than El Niño for purposes of understanding global weather patterns, as it turns out that the shifts in sea surface temperatures (SSTs) off the west coast of South America are just one part of the coupled interactions of atmosphere, oceans, and land masses. The term Southern Oscillation refers to the atmospheric component of the relationship, and El Niño represents the oceanic property in which sea surface temperatures are the main factor.

Teleconnections   ENSO occurrences are global climate events that are linked to various climatic anomalies. Not all anomalies, even in ENSO years, are due to ENSO. In fact, statistical evidence shows that ENSO can account at most for about 50% of the inter annual rainfall variance in eastern and southern Africa (Ogallo, 1994), but many of the more extreme anomalies, such as severe droughts, flooding, and hurricanes, have strong teleconnections to ENSO events. Teleconnections are defined as atmospheric interactions between widely separated regions (Glantz, 1994). Many researchers are studying the relationships between ENSO (and La Niña) events and weather anomalies around the globe to determine whether links exist. Understanding these teleconnections can help in forecasting droughts, floods, and tropical storms (hurricanes).

Estimates of the economic impacts of the 1982–83 El Niño, perhaps the strongest event in recorded history, conservatively exceeded $8 billion worldwide, from droughts, fires, flooding, and hurricanes (NOAA, 1994). Some 1000–2000 deaths have been blamed on the event and the disasters that accompanied it. Virtually every continent felt the impacts of this strong event. Incidentally, the extreme drought in the Midwest Corn Belt of the United States during 1988 has been inconclusively linked to the "cold event", or La Niña, of 1988 that followed the ENSO event of 1986–87.

ENSO and U.S. droughts   In North America, particularly the United States, the impacts of El Niño are most dramatic in the winter. El Niño produces winters that are generally mild in the northeast and central United States and wet over the south from Florida to Texas. Alaska and the north-western regions of Canada and the United States can be abnormally warm. This might be a result of the forcing caused by a Pacific–North American (PNA) pattern that is typified by a high pressure ridge over north–western North America and a low pressure trough in the south-eastern United States. This serves as an upper-level steering mechanism for moisture and temperature at the surface. Once the pattern is entrenched, regions under the ridge can expect little in the way of precipitation while those in the trough can't turn it off and are prone to frequent flooding.

Ropelewski and Halpert (1986) studied North American precipitation and temperature patterns associated with ENSO conditions and concluded the following. In the Great Basin area of the western United States, above-normal precipitation was recorded during ENSO years in 81% of the cases for the "season" that runs from April to October. In the south-eastern United States and northern Mexico, above-normal precipitation was also recorded for 81% of the cases for the "season" that began in October of the ENSO year and concluded in March of the following year. For temperature anomalies during ENSO conditions in North America, the U.S. Pacific Northwest, western Canada, and parts of Alaska showed warmer temperatures in 81% of the years while the south–eastern United States showed below-normal temperatures around 80% of the time. This would seem consistent with a typical PNA atmospheric pattern. During stronger events, the United States experiences flooding and severe storms in some regions and droughts and heat waves in other areas. Hurricane activity is usually minimal in the Atlantic Ocean, sparing the coastal areas from the Gulf of Mexico to the northeast. In the coastal west, the displacement of the jet stream can bring abnormally large amounts of rain and flooding to California, Oregon, and Washington. During the summer, heat waves and below-normal precipitation bring drought, crop failures, and even death. U.S. crop losses from the 1982–83 El Niño were projected to be in the neighbourhood of $10–12 billion (Wilhite *et al.*, 1987).

ENSO and drought around the world   During an ENSO event, drought can occur virtually anywhere in the world, although researchers have found the strongest connections between ENSO and intense drought in Australia, India, Indonesia, the Philippines, Brazil, parts of east and south Africa, the western Pacific basin islands (including Hawaii), Central America, and various parts of the United States. Drought occurs in each of the above regions at different times (seasons) during an event and in varying degrees of magnitude.

Ropelewski and Halpert also looked at the link between ENSO events and regional precipitation patterns around the globe (1987). North eastern South America from

Brazil up to Venezuela shows one of the strongest relationships. In 17 ENSO events, this region had 16 dry episodes. It is not uncommon to find the rain forests burning during these dry periods. Other areas from their study also showed a strong tendency to be dry during ENSO events. In the Pacific basin, Indonesia, Fiji, Micronesia, and Hawaii are usually prone to drought during an event. Virtually all of Australia is subjected to abnormally dry conditions during ENSO events, but the eastern half has been especially prone to extreme drought. This is usually followed by bush fires and a decimation of crops. India has also been subjected to drought through a suppression of the summer monsoon season that seems to coincide with ENSO events in many cases. Eastern and southern Africa also showed a strong correlation between ENSO events and a lack of rainfall that brings on drought in the Horn region and areas south of there. Another region they found to be abnormally dry during warm events was Central America and the Caribbean Islands.

Thus, ENSO events seem to have a stronger influence on regions in the lower latitudes, especially in the equatorial Pacific and bordering tropical areas. The relationships in the mid-latitudes aren't as pronounced, nor are they as consistent in the way wet or dry weather patterns are influenced by El Niño. The intensity of the anomalies in these regions is also more inconsistent than those of the lower latitudes. NOAA's Climate Prediction Center has short papers on the typical impacts associated with ENSO and La Niña episodes.

Can we predict ENSO?   If we can understand some of the teleconnections discussed above, it can lead us to some general predictive capabilities via numeric computer models that can help us determine and conclude when conditions are favourable for the onset of an event. Numeric models try to emulate processes (and dynamic relationships) that occur in nature using sets of numbers and equations. But once an event is underway, forecasting its duration and intensity are difficult at best.

In Ropelewski and Halpert's (1987) study on global precipitation patterns and ENSO events, they found that the consistency and magnitude of the precipitation relationships to ENSO events could serve as a practical utility for forecasting precipitation in certain regions (and seasons) once it was determined that an event was in progress. This can serve as a broad-brush approach for given regions, with the understanding that expanses within any given area will not behave in the exact same manner from event to event.

NOAA has established and now operates an array of moored buoys in the equatorial Pacific Ocean. These buoys measure temperature, currents, and winds in this region on a daily basis. The data is available to scientists around the world in real time, enabling them to use the data for both research and forecasting. This network is very valuable in that the first stages of an ENSO event occur in this region. By monitoring data from past episodes and the data from the months leading up to an episode, scientists can use numerical models (similar to but not as reliable as those used in weather forecasting) to help them predict and/or simulate ENSO events. The predictive models are becoming more sophisticated and more effective in many respects thanks in part to the expanded data sets that are available for the equatorial Pacific region. The dynamic coupled nature of the new models, has allowed for prediction of ENSO events a year or more in advance.

ENSO forecasts help countries anticipate and mitigate droughts and floods, and are very useful in agricultural planning. Countries that are in latitudes with strong El Niño connections to weather patterns, such as Brazil, Australia, India, Peru, and various African nations, use predictions of near-normal conditions, weak El Niño conditions, strong El Niño conditions, or a La Niña to help agricultural producers select crops most likely to be successful in the coming growing season. In countries or regions with a Famine Early Warning System (FEWS) in place, ENSO forecasts can play a key role in mitigating the impacts of flood or drought that can lead to famine. Famine, like drought, is a slow-onset disaster, so forewarning may enable countries to greatly reduce, if not eliminate, its worst impacts.

ENSO advisories are used to a lesser extent in planning in North America and other extratropical countries, because the links between ENSO and weather patterns are less clear in these areas. As prediction models improve, the role of ENSO advisories in planning in mid-latitude countries will increase. The Climate Prediction Center is responsible for issuing ENSO advisories. For the latest information on the status of ENSO, go to the ENSO Diagnostic Advisory.

Climate Change

- o   What is the greenhouse effect?
- o   What we know about the greenhouse effect and climate change
- o   Climate change and the hydrological cycle
- o   What we do not know?
- o   Water resources and climate change
- o   Bibliography
- o   Climate change links

What is the greenhouse effect?   The Greenhouse Effect is a naturally occurring phenomenon necessary to sustain life on earth. In a greenhouse, solar radiation passes through a mostly transparent piece of glass or plastic and warms the inside air, surface, and plants. As the temperature increases inside the greenhouse, the interior of the greenhouse radiates energy back to the outside and eventually a balance is reached.

The earth and its atmosphere simulates these greenhouse conditions. Short-wave radiation from the sun passes through the earth's atmosphere. Some of this radiation is reflected back into space, some of it is absorbed by the atmosphere, and some of it makes it to the earth's surface, where it is either reflected or absorbed. The earth, meanwhile, emits long-wave radiation toward space. Gases within the atmosphere absorb some of this long-wave radiation and re-radiate it back to the surface. It is because of this greenhouse-like function of the atmosphere that the average global temperature of the earth is 15°C (59°F). Without the atmosphere and these gases, the average global temperature would be a frigid -18°C (0°F), and life would not be possible on earth. These gases are called greenhouse gases and include carbon dioxide (CO_2), water vapour (H_2O), methane (CH_4), nitrous oxide (N_2O), chloroflourocarbons (CFCs), and ozone (O_3).

The role of greenhouse gases in the atmosphere was first discovered during the 1800s (Kellogg, 1988). By 1896, Swedish scientist Svante Arrhenius was already calculating that the earth's surface temperature would increase by 5-6°C (9–10.8°F) with a doubling or tripling of the atmospheric CO_2 content, although he did not perceive that the greenhouse gas concentrations would increase so much in such a short period of time. Such predictions received relatively little notice until the 1950s. In 1957, ROGER Revelle and Hans Suess, scientists at the Scripps Institution of Oceanography, said that by adding CO_2 into the atmosphere humans were "now carrying out a large-scale geophysical experiment." They also pointed out that the CO_2 would remain within the atmosphere for a very long time because of how slowly it is absorbed by the oceans. Since then, there has been a growing acknowledgment of the increasing concentrations of not only CO_2 but also the other greenhouse gases and their potential impact on the global climate.

Public awareness of the Greenhouse Effect and the concern that the impact of increased emissions of CO_2, CH_4, N_2O, and CFCs would raise global temperatures grew in the 1980s. During an intense drought and heat wave in 1988, the media and several scientists speculated that the drought and heat wave affecting much of the United States were evidence of climate change. In hindsight, the 1988 drought was likely within the range of normal climate variability, but the attention was focused on the Greenhouse Effect. Likewise, climate attribution researchers determined that the record-breaking warmth across much of the middle U.S. in March 2012 was much more extreme than climate change alone could explain.

What we know about the greenhouse effect and climate change   We know with certainty that the concentrations of CO_2, CH_4, N_2O, and CFCs have increased as a result of recent human activity. Carbon dioxide is responsible for more than half of the increase. Annual emissions of CO_2 between 1970 and 2004 have increased by about 80% (IPCC, 2007). Concentrations of CO_2 have been systematically monitored at the Mauna Loa Observatory in Hawaii since 1958. Scientists can determine the CO_2 concentrations from before 1958 using data collected from air bubbles within ice cores around the world. In 1750, levels of CO2 were around 270 parts per million (ppm). By 2005, these levels were up to 379 ppm, and the 10-year rate of growth in concentrations, 1.9 ppm per year, was greater than any previous 10-year period in the modern record (IPCC, 2007).

According to the congressional Office of Technology Assessment (OTA, 1993), 70–90% of the CO_2 added to the atmosphere is due to the burning of fossil fuels, and the rest is from deforestation. Methane, N_2O, and CFCs are also increasing at a similar rate. But because CFCs destroy ozone, the net warming effect from CFCs is approximately zero (but that leads to an entirely different problem: depletion of stratospheric ozone).

Scientists can use this information within large-scale models of the atmosphere called General Circulation Models (GCMs). These models are composed of mathematical equations and relationships designed to simulate global atmospheric conditions and make projections of the future climate. Although there are differences between the GCM projections, the models are in general agreement that, as a result of increasing greenhouse gas concentrations, the current best estimate is that average global temperature will increase 1.8–4.0°C (3.24–7.2°F) by 2100 (IPCC, 2007). In the past 100 years, the global average surface temperature has increased 0.60°C (1.08°F). This increase by itself is within the normal variability and, although it may be a result of climate change, it cannot be used as definitive proof that recent human activities have caused a global warming. Between 1995 and 2006, however, 11 of the 12 years ranked among the 12 warmest years on record, with records dating back to 1850 (IPCC, 2007).

Climate change and the hydrological cycle   With the projected global temperature increase, some scientists think that the global hydrological cycle will also intensify. GCMs indicate that global precipitation could increase 7–15%. Meanwhile, global evapo-transpiration could increase 5–10% (OTA, 1993). Thus, the combined impacts of increased temperature, precipitation, and evapo-transpiration will affect snowmelt, runoff, and soil moisture conditions. The models generally show that precipitation will increase at high latitudes and decrease at low and mid-latitudes. Therefore, in mid-continent regions, evapo-transpiration will be greater than precipitation and there will exist the potential for more severe, longer-lasting droughts in these areas. In addition, the increased temperatures alone will cause the water in the oceans to expand. It has been estimated that global sea levels in the 20th century rose 17cm (6.7 in), and future predictions show that sea levels could reach 22–44 cm (8.66–17.3 in) above 1990 levels by 2100 (IPCC, 2007).

Observations show that precipitation amount, intensity, frequency, and type is changing across the globe. According to the IPCC, 2007:

Pronounced long-term trends from 1900 to 2005 have been observed in precipitation amount in some places: significantly wetter in eastern North and South America, northern Europe and northern and central Asia, but drier in the Sahel, southern Africa, the Mediterranean and southern Asia. More precipitation now falls as rain rather than snow in northern regions. Widespread increases in heavy precipitation events have been observed, even in places where total amounts have decreased. These changes are associated with increased water vapour in the atmosphere arising from the warming of the world's oceans, especially at lower latitudes. There are also increases in some regions in the occurrences of both droughts and floods.

What we do not know   One of the weaknesses of GCM climate change predictions is that they cannot adequately resolve factors that might influence regional climates, such as the local effects of mountains, coastlines, lakes, vegetation boundaries, and heterogeneous soils, or how human activities might play a role. This makes it difficult to predict the impacts on regional resources. GCMs also cannot tell us whether the occurrence of extreme events will increase. Some scientists have suggested that droughts, floods, and hurricanes will become more common with climate change. Although the losses due to natural disasters have increased worldwide and within the United States, this is more the result of increased vulnerability than an increased number of events.

No one knows who will be hurt or who will benefit from a CO_2-induced climate change. Clearly, coastal regions will suffer as a result of a rise in sea levels. But for interior regions, there might be beneficial gains in agricultural production resulting from the indirect effects of a warmer climate and adequate precipitation, especially in higher latitudes across Canada and Russia. The increased CO_2 might also directly increase plant growth and productivity as well. In fact, this theory, known as the CO_2 Fertilization Effect, has led some scientists to suggest that the Greenhouse Effect might be a blessing in disguise. Laboratory experiments have shown that increased CO_2 concentrations potentially promote plant growth and ecosystem productivity by increasing the rate of photosynthesis, improving nutrient uptake and use, increasing water-use efficiency, and decreasing respiration, along with several other factors (OTA, 1993). The scientists, encouraged by these benefits, hypothesize that increased ecosystem productivity will actually help draw excess CO_2 from the atmosphere, thereby diminishing concerns about global warming (OTA, 1993). Whether any benefits would result from the CO_2 Fertilization Effect within the complex interactions of natural ecosystems is still unknown. Ecosystem productivity can only increase, however, in regions where supplies of light, water, and nutrients are plentiful (OTA, 1993).

Water resources and climate change   Although we don't know how climate change will affect regional water resources, it is clear that water resources are already stressed, independent of climate change, and any additional stress from climate change or increased variability will only intensify the competition for water resources. Current stresses on water resources around the globe include:

- Growing populations
- Increased competition for available water
- Poor water quality
- Environmental claims
- Uncertain reserved water rights
- Groundwater overdraft
- Outmoded institutions
- Aging urban water infrastructure

In all likelihood, the direct impacts of climate change on water resources will be hidden beneath natural climate variability. With a warmer climate, droughts and floods could become more frequent, severe, and longer–lasting. The potential increase in these hazards is a great concern given the stresses being placed on water resources and the high costs resulting from recent hazards. The drought of the late 1980s showed what the impacts might be if climate change leads to a change in the frequency and intensity of droughts across the United States. From 1987 to 1989, losses from drought in the United States totalled $39 billion (OTA, 1993). More frequent extreme events such as droughts and floods could end up being more cause for concern than the long-term change in temperature and precipitation averages.

The best advice to water resource managers regarding climate change is to start addressing current stresses on water supplies and build flexibility and robustness into any system. Flexibility helps to ensure a quick response to changing conditions, while robustness helps people prepare for and survive the worst conditions. With this approach to planning, water system managers will be better able to adapt to the impacts of climate change, whatever they may be, and will also be better equipped for the climate variability we have now.

Climate change links   Intergovernmental Panel on Climate Change (IPCC). The IPCC was established in 1988 and operates under the World Meteorological Organization and the United Nations Environment Program. It consists of a panel of scientists from more than 50 countries that study climate change and assess its potential impacts. Reports have been completed in 1990, 1992, 1995, 2001, and 2007.

United Nations Environment Program (UNEP). This site contains a variety of information related to the environment and sustainable development, including issues involving climate change. A section of the website is devoted to climate change. Topics in this section include information about impacts, publications, recent news, educational materials, and more.

Institute for the Study of Society and Environment (ISSE). ISSE is a multidisciplinary group of scientists working for the National Center for Atmospheric Research (NCAR) focusing on the interactions between human activities and the environment. A major research area covered by ISSE involves climate change.

Climate Impacts Group (CIG). CIG is an interdisciplinary group of researchers at the University of Washington who are focused on studying the impacts of climate change and variability on the Pacific Northwest of the United States. CIG's assessment examines climate impacts on water, forests, salmon, and coasts and the human socioeconomic and/or political systems associated with each.

Interpreting Climate Conditions, Physical Science Division, Earth System Research Laboratory, National Oceanic and Atmospheric Administration.

For more information, please contact Michael J. Hayes.

How does Drought affect our lives?

When we have a drought, it can affect our communities and our environment in many different ways. Everything in the environment is connected, just like everything in our communities is connected. Each different way that drought affects us is what we call an impact of drought.

Drought affects our lives in many different ways because water is such an important part of so many of our activities. We need water to live, and animals and plants do too. We need water to grow the food we eat. We also use water for many different things in our lives, like washing dishes, cooking, bathing, and swimming or river rafting. Water is also used to help make the electricity we use to run the lights in our houses and the video games you may like to play. When we don't have enough water for these activities because of a drought, many people and many different things will be affected in many different ways.

Drought Dominoes?

We often talk about drought's impacts as either "direct" or "indirect." What does that mean? Well, to find out, let's think about dominoes. If you set up a long line of dominoes on the floor and knock the first domino in the line over, it will cause the second domino in the line to fall and hit the third, which will fall and hit the fourth, and so on.

If those dominoes were drought impacts, the first domino you knock over might be farmers' corn crops dying. The second domino might be that the farmers would not have money to buy a new tractor from the dealer in town. The dealer would then lose money, which would be the third domino. If enough farmers lose their corn crops, the dealership might not be able to employ as many people or may even have to close down—the fourth domino. The dealership closing would cause many more impacts in the community.

The farmers' crops dying would be the "direct" impact of drought. The dealer losing money and all of the other impacts would be the "indirect" impacts of drought.

All of the impacts in the example above would be "negative" impacts. But the impacts of drought aren't always all negative. How can this be? Well, let's think about the example of the farmers we talked about earlier. The farmers who have lost their corn crops might use the money they didn't spend to buy a new tractor to hire a person to drill irrigation wells. The well-drilling business would make more money, so for them the drought might actually have a "positive" or good impact. However, the overall impact of drought in an area is almost always negative.

REFERENCES

Glantz, M. (ed.) 1994. Usable Science: Food Security, Early Warning, and El Niño. Proceedings of the Workshop on ENSO/FEWS, Budapest, Hungary, October 1993. UNEP, Nairobi; and NCAR, Boulder, Colorado.

Glantz, M.; R. Katz; and N. Nicholls (eds.). 1991. Teleconnections Linking Worldwide Climate Anomalies. Cambridge University Press, Cambridge.

Glantz, M.; R. Katz; and M. Krenz (eds.). 1987. Climate Crisis. UNEP, Nairobi; and NCAR, Boulder, Colorado.

Maunder, W.J. 1992. Dictionary of Global Climate Change. Chapman and Hall, New York.

NOAA. 1994. El Niño and Climate Prediction—Reports to the Nation on Our Changing Planet. A publication of the University Corporation for Atmospheric Research pursuant to National Oceanic and Atmospheric Administration Award No. NA27GP0232–01. UCAR, Boulder, Colorado.

Ogallo, L.A. 1994. Validity of the ENSO-Related Impacts in Eastern and Southern Africa. In M. Glantz (ed.). Usable Science: Food Security, Early Warning, and El Niño, pp. 179–184. Proceedings of the Workshop on ENSO/FEWS, Budapest, Hungary, October 1993. UNEP, Nairobi; and NCAR, Boulder, Colorado.

Philander, S.G. 1990. El Niño, La Niña, and the Southern Oscillation. Academic Press, San Diego, California.

Ropelewski, C.F.; and M.S. Halpert. 1987. Global and regional scale precipitation patterns associated with the El Niño–Southern Oscillation. Monthly Weather Review 115:1606–1626.

Ropelewski, C.F.; and M.S. Halpert. 1986. North American precipitation and temperature patterns associated with the El Niño–Southern Oscillation (ENSO). Monthly Weather Review 114:2352–2362.

Wilhite, D.A.; D.A. Wood; and S.J. Meyer. 1987. Climate-related impacts in the United States during the 1982–83 El Niño. In M. Glantz, R. Katz, and M. Krenz (eds.). Climate Crisis, pp. 75–78. UNEP, Nairobi and NCAR, Boulder, Colorado.

For more information, please contact Mark Svoboda, NDMC Climate/Water Resources Specialist.

BIBLIOGRAPHY

- IPCC (Intergovernmental Panel on Climate Change). 2007. Climate Change 2007: The Physical Science Basis. Contribution of Working Group I to the Fourth Assessment Report of the Intergovernmental Panel on Climate Change. Edited by Solomon, S., D. Qin, M. Manning, Z. Chen, M. Marquis, K.B. Averyt, M. Tignor and H.L. Miller. Cambridge University Press, Cambridge.

- Kellogg, W.W. 1988. Human impact on climate: The evolution of an awareness. In M.H. Glantz, ed. Societal Responses to Regional Climatic Change. Westview Press, Boulder, Colorado.

- OTA (Office of Technology Assessment). 1993. Preparing for an Uncertain Climate, Vol. I. OTA–O–567. U.S. Government Printing Office, Washington, D.C.

- NOAA (National Oceanic and Atmospheric Administration), Earth System Research Laboratory, Physical Sciences Division. 2012. Meteorological March Madness 2012. Hoerling, Martin.

GLOSSARY

Adaptation: The Intergovernmental Panel on Climate Change uses adaptation to refer to measures taken ahead of time to reduce vulnerability to climate change.

Taking measures ahead of time to reduce vulnerability is what the NDMC means by mitigation. See also mitigation.

Aquifer: An area that contains large amounts of water under the surface of the earth.

Climate: Day-to-day weather over a longer period of time. Climatology is the study of climate.

Climograph: A graph that shows monthly average temperature and precipitation for some location.

Dam: A structure built across a river to hold back water for a variety of reasons, including protecting areas from floods, storing water, and generating power.

Desalination: The process of removing salts and other minerals from seawater so that it can be used for drinking water.

Drought: Less rainfall than is expected over an extended period of time, usually several months or longer. Or, more formally, it is a deficiency of rainfall over a period of time, resulting in a water shortage for some activity, group, or environmental sector.

Drought index: A numerical scale that scientists use to describe the severity of a drought. Scientists take many kinds of data (like streamflow, rainfall, temperature, and snow-pack) and "blend" it into a single number, called a drought index value, to make it easier to understand the drought conditions of a particular area. Drought indices are one type of drought indicator.

Drought indicator: A way to look at one or more variables, such as precipitation, to describe available water in soil or hydrologic systems. It may be a record of a single measurement, such as rainfall at a particular rain gauge. It may also be a complex index. Drought indices (indexes) are a subset of drought indicators.

Dust bowl: An area of the U.S. Plains that included parts of Kansas, Colorado, Oklahoma, Texas, and New Mexico. The term was coined in the 1930s, when dry weather and high winds caused many dust storms throughout the United States, but particularly in this area.

El Niño: A weather phenomenon that occurs in the eastern and central equatorial Pacific Ocean. During an El Niño, the affected area's winds weaken and sea temperatures become warmer.

Erosion: A process that wears the earth's surface away, causing soil to move from one place to another. It's a natural process, but human activities can make it worse.

Groundwater: Water that is found underground in spaces in soil, sand, or rocks.

Hydroelectricity: Electricity created by channelling water through turbines in power stations located below dams.

Irrigation: An agricultural practice that involves providing water to crops through pipes, ditches, or streams.

Jet stream: Strong wind currents at high altitudes in the earth's atmosphere.

They are thousands of miles long and hundreds of miles wide, and they move weather patterns around the earth.

La Niña: A weather phenomenon that involves unusually cold ocean temperatures in the equatorial Pacific Ocean. La Niña events don't occur as often as El Niño events.

Mitigation: Actions that we can take before, or at the beginning of drought, to help reduce the impacts of drought. Mitigation includes actions as diverse as drought planning, implementing land use practices that increase the organic content and water-holding capacity of soil, and designing agricultural policy that doesn't encourage short-term economic gains at the expense of long-term productive capacity. The Intergovernmental Panel on Climate Change uses mitigation to mean reducing emissions of greenhouse gasses.

Operational definition of drought: How agencies, communities, or individuals will recognize a drought in its early stages. Do they have their own rain gauge, river flow meter, or water level meter? Do they rely on state or national climatological data? Do they have a unique way to measure soil moisture, such as the appearance of certain plants or other environmental features?

Reservoirs: Water that's collected and stored in natural or man made lakes.

Submarginal farmland: Lands with nutrient-poor soils and/or soils that have been damaged by poor cultivation practices.

Teleconnection: A relationship between two distant weather events. The weather phenomenon El Niño, for example, has been linked to a wide variety of events, including wildfires in the Australian Outback, flooding in the Peruvian Andes, and above-normal rainfall in the Greater Horn of Africa.

Water banking: A water management strategy that temporarily transfers water from those who are willing to lease it to those who are willing to pay to use it.

Water recycling: Reusing treated wastewater for purposes like agricultural and landscape irrigation. It's also referred to as water reclamation or water reuse.

Weather: The condition of the earth's atmosphere over a brief period of time, like a day or a week.

Xeriscaping: A type of landscaping around homes and businesses that uses a limited amount of water.

4

WEATHER

For many people, weather plays a crucial part in their lives. Farmer need rain so that the seeds they plant will sprout. They need sunshine for their plants to grow. Frasty nights and long droughts can cause them to lose their crops. Pilots and sailors must pay careful attention to the weather to stay safe when they travel. Foresters need to watch the weather when fighting dangerous fires.

Weather influences so much of what we do, should we go swimming to day or wear a rain coat? Is it a good day for picnicking or for ice skating? In all depends upon the weather. And whatever we decide to do. We can be sure of only two things about the weather. We are going to have it and it is going to change. Three things cause our weather (1) heat from the sun (2) moving air and (3) water.

Our weather is powered by energy from the sun. The sun heats the earth atmosphere and warms its oceans, rivers, lakes and land.

Energy from the sun fuels winds. Winds blow clouds, moving moisture around the globe. In addition, winds blow warm and cool air from one place to another.

Weather can be dry or wet, warm or cold, cloudy or clear or windy or still. Weather describes the air at a particular place during a particular time.

The suns energy powers the weather on earth. The sun appears small because it is so far away. But sun is actually huge. About log earths could fit across the middle of the sun. The sun is intensely hot. The surface of the sun (the photo sphere) gives off energy in all directions.

Energy from the sun travels to the earth as light and heat waves. The light we see is part of these waves. Some of the waves are not visible to us. We feel infrared waves as heat.

Energy from the sun reaches the earth's atmosphere. The atmosphere is the layer of gases surrounding the earth that we call air. Some energy is absorbed by gases in the atmosphere gases and clouds reflect some energy back into space. About half of the sun energy reaches the earth surface, warming the land and oceans. The ground and water give off heat, warming the air above. As the air warms, it begins to move, forming winds that carry weather. The heat is carried upwards and much of it is lost to space.

On the Earth

The heating of the earth is uneven. The amount of sunlight that a place receives changes during the day and with the seasons. Each day the sun's energy warms the land and oceans. At night the land continues to give off heat.

Some places on earth receive more of the sun's heat than others. At the equator the sun rays shine directly over the land and oceans. The heat from the sun is more concentrated along the equator so temperature higher.

At the earth's two poles sunlight travels further through the atmosphere. It is not directly over head. The sun's rays spread out over a winder area, so they have less heat when they reach land. Temperature's are lower here. Sheets of snow and ice cover the land and parts of the oceans at the poles. Even in summer you can find snow near the north pole.

In the Atmosphere

Weather takes place in the earths atmosphere. Gases in the earths atmosphere stay close to earth because of the pull of earth's gravity.

Two gases make up most of our atmosphere. Air is about 78% nitrogen and 21% oxygen. The remaining one percent is made up of small amounts of water vapour, hydrogen, carbon dioxide and other gases.

Gases is the atmosphere protect life on earth by keeping out the sun's harmful rays. They also trap some of the heat from the sun during the day keeping the earth warm at night. Without the atmosphere, there could be no life on earth.

Layers of the Atmosphere

Scientists divide the atmosphere into five layers.

The layer closest to the earth is called the troposphere. The troposphere has water vapour and clouds that produce over weather. It is 7 kms (4.3 miles) thick at the poles and 17 kms (10.5 miles) thick near the equator. Air in the troposphere is warmest near the earth's surface. The sun warms the oceans and land. Heat rises from the earth surface and warms the air above it. The temperature of the air decreases farther up in the troposphere. Water vapour and carbon dioxide in the troposphere trap the sun's heat keeping the earth warm.

Going Higher

The air above the troposphere is thinner and does not have enough oxygen to breathe. This next layer is called the stratosphere. The stratosphere contains a layer of gases called the ozone layer. Ozone absorbs harmful ultraviolet rays from the sun that would otherwise burn over skin. Jets often fly in the stratosphere because its air is stable.

And High

Above the stratosphere is mesosphere the coldest layer of the atmosphere. The layer above the mesosphere is called thermosphere. It's temperatures can reach upto 1500°C (2730°F). Space shuttles orbit here. Past the thermospheres is the exosphere. The air in the exosphere is very thin and gives away to space.

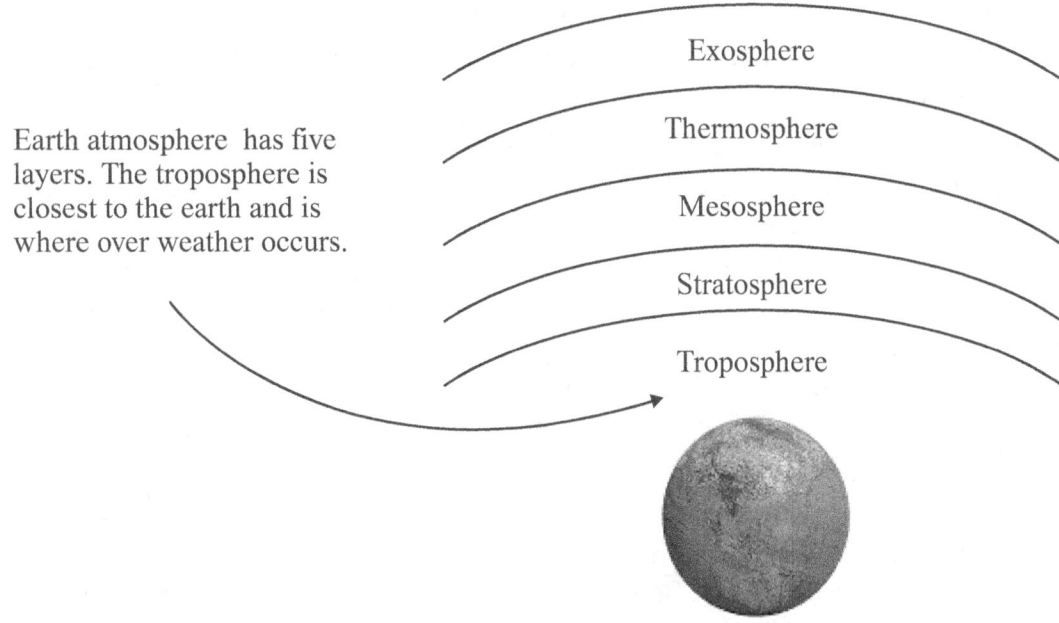

Earth atmosphere has five layers. The troposphere is closest to the earth and is where over weather occurs.

Exosphere

Thermosphere

Mesosphere

Stratosphere

Troposphere

The climate of a region also depends on the amount of rainfall it receives. Places with desert climates receive little rainfall because of the <u>dry air</u>.

All though weather changes from day to day, an area has similar patterns of weather each year. The pattern of weather is called <u>climate</u>. Near the equator, many places have hot tropical climates. Polar regions have cold polar climates.

Places with temperate climates are located between the polar regions and tropical areas. Temperate climates are warm in summer and cold in winter.

Oceans and Land

Oceans take longer to warm and cool than land. Because they do not change temperature as quickly as land, oceans moderate the climate. A warm ocean keeps the air warm in the winter. In the summer the ocean keeps the air cooler. On a cold winters day the temperature along the coast may be warmer than farther inland.

In summer, you can go to the beach to enjoy cooler weather. The climate to enjoy cooler weather. The climate of coastal lands is often cooler in summer and warmer in winter.

Ocean currents can also change the climate of the land near the coast. Ocean currents are flowing streams of cold or warm water. Winds blow ocean water. Because of the way the earth rotates, these currents move in circular paths. Norway's coast extends close to the North pole. But its climate is milder than you might expect because of the warm ocean currents that flow past.

On the summer's day an ocean breeze keeps the beach cool.

The Earth Tilts

Every year the earth revolves one time around the sun. As the earth moves, it is titled at an angle. Because the earth is titled on its axis we have seasons.

If you live in the Northern hemisphere then you enjoy long summer days in July. At this time of year the Northern half of the earth is titled more toward the sun.

The Southern hemisphere is titled away from the sun. It receives less sunlight and has cold winter days.

Six months later, the earth has completed half its around the sun. Now the northern hemisphere is farther from the sun. The days are short and it is winter. The southern hemisphere is titled more toward the sun. There the days are long and warm and it is summer.

What causes Wind

Wind is caused by the uneven heating of the earth surface. Oceans, lakes and rivers cover more than 2/3 of the earth's surface. The rest is covered by land, including mountains valleys and plains. Bodies of water and different land forms absorb the sun's energy at different rates. During the day air over the land heals up faster than air over the water. Warm air is lighter than cold air and rises. As warm air rises, cool air rushes into take its place. This movement creates wind.

Wind Patterns

Wind is described by its speed and direction. The spread and direction of wind can change from day to day. But there are large wind patterns around the globe that help people predict the weather. The land near the equator rises to higher temperature than land near the poles. This heating difference sets up global wind patterns. As air is heated, it moves north and south of the equator.

It's Raining, it's Pouring

When you think of weather, you might picture rain, snow, sleet, hail and clouds. Water falling from the sky is called precipitation. Where does precipitation come from. It is the part of the water cycle that carries water around the earth. Rainfalls when water droplets in clouds grow.

Precipitation

The heat of the sun makes some of the droplets in a cloud to evaporate. Warm air in the clouds rises with water vapour. As it rises, the air cools more droplets to condense and the cloud grows. As more and more droplets condense the cloud becomes heavier and heavier. Now the droplets are packed so closely together that light cannot shine, through and the droplets grow too heavy to stay suspended as they fall to the earth as rain, snow, sleet or hail.

Water Vapour

All air contains water vapour. The air is humid because it contains a lot of water vapour. Because warm air is rising and in motion, it can hold more water vapour than cold air. Air that has as much water vapour as it can hold is called saturated air. When it can hold is called saturated air. When it can hold no more water vapour, the vapour is saturated air begins to condense turning into water droplets. If water vapour condenses near the ground it forms fog or mist.

On a cool morning you might see fog forming above a lake, as water vapour rising from lake cools and condenses when water vapour condenses high above the ground it forms clouds.

Cloud Cover

Clouds come in three main shapes (1) cirrus (2) cumulus (3) stratus high in the sky are thin whispy cirrus clouds.

Cumulus clouds are fluffy, while clouds that can grow as they rise and bring rain.

Stratus clouds form flat blankets and are low in the sky. Clouds are classified by their shape and height in the sky.

Rain: In side a cloud, droplets condense on tiny dust particles. Tiny droplets are blown together and joined up forming larger droplets. When they grow too heavy for the cloud to carry they fall. As they fall more vapour condenses on them, making larger rain drops.

Snow: When the temperature in side a cloud is very cold, ice crystals form. Water droplets freeze on the crystals. Many crystals stick together, growing into snow flakes that fall to the ground.

Hail: Hail can form during thunder storms. Ice crystals in towering clouds are blown up and down. As air currents push the crystals up and down, water freezes on them and they grow. When hail stones fall, they can be larger than golf balls.

Sleet: On a cold day, rain drops might freeze as they fall, covering the ground with a blanket of wet, sloppy sleet.

Changing Weather

When you wake up, the sky is clear and the sun shines brightly. But the meteorologist on a TV says that a storm is heading your way. Soon the sky darkens thunder rumbles and lighting flashes. Why is the weather constantly changing? Air has weight as it presses down on the earth. This is called air pressure. High and low pressure systems are due to the movements of air around the earth. A high pressure area has more air over a location on the earth creating move weight or pressure. Low pressure areas are the opposite.

Moving Molecules

Gases in air are made of tiny particles that you cannot see, called molecules. The molecules in a gas are always moving. When they are heated the molecules get more energy and move farther apart and faster. This makes heated air less dense and it rises. A hot air balloon rises to the sky because the balloon is filled with hot air that is less dense than the surrounding air.

One source of circulating air is the hot air that rises near the equator. This hot air rises and moves away from the equator to the north or south. At the other extreme is the cold air near the north and south poles. This air has more density and slides under the warmer air from the equator. When the cold air meets the warm air they mix in a circular motion. This is caused by differences in temperature and the turning of the earth.

High and Low Pressure Areas

In low pressure areas the circular motion is counter clockwise pulling air into the centre of the area like a whirlpool. This circulating pattern of air pulls in the water vapour from around it. As the air gets to the center it has to rise. Air cools as it rises, and clouds form. Water droplets in clouds collide and join together becoming heavier until precipitation falls. Low pressure leads to cloudy weather gray skies and rain. High pressure areas have a counter clockwise circulation that pushes the air out from it, "clearing" the area of moisture and weather, high pressure areas usually have clear skies and good weather.

Air travels in large masses of the same temperature and moisture. The place where two air masses meet is called a front. Symbol on a weather map show fronts.

Blue triangles: These show a cold front where a cold air mass push's against a warm one and replaces it. The warm air quickly rises to the sky, farming clouds and wind. A cold front can produce clouds and storms followed by clear, cooler weather.

Red semi circles: These show a warm front where a warm air mass pushes over a cold air mass. Warm fronts move slowly bringing cloudy weather and rain.

Red semi circles alternating with blue triangles: An occluded front shows where a fast moving cold front over takes a warm front. The cold air lifts up the warm air mass pushing it higher in the sky. Heavy rains and thunder storms may occur.

Climate Change

Many scientists are concerned that climate change will effect our weather. In the past the climate of different regions of the earth has changed, from warm to cold and cold to warm. Places that were once dry became wetter and wet places became drier.

During past ice ages parts of North America, Europe and Asia were covered with huge sheets of ice. There have been at least four ice ages in the past, the most recent one ending about 11,000 years ago. Around 65 million years ago, a changing climate may have led to the disappearance of the dinosaurs. An asteroid might have collided with the earth, changing the temperature. The earth has seen many natural changes in climate.

Global Warming and Weather

In the last few decades, people have caused climates to change at a fast pace. We use fuels such as gasoline and coal in our cars, factories and to heat buildings.

Burning fuels release carbon dioxide and other gases into the atmosphere. When people burn and clear forests to make large farms, the atmosphere also changes, gases such as carbon dioxide trap heat.

Because these are more of there gases now in the atmosphere, more heat is trapped close to earth. The temperature of the earth rises causing global warming.

Scientists say that earth will warm by as much as 5°F in the next one hundred years and that global heating will cause more powerful storms and other great changes in the earth's weather.

A warm atmosphere may cause more severe storms. Ice caps might melt and oceans warm causing flooding in coastal areas. Other places may have droughts.

People are not certain how our weather will change as a result of global warming, but scientists think that powerful storms might become more common.

Tornadoes

These are by far the most extreme of the family of whirls and are often highly destructive.

Although hurricanes (tropical cyclones) affects for greater areas, the damage caused by Tornadoes is highly concentrated and generally more severe. (The highest reliably known wind speed is 307 mph recorded in the Oklahoma city Tornado of 1999). The diameter at the base varies considerably ranging from 300 to 6600 feet.

Earths weather is driven by the intense heat of the sun. The sun's energy travels through space in the form of visible light waves and visible ultraviolet and infra red rays.

About one third of the energy reaching earth's atmosphere is reflected back into space. The remaining two thirds is absorbed during a process called insolation (from incoming solar radiation).

The atmosphere lets sun light pass through. Sun light heats the ground which in turn warms the air near the surface. But the atmosphere prevents most of the heat from escaping into space. This is called the green house effect, because the glass windows in a green house trap heat in the same way. Insolation and the green house effect strike a balance and make our planet liveable. If earth's average temperature were to drop by a few degrees, the ice ages would return and glaciers would cover North America and Europe.

If the temperature were to increase by a few degrees the polar ice caps would melt and the oceans would flood, low–lying coastal lands.

INDIA – CLIMATE AND VEGETATION

Climatic Regions and Factors Influencing the Climate

India has a great variety in climate, as also in relief. The climate of India ranges from dry desert type in Rajasthan, to humid climate near sea coasts and wet climate in Meghalaya in the North–East. These variations are shown in the map enclosed. Inspite of great local variations, the climate of India is broadly described as "Tropical Monsoon type climate".

Three factors responsible for this broad categorization of climate are:

1. The tropic of cancer divides India into two equal halves. While the southern part lies in the tropical zone, the northern part is in sub–tropical zone.

2. India lies at the head of the Indian Ocean. The moist winds blowing over the Indian Ocean blow towards the mainland.

3. The Himalayas protect India from cold polar winds blowing from the north and impart a distinctly tropical touch to the climate of India.

Hence, it is apt to describe the climate of India as Tropical Monsoon climate. Other factors including the climate of India are the shifting jet streams or upper air currents blowing in a narrow zone over the northern plains in winter and summer. The general relief of the Indian landmass is more pronounced due to altitude and direction of mountain ranges and lastly Indian Ocean and the sea also play an important role.

The climate of India is affected by location, altitude and distance from the sea and its relief topograph. This explains why we experience regional and seasonal differences in climate. Low temperature during winter in Northern plains causes high pressure areas on land. Because winds always blow from high pressure areas to low pressure areas, winds from the Northern plains blow towards the seas. Similarly the temperature is very high over plains in summer. The warm air rises up creating an area of low pressure. This causes moisture laden air from the sea to blow towards land (i.e.) on store breeze.

Pressure and Temperature

The pressure and temperature map of India during January and July is enclosed. It can be seen from the maps that in January pressure goes on increasing from ocean to land and vice–versa in July. Other climatic factors also play their role in causing rainfall. That is why Chennai gets rainfall in winter from north–east monsoon. There are dry winds blowing from land to sea but they pick up moisture while passing over Bay of Bengal and shed. This moisture over the coromandal coast.

Normal Dates of Onset of Monsoon

Season of advancing monsoon is also known as the season of south–west monsoon or the rainy season. The season is marked by the onset and advance of monsoon. During this season, winds blow from both Arabian sea and Bay of Bengal towards the land. They carry moisture and as they strike the mountain barriers, they are forced to rise up. The vapour condenses and causes rainfall. The monsoon appears at different dates at different places. While Kerala on west-coast is the first to receive rainfall (on 1st June) North – Western Rajastan is the last to receive rainfall (by around 15th July). The triangular shape of Peninsular India and the Himalayan barriers deflect the winds from west to east. This position can best be understood with help of map enclosed. The north-east receives monsoon rains by around 5th June. The moisture laden winds then go onwards and reach Delhi by around 15th June. The distribution of rainfall is highly uneven over the landmass North–Eastern parts on south-facing slopes (Assam, Arunachal, Meghalaya and Sikkim) Western parts of Western Ghats—including the coastal plains and the deltaic region of West Bengal—receive heavy rainfall of more than 200 cm in a year.

Rainfall

Mawsynram in Meghalaya gets 1142 cm the Worlds heaviest rainfall. These areas receive between 100-200 cm in a year. There areas include parts of North-West, Bihar, Bengal and Eastern U.P. which receive rainfall in decreasing order from south-west monsoons in the South Eastern Coasts of Tamilnadu. The eastern slopes of Western Ghats also fall in the moderate rainfall region. The interior parts of the Northern plains get rainfall between 50–100 cms. These areas get less than 50 cm rainfall in a year. These areas are scattered from rain – shadow areas to north of Himalayas, such as Lahaul–spit in Himachal Pradesh and Kashmir. On account of parallel direction of Aravallis to Arabian branch of monsoon winds, the Thar Desert gets less than 25 cm annual rainfall. Southern parts of Haryana and south–western Punjab also get scanty rainfall.

The map showing wind direction, heavy rainfall, medium rainfall and low rainfall areas is enclosed.

Natural Vegetation

The natural vegetation of India is divided into Tropical deciduous forests, Evergreen forests, mountain forests, Thorn forests and Tidal or delta forests.

REFERENCES

1. www.weatherwizkids.com. Learn about hurricanes, tornadoes clouds, winter storms and more from a meteorologist.

2. National weather service www.nws.noaa.goc/out look.tab.php. Read weather maps to find out about weather across the USA.

3. www.exploratorium.edu/spaceweather/index.html Learn how weather in space affect earth.

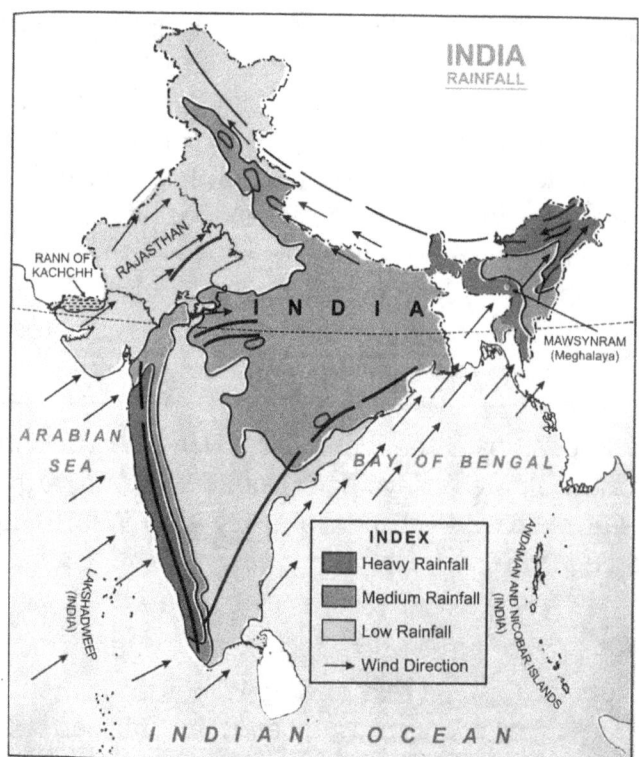

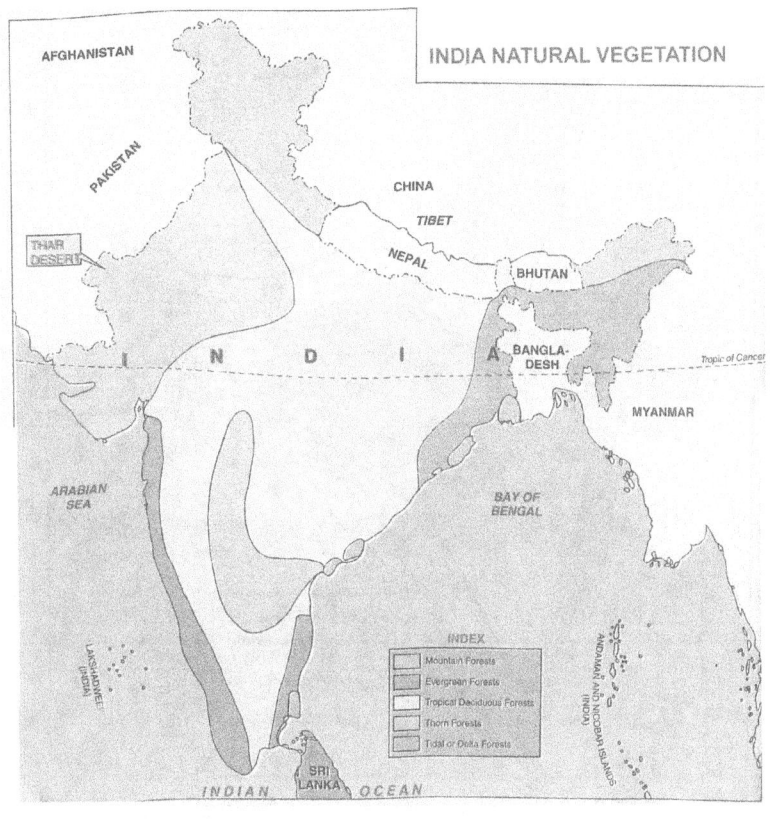

INDIA NATURAL VEGETATION

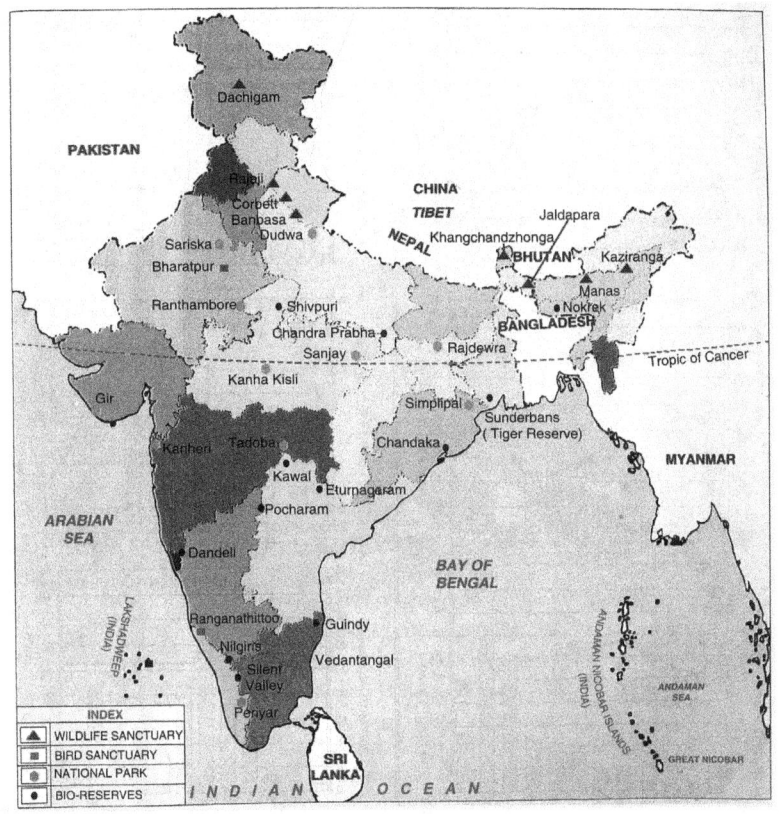

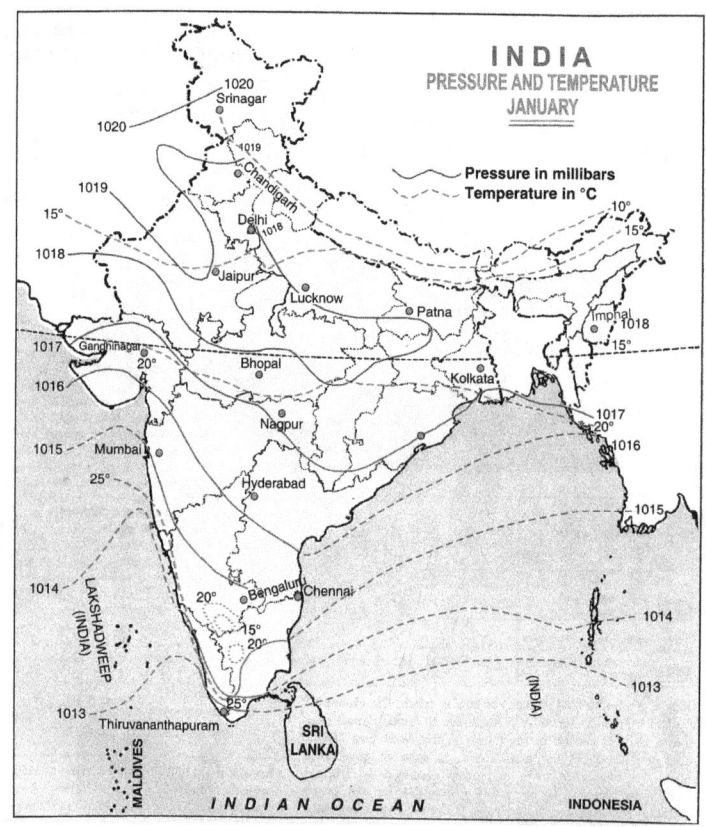

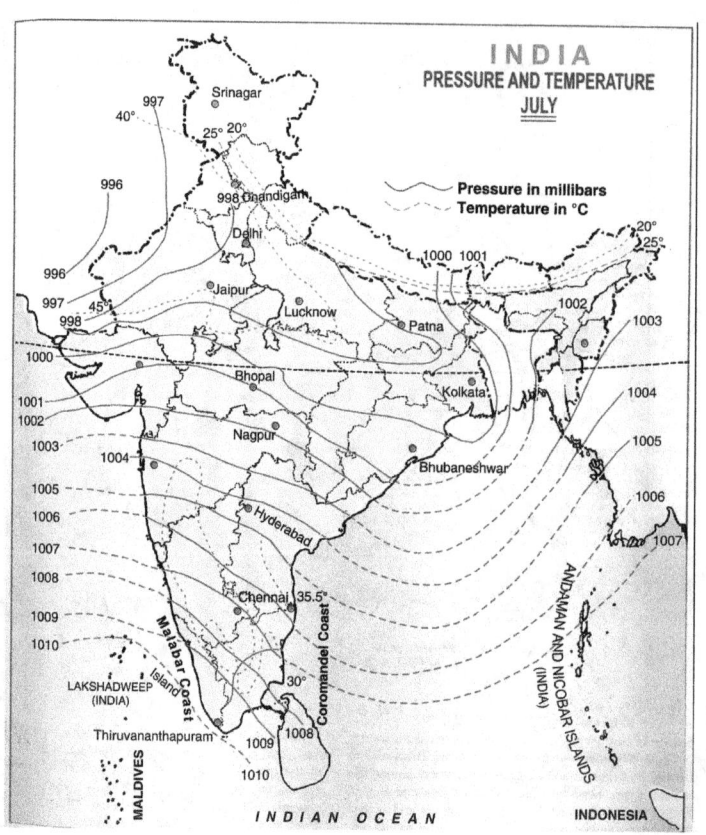

INDIA – CLIMATE AND VEGETATION

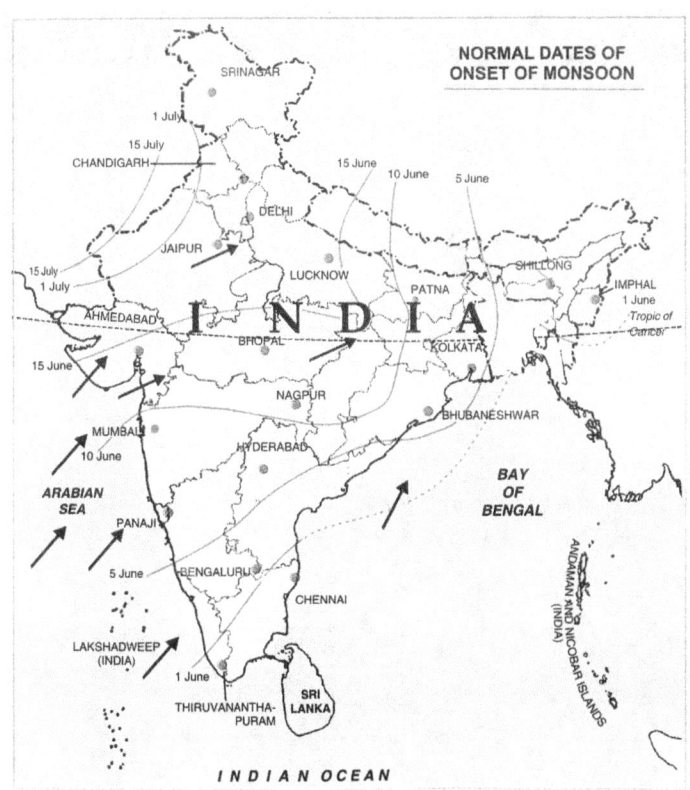

Global warming and the greenhouse effect

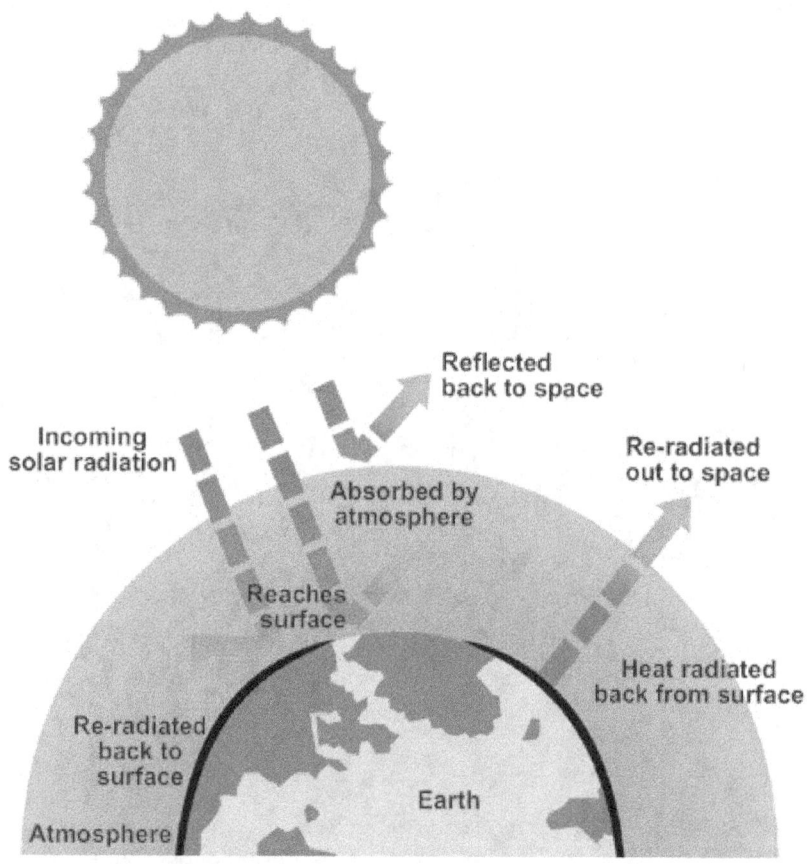

METEOROLOGICAL DROUGHT

Meteorological drought is defined as the deficiency of precipitation from expected or normal levels over an extended period of time. Meteorological drought usually precedes other kinds of drought.

Indian Meteorological Department (IMD) defines meteorological drought based on rainfall deficiency (S.W. Monsoon) (June–Sept) on subdivision wise basis. The meteorological droughts are classified into (a) moderate and (b) severe based on rainfall deficiency (i.e.) 26 to 50% and more than 50% respectively and 25% and less as normal.

Season's Rainfall

When the rainfall deficiency for the monsoon season of June to September for the country as a whole is within 10% of its long period average, it is categorized as a normal monsoon. When the monsoon rainfall deficiency exceeds 10% and it is covered more than 20% of area and less than 40% area of the country is under drought conditions, then the year is termed as All India drought year: when the rainfall deficiency is more than 10% and when the spatial coverage of drought is more than 40% it is called as all India severe drought year.

> **Example** South West monsoon rainfall for country as a whole was 19% below normal making 2002 year an All India drought year.

The weekly / seasonal rainfall distribution on regional scale It is common for drought conditions to exist on a sub-divisional scale or district level even when the monsoon is normal for the country as a whole. The performance of monsoon rainfall for much small areas the country is monitored by evaluating the departures from the normal for each meteorological subdivision and district. The rainfall is classified as excess, normal, deficient or scanty as per the following criteria.

Percentage departure of realized rainfall from normal rainfall is:

Excess   – +20% of the normal or more

Normal   – +19% to (–) 19% of normal

Deficient – (–) 20% to 59% of normal

Scanty   – (–) 60% to (–) 99% of normal or less

No rain   – (–) 100%

Rainfall Distribution on all India Scale

Normal: Percentage departure of realized rainfall is within

(±) 10% of the long period average

Below normal	Percentage departure of realized rainfall is < (less) 10% of the long period average
Above normal	Percentage departure of realized rainfall is > (greater) 10% of the long period average

India – Specific Classification for Drought

How IMD declares a year as a drought year In India, a year is considered to be a drought year in case, the area affected by moderate and severe drought, either individually or together, is 20% – 40% of the total area of the country and seasonal rainfall deficiency during south west monsoon (June – September) for the country as a whole is atleast 10% or more.

When the spatial coverage of drought is more than 40% it will be called as All India severe drought year. The country experienced all India drought years in 1972,

1979, 1987 and 2002. An rainfall deficiency of 49% in the month of July on all India basis, is a case of worst meteorological drought. Only two occasions in the past (1911 & 1918) it was over 45% and both the years had a major drought.

July rainfall is critical to agriculture. July is the rainest month of the monsoon season that registers one third of the seasonal rainfall. All though long range prediction of month wise rainfall distribution at the beginning of monsoon season is beyond the current state-of-art, efforts will be made to develop a model for prediction of July rainfall using meteorological data upto June and so as to enable a mid-season condition.

Meteorological drought need not necessarily result in agricultural drought, since timely rainfall during critical crop phases may save the crop or irrigation water may be available. However rainfall being the ultimate source of water in the event of an extreme rainfall deficiency its agricultural and hydrological impacts are enveritable.

Droughts are also classified according to the timing of rainfall deficiency during a particular rainfall season, usually June to September in India.

An analysis of droughts in recent decades indicates that in 1972, 1987 and 2002 rainfall deficiency was the maximum during July. These are called early season droughts and provide sufficient lead – time to mitigate the impact of the drought.

In 1965 and 1979 rainfall deficiency in September was the major cause for drought. These droughts are called late season droughts.

Mid – season droughts occur in association with the breaks in the south west monsoon. If the drought conditions occur during the vegetative phase of the crop growth, it might result in stunted growth, low leaf area development and even reduced plant population.

Further in India around 68%–70% of the country is prone to drought in varying degrees. Of the entire area, 35% of the area which receives rainfall between 750 mm and 1125 mm is considered drought prone, while another 33% which receives less than 750 mm of the rainfall, is called chronically drought prone.

A further classification of Indias regions into arid (19.6%) semi arid (37%) and sub-humid areas (21%) comprise three fourths of Indian drought prone total area of 329 million hectares. The drought prone areas of the country are confined to peninsular and western India, primarily arid, semi aid and sub humid areas.

1. Arid zone (19.6%)	-	Mean annual precipitation (MAP) OF 100 – 400 mm (Water deficit throughout the year) Ex: Rajasthan, Parts of Hariyana and Gujarat. (Droughts are severe in this zone)

		MAP of 400-600 mm (Water surplus in some months and deficit in other months) Ex: Parts of Hariyana, Punjab West, U.P. West Madhya Pradesh and also most of the peninsular parts of the Western Ghats. (Droughts can be moderate to severe in this zone)
2.	Semi-arid zone (37%) -	
3.	Dry sub-humid zone (21%) -	MAP of 600 – 900 mm in India, parts of northern plains, central high lands eastern plateau, Himalayas (Droughts are moderate in this zone)

4. Humid and pre-humid regions such as Assam and other north – east states rarely face drought.

In India IMD collects its rainfall data, using a network of 2800 rain gauge stations distributed across 36 meteorological subdivisions of the country.

What are the various classifications of rainfall intensity as per IMD

S.No	Description of Rainfall	Quantity of Rainfall in mm
1.	No rain	Rainfall amount realized in a day is 0.00 mm
2.	Trace	do is 0.01 to 0.04 mm
3.	Very light rain	do is 0.1 to 2.4 mm
4.	Light rain	do is 2.5 to 7.5 mm
5.	Moderate rain	do is 7.6 to 35.5 mm
6.	Rather heavy	do is 35.6 to 64.4 mm
7.	Heavy rain	do is 64.5 to 1.24.4 mm
8.	Very heavy rain	do is 124.5 to 244.4 mm
9.	Extremely heavy rain	do more than or equal to 244.5 mm
10.	Exceptionally heavy rain	do when exceeds 12 mm
11.	Rainy day	Rainfall amount realized in a day is 2.5 mm or more

In India, the monsoon rainfall accounts for about 75–80% of the total annual rainfall, in large areas of central and North West India, the monsoon contribution to the annual rainfall is 90% or more. Thus there is a pressing need to understand the Indian monsoon and forecast, its inter annual variability on long range scale.

What is the accuracy of the long range forecast for monsoon rainfall by IMD   The monsoon forecast in India is being done with reasonable accuracy. The success rate of IMD forecasts since 1988 has been high. During the last 21 years (1988–2008) IMD forecasts were qualitatively correct in 19 years (i.e. 90% of the years).

The exception was during years 2002 and 2004 both of which were drought years. However in some years (1994, 1997, 1999, 2002, 2004 & 2007), the forecast error (difference between actual rainfall and forecast rainfall) was more than 10%.

The 2002 drought was due to exceptionally low rainfall during the month of July (46% of long term period) caused by unexpected sudden warming of sea surface over equatorial central pacific that started in the month of June. It may be mentioned that the exceptionally deficient rainfall of July 2002 was not predicted by any prediction group in India or abroad.

It is not possible to have 100% success for forecasts based on statistical models. The problems with statistical models are inherent in this approach and are being faced by forecast world wide.

What is the relationship between monsoon and ENSO?   Both monsoon and ENSO (EL NiNO Southern Oscillation) are ocean–atmosphere couple Phenomena. This is a general inverse relationship between monsoon and ENSO.

The warm phase of ENSO is generally associated with weaker than normal monsoon and vice versa.

During the period (1885–2007) there were 36 years of warm ENSO (EL nino) and 25 years of cold ENSO (LANINA).

During the 15 of the 35 EL nina years 42% Indian summer monsoon rainfall (ISMR) was below normal and 9 of the 25 LANINA years 36% ISMR was above normal.

This shows that there is no one to one correspondence between ENSO and ISMR (Indian summer monsoon rainfall).

Monsoons

The climate in India is broadly described as a tropical monsoon type. The four seasons are winter (January – February) a hot summer period (March–May) a rainy south–western monsoon period (June–September) and a north – eastern monsoon period (October–December).

In addition, a number of micro–climatic patterns occur. The Kashmir Valley and some other higher altitude regions experience a typical temperate climate while still higher areas, such as Ladakh, Lahul and spiti, have a typical cold – arid desert climate.

India's climate is formed by the north east monsoon (winter monsoon) winds, which blow from land to sea and the south–west monsoon (Summer monsoon) winds which blow from sea to land after crossing the Indian ocean, the Arabian sea and the bay of Bengal.

Most rainfall in India is caused by the south–west monsoon. The south–western summer monsoons occur from June through September. The Great Indian Desert (Thar Desert) and adjoining areas of the northern and central Indian sub- continent heat up considerably during the hot summers. This causes a low pressure area over the northern and central Indian subcontinent. To fill up this void, the moisture laden winds from the Indian ocean rush into the subcontinent.

These winds rich in moisture, are drawn towards the Himalayas, creating winds blowing storm clouds towards the sub–continent. However the Himalayas act like a high wall and do not allow the winds to pass into Central Asia, forcing them to rise with the gain in the altitude of the clouds, the temperature drops and precipitation occurs. Some areas of the sub-continent receive upto 10,000 mm of rain.

A delay of a few days in the arrival of the monsoon can and does, badly affect the economy, as evidenced in the numerous droughts in India in the 90s. First of June is regarded as the date of onset of the monsoon in India in which is the average date on which the monsoon strikes State of Kerala over the years (fig).

North–East Monsoon (Retreating Monsoon)

Around September, with the sun fast retreating south, the northern land mass of the Indian subcontinent begin to cool off rapidly.

With this air pressure begins to build over northern India. The Indian Ocean and its surrounding atmosphere still holds its heat.

This causes the cold wind to sweep down from the Himalayas and Indo-gangetic plain towards the vast spans of the India Ocean, South of the Deccan Peninsula.

This is known as the north – east monsoon or retreating monsoon. While travelling towards the Indian Ocean, the dry cold wind picks up some moisture from the Bay of Bengal and pours it over Peninsular India. Cities like Chennai, which gets less rain from the south – west monsoon, receives rain from the retreating monsoon. About 50 - 60% of the rain received by the State of Tamilnadu is from the north – east monsoon. It is worth noting that north–east monsoon (or retreating monsoon) is not able to bring as much rain as the south–west monsoon.

Factors influencing the occurrence and conditions of drought in India   In India, the occurrence and conditions of drought are influenced by a number of factors

    (a) Rainfall and cropping patterns are different across many geographical regions.

    (b) It is not just the deficiency of rainfall, but also the uneven distribution of rainfall across the season, duration of rainfall deficiency and its impacts on different regions of the country that characterize drought conditions.

The facts presented below show seasonal and regional characteristics of drought conditions in India

1. **Seasonal distribution of rains in India**

   India receives most of its rainfall (73%) from the south west monsoon or summer monsoon (the rainfall received between June and September). The performance of Indian economy is vitally linked with the rainfall that occurs during these months. The summer monsoon appears at different dates at different places. While Kerala on the West Coast of India is the first to receive rainfall (on 1st June) and gradually proceeds towards north west region covering the entire country. Rajasthan is the last to receive rainfall (by around July 15th). The triangular shape of peninsular India and the Himalayan barriers deflect the winds from west to east. This position can best be understood with the help of the map enclosed. The north – east receives monsoon rains by around 5th of June. The moisture Laden winds then go onwards and reach Delhi by around 15th June. Monsoon starts its withdrawal during the first week of September from the west and north and gradually recedes from the entire country.

   The distribution of rainfall is highly uneven over the landmass. Due to this pattern of onset and withdrawal, the north – west region receives less than a month of rainy season due to late arrival and early cessation of monsoon conditions conversely Kerala and north eastern parts of India receive more than four (4) months of rainfall due to the early arrival and later withdrawal of the monsoon.

   Coastal areas of Peninsular India and Tamilnadu, in particular, also receive rains from October to December primarily due to periodic cyclonic disturbances in the Bay of Bengal (north east monsoon or post–monsoon). It also receives some rainfall during other months. The broad seasonal distribution of rainfall is presented in Table below.

   Table 5.1   Seasonal Distribution of Rains in India

Season	Months	Percentage of Distribution
Pre monsoon	March–May	10.4
South west monsoon	June–September	73.4
Post monsoon	October–December	13.3
Winter rains	January–February	2.9

   Source: Indian Meteorological, Dept. Govt. of India.

2. **Uneven distribution across monsoon**

   Uneven distribution of rain also occurs across the monsoon season (done by analyzing the number of meteorological subdivisions that receive deficient / scanty rainfall during mid July, mid August and mid September during major drought years as put up below.

Table 5.2 Sub-division Wise Distribution of Rains During Major Drought Occasions (number of sub-divisions received deficient rainfall)

Drought Year	Mid – July	Mid – August	Mid – September
1966	19	14	16
1972	13	21	21
1979	17	15	15
1987	25	25	21
2002	25	25	21

Source: IMD cited in drought 2002 – A Report, Ministry of Agriculture, Govt. of India.

1987 and 2002 were the worst years with regard to deficiency and season.

A comparison of the 2002 monsoon with three recent major droughts (1972 1979 and 1987) is shown in the table below. Rainfall variation in 2002 was much higher compared to the earlier drought years. A deficiency of 51% rainfall in July could not be compensated by almost average rainfall in August 4% and September 10%.

Table 5.3 Month-Wise Rainfall Distribution (percentage departure for the country as a whole in recent major drought years)

Year	June	July	August	September	June – September
1972	–27	–31	–14	–24	–24
1979	–15	–16	–19	–28	–19
1987	–22	–29	–4	–25	–19
2002	+4	–51	–4	–10	–19

Source: IMD cited in drought 2002 – A Report, Ministry of Agriculture, Govt. of India.

3. **Duration of drought**

Drought conditions are prolonged due to poor rainfall in consecutive years. Prolonged drought conditions do not allow people the opportunity to recover from the impact of drought and it depletes their capacity to cope with the impact.

Table below shows that drought in 1965 and 1966 became severe due to failure of rainfall during successive years, though deficiency did not hit the same area in both the years. During 1985-87, there were progressive reductions in rainfall in the same regions, with the result that 1987 experienced a multi-year drought. Year 2002 was preceded by normal monsoon years. However after 1998 the rainfall was less than normal, though the deficiency was moderate. The subdued rainfall in 1999 - 2001 reduced the availability of water and moisture and aggravated the impact of the 2002 drought when the rainfall deficiency was very high.

Table 5.4 Percentage Departure of Rainfall from Normal for Country as a Whole (SW monsoon)

Year	Percentage Departure from Normal
1965	−18
1966	−16
1972	−24
1979	−19
1985	−7
1986	−13
1987	−19
1999	−4
2000	−5
2001	−8
2002	−19

Why does drought recur in India?   Conditions for onset of drought in India vary across agro-climatic zones. In the semi arid regions even a 400 mm rainfall would be adequate for the growth of crops, while in high rainfall regions of Assam even an annual rainfall of 1000 mm would create conditions for drought.

Drought is a recurrent climatic phenomenon in India and is caused due to the country's peculiar physical and climatic characteristics as well as resulting economic and agricultural impacts. These include

(a) High average rainfall of around 1150 mm no other country has such a high annual average; however there is <u>considerable annual variation</u>

(b) About 73% of the total annual rainfall is received in less than 100 days during the south – west monsoon and the <u>geographic spread is uneven</u>

(c) Even though India receives abundant rainfall as a whole, <u>disparity in its distribution</u> over different parts of the country is so great that some parts suffer from perennial dryness.

In other parts, however, the rainfall is so excessive that only a small fraction can be utilized. The variability in rainfall as compared to long period average (LPA) exceeds 30% in large areas of the country and is over 40 – 50% in parts of drought prone Saurashtra, Kutch & Rajastan.

(d) About 33% of the cropped area in the country receives less than 750 mm rain annually making such areas hot spots of drought.

(e) Inadequacy of rains coupled with adverse land – man ratio compels the farmers to practice rain – fed agriculture in large parts of the country.

(f) Irrigation, using ground water aggravates the situation in the long term as groundwater withdrawal exceeds replenishment; in the peninsular region availability of surface water itself becomes scare in the years of rainfall insufficiency.

(g) Per capita water availability in the country is steadily declined.

(h) Traditional water harvesting systems have been largely abandoned.

Considerations being followed by various countries to define the drought in their countries   Drought is a recurrent feature of climate and it is an insidious (developing without being noticed with a harmful effect) often referred to a creeping phenomenon and its impacts vary from region to region. Therefore it is difficult for the people to understand unless the impacts are felt.

It is also equally difficult to define, because what may be considered a drought in say Bali (six days without rain) would certainly not be considered a drought in Libya (annual rainfall less than 180 mm) and similarly in India as per IMD a period of drought as a year or season in which the total rainfall is less than 75% of the normal. Meteorological drought is a situation when there is significant (more than 25%) decrease from normal precipitation over an area (A symposium on droughts in Asiatic monsoon region held at Delhi on 14 to 16th December, 1972).

In U.S.A. according to Conard (1944) a period of consecutive 20 days or more without 6.4 mm precipitation in 24 hours during season March to September is considered as drought situation (Ibid p4).

In Australia according to Gibbs and Maher (1967) the rainfall is the best single index of drought and use of rainfall declines demonstrate temporal and spatial distribution (Ibid p4).

In USSR drought is defined as period of ten days with a total rainfall not exceeding 5 mm (Report on national commission on agriculture 1976 p35).

The British rainfall organization in U.K. defines "absolute drought" when at least 15 consecutive days none of which receive at least 0.25 mm of rainfall and partial drought when at least 29 days during which mean rainfall does not exceed 0.25 mm per day.

Urgent need for developing a better drought monitoring and early warning systems in India   In most general sense drought originates from a deficiency of precipitation over an extended period of time usually a season or more, resulting in a water shortage for some activity, group or environment sector and causing extensive damage to crops resulting in loss of yield, due to unusual dryness of soil hot dry winds and high temperatures.

The impacts of drought are largely non-structural and spread over large geographical area than are damages from other natural hazards. The percentage of area affected by serious drought has doubled from the 1970's to the early 2000.

Rising global temperatures appear to be a major factor, provoking more frequent and intense droughts in sub-tropical areas of Asia (north India also) and Africa. It is reported that the northern himisphere has recorded an increase of 1°C in the past 100 years and the fluctuations is quite rapid during 1850 and 2000 and upward trend still persists. The climate models predict a global warming of about 1.4°C to 5.8°C between 1990 and 2100 and the sea level to rise by 9 to 88 cms. With increase in global temperatures the world is likely to experience more hot days and heat waves and fewer frost days and cold spells. Global warming happens when green house gases collect in the earths atmosphere. They let heat from sun through, but as it bounces back, it gets trapped close to the earth, making the planet heat up. Carbondiaoxide and Methane gases are released into the air as pollution. They are known as green house gases and can stop the suns heat escaping from the atmosphere.

The pattern of earth temperature as recorded from balloon and satellite measurements tend to exceed during the past few decades, indicating the pattern of green house warming. Even a small rise in temperature is likely to be accompanied by many other changes in cloud cover and wind pattern. The period of unusual dryness (i.e. drought) is a normal feature of climate and weather system in semi-arid and arid regions of the tropics of the world and it covers more than one third of the land surface vulnerable to drought and desertification. Droughts make forests fires more likely.

Most of the earth surface has become desert since the dawn of the civilization and more areas are in the process of desertification. The existing 18, 16 and 6 million square kilometers area of degraded lands in Africa, Asia and Australia respectively are on the increase due to rainfall variation and human activities. The variability in rainfall is significant in the tropical areas. Example being South India.

Every few years a change in sea temperature in Pacific called EL NINO a weather phenomenon with negative consequences for the monsoon rains affects weather around the world and causes droughts. Droughts are disastrous for people animals and plants. Droughts can also cause wars, when people are forced to leave their lands and flock into other areas.

Several years of drought, dried out farm soil in the Central States of United States of America (USA), such as Okkloma and Kansas, the dust bowl, was a great drought disaster that hit the USA in the year 1930's. It blow away in huge dust storms and farmers could not grow their crops. Hundreds of thousands of people had to leave the area. Many trekked west in search of new lives and jobs.

Tropic of cancer divides India into two equal halves. While the Southern part lies in the tropical zone, the northern part is in sub-tropical zone. India lies at the head of the Indian ocean. The moist winds blowing over the Indian Ocean blow towards the main land. Himalayas protect India from cold polar winds blowing from the north and impart a distinctly tropical touch to the climate of India.

Mawsynram in Meghalaya gets 1142 cms the Worlds heaviest rainfall. Over India, the monsoon rainfall (June – September) accounts for about 75-80% of the total annual rainfall, in large areas of central and northwest India, the monsoon contribution to the annual rainfall is 90% or more. Thus there is a pressing need to understand the Indian monsoon and forecast its inter annual variability on large range scale.

A delay of a few days, in the arrival of the monsoon can and does, badly affects the economy as evidenced in the numerous droughts in India in the 90s. June 1st is regarded as the date of onset of the monsoon in India which is the average date on which the monsoon strikes Kerala state over the years.

Increased climate variability has made rainfall patterns more inconsistent and unpredictable in the country increasing the recurrence of drought or drought like situation. Apart from decrease in agricultural production drought has other multifarious long dawn impacts like shortage of drinking water, fodder, less water in dams / reservoirs for power generation etc. which has severe impact on the economy of the country affecting its growth.

Today the world is facing a greater water crisis than ever. Droughts of lesser magnitude are resulting in greater impact. Even in years with normal precipitation water shortage has become wide spread on both developing and developed nations in humid as well as arid climates. When faced with severe drought Governments become eager to act. Unfortunately this eagerness usually wanes when precipitation returns to normal.

Therefore, there is an urgent need to develop better drought monitoring and early warning systems in India. A critical component of national drought strategies should be a comprehensive drought monitoring system, that can provide early warning of droughts onset and END, determine its severity and deliver that information to a broad group of users in a timely manner. With this information the impacts of drought can be reduced or avoided in many cases.

Droughts are commonly classified by type as meteorological, agricultural and hydrological and socio-economical droughts differ from one another in three essential characteristics, intensity, duration and spatial coverage.

Limitations of percentage of normal precipitation used by India to characterize meteorological drought are as follows   At present in India, percentage of normal precipitation method is being used to characterize the meteorological droughts. It is a simple method to detect the meteorological droughts. It is calculated by dividing actual precipitation typically by a 30 years mean and multiplying it by 100% for each rainfall station. This method is effective in single region or season. **The data is not normalized.**

It is lagging in its detection of drought over months because the precipitation data collected is not true to its ground realities. The data being used will show as more in quantity than actual precipitation being put to use by the people since the temperature element of climate has not been considered. The deficiency of precipitation is being resulted after some time than actually felt; resulting in shortage of drinking water supply and the people will suffer.

Percentage of normal method is effective in single region or season. The intensities of meteorological drought can not be compared, to identify the order of severity felt village or mandal to later to mitigation activities.

Percentage normal, cannot determine the frequency of the departures from normal or compare with different locations. Also it cannot identify specific impact of drought or the inhibition factor for drought risk mitigation plans. Further the date of onset of drought and its end along with intensity at different scales are not provided in this method. This method is misleading because the median rainfall is different than the mean. For instance, a few years of extremely heavy rain can significantly increase the mean rainfall, meaning that more years will be less than "100 percent" of the mean, even though they won't signify droughts.

"Deficit of rainfall over a period of time at a certain location could lead to various degrees of drought conditions affecting water resources, agriculture and socio-economic activities. Since rainfall varies significantly among different regions, the concept of drought may differ from places to places" (W.U. Man – chi, Sep. 13, 2013).

As such effective drought early warning systems must integrate precipitation and other climatic parameters. The percentage average rainfall will only indicate the numerical magnitude of variation (i.e. 40 mm less or 10 mm more etc.) without stating how much less than what (is it 40 mm less than 80 mm, 40 mm less than 200 mm). **The standardized precipitation index recently developed shows the statistical magnitude of deviation from the average and therefore better portrays the seriousness of the shortage.**

Standardized precipitation index   At the international workshop on indices and early warning systems for drought held at Lincoln Nebraska, USA on 8 to 11 December, 2009, for which forty four participant countries attended including India, sponsored by World Meteorological Organization, the following indices were identified for characterization of meteorological drought during session No.1, are standardized precipitation index, percentage of normal precipitation, a soil moisture index, percentile ranking methods (deciles and quantiles) palmer drought severity index (PDSI) and K-index.

During session No.2 the participants reached a consensus that the WMO should recommend the standardized precipitation index for widespread use in countries wanting to track meteorological droughts.

The recommendation to use the spi was approved by world meteorological organization congress in June 2011. The recommendation of WMO to use spi in place of percentage normal precipitation is a better drought indice to characterize meteorological droughts, is indeed a blessing in disguise for Indian conditions.

The Strengths and Weaknesses of Spi can be Summarized as Follows

Strengths

1. It is flexible, it can be computed for multiple time scales.
2. Shorter time scales spi's for example 1, 2 & 3 months spi's can provide early warning of drought and help assess drought severity.
3. It is spatially consistent. It allows for comparisons between different locations in different climates.
4. Its probabilistic nature gives it historical context, which is well suited for decision making.

Weaknesses

1. It is based only on precipitation
2. No soil water – balance component, thus no ratios of evapo-transpiration/ potential evapo-transpiration ET / PET) can be calculated.

SPI Methodology

1. The spi calculation for any location is based on the long term precipitation record for a desired period. This long term record is fitted to a probability distribution, which is then transformed into a normal distribution so that the mean spi for the location and desired period is zero (Edwards and McKee 1997).
2. Positive spi values indicate greater than median precipitation, and negative values indicate less than median precipitation.
3. Drought according to the spi, starts when the spi value is equal or below (–) 1.0 and ends when the value becomes positive.

The spi was designed to quantify the precipitation deficit for multiple time scales. These time scales reflect the impact of drought on the availability of the different water resources. Soil moisture conditions respond to precipitation anomalies on a relatively short scale. Ground water, stream flow and reservoir storage reflect the long – term precipitation anomalies. For these reasons McKee and Others (1993) originally calculated the spi for 3, 6, 12, 24 and 48 month time scales.

McKee and Others (1993) used the classification system shown in the spi value table below, to define drought intensities resulting from the spi. They also defined the criteria for a drought event for any of the timescales. A drought event occurs any time the spi is continuously negative and reaches an intensity of (–) 1.00 or less. The event ends when the spi becomes positive. Each drought event, therefore has a duration defined by its beginning and end, and an intensity for each month that the event continues. The positive sum of the spi for all the months within a drought event can be termed as the droughts "magnitude".

Table 5.5 SPI Values

2.0 +	Extremely wet
1.5 to 1.99	Very wet
1.0 to 1.49	Moderately wet
(–) 0.99 to 0.99	Near normal
(–) 1.0 to (–) 1.49	Moderately dry
(–) 1.5 to (–) 1.99	Severely dry
(–) 2 and less	Extremely dry

The standardization allows the spi to determine the rarity of a current drought, as well as the probability of the precipitation necessary to end it (Mc Kee and Others 1993). It also allows the user to confidently compare historical and current droughts between different climatic and geographic locations when assessing how rare or frequent a given drought event is.

Some key points

1. Because the spi is normalized, wetter and drier climates can be represented in the same way, thus, wet periods can also be monitored using the spi, it must be stressed that the spi is not suitable for climate change analysis because temperature is not an input parameter.
2. The spi was designed to quantify the precipitation deficit for multiple time scales.
3. These time scales reflect the impact of drought on the availability of different water resources which was the initial intent of the spi's creators.
4. Soil moisture conditions respond to precipitation anomalies on a relatively short time scale. Ground water, stream flow and reservoir storage reflect the longer – term precipitation anomalies.

So, for example, one may want to look at a 1 or 2 month spi for meteorological drought, anywhere from 1 month to 6 month spi for agricultural drought and something like 6 month upto 24 month spi or more for hydrological drought analyses and applications.

How it works

1. Precipitation is normalized using a probability distribution function so that values of spi are actually seen as standard deviations from the median.
2. A normalized distribution allows for estimation of both dry and wet periods.

3. Accumulated values can be used to analyse drought severity (magnitude).

4. At least 30 years of continuous monthly precipitation data are needed but longer – term records would be preferable.

5. spi time scale intervals shorter than 1 month and longer than 24 months may be unreliable.

6. It is spatially in variant in its interpretation.

7. Its probability – based nature (Probability of observed precipitation transformed into an index) makes it well suited to risk management and triggers for decision – making.

How to obtain the program   The program is available in a windows / p.c version and can be down loaded or it is available in this book as an attachment. The latest spi programme (spi_sl_6.exe) can be found at htt://drought.unl.edu/monitoringTools/ downloadable spi program.aspx. The program can calculate upto 6 spi time windows at one time for any given location. It was compiled in C++ for p.c.

Mapping Capabilities

Many countries are calculating and mapping the spi and other drought indices or meteorological parameters on a regular basis. Below is an over view of the methods often employed in mapping drought indicators.

There are multiple ways, such as the standard drought indicators and indices, to map meteorological variables. Most drought – related data originate as point ('station – based' or 'site – specific') data. These data serve their purposes, but it is often in map form that the data best communicate a massage based on a geographic context to the decision – maker trying to understand drought severity and spatial extent. The point data can be placed on a map, and derivative products or characteristics of that site can be provided for additional information. This could include, for example, a time series plot of the indicator or index. The limitation of this level of spatial detail is that information on what is taking place between points is not available.

A variety of techniques can be used to generate a continuous map of meteorological drought. One such technique generates an interpolated surface of estimated values at locations between sites based on mathematical relationships of the indicator or index between the original point data. Often this produces a map that appears "natural", but is still based on the data from specific points and is only as accurate as the original data and the interpolation technique. No single interpolation method can be applied to all situations and the most commonly used interpolation techniques include kriging, spline and inverse distance weighting (IDW).

Each interpolation technique has its advantages and disadvantages. Some techniques are more exact than others, but take longer to produce the desired output. The kriging method, which has its origins in geological applications and the mining industry assumes that there is a relationship between points that is non-random and changes over space. The spline method is used when minimizing the overall surface curvature is important. Inverse distance weighting (IDW) is used when the data points are scattered but dense enough to represent local variations. The data, as the name implies, are weighted to favour data closer in proximity to the point being processed.

Another technique that has been used for meteorological drought monitoring and mapping is to map point data to grid cells. These data can also originate from airborne, radar and satellite sources. These gridded data products appear less 'natural' than interpolated products but they are easier to use the comparative purpose because of the common grid cell sizes. These cells can vary in size from degrees down to meters depending on the source and application needed. They also vary in their temporal frequency, with return periods ranging from daily (or multiple times a day) to weekly or longer. In the United States, gridded products for monitoring meteorological drought are becoming much more common, while in other regions, particularly Africa, there has been a long history of using gridded information to determine drought conditions. The famine early warning system (FEWS) and similar networks have utilized gridded data in their analyses. Multiple examples of gridded meteorological drought products exists in Australia, China, the United Kingdom and United States.

To develop a gridded map product, the point data are aggregated to a grid cell resolution selected for that product using a mathematical relationship. An interpolated surface is then created between the grid cells (not the point data). For example, the high plains regional climate center, in partnership with the national drought mitigation center, is mapping the standardized precipitation index on a daily basis at state, regional and national scales across the united states.

Spi maps are generated by using the Grid analysis and display system (Gr ADS). The discrete station spi data are interpolated using a cressman objective analysis with radii of influence of 10, 7, 4, 2, 1. The grid resolution is 0.4 degrees. Gridded contour maps are generated at national, regional and state levels for the high plains region. For national maps, north polar stereographic (rps) projection is applied. Regional and state maps use a latitude longitude (dat/con) projection with aspect ratio maintained. This interface and resultant products can be found at http://www. hprcc.unl.edu/maps/current.

Successful mapping of meteorological drought deports on the quality of the data. Drought indicator and index data quality is determined by several factors including the availability of the data, the timing of the recording, the quality of the historical data at a station, the transmission of data in near real time, the maintenance of the station network and the ability to measure precipitation in cold temperatures, particularly in northern or alpine locations. Some of these issues are released to the ability to provide the data in a timely fashion, which can be very important with meteorological drought. Finally the data density plays a huge role in the spatial resolution that can be achieved in mapping drought.

One of the highest challenges in mapping meteorological drought is to try and match the spatial resolution that decision – makers need and demand with the information that is available to-day. This limitation is related to the density of point data, which might not be available at the resolution desired by the decision makers. Because of this challenge, the promise of potential remote sensing based products is encouraging. Some remote sensing products can provide data at spatial resolutions in regions where in situ point data are relatively sparse and unreliable. Most satellite products are already incorporated into a grid cell (or pixel) as described above. In the United States, some products are now being developed to utilize a combination of station – based and remote – sensing data. The station data are used to help refine the remote sensing data, and the resulting 'hibrid' maps have a higher level of accuracy.

Topographical issues, particularly those related to mountains and rapid terrain changes, represent a real challenge when mapping meteorological drought. There are two reasons for this. First, data density tends to be lower in mountainous regions. Second, because interpolation methodologies are often based on correlations, the relationship between adjoining regions with regard to precipitation tends to be particularly discontinuous in areas where the terrain changes rapidly and significantly. As a result, smoothed interpolated surfaces on a finished map product may not realistically match natural variability especially the indicators and indices related to precipitation.

Because of all the complexities involved in meteorological drought data and the characteristics of mapping techniques, it is important that a decision – maker understand these factors when interpreting maps of drought severity and spatial extent.

Source: World Meteorological Organization 2012 standardized precipitation index users guide (M.Svode M. Hayes and D. Wood) (WMO – No 1090) Geneva. Switzerland).

Vide attachment: standardized precipitation index user guide.

STANDARDIZED PRECIPITATION INDEX

USER GUIDE

EDITORIAL NOTE

METEOTERM, the WMO terminology database, may be consulted at: http://www. wmo.int/pages/prog/ lsp/meteoterm_wmo_en.html. Acronyms may also be found at: http://www.wmo.int/pages/themes/ acronyms/ index_en.html.

WMO-No. 1090

© World Meteorological Organization, 2012

The right of publication in print, electronic and any other form and in any language is reserved by WMO. Short extracts from WMO publications may be reproduced

Chair, Publications Board

World Meteorological Organization (WMO)

7 bis, avenue de la Paix   Tel.: +41 (0) 22 730 84 03

P.O. Box 2300   Fax: +41 (0) 22 730 80 40

CH-1211 Geneva 2, Switzerland   E-mail: publications@wmo.int

ISBN 978-92-63-11091-6

NOTE

CONTENTS

STANDARDIZED PRECIPITATION INDEX USER GUIDE

PREFACE

Over the years, there has been much discussion on what drought indices should be used in a particular climate and for what application. Many drought definitions and indices have been developed and attempts have been made to provide some guidance on this issue.

With this in mind, the Interregional Workshop on Indices and Early Warning Systems for Drought was organized and held at the University of Nebraska-Lincoln, United States of America, from 8 to 11 December 2009. It was jointly sponsored by the School of Natural Resources (SNR) of the University of Nebraska, the United States National Drought Mitigation Center (NDMC), the World Meteorological Organization (WMO), the United States National Oceanic and Atmospheric Administration (NOAA), the United States Department of Agriculture (USDA) and the United Nations Convention to Combat Desertification (UNCCD). The workshop brought together 54 participants representing 22 countries from all over the world. They reviewed the drought indices currently in use in different regions of the world to explain meteorological, agricultural and hydrological droughts; assessed the capacity for collecting information on the impacts of drought; reviewed the current and emerging technologies for drought monitoring, and discussed the need for consensus standard indices to describe different types of droughts.

The experts at the meeting elaborated and approved the Lincoln Declaration on Drought Indices, which recommended that the Standardized Precipitation Index (SPI) be used by all National Meteorological and Hydrological Services (NMHSs) around the world to characterize meteorological droughts, in addition to other drought indices that were in use in their service.

The Lincoln Declaration also recommended the development of a comprehensive SPI user manual. In June 2011, the Sixteenth World Meteorological Congress adopted a resolution that endorsed both of these recommendations. The Congress also requested that the SPI manual be published and distributed in all official languages of the United Nations.

The full Lincoln Declaration on Drought Indices can be found on the WMO website at http://www.wmo.int/pages/prog/wcp/m/meetings/wies09/documents/Lincoln_Declaration_Drought_ Indices.pdf.

WMO would like to thank Mark Svoboda, Michael Hayes and Deborah A. Wood of the National Drought Mitigation Center (NDMC) at the University of Nebraska for preparing this user guide on the Standardized Precipitation Index[1]. We hope that it will help countries and institutions to understand how to calculate and use the SPI in order to develop or further enhance their own drought monitoring and early warning capabilities.

For any questions or comments on the content of this guide, including any suggestions for improvement, please email the WMO Agricultural Meteorology Division at agm@wmo.int.

BACKGROUND

Drought is an insidious natural hazard that results from lower levels of precipitations than what is considered normal. When this phenomenon extends over a season or a longer period of time, precipitation is insufficient to meet the demands of human activities and the environment. Drought must be considered a relative, rather than absolute, condition. There are also many different methodologies for monitoring drought. Droughts are regional in extent and each region has specific climatic characteristics. Droughts that occur in the North American Great Plains will differ from those that occur in Northeast Brazil, Southern Africa, Western Europe, Eastern Australia, or the North China Plain. The amount, seasonality and form of precipitation differ widely between each of these locations.

Temperature, wind and relative humidity are also important factors to include in characterizing drought. Drought monitoring also needs to be application-specific because drought impacts will vary between sectors. Drought means different things to different users such as water managers, agricultural producers, hydroelectric power plant operators and wildlife biologists. Even within sectors, there are many different perspectives of drought because impacts may differ markedly. Droughts are commonly classified by type as meteorological, agricultural and hydrological, and differ from one another in intensity, duration and spatial coverage.

INTRODUCTION TO THE STANDARDIZED PRECIPITATION INDEX

Over the years, many drought indices were developed and used by meteorologists and climatologists around the world. Those ranged from simple indices such as percentage of normal precipitation and precipitation percentiles to more complicated indices such as the Palmer Drought Severity Index. However, scientists in the United States realized that an index needed to be simple, easy to calculate and statistically relevant and meaningful. Moreover, the understanding that a deficit of precipitation has different impacts on groundwater, reservoir storage, soil moisture, snowpack

---

[2] This guide should be cited as: World Meteorological Organization, 2012: Standardized Precipitation Index User Guide (M. Svoboda, M. Hayes and D. Wood). (WMO-No. 1090), Geneva.

and stream flow led American scientists McKee, Doesken and Kleist to develop the Standardized Precipitation Index (SPI) in 1993.

The SPI (McKee and others, 1993, 1995) is a powerful, flexible index that is simple to calculate. In fact, precipitation is the only required input parameter. In addition, it is just as effective in analysing wet periods/cycles as it is in analysing dry periods/cycles. The program can run in both Windows and UNIX environments. This SPI user guide describes the Windows version.

Ideally, one needs at least 20-30 years of monthly values, with 50-60 years (or more) being optimal and preferred (Guttman, 1994). The program can be run with missing data, but it will affect the confidence of the results, depending on the distribution of the missing data in relation to the length of the record. More information on usage can be found in section 6, Computational methodology.

Climatologists would prefer to see serially complete data sets, which means there should be no missing data. However, it is more than likely that data sets would only have 90% or even 85% complete records. In reality, many users don't have this luxury and may have to settle for less (75-85% complete) unless they look to estimation techniques to fill in the gaps in the record. Of course, long and pristine data records are neither practical nor typical in many cases, so the user needs to be aware of the statistical limitations of extreme events when dealing with shorter periods of records for various locations. In the end, users have to make a subjective decision as to what tolerance of missing data they are willing to incorporate into the SPI calculations and analyses. Depending on the confidence and method of calculation, the use of estimated data is acceptable. Naturally, the fewer estimated data used the better.

DESCRIPTION OF THE STANDARDIZED PRECIPITATION INDEX

Overview: The SPI is based on the probability of precipitation for any time scale. The probability of observed precipitation is then transformed into an index. It is being used in research or operational mode in more than 70 countries.

Who uses it: Many drought planners appreciate the SPI's versatility. It is also used by a variety of research institutions, universities, and National Meteorological and Hydrological Services across the world as part of drought monitoring and early warning efforts.

Strengths: Precipitation is the only input parameter. The SPI can be computed for different time scales, provide early warning of drought and help assess drought severity. It is less complex than the Palmer Drought Severity Index and many other indices.

Weaknesses: It can only quantify the precipitation deficit; values based on preliminary data may change, and values change as the period of record grows.

Developed by T.B. McKee, N.J. Doesken and J. Kleist, Colorado State University, 1993.

The SPI was designed to quantify the precipitation deficit for multiple timescales. These timescales reflect the impact of drought on the availability of the different water resources. Soil moisture conditions respond to precipitation anomalies on a relatively short scale. Groundwater, streamflow and reservoir storage reflect the longer-term precipitation anomalies. For these reasons, McKee and others (1993) originally calculated the SPI for 3-, 6-,12-, 24- and 48-month timescales.

The SPI calculation for any location is based on the long-term precipitation record for a desired period. This long-term record is fitted to a probability distribution, which is then transformed into a normal distribution so that the mean SPI for the location and desired period is zero (Edwards and McKee, 1997). Positive SPI values indicate greater than median precipitation and negative values indicate less than median precipitation. Because the SPI is normalized, wetter and drier climates can be represented in the same way; thus, wet periods can also be monitored using the SPI.

McKee and others (1993) used the classification system shown in the SPI value table below (Table 1) to define drought intensities resulting from the SPI. They also defined the criteria for a drought event for any of the timescales. A drought event occurs any time the SPI is continuously negative and reaches an intensity of −1.0 or less. The event ends when the SPI becomes positive. Each drought event, therefore, has a duration defined by its beginning and end, and an intensity for each month that the event continues. The positive sum of the SPI for all the months within a drought event can be termed the drought's "magnitude".

Table 5.6   SPI Values

2.0+	Extremely wet
1.5 to 1.99	Very wet
1.0 to 1.49	Moderately wet
−0.99 to 0.99	Near normal
−1.0 to −1.49	Moderately dry
−1.5 to −1.99	Severely dry
−2 and less	Extremely dry

Based on an analysis of stations across Colorado in the United States, McKee determined that the SPI indicates mild drought 24% of the time, moderate drought 9.2% of the time, severe drought 4.4% of the time and extreme drought 2.3% of the time (McKee and others, 1993). Because the SPI is standardized, these percentages are expected from a normal distribution of the SPI. The 2.3% of SPI values within the "extreme drought" category is a percentage that is typically expected for an "extreme" event. In contrast, the Palmer Drought Severity Index reaches its "extreme" category more than 10% of the time across portions of the central Great Plains in the United States. This standardization allows the SPI to determine the rarity of a current drought

(Table 2), as well as the probability of the precipitation necessary to end it (McKee and others, 1993). It also allows the user to confidently compare historical and current droughts between different climatic and geographic locations when assessing how rare, or frequent, a given drought event is.

Table 5.7   Probability of Recurrence

SPI	Category	Number of Times in 100 Years	Severity of Event
0 to –0.99	Mild dryness	33	1 in 3 years
–1.00 to –1.49	Moderate dryness	10	1 in 10 years
–1.5 to –1.99	Severe dryness	5	1 in 20 years
< –2.0	Extreme dryness	2.5	1 in 50 years

Some key points:

- Because the SPI is normalized, wetter and drier climates can be represented in the same way; thus, wet periods can also be monitored using the SPI. However, it must be stressed that the SPI is not suitable for climate change analysis because temperature is not an input parameter.
- The SPI was designed to quantify the precipitation deficit for multiple timescales.
- These timescales reflect the impact of drought on the availability of the different water resources, which was the initial intent of the SPI's creators.
- Soil moisture conditions respond to precipitation anomalies on a relatively short timescale. Groundwater, stream flow and reservoir storage reflect the longer-term precipitation anomalies. So, for example, one may want to look at a 1- or 2-month SPI for meteorological drought, anywhere from 1-month to 6-month SPI for agricultural drought, and something like 6-month up to 24-month SPI or more for hydrological drought analyses and applications.

STRENGTHS AND WEAKNESSES

The strengths and weaknesses of the SPI can be summarized as follows:

Strengths

- It is flexible: it can be computed for multiple timescales
- Shorter timescale SPIs, for example 1-, 2- or 3-month SPIs, can provide early warning of drought and help assess drought severity.
- It is spatially consistent: it allows for comparisons between different locations in different climates.

○ Its probabilistic nature gives it historical context, which is well suited for decision-making.

Weaknesses

○ It is based only on precipitation

○ No soil water-balance component, thus no ratios of evapo-transpiration/ potential evapo-transpiration (ET/PET) can be calculated.

○ A new variation of the index by Vicente-Serrano and others (2010) attempts to address the PET issue by including a temperature component in the calculation of their new index called the Standardized Precipitation and Evapo-transpiration Index (SPEI). The inputs required to run the program are precipitation, mean temperature and latitude of the site(s). Further information on the SPEI is available at http://sac.csic.es/spei/index.html.

INTERPRETATION: SPATIAL AND TEMPORAL FLEXIBILITY DESCRIBED

There is no single definition of drought (Wilhite and Glantz, 1985). We can generally group them into meteorological, agricultural, hydrological and socioeconomic droughts. Drought is a very complex hazard to define and detect. It spans multiple sectors and timescales. Just as there is no single definition of drought, there is no single drought index that meets the requirements of all applications.

That said, a real strength of the SPI is its ability to be calculated for many timescales, which makes it possible to deal with many of the drought types described above. The ability to compute the SPI on multiple timescales allows for temporal flexibility in the evaluation of precipitation conditions in relation to water supply.

As mentioned earlier, the SPI was designed to quantify the precipitation deficit for multiple timescales, or moving averaging windows. These timescales reflect the impacts of drought on different water resources needed by various decision-makers. Meteorological and soil moisture conditions (agriculture) respond to precipitation anomalies on relatively short timescales, for example 1-6 months, whereas stream flow, reservoirs, and groundwater respond to longer-term precipitation anomalies of the order of 6 months up to 24 months or longer. So, for example, one may want to look at a 1- or 2-month SPI for meteorological drought, anywhere from 1-month to 6-month SPI for agricultural drought, and something like 6-month up to 24-month SPI or more for hydrological drought analyses and applications.

The SPI can be calculated from 1 month up to 72 months. Statistically, 1–24 months is the best practical range of application (Guttman, 1994, 1999). This 24-month cutoff is based on Guttman's recommendation of having around 50–60 years of data available. Unless one has 80–100 years of data, the sample size is too small and the statistical confidence of the probability estimates on the tails (both wet and dry extremes) becomes weak beyond 24 months. In addition, having only the minimum 30 years of data (or less) shortens the sample size and weakens the confidence. Technically, one could run the SPI on less than 30 years of data bearing in mind, however, the statistical limitations and weaker confidence pointed out above.

Short-Versus Long-Term Standardized Precipitation Index Values

1-month SPI   A 1-month SPI map is very similar to a map displaying the percentage of normal precipitation for a 30-day period. In fact, the derived SPI is a more accurate representation of monthly precipitation because the distribution has been normalized. For example, a 1-month SPI at the end of November compares the 1-month precipitation total for November in that particular year with the November precipitation totals of all the years on record. Because the 1-month SPI reflects short-term conditions, its application can be related closely to meteorological types of drought along with short-term soil moisture and crop stress, especially during the growing season. The 1-month SPI may approximate conditions represented by the Crop Moisture Index, which is part of the Palmer Drought Severity Index suite of indices.

Interpretation of the 1-month SPI may be misleading unless climatology is understood. In regions where rainfall is normally low during a month, large negative or positive SPIs may result even though the departure from the mean is relatively small. The 1-month SPI can also be misleading with precipitation values less than the normal in regions with a small normal precipitation total for a month. As with a percent of normal precipitation map, useful information is contained in the 1-month SPI maps, but caution must be observed when analysing them.

Note: In theory, the SPI can be calculated on a sub-monthly basis, but in practice this is not recommended. It is highly recommended that the user look at a minimum averaging window of 4 weeks. One could compute a 1-week SPI, but the reality is that one will likely encounter many dry day events (0.00 rainfall even in non-arid climates), which makes the SPI behave erratically (Wu and others, 2006), therefore this approach is not recommended. However, updating the SPI every day or every week for a 1-month up to a 24-month time frame is acceptable. This "moving window" approach does not compromise the program as it is still looking at a minimum of 4 weeks of data each day it moves.

3-month SPI   The 3-month SPI provides a comparison of the precipitation over a specific 3-month period with the precipitation totals from the same 3-month period for all the years included in the historical record. In other words, a 3-month SPI at the end of February compares the December–January– February precipitation total in that particular year with the December–February precipitation totals of all the years on record for that location. Each year data is added, another year is added to the period of record, thus the values from all years are used again. The values can and will change as the current year is compared historically and statistically to all prior years in the record of observation.

A 3-month SPI reflects short- and medium-term moisture conditions and provides a seasonal estimation of precipitation. In primary agricultural regions, a 3-month SPI might be more effective in highlighting available moisture conditions than the slow-responding Palmer Index or other currently available hydrological indices. A 3-month SPI at the end of August in the United States Corn Belt would capture precipitation trends during the important reproductive and early grain-filling stages for both corn and soybeans. Meanwhile, the 3-month SPI at the end of May gives an indication of soil moisture conditions as the growing season begins.

It is important to compare the 3-month SPI with longer timescales. A relatively normal or even a wet 3-month period could occur in the middle of a longer-term drought that would only be visible over a long period. Looking at longer timescales can prevent misinterpretation believing that a drought might be over when in fact it is just a temporary wet period. Continuous and persistent drought monitoring is essential to determine when droughts begin and end. This helps avoid "false alarms" when going into and coming out of drought. Having a set of "triggers" in place, which are tied to actions within a drought plan, can help ensure this.

As with the 1-month SPI, the 3-month SPI may be misleading in regions where it is normally dry during any given 3-month period. Large negative or positive SPIs may be associated with precipitation totals not very different from the mean. This caution can be explained with the Mediterranean climate of California and around northern Africa and southern Europe, where very little rain falls or is expected over distinct periods of the year. Because these periods are characterized by little rain, the corresponding historical totals will be small, and relatively small deviations on either side of the mean could result in large negative or positive SPIs. Conversely, this time period can be a good indicator for some monsoon regions around the world.

6-month SPI   The 6-month SPI compares the precipitation for that period with the same 6-month period over the historical record. For example, a 6-month SPI at the end of September compares the precipitation total for the April–September period with all the past totals for that same period.

The 6-month SPI indicates seasonal to medium-term trends in precipitation and is still considered to be more sensitive to conditions at this scale than the Palmer Index. A 6-month SPI can be very effective in showing the precipitation over distinct seasons. For example, a 6-month SPI at the end of March would give a very good indication of the amount of precipitation that has fallen during the very important wet season period from October through March for certain Mediterranean locales. Information from a 6-month SPI may also begin to be associated with anomalous streamflows and reservoir levels, depending on the region and time of year.

9-month SPI   The 9-month SPI provides an indication of inter-seasonal precipitation patterns over a medium timescale duration. Droughts usually take a season or more to develop. SPI values below −1.5 for these timescales are usually a good indication that dryness is having a significant impact on agriculture and may be affecting other sectors as well. Some regions may find that the pattern displayed by the map of the Palmer Index is closely related the 9-month SPI maps. For other areas, the Palmer Index is more closely related to the 12-month SPI. This time period begins to bridge a short-term seasonal drought to those longer-term droughts that may become hydrological, or multi-year, in nature.

12-month up to 24-month SPI   The SPI at these timescales reflects long-term precipitation patterns. A 12-month SPI is a comparison of the precipitation for 12 consecutive months with that recorded in the same 12 consecutive months in all previous years of available data. Because these timescales are the cumulative result of shorter periods that may be above or below normal, the longer SPIs tend to gravitate toward zero unless a distinctive wet or dry trend is taking place. SPIs of these timescales are usually tied to streamflows, reservoir levels, and even groundwater levels at longer timescales. In some locations, the 12-month SPI is most closely related with the Palmer Index, and the two indices can reflect similar conditions.

COMPUTATIONAL METHODOLOGY

The SPI is determined by normalizing the precipitation for a given station after it has been fitted to a probability density function as described by McKee and others (1993, 1995), Edwards and McKee (1997), and Guttman (1998). A full description of the SPI computational procedure can be found in McKee and others (1993, 1995) and Edwards and McKee (1997). The basics, as taken from Edwards (1997), are described below.

SPI Methodology

- The SPI calculation for any location is based on the long-term precipitation record for a desired period. This long-term record is fitted to a probability distribution, which is then transformed into a normal distribution so that the mean SPI for the location and desired period is zero (Edwards and McKee, 1997).

- Positive SPI values indicate greater than median precipitation, and negative values indicate less than median precipitation.

- Drought, according to the SPI, starts when the SPI value is equal or below −1.0 and ends when the value becomes positive.

How it Works

- Precipitation is normalized using a probability distribution function so that values of SPI are actually seen as standard deviations from the median.

- A normalized distribution allows for estimation of both dry and wet periods.

- Accumulated values can be used to analyse drought severity (magnitude).

○ At least 30 years of continuous monthly precipitation data are needed but longer-term records would be preferable.

○ SPI timescale intervals shorter than 1 month and longer than 24 months may be unreliable.

○ It is spatially invariant in its interpretation.

○ Its probability-based nature (probability of observed precipitation transformed into an index) makes it well suited to risk management and triggers for decision-making.

HOW TO OBTAIN THE PROGRAM

The program is available in a Windows/PC version and can be downloaded for free. The latest SPI program (SPI_SL_6.exe), sample files such as those described below and instructions for Windows/PC use can be found at http://drought.unl.edu/ MonitoringTools/ DownloadableSPIProgram.aspx.

The program can calculate up to 6 SPI time windows at one time for any given location. It was compiled in C++ for PC and all libraries are included.

HOW TO RUN THE PROGRAM IN WINDOWS

To run the program in Windows, simply follow the steps indicated below:

1. Set up an input file as in the following sample containing precipitation data from Falls City, Nebraska:

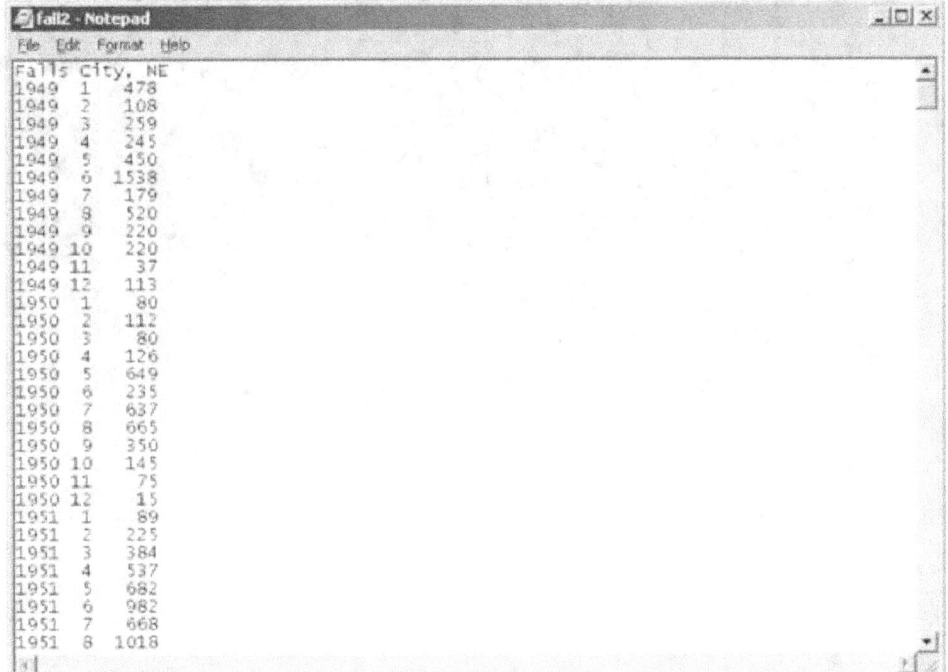

All input files must follow this format, which features three columns indicating respectively the year, month and monthly precipitation value. A header, usually the name of the station, must be included at the top of the input file otherwise the program will produce an empty output file. The precipitation total must NOT include decimals and can be in inches or millimetres.

Note: Pay attention to column spacing and missing data in the input file. If the monthly precipitation value is missing for a particular month or months, one must use −99 for the missing data value. Do not use a blank in the precipitation column. Zero is a valid value for typically dry months in arid regions or for those locations having distinct wet or dry seasons. Ideally, one would want at least 30 years of monthly/weekly data in order to have some confidence in the statistics, but that would be the case with most indices when assessing any drought climatology for a given location or region.

The input files can be generated with Excel or any text editor, but must be renamed with a .cor extension before executing the program.

2. Right click on the SPI_SL_6.exe file and save it. Then execute (double click) the program and follow the instructions in the pop-up window.

3. Choose the number of SPI timescales to be computed:

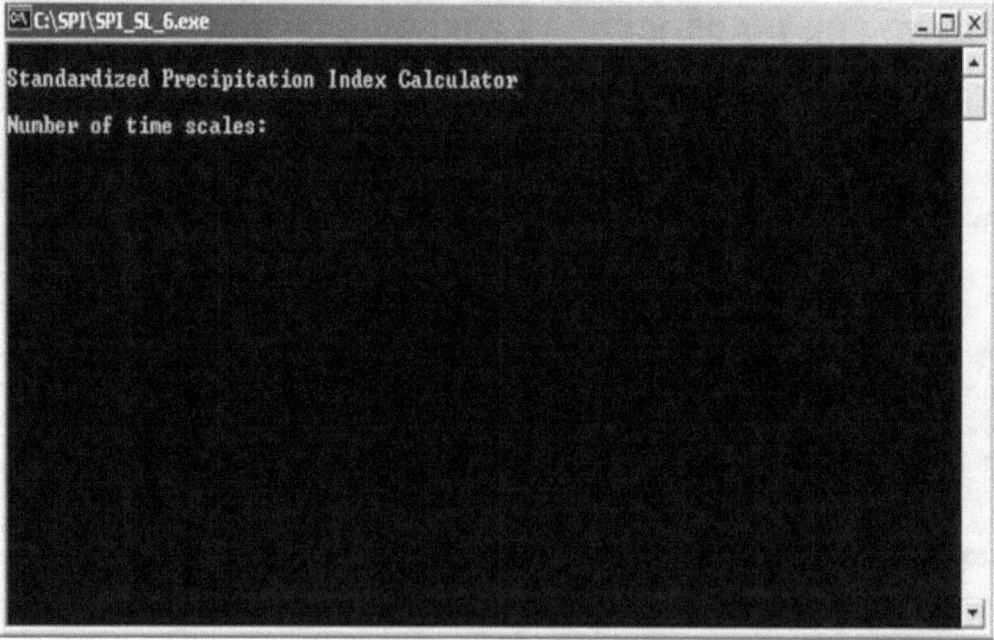

4. Specify the SPI timescales to be computed. In the example below, the user will generate five SPI timescales or windows: 1-month, 3-month, 6-month, 9-month and 12-month SPIs:

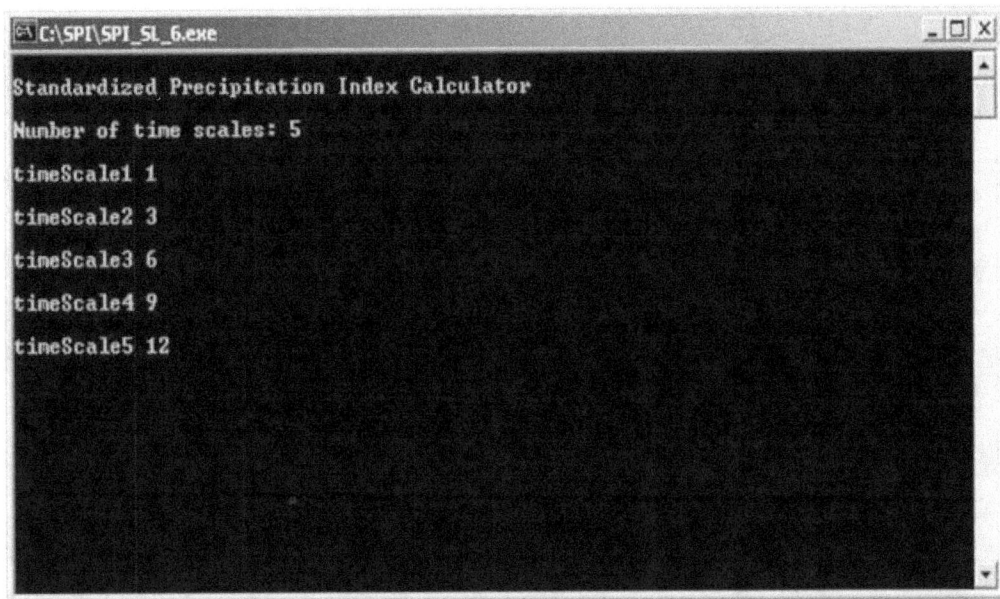

5. Enter an input and an output filename. It is recommended to adopt a naming system that reflects the SPI analyses to be carried out in order to keep the results of each analysis separate:

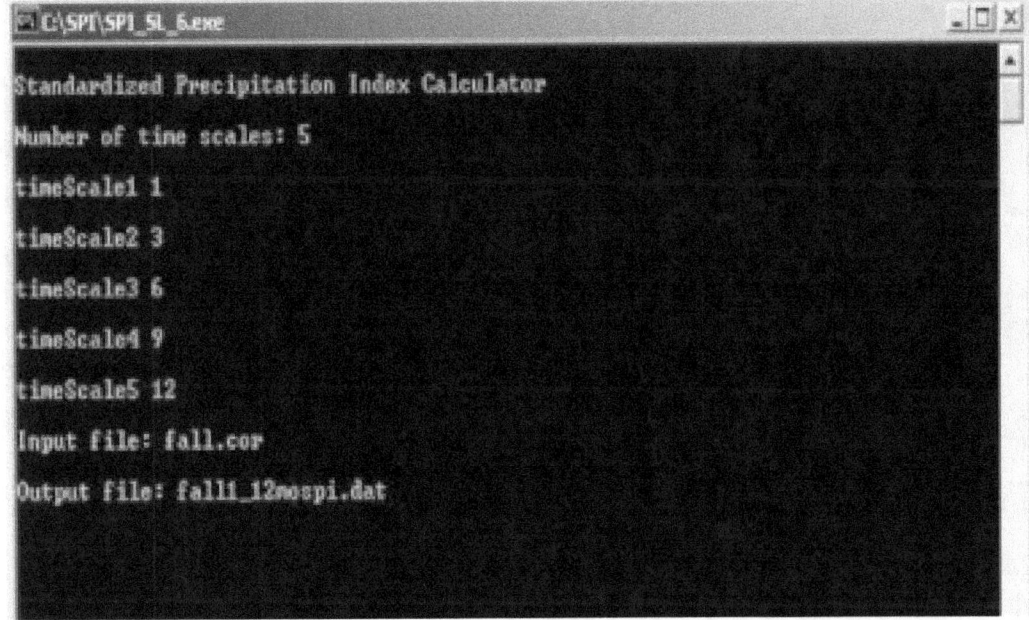

The output file can be given any name but must have a .dat extension. It will be placed in the same folder where the executable file is located.

The results can be processed with Microsoft Notepad or any other text or word processing software. These files are saved as MS_DOS ASCII text files.

The output data can then be plotted, graphed or mapped by any means.

A sample output file for Fall City, Nebraska, is shown and described below. The input file was set up to analyse the 1-, 3-, 6-, 9- and 12-month SPIs. The corresponding values appear in the third, fourth, fifth, sixth and seventh columns:

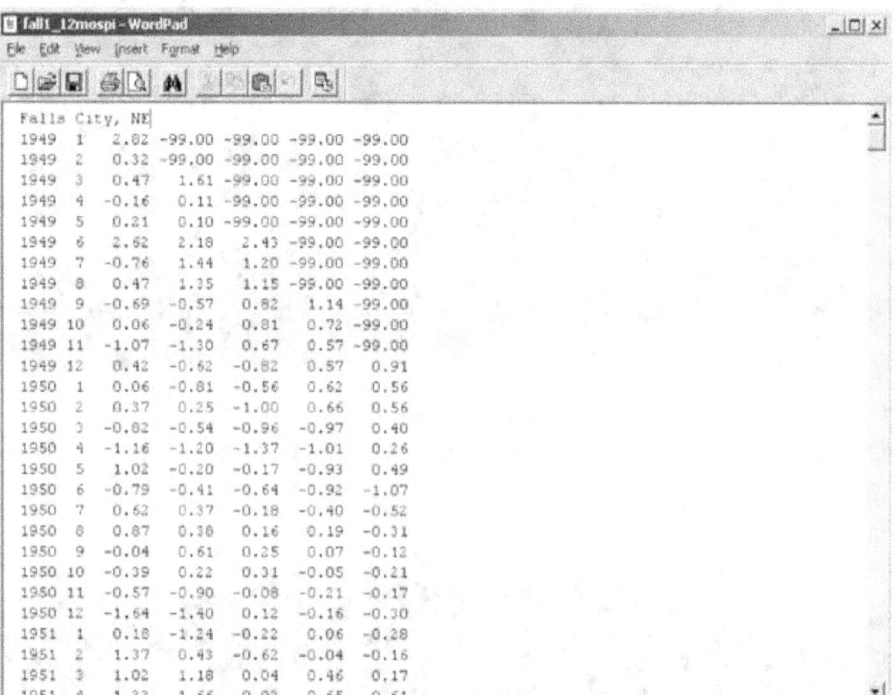

Note: The −99.00 value does not indicate missing data in this case. It simply reflects that, for example, in the fourth column, one cannot have a 3-month SPI value until one is 3 months into the period of record. The same applies to the last column where one does not see a 12-month SPI until December 1949, or the twelfth month available for calculation. This becomes the first 12-month SPI generated.

MAPPING CAPABILITIES

Many countries are calculating and mapping the SPI and other drought indices or meteorological parameters on a regular basis. Below is an overview of the methods often employed in mapping drought indicators.

There are multiple ways, such as the standard drought indicators and indices, to map meteorological variables. Most drought-related data originate as point ("station-based" or "site- specific") data. These data serve their purposes, but it is often in map form that the data best communicate a message based on a geographic context to the decision-maker trying to understand drought severity and spatial extent. The point data can be placed on a map, and derivative products or characteristics of that site can be provided for additional information. This could include, for example, a time series plot of the indicator or index. The limitation of this level of spatial detail is that information on what is taking place between points is not available.

A variety of techniques can be used to generate a continuous map of meteorological drought. One such technique generates an interpolated surface of estimated values at locations between sites based on mathematical relationships of the indicator or index between the original point data. Often this produces a map that appears "natural", but is still based on the data from specific points and is only as accurate as the original data and the interpolation technique. No single interpolation method can be applied to all situations, and the most commonly used interpolation techniques include Kriging, Spline, and Inverse Distance Weighting (IDW).

Each interpolation technique has its advantages and disadvantages. Some techniques are more exact than others, but take longer to produce the desired output. The Kriging method, which has its origins in geological applications and the mining industry, assumes that there is a relationship between points that is non-random and changes over space. The Spline method is used when minimizing the overall surface curvature is important. Inverse Distance Weighting (IDW) is used when the data points are scattered but dense enough to represent local variations. The data, as the name implies, are weighted to favour data closer in proximity to the point being processed.

Another technique that has been used for meteorological drought monitoring and mapping is to map point data to grid cells. These data can also originate from airborne, radar, and satellite sources. These gridded data products appear less "natural" than interpolated products, but they are easier to use for comparative purposes because of the common grid cell sizes. These cells can vary in size from degrees down to meter(s) depending on the source and application needed. They also vary in their temporal frequency, with return periods ranging from daily (or multiple times a day) to weekly or longer. In the United States, gridded products for monitoring meteorological drought are becoming much more common, while in other regions, particularly Africa, there has been a long history of using gridded information to determine drought conditions. The Famine Early Warning System (FEWS) and similar networks have utilized gridded data in their analyses. Multiple examples of gridded meteorological drought products exist in Australia, China, the United Kingdom and the United States.

To develop a gridded map product, the point data are aggregated to a grid cell resolution selected for that product using a mathematical relationship. An interpolated surface is then created between the grid cells (not the point data). For example, the High Plains Regional Climate Center, in partnership with the National Drought Mitigation Center, is mapping the Standardized Precipitation Index on a daily basis at state, regional, and national scales across the United States.

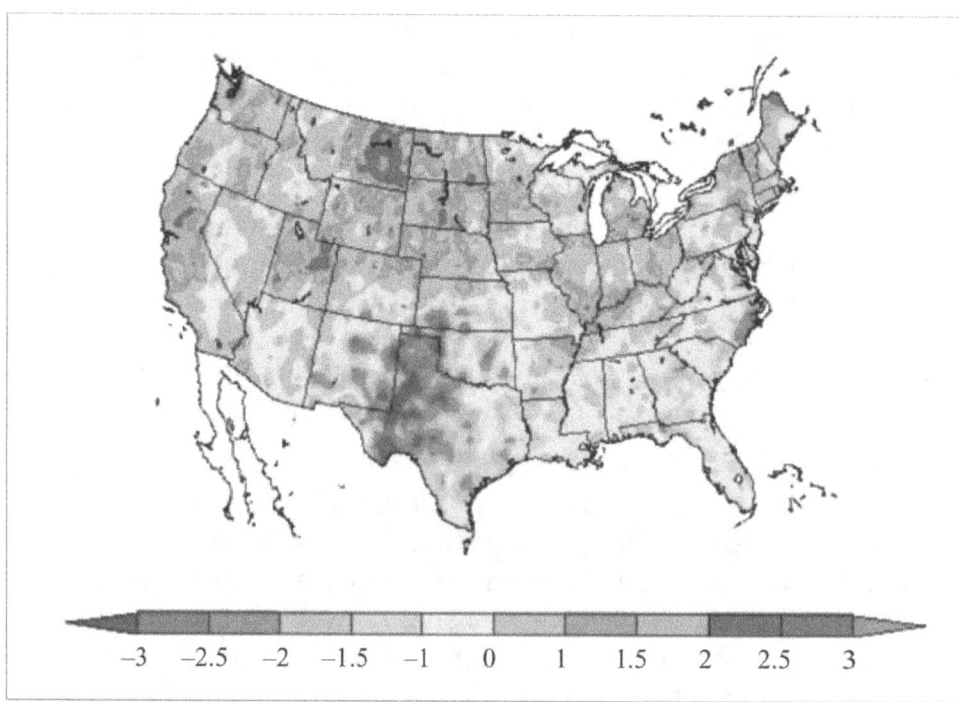

Source: United States High Plains Regional Climate Center

Example of 3-month SPI (1 May 2011 – 31 July 2011)

SPI maps are generated by using the Grid Analysis and Display System (GrADS). The discrete station SPI data are interpolated using a Cressman objective analysis with radii of influence of 10,7,4,2,1. The grid resolution is 0.4 degrees. Gridded contour maps are generated at national, regional and state levels for the High Plains region. For national maps, north polar stereographic (nps) projection is applied. Regional and state maps use a latitude/longitude (lat/lon) projection with aspect ratio maintained. This interface and resultant products can be found at: http://www. hprcc.unl.edu/maps/current/.

Successful mapping of meteorological drought depends on the quality of the data. Drought indicator and index data quality is determined by several factors, including the availability of the data, the timing of the recording, the quality of the historical data at a station, the transmission of data in near real time, the maintenance of the station network and the ability to measure precipitation in cold temperatures, particularly in northern or alpine locations. Some of these issues are related to the ability to provide the data in a timely fashion, which can be very important with meteorological drought. Finally, the data density plays a huge role in the spatial resolution that can be achieved in mapping drought.

One of the biggest challenges in mapping meteorological drought is to try and match the spatial resolution that decision-makers need and demand with the information that is available today. This limitation is related to the density of point data, which might not be available at the resolution desired by the decision-makers.

Because of this challenge, the promise of potential remote sensing-based products is encouraging. Some remote-sensing products can provide data at spatial resolutions in regions where in situ point data are relatively sparse and unreliable. Most satellite products are already incorporated into a grid cell (or "pixel") as described above. In the United States, some products are now being developed to utilize a combination of station-based and remote-sensing data. The station data are used to help refine the remote-sensing data, and the resulting "hybrid" maps have a higher level of accuracy.

Topographical issues, particularly those related to mountains and rapid terrain changes, represent a real challenge when mapping meteorological drought. There are two reasons for this. First, data density tends to be lower in mountainous regions. Second, because interpolation methodologies are often based on correlations, the relationship between adjoining regions with regard to precipitation tends to be particularly discontinuous in areas where the terrain changes rapidly and significantly. As a result, smoothed interpolated surfaces on a finished map product may not realistically match natural variability, especially the indicators and indices related to precipitation.

Because of all the complexities involved in meteorological drought data and the characteristics of mapping techniques, it is important that a decision-maker understands these factors when interpreting maps of drought severity and spatial extent.

REFERENCES

Edwards, D. C. and T. B. McKee, 1997: Characteristics of 20th century drought in the United States at multiple time scales. Climatology Report 97-2, Department of Atmospheric Science, Colorado State University, Fort Collins, Colorado.

Guttman, N.B., 1994: On the sensitivity of sample L moments to sample size. Journal of Climate, 7(6):1026 –1029.

Guttman, N.B., 1998: Comparing the Palmer drought index and the Standardized Precipitation Index.

Journal of the American Water Resources Association, 34(1):113–121.

——— , 1999: Accepting the Standardized Precipitation Index: a calculation algorithm. Journal of the American Water Resources Association, 35(2):311–322.

McKee, T.B., N.J. Doesken and J. Kleist, 1993: The relationship of drought frequency and duration to time scale. In: Proceedings of the Eighth Conference on Applied Climatology, Anaheim, California,17–22 January 1993. Boston, American Meteorological Society, 179–184.

——— , 1995: Drought monitoring with multiple timescales. In: Proceedings of the Ninth Conference on Applied Climatology, Dallas, Texas, 15–20 January 1995. Boston American Meteorological Society, 233–236.

Vicente-Serrano, S.M., S. Beguería and J.I. López-Moreno, 2010: A multi-scalar drought index sensitive to global warming: the Standardized Precipitation Evapotranspiration Index – SPEI. Journal of Climate, 23(7):1696 –1718, doi:

10.1175/2009JCLI2909.1.

Wilhite, D.A. and M.H. Glantz, 1985: Understanding the drought phenomenon: the role of definitions. Water International, 10:111–120.

Wu, H., M.D. Svoboda, M.J. Hayes, D.A. Wilhite and F. Wen, 2007: Appropriate application of the Standardized Precipitation Index in arid locations and dry seasons. International Journal of Climatology, 27(1):65–79.

OTHER ONLINE RESOURCES

http://drought.unl.edu/MonitoringTools/DownloadableSPIProgram.aspx

http://drought.mssl.ulc.ac.uk/spi.html

http://www,wrcc.dri.edu/spi/spi.html

http://ccc.atmos.colostate.edu/standardizedprecipitation.php

http://www.wmo.int/drought

Case Study of Anantapur District / precipitation received from 1900 to 2002 (period 102 years) located in Andhra Pradesh State India using standardized precipitation index to characterize Meteorological drought events.

The normal rainfall of Anantapur District is 538 mm and 75% confidence limit is 430 mm and 50% confidence limit is 510 mm. The district is characterized by alternative year drought occurrence. The District receives its rainfall from South West monsoon to a quantity of 57% (June to September) and from the north east monsoon – 28% (Oct – Dec) and hot weather period (March to May) – 14.5% and cold weather period (January to February) 0.5%.

The rainfall deficit with respect to potential evapotranspiration was as follows for the period (1946-1975) – South West Monsoon.

Month	Rainfall mm	P.E. mm	Percentage
June	45.59	244.00	81.32
July	63.60	193.9	67.00
August	69.80	177.6	61.00
September	129.94	166.8	22.00

The District lies in the agro-climate zone namely "scarce rainfall zone of Rayalaseema" (i.e.) average rainfall being in the zone 500-750 mm. In light of the above as an exercise spi has been attempted to find for the district having 101 rainfall stations, considering the average monthly rainfall over the district for a period of 102 years i.e. from 1900 to 2002. Standardized precipitation index values have been worked out for 102 years and enclosed in this book, for time scales of 1 month 3 months 6 months 9 months and 12 months.

Based on the spi values arrived, considering the values for the years 2001 & 2002, individual graphs have been drawn separately for each time scale, wherein it can be seen the time of onset of drought, end of drought, duration and magnitude.

For those who are interested to find the above details for the rest of years, based on the values arrived for spi, graphs can be drawn and particulars obtained.

Based on an analysis of stations across Anantapur District for 102 years, it is determined that the spi indicates the rarity of a current drought as per tables below

Table 5.8:   Probability of Occurrence

Spi Value Intensity	Spi 1st month	Spi 3rd month	Spi 6th month	Spi 9th month	Spi 12th month	Severity of Event
Moderate dryness (–) 1.0 to (–) 1.49	1 in 1 year	1 in 1 year	1 in 1 year	1 in 1 year	1 in 1 year	1 in 1 year
Severe dryness (–) 1.5 to (–) 1.99	1 in 3 years	1 in 2 years	1 in 1 year	1 in 2 years	1 in 1 year	1 in 2 years
Extreme dryness < (–) 2.00 less than	1 in 4 years	1 in 4 years	1 in 4 years	1 in 3 years	1 in 4 years	1 in 4 years

From the table it can be seen that moderate drought is occurring at alternative year and sever drought in one year for every 2 years, similarly extreme drought one year for every 4 years. This type of analysis of drought intensity is helpful to evolve a better drought management system in the district.

The years 2002 and 2004, saw some of India's most significant droughts and nothing in IMDs weather models indicated any sign that such rainfall deficits were imminent. In 2009 even after junking the mathematical techniques it had used for decades and adopting new climate variables, the IMD failed to anticipate a drought India's biggest and most wide spread in 40 years.

It is glaringly incongruous because the IMD does not like rocking the boat. Between 2002 and 2013, when India saw four drought years, the IMD never gave any indication in its April forecast that the country should be preparing for a drought.

Source: The economic times (Agriculture) by Jacob Koshy science writer to new paper (The forecast game between skymet (Private agency) and IMD on monsoon of India April 26, 2015 3.00 am 1st)

IMD – South West monsoon 2002 end of season report Chapter – 2 (Rainfall and Drought)

It may be mentioned here that long range forecasts of the Indian south – west monsoon rainfall are generated by many other national and global centres which employ different prediction techniques. None of these forecasts indicated a large deficiency of the monsoon rainfall that was actually observed. This calls for a through scientific investigation. IMD while issuing the long range forecast for 2002 monsoon, had indicated the existence of ELNINO and its adverse influence on the monsoon performance only.

It is interesting to note that the spi values for the year 2002 for 1st month scale are as follows (1900-2002 year 102)

1st month – scale Time of onset of drought	-	2nd week of June 2002 ending at last week of Sept. 2002 with a magnitude of (–) 3.02 indicating extremely dryness.
3rd month – scale Onset of drought	-	Last week of July 2002 and extended to 2003 year also with magnitude (–) 4.36 indicating extremely dryness.
6th month – scale Onset of drought	-	August 2nd week extending to 2003 year also with a magnitude of (–) 3.81 indicating extremely dryness.
9th month – scale Onset of drought	-	1st week of July extending to 2003 year also with a magnitude of (–) 3.36 indicating extremely dryness.
12th month – scale Onset of drought	-	1st week of July extending to 2003 year also with a magnitude of (–) 3.72 indicating extremely dryness.

Note: The magnitude mentioned at 3, 6, 9 and 12 months are to the extent of 2002 year only, the magnitude will increase as the drought has extended to 2003 year.

Drought climatology based on spi is not biased by the aridity of the region and hence is better index for drought monitoring (ibid, p.10).

As the all India rainfall is significantly, normally distributed, the years of all India drought (moderate and above) identified both percentage departure from normal and spi were nearly same. However there were significant differences in the district – wise climatology based on these two drought indices. The district - wise drought climatology over India based on PN was found to be highly biased by the aridity of region. Highest probability for droughts of moderate intensity was observed over many districts from North West India and neighbouring Central India and interior parts of South Peninsula. The lowest probability for droughts of moderate intensity was mainly observed over several districts along West Coast of the North Peninsula and Eastern and Northern parts of the country.

The highest probability of severe droughts was also observed over the North Western part of the country. On the other hand district wise climatology of the drought based on the spi was not biased by the aridity of the region.

Therefore spi is a better index than PN for monitoring the district wise drought indices over India. Further as the spi is normalized index, it can be used to represent the excess rainfall or wet conditions in the same way as it is used to represent the drought / dry conditions and wet periods can be monitored using the spi. As seen in the previous paragraph where both PN and spi are suitable for the seasonal drought monitoring on all India scale. Spi is more suitable for district – wise drought monitoring. This is mainly because of the higher C.V. of the district – wise rainfall than all India rainfall. The C.V. also increased with decrease in the spatial and temporal scale for region of reference because of more frequent extreme events at these scales.

Therefore, the spi will also be a better index than PN in monitoring wet and dry incidences at intra – seasonal scales such as break and active events over India.

Spi is also more suitable as it allows drought severity at two or more locations to be compared with each other regardless of climate difference between them. As the variability of spi is nearly same as that of the precipitation anomaly, prediction models can be developed for spi and hence it is suitable for drought prediction. Dynamical model forecasts of rainfall can be altered to produce spatial maps of spi for drought prediction.

Source: ibid, p12.

Concerns have been raised about the applicability of spi as a measure of changes in drought associated with climate change, as it does not deal with changes in evapotranspiration.

Analysis of drought over India based on spi   A few studies (Hughes and Saunders 2002 Hayes et.al 1999, Mihajlovic 2006) have been done on spi based drought monitoring on a monthly / seasonal time scale. Keeping this in mind, an attempt has been made to analyse drought over India on spi (Table).

The main objective was to see how effective spi was in diagnosing drought intensity over a longer period of time rainfall data (All India seasonal rainfall (June-Sept) used for the study was from 1875 to 2009.

Computation of spi involved fitting a gamma probability density function to a given frequency distribution of precipitation totals. The alpha and beta, shape and scale parameters of the gamma distribution were estimated for a suitable time scale for each year. Alpha and Beta parameters were then used to find the cumulative probability of an observed precipitation amount, which was then transformed into the standardised normal distribution.

Thus spi could be said to be normalized in space and time scale. spi as a drought index is versatile, as it can be calculated on any time scale so it is suitable for agriculture and hydrological applications. This versatility is also critical for monitoring the temporal dynamics of a drought, including its development and decline. These aspects of a drought have always been difficult to track with other indices, further, as spi values are normally distributed, the frequencies of extreme and severe drought events for any location and time scale are consistent.

Table 5.9 Drought intensity over India based on standardized Precipitation Index (spi)

Drought Year	Seasonal Rainfall (June – September) cm	Spi Value	Drought Intensity as per Spi Value
1877	58.7	(–) 3.47	ED
1899	62.1	(–) 3.01	ED
1901	76.5	(–) 1.21	MD
1904	77.6	(–) 1.08	MD
1905	72.7	(–) 1.66	SD
1911	75.1	(–) 1.38	MD
1918	66.1	(–) 2.48	ED
1920	73.3	(–) 1.59	SD
1941	76.3	(–) 1.23	MD
1951	71.5	(–) 1.81	SD
1965	72.0	(–) 1.75	SD

1966	76.4	(–) 1.22	MD
1972	67.00	(–) 2.37	ED
1974	77.4	(–) 1.11	MD
1979	71.4	(–) 1.82	SD
1982	75.2	(–) 1.36	MD
1986	76.8	(–) 1.18	MD
1987	70.9	(–) 1.88	SD
2002	71.3	(–) 1.83	SD
2004	76.6	(–) 1.20	MD
2009	69.8	(–) 2.02	ED

ED: Extreme drought – spi more than (–) 2.00

SD: Severe drought – spi (–) 1.50 to (–) 1.99

MD: Moderate drought – spi 1.0 to (–) 1.49

The analysis revealed that out of 135 years (1875 – 2009) spi diagnosed five years (1877, 1899, 1918, 1972 and 2009) as all India extreme drought years when the spi value exceeded (–) 2.00. This result was in agreement with the analysis of Mooley (1994) who while analyzing data from (1871 to 1966) found that in the years 1877, 1899, 1918, 1972 and 1987. Phenomenal all India droughts affected the country. Mooley (1994) defined phenomenal all India drought as a phenomenon occurring when percent departure of monsoon rainfall was ≤ (–) 2.0 SD (i.e (–) 20%) and the percentage area under deficient monsoon rainfall equal to or more than mean +2 SD (i.e. 47.7%).

Therefore phenomenal drought years identified by Mooley (1994) have been effectively diagnosed as extreme drought years that affected the country. Further, when spi was used to examine whether any trend existed in drought over the country. No trend was found.

However it should be mentioned that despite the current optimism about spi, it cannot solve all moisture monitoring concerns. Rather it can be considered as a tool that can be used in coordination with other tools such as the aridity anomaly index or remote sensing data, to detect the development of drought and monitor their intensity and duration. This will further improve the timely identification of emerging drought conditions that can trigger appropriate responses by the policy makers.

Table 5.10:   Probabilities of Occurrence of Drought in India (Temporal variation of spi in India period 1875 – 2004) 130 years

S. No.	Name of the Sub-Division	Moderate	Severe	Total	Drought Probabilities Total Percentage
1.	Andaman and Nicobar islands	17	0	17	13%
2.	Arunachal Pradesh	7	1	8	6%
3.	Assam and Meghalaya	5	0	5	4%
4.	Nagaland Manipur Mizoram & Tripura	12	0	12	9%
5.	Sub-Himalayan West Bengal	7	0	7	5%
6.	Gangetic West Bengal	2	0	2	1%
7.	Orissa	5	0	5	4%
8.	Bihar	12	0	12	9%
9.	Jharkhand	6	0	6	4%

S. No.	Name of the Sub-Division	Moderate	Severe	Total	Drought Probabilities Total Percentage
10.	East Uttar Pradesh	13	1	14	10%
11.	West Uttar Pradesh	13	1	14	10%
12.	Uttarakhand	16	2	18	13%
13.	Haryana Delhi and Chandigarh	21	4	25	19%
14.	Punjab	20	4	24	18%
15.	Himachala Pradesh	20	3	23	17%
16.	Jammu & Kashmir	21	6	27	20%
17.	West Rajastan	22	12	34	25%
18.	East Rajastan	18	5	23	17%
19.	West Madhya Pradesh	14	0	14	10%

S. No.	Name of the Sub-Division	Moderate	Severe	Total	Drought Probabilities Total Percentage
20.	East Madhya Pradesh including Chhattisgarh	12	0	12	9%
21.	Gujarat Region	17	11	28	21%
22.	Saurastra and Kutch	16	15	31	23%
23.	Kokan & Goa	9	0	9	7%
24.	Madhya Maharastra	7	2	9	7%
25.	Marathwada	17	1	18	13%
26.	Vidarbha	16	1	17	13%
27.	Coastal Andhra Pradesh	13	0	13	10%
28.	Telangana	18	0	18	13%
29.	Rayalaseema	20	2	22	16%

S. No.	Name of the Sub-Division	Moderate	Severe	Total	Drought Probabilities Total Percentage
30.	Tamilnadu & Pandicherry	12	0	12	9%
31.	Coastal Karnataka	5	0	5	4%
32.	North interior Karnataka	10	0	10	7%
33.	South interior Karnataka	9	0	9	7%
34.	Kerala	10	0	10	7%
35.	Lakshdweep	10	3	13	10%

Based on probabilities of occurrence of drought (percentage) entire country has been divided into chronically drought prone areas (probability of occurrence of drought more than 20%) frequently drought prone areas (probability of occurrence of drought 10-20%) and least drought areas (probability of occurrence of drought less than 10%) (IMD 2005).

As per data shown in the table above reveal that the arid west namely West Rajasthan (34 cases) and Saurastra and Kuch (31 cases) have the highest occurrences of drought. The adjoining Gujarat region which mostly belongs to a semiarid climate also experiences high incidences of drought (28 cases) other areas recording large incidences of drought are Haryana Delhi and Chandigarh, Punjab, Himachal Pradesh and East Rajasthan in North West of India and Rayalaseema in the South Peninsula. The sub humid and humid areas of east and northeast India (Arunachala Pradesh, Assam and Meghalaya, Orissa, Gangetic West Bengal and Jharkhand) for obvious reasons have the lowest occurrences of drought.

REFERENCES

WMO 1994 on the Frontline Public Weather Services, WMO, No.816 Geneva.

Mooley DA 1994 origin incidence and impact of droughts over India and remedial measures for their mitigation. Sadhna (Academy proceedings in Engineering Services) India Academy of Sciences, 1994.

Naresh Kumar M. C.S. Murthy MVR Sesha Sai and P.S. Roy on the standardized precipitation index (spi) for drought intensity assessment. Meteorological application 16(3): 381-389 published online in Wiley inter science. htt:// onlinelibrary.wiley.com/doi/10.1002/met.136/pdf.

Hayes M.T. M.D. Srohoda DA Wihite and O.V. Vanjakho 1999 monitoring the 1996 drought using the standardized precipitation index. Bulletin of Meteorological Society, 8: 429-438.

NRSC 2009 – Drought report – National remote sensing centre ISRO, Hyderabad.

Thornth Waite C.W. 1948 – An approach toward a rational classification of climate Geographic Review, 38: 55-94.

6

STANADARDIZED PRECIPITATION AND EVAPOTRANSPIRATION INDEX (SPEI)

CHAPTER OUTLINE

- Dealt about "SPEI" a multi-scaler drought index sensitive to global warming – performance of drought indices around the world for ecological, agricultural and hydrological applications compared with SPEI – "SPEI" worked to be best. Application of "SPEI" for data at Anantapur Dist. A.P. enclosed with results.

Standardized precipitation index (SPI) developed by T.B. Makee N.J. Doesken and J. Kleist Colorado State University 1993. The SPI is a powerful, flexible index that is simple to Calculate. Infact, precipitation is the only required input parameter. In addition, it is just as effective in analyzing Wet periods / cycles as it is in analyzing dry periods / cycles. Ideally one needs at least 20–30 years of monthly values, with 50–60 years (or more) being optimal and preferred. The SPI was designed to quantify the precipitation deficit for multiple time scales. One may want to look at a 1 or 2 month SPI for meteorological drought anywhere from 1 month to 6 month SPI for agricultural drought and something like 6 month up to 24 month or more for hydrological drought analysis and applications. It is developed to improve drought detection and monitoring capabilities.

Concerns have been raised about the application of SPI as a measure of changes in drought associated with climate change, as it does not deal with changes in evapotranspiration.

Because SPI is based only on precipitation and not on soil moisture conditions, the SPI is just as effective during winter months.

SPI by itself cannot identify regions that may be more "drought prove" than others. Because of characteristics associated with the normal distribution severe and extreme droughts measured by SPI occur with same frequency for all locations across the country over a long time period.

Before the SPI is applied in a specific situation, a knowledge of the climatology for that region is necessary.

At the shorter time scales (1, 2, or 3 months) the SPI is very similar to the percent of normal representation of precipitation, which can be misleading in regions with normally low seasonal precipitation totals (i.e. Seasonal rainfall regime). Understanding the climatology of these regions improves the interpretation of the SPI values.

Even with the current optimism about the SPI, it must be remembered that the SPI cannot solve all the moisture monitoring concerns. Rather, it is one tool that can be used in coordination with other tools, such as moisture anomaly index, PDSI or remote sensing data to detect the development of droughts and monitor their intensity and duration.

Therefore a new variation of the index by Vicente – Serrano and others (2010) attempts to address the PET issue by including a temperature component in the calculation of their new index called **standardized precipitation and Evapo–transpiration index (SPEI).** The inputs required to run the program are precipitation, mean temperature and latitude of the site. Further information on the SPEI is available at http://sac.csic.es/spei/index.html.

The main criticism of the SPI is that its calculation is based only on precipitation data. The index does not consider other variables that can influence droughts, such as temperature, evapo–transpiration wind speed and soil water holding capacity. Empirical studies have shown that temperature rise markedly affects the severity of droughts – For example Abramopouloz et al. (1988) used a general circulation model experiment to show that evaporation and transpiration can consume upto 80% of rainfall. There has been a general temperature increase (0.5°–2°C) during the last 150 years (Jones and Moberg, 2003) and climate change, models predict a marked increase during the 21st century (IPCC 2007).

It is expected that this will have dramatic consequences for drought conditions with an increase in water demand due to evapotranspiration (Sheffield and wood 2008).

Dubrov Sky et al. (2008) recently showed that the drought effects of warming predicted by global climate models can be seen in the PDSI where as SPI (which is based only on precipitation data) does not reflect expected changes in drought conditions.

Therefore the use of drought indices which include temperature data in their formulation (such as the PDSI) is preferable, especially for applications involving future climate scenarios. However the PDSI (Palmer drought severity index) lacks the multi–scalar character essential for both assessing drought in relation to different hydrological systems and differentiating among different drought types. Therefore this necessity made to formulate a new drought index – standardized precipitation and evapotranspiration index (SPEI) based on precipitation and potential evapotranspirfation (PET). The SPEI combines the sensitivity of PDSI to changes in evaporation demand (caused by temperature fluctuations and trends) with the simplicity of calculation and the multi temporal nature of SPI.

The new index is particularly suited to detecting monitoring and exploring the consequences of global warming on drought conditions.

A reduction in precipitation due to climate change will affect the severity of droughts. The SPEI is very easy to calculate and it is based on the SPI calculation procedure. The SPI is calculated using monthly [or weekly] precipitation as the input data. The SPEI uses the monthly (or weekly) difference between precipitation and PET. This represents a simple climatic water balance (Thornth Waite, 1948) which is calculated at different time scales to obtain the SPEI.

The log–logistic distribution was used to calculate the SPEI, as the Pearson III gamma distributions were used to calculate the SPI. Only when the index was computed at short time scales for some few very low precipitation PET values (mainly arid locations with a highly variable climatology) were many problems experienced. These were minor and already known for SPI calculations when the two parameter gamma distribution is used (wee *et. al* 2007). However the use of three parameter distributions to calculate the SPEI reduced this problem noticeably.

SPI cannot identify the role of temperature increase in future drought conditions and independently of global warming scenarious cannot account for the influence of temperature variability and the role of heat waves, such as that which affected central Europe in 2003.

The SPEI can account for the possible effects of temperature variability and temperature extremes beyond the context of global warming.

Therefore, given the minor additional data requirements of the SPI, use of the former is preferable for the identification analysis and monitoring of droughts in any climate region of the world.

In summary, the SPEI fulfils the requirements of a drought index, as indicated by Nkemdivim and Weber (1999) since its multi–scaler character enables it to be used by different scientific disciplines to detect, monitor and analyze droughts. SPEI can measure drought severity according to its intensity and duration and can identify the onset and end of drought episodes. The SPEI allows comparison of drought severity through time and space, since it can be calculated over a wide range of climates as can the SPI.

Moreover, Keyantash and Dracup (2002) indicated that drought indices must have statistically robust and easily calculated, and have a clear and comprehensible calculation procedure. All these requirements are met by the SPEI.

However, a critical advantage of the SPEI over the most widely used drought indices that consider the effect of PET on drought severity is that its multi scalar characteristics enable identification of different drought types and impacts in the context of global warming.

In view of the above the author has attempted to work out SPEI values for Anantapur district of Andhra Pradesh using precipitation & mean temperature from year 1900–2002 and out put values have been enclosed.

A graph for SPEI – 1 month scale has been plotted from the values obtained and the probability of recurrence of drought has been found. Similarly for 3 months scale and 6–months, 9–month scale and 12th month scales graph for 102 years versus SPEI values are separately plotted and the following probability of recurrences of drought is exhibited in the Table below for Anantapur District of A.P. India.

Description SPEI Value and Intensity	SPEI one Month Scale	SPEI 3 Months Scale	SPEI 6 Months Scale	SPEI 9 Months Scale	SPEI 12 Months Scale	Severity of Event
Moderate dry (–) 1 to (–) 1.49	1 in 1 year	1 in 1 year	1 in 1 year	1 in 1 year	1 in 1 year	1 in 1 year
Severely dry (–)1.5 to (–) 1	1 in 2 years	1 in 1 year	1 in 1 year	1 in 1 year	1 in 1 year	1 in 1 year
Extremely dry Less than (–) 2	1 in 13 years	nil	nil	nil	nil	nil

As an example individual graphs are prepared for years 2000 and 2001 for 1 month 3 months, 6 months, 9 months and 12 months scales and based on the SPEI values the time of onset end and duration along with magnitude is found out. The following are the description of the graphs of 2000 & 2001 years.

SPEI – 1st month year 2000	SPEI – 1st month year 2001
Time of on set – end of September	Time of onset 2nd week of October
End of drought – end of March 2001	End of February 2002 year
Magnitude = (–) 4.124	Magnitude (–) 4.376

3 months scale 2000	3 months scale 2001
On set of drought – 1st Week of September	Onset of drought 1st week of September
End of drought – end of February 2001	End of drought January
Magnitude = (–) 3.856	Magnitude (–) 3.870

6 months scale 2000	6 months year 2001
On set of drought – 1st week of August	Onset of drought 1st week of July
End of drought – 1st week of January 2001	End of drought 1st week of January 2002
Magnitude – (–)2.11	Magnitude (–) 2.307

9 months scale 2000	9 months year 2001
On set of drought – 4th week of June	Onset of drought 3rd week of May
End of drought – 4th week of November	End of drought 1st week of November
Magnitude – (–)0.510	Magnitude (–) 0.741

12 months scale 2000	12 months year 2001
On set of drought – end of June	Onset of drought 5th week of May
End of drought – 4th week of November	End of drought 1st week of November
Magnitude – (–)0.464	Magnitude – (–) 0.695

With the aid of SPEI values for 1900–2002 years of Anantapur Dist, A.P. shows that there are droughts moderate and severe occurring 1 in 1 yr and there are no extreme droughts in the district.

SPEI values of year 2000 & 2001 indicate a moderate agricultural drought as the drought being set just before the monsoon period or at the middle of crop season.

The summary at 5th South Asian climate outlook forum during April, 2014 at Pune India, was conducted and compiled by Global Water partnership South Asia with the support from WMO/GWP integrated drought management programme (IDMP), India has reported during March 2014 that "Meteorological and hydrological drought indices (such as SPI & SWI) and remote sensing drought indices (such as NDVI, NDWI, VCI, TCI etc.) are used by IARI WTC to predict drought and it is being successfully tested in North and North Western India".

"The IARI (Indian Agriculture Research Institute) WTC (Water Technology Centre) uses the National Agricultural Drought Assessment System (NA DAMS) provided by the National Remote Sensing Centre (NRSC) and is in process of developing drought early warning system in collaboration with the national drought mitigation centre of the University of Nebraska – Lincoln, USA".

"The Department of Agriculture and Cooperation of the Ministry of Agriculture (MOA) used several criteria for monitoring drought among which the most important is "moisture adequacy index" (MAI) which has been provided by the institute of ICAR".

Note: In this connection it to suggest that IMD or Govt. of India or State Governments to attempt the usage of SPEI for Meteorological and Hydrological droughts where in the component of PET is considered instead of SPI using a single component of precipitation.

A MULTI–SCALAR DROUGHT INDEX SENSITIVE TO GLOBAL WARMING : THE STANDARDIZED PRECIPITATION EVAPOTRANSPIRATION INDEX – SPEI

Sergio M. Vicente–Serrano[1], Santiago Beguería[2], Juan I. López–Moreno[1]

[1]Instituto Pirenaico de Ecología—CSIC, Campus de Aula Dei, P.O. Box 13034, Zaragoza 50080, Spain.

[2]Estación Experimental de Aula Dei— CSIC, Campus de Aula Dei, P.O. Box 13034, Zaragoza 50080, Spain. e–mail: svicen@ipe.csic.es

ABSTRACT

We propose a new climatic drought index: the Standardized Precipitation Evapotranspiration Index (SPEI). The SPEI is based on precipitation and temperature data, and has the advantage of combining a multi–scalar character with the capacity to include the effects of temperature variability on drought assessment. The procedure to calculate the index is detailed, and involves a climatic water balance, the accumulation of deficit/surplus at different time scales, and adjustment to a Log–logistic probability distribution. Mathematically, the SPEI is similar to the Standardized Precipitation Index (SPI), but includes the role of temperature. As the SPEI is based on a water balance, it can be compared to the self–calibrated Palmer Drought Severity Index (sc–PDSI). We compared time series of the three indices for a set of observatories with different climate characteristics, located in different parts of the world. Under global warming conditions only the sc–PDSI and SPEI identified an increase in drought severity associated with higher water demand due to evapotranspiration. Relative to the sc–PDSI, the SPEI has the advantage of being multi–scalar, which is crucial for drought analysis and monitoring.

Key–words: drought, drought index, Log–logistic distribution, global warming, Standardized Precipitation Index, Palmer Drought Severity Index, precipitation, evapotranspiration.

INTRODUCTION

Drought is one of the main natural causes of agricultural, economic and environmental damage (Burton *et al.* 1978; Wilhite and Glantz, 1985; Wilhite, 1993). Droughts are apparent after a long period without precipitation, but it is difficult to determine their onset, extent and end. Thus, it is very difficult to objectively quantify their characteristics in terms of intensity, magnitude, duration and spatial extent. For this reason, much effort has been devoted to developing techniques for drought analysis and monitoring. Among these, objective indices are the most widely used, but subjectivity in the definition of drought has made it very difficult to establish a unique and universal drought index (Heim, 2002). A number of indices were developed during the 20th century for drought quantification, monitoring and analysis (Du Pisani *et al.* 1998; Heim, 2002; Keyantash and Dracup, 2002).

In recent years there have been many attempts to develop new drought indices, or to improve existing ones (González and Valdés, 2004; Keyantash and Dracup, 2004; Wells *et al.* 2004; Tsakiris *et al.* 2007). Most studies related to drought analysis and monitoring systems have been conducted using either i) the Palmer Drought Severity Index (PDSI) (Palmer, 1965), based on a soil water balance equation, or ii) the Standardised Precipitation Index (SPI; McKee *et al.* 1993), based on a precipitation probabilistic approach.

The PDSI was a landmark in the development of drought indices. It enables measurement of both wetness (positive value) and dryness (negative values), based on the supply and demand concept of the water balance equation, and thus incorporates prior precipitation, moisture supply, runoff and evaporation demand at the surface level. The calculation procedure has been explained in a number of studies (e.g., Karl, 1983 and 1986; Alley, 1984). Nevertheless, the PDSI has several deficiencies (Alley, 1984; Karl, 1986; Soulé, 1992; Akimremi *et al.* 1996; Weber and Nkemdirim, 1998), including the strong influence of calibration period, its limited utility in areas other than that used for calibration, problems in spatial comparability, and subjectivity in relating drought conditions to the values of the index. Many of these problems were solved by development of the self–calibrated PDSI (sc–PDSI) (Wells *et al.* 2004), which is spatially comparable and reports extreme wet and dry events at frequencies expected for rare conditions. Nevertheless, the main shortcoming of the PDSI has not been resolved. **This relates to its fixed temporal scale (between 9 and 12 months), and an autoregressive characteristic whereby index values are affected by the conditions up to four years in the past (Guttman, 1998).**

It is commonly accepted that drought is a multi–scalar phenomenon. McKee *et al.* (1993) clearly illustrated this essential characteristic of droughts through consideration of usable water resources including soil moisture, ground water, snowpack, river discharges, and reservoir storages. The time period from the arrival of water inputs to availability of a given usable resource differs considerably. Thus, the time scale over which water deficits accumulate becomes extremely important, and functionally separates hydrological, environmental, agricultural and other droughts. For example, the response of hydrological systems to precipitation can vary markedly as a function of time (Changnon and Easterling, 1989; Elfatih *et al.*, 1999; Pandey and Ramasastri, 1999). This is determined by the different frequencies of hydrologic/climatic variables (Skøien *et al.* 2003). For this reason, drought indices must be associated with a specific timescale to be useful for monitoring and management of different usable water resources. This explains the wide acceptance of the SPI, which is comparable in time and space (Guttman, 1998; Hayes *et al.* 1999), and can be calculated at different time scales to monitor droughts with respect to different usable water resources. A number of studies have demonstrated variation in response of the SPI to soil moisture, river discharge, reservoir storage, vegetation activity, crop production and piezometric fluctuations at different time scales (e.g. Szalai *et al.* 2000; Sims *et al.* 2002; Ji and Peters, 2003; Vicente–Serrano and López–Moreno, 2005; Vicente–Serrano *et al.* 2006; Patel *et al.* 2007; Vicente–Serrano, 2007; Khan *et al.* 2008).

The main criticism of the SPI is that its calculation is based only on precipitation data. The index does not consider other variables that can influence droughts, such as temperature, evapotranspiration, wind speed and soil water holding capacity. Nevertheless, several studies have shown that precipitation is the main variable determining the onset, duration, intensity and end of droughts (Chang and Cleopa, 1991; Heim, 2002). Thus, the SPI is highly correlated with the PDSI at time scales of 6 to 12 months (Lloyd–Hughes and Saunders, 2002; Redmond, 2002). Low data requirements and simplicity explain the wide use of precipitation– based indices, such as the SPI, for drought monitoring and analysis.

Precipitation–based drought indices including the SPI rely on two assumptions: i) the variability of precipitation is much higher than that of other variables, such as temperature and potential evapotranspiration (PET), and ii) the other variables are stationary (i.e. they have no temporal trend). In this scenario the importance of these other variables is negligible, and droughts are controlled by the temporal variability in precipitation. However, some authors have warned against systematically neglecting the importance of the effect of temperature on drought conditions. For example, Hu and Willson (2000) assessed the role of precipitation and temperature in the PDSI, and found that the index responded equally to changes of similar magnitude in both variables. Only where the temperature fluctuation was smaller than that of precipitation was variability in the PDSI controlled by precipitation. Empirical studies have shown that temperature rise markedly affects the severity of droughts. For example, Abramopoulos et al. (1988) used a general circulation model experiment to show that evaporation and transpiration can consume up to 80% of rainfall. In addition, they found that the efficiency of drying due to temperature anomalies is as high as that due to rainfall shortage. The role of temperature was evident in the devastating central European drought during the summer of 2003. Although previous precipitation was lower than normal, the extremely high temperatures over most of Europe during June and July (more than 4°C above the average) caused the greatest damage to cultivated and natural systems, and dramatically increased evapotranspiration rates and water stress (Rebetez et al. 2006). Some studies have also found that the PDSI explains the variability in crop production and the activity of natural vegetation better than does the SPI (Mavromatis, 2007; Kempes et al. 2008).

There has been a general temperature increase (0.5–2°C) during the last past 150 years (Jones and Moberg, 2003), and climate change models predict a marked increase during the 21st century (IPCC, 2007). It is expected that this will have dramatic consequences for drought conditions, with an increase in water demand due to evapotranspiration (Sheffield and Wood, 2008). Dubrovsky et al. (2008) recently showed that the drought effects of warming predicted by global climate models can be clearly seen in the PDSI, whereas the SPI (which is based only on precipitation data) does not reflect expected changes in drought conditions. Therefore, the use of drought indices which include temperature data in their formulation (such as the PDSI) is preferable, especially for applications involving future climate scenarios. However, the PDSI lacks the multi–scalar character essential for both assessing drought in relation to different hydrological systems, and differentiating among different drought types. We therefore formulated a new drought index (the Standardized Precipitation Evapotranspiration Index; SPEI) based on precipitation and PET. The SPEI combines the sensitivity of PDSI to changes in evaporation demand (caused by temperature fluctuations and trends) with the simplicity of calculation and the multi-temporal nature of the SPI. **The new index is particularly suited to detecting, monitoring and exploring the consequences of global warming on drought conditions.**

PROBLEM OVERVIEW

As an illustrative example, Figure 1 shows the evolution of the sc–PDSI and the SPI at different time scales from 1910 to 2007 at the Indore observatory (India). The sc–PDSI was devised by Wells *et al.* (2004) to address the shortcomings of the PDSI. For calculations we used the software developed by Wells (2003; and available at http://greenleaf. unl.edu/downloads). The time series of monthly precipitation and monthly mean temperature were obtained from the Global Historical Climatology Network (GHCN– Monthly) database (http://www.ncdc.noaa.gov/oa/climate/ghcn–monthly/). The water field capacity at Indore, needed to derive the sc–PDSI, was obtained from a global digital format data set of water holding capacity, described by Webb *et al.* (1993). The SPI was calculated according to a Pearson III distribution and the L-moment method to obtain the distribution parameters, following Vicente–Serrano (2006). Figure 1 shows that the sc–PDSI has a unique time scale, in which the longest and most severe droughts were recorded in the decades 1910, 1920, 1950, 1960 and 2000. These episodes are also clearly identified by the SPI at long time scales (12–24 months). This provides evidence about the suitability of identifying and monitoring droughts using an index that only considers precipitation data. Moreover, this example shows the advantage of the SPI over the sc–PDSI, since the different time scales over which the SPI can be calculated allows the identification of different drought types. At the shortest time scales the drought series show a high frequency of drought, and moist periods of short duration. In contrast, at the longest time scales the drought periods are of longer duration and lower frequency. **Thus, short time scales are mainly related to soil water content and river discharge in headwater areas, medium time scales are related to reservoir storages and discharge in the medium course of the rivers, and long time–scales are related to variations in groundwater storage. Therefore, different time scales are useful for monitoring drought conditions in different hydrological sub-systems.**

Climatic change processes result in two main predictions with implications for the duration and magnitude of droughts (IPCC 2007): i) precipitation will decrease in some regions, and ii) an increase in global temperature, which will be more intense in the northern hemisphere, will cause an increase in the evapotranspiration rate.

A reduction in precipitation due to climate change will affect the severity of droughts. Current climate change A2 scenarios for the end of the 21st century (IPCC, 2007) show a maximum reduction of 15% in total precipitation in some regions. The influence of a reduction in precipitation on future drought conditions is identified by both the sc–PDSI and the SPI.

Figure 2 shows the evolution of the sc–PDSI and the 18–month SPI at the Albuquerque (New Mexico, USA) observatory between 1910 and 2007. Both indices were calculated using a hypothetical progressive precipitation decrease of 15% during this period. Both the modeled SPI and sc–PDSI series showed an increase in the duration and magnitude of droughts at the end of the century relative to the series computed with real data. As a consequence of the precipitation decrease, droughts recorded in the decades of 1970 to 2000 increased in maximum intensity, total magnitude and duration. In contrast, the humid periods showed the opposite behavior. Therefore, both indices have the capacity to record changes in droughts related to changes in precipitation.

However, climate change scenarios also show a temperature increase during the 20th century. In some cases, such as the A2 greenhouse gas emissions scenario, the models predict a temperature increase that might exceed 4°C with respect to the 1960–1990 average (IPCC, 2007). This increase will have consequences for drought conditions, which are clearly identified by the PDSI (Mavromatis, 2007; Dubrovsky *et al.* 2008). Figure 3 shows the evolution of the sc–PDSI in Albuquerque, computed with real data between 1910 and 2007, but also considers a progressive increase of 2–4°C in the mean temperature series. The differences between the sc–PDSI using real data and the two modelled series are also shown. This simple experiment clearly shows an increase in the duration and magnitude of droughts at the end of the century, which is directly related to the temperature increase. A similar pattern could not be identified using the SPI, demonstrating the shortcomings of this widespread index in addressing the consequences of climate change.

METHODOLOGY

We describe here a simple multi–scalar drought index (the SPEI) that combines precipitation and temperature data. The SPEI is very easy to calculate, and it is based on the original SPI calculation procedure. The SPI is calculated using monthly (or weekly) precipitation as the input data. The SPEI uses the monthly (or weekly) difference between precipitation and PET. This represents a simple climatic water balance (Thornthwaite, 1948) which is calculated at different time scales to obtain the SPEI.

The first step, calculation of the PET, is difficult because of the involvement of numerous parameters including surface temperature, air humidity, soil incoming radiation, water vapor pressure and ground–atmosphere latent and sensible heat fluxes (Allen *et al.* 1998). Different methods have been proposed to indirectly estimate the PET from meteorological parameters measured at weather stations. According to data availability, such methods include physically based methods (e.g. the Penman–Monteith method; PM) and models based on empirical relationships, where PET is calculated with fewer data requirements. **The PM method has been adopted by the International Commission for Irrigation (ICID), the Food and Agriculture Organization of the United Nations (FAO), and the American Society of Civil Engineers (ASCE) as the standard procedure for computing PET. The PM method requires large amounts of data since its calculation involves values for solar radiation, temperature, wind speed and relative humidity. In the majority of regions of the world this meteorological data is not available.** Accordingly, alternative empirical equations have been proposed for PET calculation where data are scarce (Allen *et al.* 1998). Although some methods in general provide better results than others for PET quantification (Droogers and Allen, 2002), the purpose of including PET in the drought index calculation is to obtain a relative temporal estimation, and therefore the method used to calculate the PET is not critical. Mavromatis (2007) recently showed that the use of simple or complex methods to calculate the PET provide similar results when a drought index such as the PDSI is calculated. **Therefore, we followed the simplest approach to calculate PET (Thornthwaite, 1948), which has the advantage of only requiring data on monthly mean temperature. Following this method, the monthly PET (mm) is obtained by:**

$$PET = 16K\left(\frac{10T}{I}\right)^m$$

Where T is the monthly mean temperature in °C; I is a heat index, which is calculated as the sum of 12 monthly index values i, the latter being derived from mean monthly temperature using the formula:

$$i = \left(\frac{T}{5}\right)^{1.514}$$

M is a coefficient depending on I: $m = 6.75E^{-7}I^3 - 7.71E^{-5}I^2 + 1.79E^{-2}I + 0.492$ and K is a correction coefficient computed as a function of the latitude and month by:

$$K = \left(\frac{N}{12}\right)\left(\frac{NDM}{30}\right)$$

Where NDM is the number of days of the month and N is the maximum number of sun hours, which is calculated according to:

$$N = \left(\frac{24}{\pi}\right)\varpi_s,$$

Where ϖ_s is the hourly angle of sun rising, which is calculated according to:

$\varpi_s = \arccos(-\tan\phi\tan\delta)$

where ϕ is the latitude in radians and δ is the solar declination in radians, calculated according to:

$$\delta = 0.4093 \, \text{sen} \left(\frac{2\pi J}{365} - 1.405 \right)$$

where J is the average Julian day of the month.

With a value for PET, the difference between the precipitation (P) and PET for the month i is calculated according to:

$D_i = P_i - PET_i,$

which provides a simple measure of the water surplus or deficit for the analyzed month. Tsakiris *et al.* (2007) proposed the ratio of P to PET as a suitable parameter for obtaining a drought index that accounts for global warming processes. This approach has some shortcomings, since the parameter is not defined when PET = 0 (which is common in many regions of the world during winter), and the P/PET quotient reduces dramatically the range of variability and de–emphasizes the role of temperature in droughts.

The calculated Di values are aggregated at different time scales, following the same procedure as that for the SPI. The difference Dk_{ij} in a given month j and year i depends on the chosen time scale, k. For example, the accumulated difference for one month in a particular year, i with a 12–month time scale is calculated according to:

$$X_{i,j}^k = \sum_{l=13-k+j}^{12} D_{i-1,l} + \sum_{l=1}^{j} D_{i,l} \, , \, \text{if } j < k, \text{ and}$$

$$X_{i,j}^k = \sum_{l=j-k+1}^{j} D_{i,l} \, , \, \text{if } j \geq k,$$

where $D_{i,l}$ is the P–PET difference in the 1st month of year i, in mm.

For calculation of the SPI at different time scales a probability distribution of the gamma family is used (the two parameter gamma or three parameter Pearson III distributions), since the frequencies of precipitation accumulated at different time scales are well modeled using these statistical distributions. **While the SPI can be calculated using a two parameter distribution such as the gamma distribution, a three parameter distribution is needed to calculate the SPEI, since in two parameter distributions the variable (x) has a lower boundary of zero (0 > x < ∞), whereas in three parameter distributions x can take values in the range (γ > x < ∞, where g is the parameter of origin of the distribution); consequently, x can have negative values, which are common in D series.**

We tested the most suitable distribution to model the Di values calculated at different time scales. For this purpose, L–moment ratio diagrams were used because they allow comparison of the empirical frequency distribution of D series computed at different time scales with a number of theoretical distributions (Hosking, 1990). L–moments are analogous to conventional central moments, but are able to characterize a wider range of distribution functions, and are more robust in relation to outliers in the data.

To create the L–moment ratio diagrams, L–moment ratios (L–skewness, τ3; and L–kurtosis, τ4) must be calculated τ3 and τ4 are calculated as follows:

$$\tau_3 = \frac{\lambda_3}{\lambda_2}$$

$$\tau_4 = \frac{\lambda_4}{\lambda_2}$$

where λ_2, λ_3 and $\lambda4$ are the L–moments of the D series, obtained from probability–weighted moments (PWMs) using the formulae:

$\lambda_1 = w_0$

$\lambda_2 = w_0 - 2w_1$

$\lambda_3 = w_0 - 6w_1 + 6w_2$

$\lambda_3 = w_0 - 6w_1 + 6w_2$

$\lambda_4 = w_0 - 12w_1 + 30w_2 - 20w_3$

The PWMs of order s are calculated as:

$$w_s \frac{1}{N} \sum_{i=1}^{N} (1 - F_i)^s D_i$$

where F_i is a frequency estimator calculated following the approach of Hosking (1990):

$$F_i = \frac{i - 0.35}{N}$$

where i is the range of observations arranged in increasing order, and N is the number of data points. τ_3 and τ_4 were calculated from the D series of 11 observatories between 1910 and 2007 in different regions of the world, under varying conditions that included tropical (Tampa, Sao Paulo), monsoon (Indore), Mediterranean (Valencia), semiarid (Albuquerque), continental (Wien), cold (Punta Arenas), and oceanic (Abashiri) climates (Figure 4). The dataset was obtained from the Global Historical Climatology Network (GHCN–Monthly) database (http://www.ncdc.noaa.gov/oa/climate/ghcn–monthly).

Figure 5 shows the L–moment diagrams for the D series accumulated for time scales of 3– and 18– months for the 11 selected observatories. For each observatory 12 points are shown, each corresponding to a 1–month series. The empirical L–moment ratios for the analyzed D series at different time scales could be adjusted by different candidate distributions (e.g. Pearson III, Log–normal, General Extreme Value, Log–logistic) because the empirical statistics oscillate around these curves. According to the Kolmogorov–Smirnoff test none of these four distributions can be rejected in the different monthly series and time scales for the 11 observatories analyzed. Figure 6 shows the curves of the four distributions and the empirical frequencies for the D series calculated at the time scales of 1, 3, 6, 12, 18 and 24 months for the Albuquerque observatory (New Mexico, USA). It is evident that the four distributions adapt well to the empirical frequencies of the D series, independently of the time scale analyzed. Figure 7 shows the modelled accumulated probabilities, F(x), for the Albuquerque observatory for time scales of 1, 6, 12 and 24 months, using the four distributions and the empirical cumulative probabilities. This figure shows the high degree of similarity among the four curves. Independently of the probability distribution selected, the modelled F(x) values adjust very well to the empirical probabilities. This was also observed for the other analyzed observatories. Therefore, selection of the most suitable distribution to model the D series is difficult, given the similarity among the four distributions. We therefore based our selection on the behavior at the most extreme values. Given the marked decrease in the curves that adjust the lower values for the Pearson III, Lognormal and General Extreme Value distributions, we found extremely low cumulative probabilities for very low values corresponding to less than 1 occurrence in 1,000,000 years, mainly at the shortest time scales. Also, in some cases we found values of D which were below the origin parameter of the distribution, which implies that f(x) and F(x) cannot be defined for these values. In contrast, the Log–logistic distribution showed a more gradual decrease in the curve for low values, and more coherent probabilities were obtained for very low values of D, corresponding to 1 occurrence in 200 to 500 years. Additionally, no values were found below the origin parameter of the distribution. **These results suggested selection of the Log–logistic distribution for standardizing the D series to obtain the SPEI.** The probability density function of a three parameter Log–logistic distributed variable is expressed as:

$$f(x) = \frac{\beta}{\alpha} \left(\frac{x-y}{\alpha} \right)^{\beta-1} \left(1 + \left(\frac{x-y}{\alpha} \right)^{\beta} \right)^{-2}$$

where α, β and γ are scale, shape and origin parameters, respectively, for D values in the range ($\gamma > D < \infty$).

Parameters of the Log–logistic distribution can be obtained following different procedures. Among them, the L–moment procedure is the most robust and easy approach (Ahmad et al. 1988). When L–moments are calculated, the parameters of the Pearson III distribution can be obtained following Singh et al. (1993):

$$\beta = \frac{2w_1 - w_0}{6w_1 - w_0 - 6w_2}$$

$$\alpha = \frac{(w_0 - 2w_1)\beta}{\Gamma(1+1/\beta)\Gamma(1-1/\beta)}$$

$$\gamma = w_0 - \alpha\Gamma(1+1/\beta)\Gamma(1-1/\beta)$$

where $\Gamma(\beta)$ is the gamma function of β.

The Log–logistic distribution adapted very well to the D series for all time scales. Figure 8 shows the probability density functions for the Log–logistic distribution obtained from the D series at different time scales for the Albuquerque observatory.

The Log–logistic distribution can account for negative values, and is capable of adopting different shapes to model the frequencies of the D series at different time scales.

The probability distribution function of the D series according to the Log–logistic distribution is given by:

$$f(x) = \left[1 + \left(\frac{\alpha}{x - \gamma} \right)^{\beta} \right]^{-1}$$

F(x) values for the D series at different time scales adapt very well to the empirical F(x) values at the different observatories, independently of the climate characteristics and the time scale of the analysis. Figure 9 shows an example of the results for the 3– and 12–month series of Albuquerque, Sao Paulo and Helsinki, but similar observations were made for the other observatories and time–scales. This demonstrates the suitability of the Log–logistic distribution to model F(x) values from the D series in any region of the world.

With F(x) the SPEI can easily be obtained as the standardized values of F(x). For example, following the classical approximation of Abramowitz and Stegun (1965):

$$SPEI = W - \frac{C_0 + C_1 W + C_2 W^2}{1 + d_1 W + d_2 W^2 + d_3 W^3}$$

where

$$W = \sqrt{-2 \ln(P)} \text{ for } P \leq 0.5,$$

and P is the probability of exceeding a determined D value, $P = 1 \square F(x)$. If $P > 0.5$, P is replaced by $1 - P$ and the sign of the resultant SPEI is reversed. The constants are: $C_0 = 2.515517$, $C_1 = 0.802853$, $C_2 = 0.010328$, $d_1 = 1.432788$, $d_2 = 0.189269$, $d_3 = 0.001308$. The average value of SPEI is 0, and the standard deviation is 1. The SPEI is a standardized variable, and it can therefore be compared with other SPEI values over time and space. An SPEI of 0 indicates a value corresponding to 50% of the cumulative probability of D, according to a Log–logistic distribution.

RESULTS

Current Climatic Conditions

Figure 10 shows the sc–PDSI, and the 3–, 12– and 24–monthly SPIs and SPEIs for Helsinki between 1910 and 2007. According to the sc–PDSI, the main drought episodes occurred in the decades of 1930, 1940, 1970 and 2000. These droughts are also clearly identified by the SPI and the SPEI. Few differences were apparent between the SPI and the SPEI series, independently of the time scale of analysis. This result shows that under climate conditions in which low interannual variability of temperature dominates, both drought indices respond mainly to the variability in precipitation.

Figure 11 shows the results for the Sao Paulo (Brazil) observatory, in which the sc–PDSI identified drought episodes in the decades of 1910, 1920, 1960 and 2000. In contrast, these episodes were not clearly evident with the SPI, especially at longer time scales. Thus the SPI identified droughts in the decades of 1910, 1950 and 1960, but not the long and severe drought of 2000. In contrast the SPEI identified all four drought periods. The mean temperature increased markedly at Sao Paulo between 1910 and 2007 (0.29°C per decade), and this increase would have produced a higher water demand by PET at the end of the century. This would have affected drought severity, which was clearly recorded by sc–PDSI in the 2000 decade. **The role of temperature increase on drought conditions was not recognized using the precipitation–based SPI drought index, but was identified for the 2000 drought using the SPEI index.**

Figure 12 shows the correlation between the 1910–2007 series for sc–PDSI and the 1– to 24– monthly SPI and SPEI, for each of the observatories shown in Figure 4. As indicated in previous reports, there is strong agreement between the sc–PDSI and the SPI, with maximum values that oscillate between 0.6 and 0.85 at time scales between 5 and 24 months. A similar result was found for the SPEI, although in general the correlations increased with respect to the SPI, mainly for observatories affected by warming processes during the 20th century, including Valencia (0.32°C per decade), Albuquerque (0.2°C per decade) and Sao Paulo. The correlation between the SPI and the SPEI was high for the different series, independently of the time scale analyzed; the exceptions were Valencia and Albuquerque, where correlations decreased at the longest time scales. These results are in agreement with the hypothesis that the main explanatory variable for droughts is precipitation. Therefore, under the current climate conditions inclusion of a variable to quantify PET in the SPEI and the sc–PDSI does not provide much additional information. This is particularly obvious at those observatories where the evolution of temperature was stationary during the analysis period. However, some of the results presented in Figure 12 indicate that this hypothesis may not hold over long time scales under global warming conditions, since differences were found between the SPI and the SPEI for the three observatories where temperature increased over the analysis period.

Global Warming Effects

In the two scenarios (i.e. temperature increases of 2°C and 4°C), the D series obtained at the 11 observatories showed a similar statistical behavior to that observed under real climate conditions. Figure 13 shows the L–moment ratio diagrams for the D series at the same 11 observatories, but with the addition of a progressive temperature increase of 2°C and 4 °C between 1910 and 2007, in relation to the original series from which the PET was calculated. The L–moment ratio diagrams show small changes from those obtained for the original series. The empirical L–moment ratios show that the Log–logistic distribution is also suitable to model the D series at the various observatories, independent of the time scale involved and the magnitude of the temperature increase. Therefore, global warming does not affect the choice of model for determining the SPEI. The modeled F(x) values from the Log–logistic distribution also showed a good fitting of the empirical F(x) values under a temperature increase of 2°C and 4°C at the various observatories, independently of the region of the world analyzed (Figure 14).

Figure 15 shows the evolution of the sc–PDSI obtained using the original and the modeled series for the Valencia observatory (Spain). The 18–month SPI and SPEI obtained with that series are also shown. Using the original data, the sc–PDSI identified the most important droughts in the decades of 1990 and 2000. With a progressive temperature increase of 2°C and 4°C, the droughts increased in magnitude and duration at the end of the century. The SPI did not identify those severe droughts associated with a marked temperature increase, and it did not take into account the role of increased temperature in reinforcing drought conditions, as was shown by the sc–PDSI. In contrast, the main drought episodes were identified by the SPEI, with similar evolution to that observed for the sc–PDSI. Moreover, if temperature increased progressively by 2°C or 4°C, the reinforcement of drought severity associated with higher water demand by PET was readily identified by the SPEI, with the time series showing a high similarity to the sc–PDSI observed under warming scenarios. The same pattern was observed for the other analyzed observatories. **Figure 16 shows the evolution at the Abashiri (Japan) observatory, where no temperature increase occurred during the 1910–2007 period. The SPI and SPEI series were similar, both identifying the main drought episodes in the decades of 1920, 1950, 1980, 1990 and 2000.** There was also a high degree of similarity with the sc–PDSI series during the same period. If the temperature was increased by 2°C and 4°C during the same period, the sc–PDSI showed reinforcement of drought severity at the end of the century. This was also observed with the SPEI. Therefore, the sc–PDSI and SPEI series were similar under the simulated warming conditions.

Thus, under the progressive temperature increase predicted by current climate change models, the relationship between the sc–PDSI and the SPI was dramatically reduced. Figure 17 shows the correlations between the sc–PDSI, the SPI and the SPEI under the two considered scenarios of temperature increase. With a temperature increase of 2°C, the correlation coefficients between the sc–PDSI and the SPI decreased noticeably in comparison to the sc– PDSI calculated from the original series. The correlation values for the original series were 0.65–0.80 for the various observatories, but under a scenario of 2°C temperature increase the correlation values decreased to 0.52–0.75. However, the correlations between the sc–PDSI and the SPEI for a temperature increase of 2°C were similar, and higher than that calculated using the original series; this implies that the SPEI also accounts for the effect of warming processes on drought severity. In contrast, the correlation values between the SPI and the SPEI decreased noticeably under a scenario of 2°C temperature increase. This occurred mainly at the longest time scales, where deficits due to PET accumulate, and also in the observatories located in tropical (Sao Paulo, Indore), Mediterranean (Valencia, Kimberley) and semi–arid (Albuquerque, Lahore) climates. In these regions of high mean temperature, an additional temperature increase of 2°C would markedly increase water losses by PET. In cold areas (e.g. Abashiri and Helsinki) the relationship between SPI and SPEI under a scenario of 2°C temperature increase did not change noticeably in relation to the original series, since PET would remain relatively low.

With a temperature increase of 4°C (Figure B), the correlation between the sc–PDSI and the SPI decreased even more than for a 2°C increase (0.40–0.70), while that between the sc–PDSI and the SPEI remained generally unchanged, and at some observatories values were higher than the indices calculated from the temperature series and for a temperature increase of 2°C. With a temperature increase of 4°C, the correlations between SPI and SPEI decreased markedly for the majority of observatories, particularly those located in warm climates. This suggests that if precipitation does not change from the present conditions, temperature will play a major role in determining future drought severity.

Intensification of drought severity due to global warming is correctly identified by the sc–PDSI, which is based in a complex and reliable water balance widely accepted by the scientific community. Our results confirm that the increase in water demand due to PET in a global change context will affect the future occurrence, intensity and magnitude of droughts. This suggests that the SPI is sub–optimal for the analysis and monitoring of droughts under a warming scenario. **However, given the fixed time scale of the sc–PDSI, the SPEI offers advantages since it provides similar patterns to that of the sc–PDSI but accounts for different time scales, which is essential for the monitoring of different drought types and assessment of the potential impact of droughts on different usable water sources.** Figure 18 compares the SPEI and the sc–PDSI under a 4°C temperature increase scenario throughout the analysis period at the Tampa (Florida, USA) observatory. Under this warming scenario, the sc–PDSI shows quasi–continuous drought conditions between 1970 and 2000, with some minor humid periods. The persistent drought conditions during this period are also clearly identified by the SPEI, independent of the analysis time scale. Thus, the sc–PDSI provides the same information as the SPEI at time scales of 7 to 10 months (R values between 0.850 and 0.857), but **Figure 18 clearly shows that the SPEI also provides information about drought conditions at shorter and longer time scales.**

DISCUSSION AND CONCLUSIONS

We have described a multi–scalar drought index (the Standardized Precipitation Evapotranspiration Index; SPEI) that uses precipitation and temperature data and is based on a normalization of the simple water balance developed by Thornthwaite (1948). We assessed the properties and advantages of this index in comparison to the two most widely used drought indices: the self–calibrated Palmer Drought Severity Index (sc–PDSI) and the Standardized Precipitation Index (SPI). A multi–scalar drought index is needed to take into account deficits which affect different usable water sources, and to distinguish different types of drought. This has been demonstrated in a number of studies that have shown how different usable water sources respond to the different time scales of a drought index (e.g. Szalai *et al.* 2000; Vicente–Serrano and López–Moreno, 2005; Vicente–Serrano, 2007).

Under climatic conditions with low temporal variability in temperature the SPI is superior to the sc–PDSI, since it identifies different drought types because of its multi–scalar character. Both indices have the capacity to identify an intensification of drought severity related to reduced precipitation in a climatic change context. Both indices similarly record the effect of a reduction in precipitation on the drought index. Nevertheless, we have demonstrated that global warming processes predicted by GCMs (IPCC, 2007) have important implications for evapotranspiration processes, increasing the influence of this parameter on drought severity. We have shown that this is readily identified by the PDSI, in line with recent results of Dubrovsky *et al.* (2008), but this behavior is not well recorded by the SPI, given the unique use of precipitation data in its calculation.

There is some scientific debate about which are the most important climate parameters that determine drought severity (e.g., precipitation, temperature, evapotranspiration, wind speed, relative humidity, solar radiation, etc.). There is general agreement on the importance of precipitation in explaining drought variability, and the need to include this variable in the calculation of any drought index. However, inclusion of a variable that accounts for climatic water demand (such as evapotranspiration) is not always accepted, since its role in drought conditions is not well understood or it is underestimated. Various studies have shown that precipitation is the major variable defining the duration, magnitude and intensity of droughts (Alley, 1984; Chang and Cleopa, 1991). Oladipo (1985) compared different drought indices and concluded that indices using only precipitation data provided the best option for identifying climatic droughts. Nevertheless, Hu and Willson (2000) demonstrated that evapotranspiration plays a major role in explaining drought variability in drought indices based on soil water balances, such as the PDSI, and that this is comparable to the role of precipitation under some circumstances. It is not well understood how evapotranspiration processes can affect different usable water resources, and how the different time scales can determine water deficits. However, it is widely recognized that evapotranspiration determines soil moisture variability, and consequently vegetation water content, which directly affects agricultural droughts commonly recorded using short time scale drought indices. **Thus, drought indices that only use evapotranspiration data to monitor agricultural droughts have shown better results than precipitation–based drought indices (Naramsimhan and Srinivasan, 2005).** Soil water losses due to evapotranspiration will also affect runoff, and these deficits will affect river discharge and groundwater storage. However, PET can also cause large losses from water bodies such as reservoirs (Wafa and Labib, 1973; Snoussi *et al.* 2002), which commonly have a low temporal inertia and are well monitored by long time scale drought indices (Szalai *et al.* 2000; Vicente–Serrano and López–Moreno, 2005). **Therefore, although it is very complex to determine the influence of evapotranspiration on drought conditions, it seems reasonable** to include this variable in the calculation of a drought index. The need for this increases under increasing temperature conditions, and also because the role of different climate parameters in explaining water resource availability is not constant in space. For example, Syed *et al.* (2008) have shown that precipitation dominates terrestrial water storage variation in the tropics, but evapotranspiration is most effective in explaining the variability at mid–latitudes.

Where temporal trends in temperature are not apparent, we found little difference between the values obtained using a precipitation drought index, such as the SPI, and other indices that include PET values, such as the sc–PDSI and the SPEI. Given that drought is considered an abnormal water deficit with respect to average conditions, the onset, duration and severity of drought could be determined from precipitation data. The inclusion of PET to calculate the SPEI only affects the index when PET differs from average conditions, for example under global warming scenarios. The same pattern has been observed in the sc–PDSI.

We detailed the procedure for calculating the SPEI. This is based on the method used to calculate the SPI, but with modifications to include PET. The Log–logistic distribution was chosen to model D (P–PET) values, and the resulting cumulative probabilities were transformed into a standardized variable. The distribution adapted very well to climate regions with different characteristics, independently of the time scale used to compute the deficits. **Therefore, the Log–logistic distribution was used to calculate the SPEI, as the Pearson III or gamma distributions were used to calculate the SPI. Only when the index was computed at short time scales for some few very low precipitation PET values (mainly arid locations with a highly variable climatology) were any problems experienced. These were minor and already known for SPI calculations when the two parameter gamma distribution is used (Wu *et al.*, 2007). However, the use of three parameter distributions to calculate the SPEI reduced this problem noticeably.**

We showed that under warming climate conditions the sc–PDSI decreases markedly, indicating more frequent and severe droughts. Thus, according to the sc–PDSI, temperature could play an important role in explaining drought conditions under global warming. This is consistent with the results of a number of studies which show an increase in future drought severity caused by temperature increase (Beniston *et al.* 2007; Sheffield and Wood, 2008). The increase in severity will be proportional to the magnitude of the temperature change, and in some regions the observed temperature increase over the past century has already had an impact on the sc–PDSI values. This phenomenon can also be assessed using the SPEI, which was very similar to the sc–PDSI under the two temperature increase scenarios tested. This suggests that the SPEI should be used in preference to the sc–PDSI, given the former index's simplicity, lower data requirements and multi–scalar properties.

The SPI can not identify the role of temperature increase in future drought conditions, and independently of global warming scenarios can

not account for the influence of temperature variability and the role of heat waves, such as that which affected central Europe in 2003. The SPEI can account for the possible effects of temperature variability and temperature extremes beyond the context of global warming. Therefore, given the minor additional data requirements of the SPEI relative to the SPI, use of the former is preferable for the identification, analysis and monitoring of droughts in any climate region of the world.

In summary, the SPEI fulfils the requirements of a drought index, as indicated by Nkemdirim and Weber (1999), since its multi–scalar character enables it to be used by different scientific disciplines to detect, monitor and analyze droughts. Like the sc–PDSI and the SPI, the SPEI can measure drought severity according to its intensity and duration, and can identify the onset and end of drought episodes. The SPEI allows comparison of drought severity through time and space, since it can be calculated over a wide range of climates, as can the SPI. Moreover, Keyantash and Dracup (2002) indicated that drought indices must be statistically robust and easily calculated, and have a clear and comprehensible calculation procedure. **All these requirements are met by the SPEI. However, a crucial advantage of the SPEI over the most widely used drought indices that consider the effect of PET on drought severity is that its multi–scalar characteristics enable identification of different drought types and impacts in the context of global warming.**

Software has been created to automatically calculate the SPEI over a wide range of time scales. The software is freely available in the web repository of the Spanish National Research Council: <u>http://digital.csic.es/handle/</u> 10261/10002.

ACKNOWLEDGEMENTS

This work has been supported by the research projects CGL2006–11619/HID, CGL2008–01189/BTE, and CGL2008–1083/CLI financed by the Spanish Commission of Science and Technology and FEDER, EUROGEOSS (FP7–ENV–2008–1–226487) and ACQWA (FP7– ENV–2007–1–212250) financed by the VII Framework Programme of the European Commission, "Las sequías climáticas en la cuenca del Ebro y su respuesta hidrológica" and "La nieve en el Pirineo Aragonés: distribución espacial y su respuesta a las condiciones climáticas" Financed by "Obra Social La Caixa" and the Aragón Government and the "Programa de grupos de investigación consolidados" financed by the Aragón Government.

REFERENCES

Abramopoulos, F., C. Rosenzweig, and B. Choudhury, 1988: Improved ground hydrology calculations for global climate models (GCMs): Soil water movement and evapotranspiration. *Journal of Climate,* 1, 921–941.

Abramowitz, M. and I.A. Stegun, 1965: *Handbook of Mathematical Functions.* Dover Publications, New York.

Ahmad, M.I., C.D. Sinclair, and A. Werrity, 1988: Log–logistic flood frequency analysis. *Journal of Hydrology,* 98, 205–224.

Akinremi, O.O., S.M. McGinn, and A.G. Barr, 1996: Evaluation of the Palmer Drought Index on the canadian praires. *Journal of Climate,* 9, 897–905.

Allen, R.G., L.S. Pereira, D. Raes, M. Smith, 1998: *Crop evapotranspiration: guidelines for computing crop requeriments.* Irrigation and drainage paper 56. FAO. Roma. Italia.

Alley, W.M., 1984: The Palmer drought severity index: limitations and applications. *Journal of Applied Meteorology,* 23, 1100–1109.

Beniston, M., D.B. Stephenson, O.B. Christensen, C.A.T. Ferro, C. Frei, S. Goyette, K. Halsnaes, *et al.* 2007: Future extreme events in European climate: An exploration of regional climate model projections. *Climatic Change,* 81 (SUPPL. 1), 71–95.

Burton, I., R.W. Kates, and G.F. White, 1978: *The environment as hazard.* Oxford University Press. Nueva York, 240 pp.

Chang, T.J., and X.A. Cleopa, 1991: A proposed method for drought monitoring. *Water Resources Bulletin,* 27, 275–281.

Changnon, S.A., and W.E. Easterling, 1989: Measuring drought impacts: the Illinois case. *Water Resources Bulletin,* 25, 27–42.

Droogers, P., and R.G. Allen, 2002: Estimating reference evapotranspiration under inaccurate data conditions. *Irrigation and Drainage Systems,* 16, 33–45.

Du Pisani, C.G., H.J. Fouché, and J.C. Venter, 1998: Assessing rangeland drought in South Africa. *Agricultural Systems,* 57, 367–380.

Dubrovsky, M., M.D. Svoboda, M. Trnka, M.J. Hayes, D.A. Wilhite, Z. Zalud, and P. Hlavinka, 2008: Application of relative drought indices in assessing climate–change impacts on drought conditions in Czechia. *Theoretical and Applied Climatology,* 96, 155–171.

Elfatih, A., B. Eltahir, and P.J.F. Yeh, 1999: On the asymmetric response of aquifer water level to floods and droughts in Illinois, *Water Resources Research,* 35, 1199–1217.

González, J., and J.B. Valdés, 2006: New drought frequency index: Definition and comparative performance analysis, *Water Resources Research,* 42, W11421, doi:10.1029/2005WR004308

Guttman, N.B., 1998: Comparing the Palmer drought index and the Standardized Precipitation Index. *Journal of the American Water Resources Association,* 34, 113–121.

Hayes, M., D.A. Wilhite, M. Svoboda, and O. Vanyarkho, 1999: Monitoring the 1996 drought using the Standardized Precipitation Index. *Bulletin of the American Meteorological Society,* 80, 429–438.

Heim, R.R., 2002: A review of twentieth–century drought indices used in the United States. *Bulletin of the American Meteorological Society,* 83, 1149–1165.

Hosking, J.R.M., 1990: L–Moments: Analysis and estimation of distributions using linear combinations of order statistics. *Journal of Royal Statistical Society* B, 52, 105–124.

Hu, Q., and G.D. Willson, 2000: Effect of temperature anomalies on the Palmer drought severity index in the central United States. *International Journal of Climatology,* 20, 1899–1911.

IPCC, (2007): *Climate Change* 2007: *The Physical Science Basis.* Contribution of Working Group I to the Fourth Assessment. Report of the Intergovernmental Panel on Climate Change [Solomon, S., D. Qin, M. Manning, Z. Chen, M. Marquis, K.B. Averyt, M. Tignor and H.L. Miller (eds.)]. Cambridge University Press, Cambridge, United Kingdom and New York, NY, USA, 996 pp.

Ji, L. and A.J. Peters, 2003: Assessing vegetation response to drought in the northern Great Plains using vegetation and drought indices. *Remote Sensing of Environment,* 87, 85–98.

Jones, P.D. and A. Moberg, 2003: Hemispheric and large–scale surface air temperature variations: An extensive revision and an update to 2001. *Journal of Climate,* 16, 206–223.

Karl, T.R., 1983: Some spatial characteristics of drought duration in the United States. *Journal of Climate and Applied Meteorology,* 22, 1356–1366.

Karl, T.R., 1986: The sensitivity of the Palmer Drought Severity Index and the Palmer z– Index to their calibration coefficients including potential evapotranspiration. *Journal of Climate and Applied Meteorology,* 25, 77–86.

Kempes, C.P., O.B. Myers, D.D. Breshears, and J.J. Ebersole, 2008: Comparing response of Pinus edulis tree–ring growth to five alternate moisture indices using historic meteorological data. *Journal of Arid Environments,* 72, 350–357.

Keyantash, J. and J. Dracup., 2002: The quantification of drought: an evaluation of drought indices. *Bulletin of the American Meteorological Society,* 83, 1167–1180.

Keyantash, J. A., and J.A. Dracup, 2004: An aggregate drought index: Assessing drought severity based on fluctuations in the hydrologic cycle and surface water storage. *Water Resources Research,* 40, W09304, doi:10.1029/2003 WR002610.

Khan, S., H.F. Gabriel, and T. Rana, 2008: Standard precipitation index to track drought and assess impact of rainfall on watertables in irrigation areas. *Irrigation and Drainage Systems,* 22, 159–177.

Lloyd–Hughes, B. and M.A. Saunders, 2002: A drought climatology for Europe. *International Journal of Climatology,* 22, 1571– 1592.

Mavromatis, T., 2007: Drought index evaluation for assessing future wheat production in Greece. *International Journal of Climatology,* 27, 911–924.

McKee, T.B.N., J. Doesken, and J. Kleist, 1993: The relationship of drought frequency and duration to time scales. *Eight Conf. On Applied Climatology.* Anaheim, CA, Amer. Meteor. Soc. 179–184.

Narasimhan, B. and R. Srinivasan, 2005: Development and evaluation of Soil Moisture Deficit Index (SMDI) and Evapotranspiration Deficit Index (ETDI) for agricultural drought monitoring. *Agricultural and Forest Meteorology*, 133, 69–88.

Nkemdirim, L. and L. Weber, 1999: Comparison between the droughts of the 1930s and the 1980s in the southern praires of Canada. *Journal of Climate*, 12, 2434–2450.

Oladipo, E.O., 1985: A comparative performance analysis of three meteorological drought indices. *Journal of Climatology*, 5, 655–664.

Palmer, W.C., 1965: *Meteorological droughts*. U.S. Department of Commerce Weather Bureau Research Paper 45, 58 pp.

Pandey, R.P. and K.S. Ramasastri, 2001: Relationship between the common climatic parameters and average drought frequency. *Hydrological Processes*, 15, 1019–1032, 2001.

Patel, N.R., P. Chopra, and V.K. Dadhwal, 2007: Analyzing spatial patterns of meteorological drought using standardized precipitation index. *Meteorological Applications*, 14, 329–336.

Rebetez, M., H. Mayer, O. Dupont, D. Schindler, K. Gartner, J.P. Kropp, and A. Menzel, 2006: Heat and drought 2003 in Europe: A climate synthesis. *Annals of Forest Science*, 63, 569–577.

Redmond, K.T., 2002: The depiction of drought. *Bulletin of the American Meteorological Society*, 83, 1143–1147.

Sheffield, J. and E.F. Wood, 2008: Projected changes in drought occurrence under future global warming from multi–model, multi–scenario, IPCC AR4 simulations. *Climate Dynamics*, 31, 79–105.

Sims, A.P., D.S. Nigoyi, and S. Raman, 2002: Adopting indices for estimating soil moisture: A North Carolina case study. *Geophysical Research Letters*, 29, 1183, doi:10.1029/2001GL013343.

Singh, V.P. and F.X.Y. Guo, 1993: Parameter estimation for 3–parameter log–logistic distribution (LLD3) by Pome. *Stochastic Hydrology and Hydraulics*, 7, 163–177.

Snoussi, M., S. Haïda, and S. Imassi, 2002: Effects of the construction of dams on the water and sediment fluxes of the Moulouya and the Sebou rivers. Morocco. *Regional Environmental Change*, 3, 5–12.

Skøien, J.O., G. Blösch, and A.W. Western, 2003: Characteristic space scales and timescales in hydrology. *Water Resources Research*, 39, 1304, doi:10.1029/2002WR001736, 2003.

Soulé, P.T., 1992: Spatial patterns of drought frequency and duration in the contiguous USA based on multiple drought event definitions. *International Journal of Climatology*, 12, 11–24.

Syed, T.H., J.S. Famiglietti, M. Rodell, J. Chen, and C.R. Wilson, 2008: Analysis of terrestrial water storage changes from GRACE and GLDAS, *Water Resources Research*, 44, W02433, doi:10.1029/2006WR005779.

Szalai, S., C.S. Szinell, and J. Zoboki, 2000: Drought monitoring in Hungary. In Early warning systems for drought preparedness and drought management. *World Meteorological Organization.* Lisboa: 182–199.

Thornthwaite, C.W., 1948: An approach toward a rational classification of climate. *Geographical Review,* 38, 55–94.

Tsakiris, G., D. Pangalou, and H. Vangelis, 2007: Regional drought assessment based on the Reconnaissance Drought Index (RDI). *Water Resources Management,* 21, 821–833.

Vicente Serrano, S.M. and J.I. López–Moreno, 2005: Hydrological response to different time scales of climatological drought: an evaluation of the standardized precipitation index in a mountainous Mediterranean basin. *Hydrology and Earth System Sciences,* 9, 523–533.

Vicente–Serrano, S.M., 2006: Differences in spatial patterns of drought on different time scales: an analysis of the Iberian Peninsula. *Water Resources Management,* 20, 37–60.

Vicente–Serrano, S.M., J.M. Cuadrat, and A. Romo, 2006: Early prediction of crop productions using drought indices at different time scales and remote sensing data: application in the Ebro valley (North–east Spain). *International Journal of Remote Sensing,* 27, 511–518.

Vicente–Serrano, S.M., 2007: Evaluating the Impact of Drought using Remote Sensing in a Mediterranean, Semi–Arid Region, Natural Hazards, 40, 173–208.

Wafa, T.A., and A.H. Labib, 1973: Seepage losses from lake Nasser. p: 287–291 in Man Made Lakes: Their Problems and Environmental Effects. Eds. W.C. Ackermann, G.F. White y E.B. Worgthington. Geophysical Monograph 17, American Geophysical Union. *Washingtong,* DC, USA: 847 pp.

Webb, R.S., C.E. Rosenzweig, and E.R. Levine, 1993: Specifying land surface characteristics in general circulation models: soil profile data set and derived water–holding capacities. *Global Biogeochemical Cycles,* 7, 97–108.

Weber, L., and L.C. Nkemdirim, 1998: The Palmer drought severity index revisited. Geografiska Annaler, 80A, 153–172.

Wells, N., 2003: PDSI Users Manual Version 2.0. National Agricultural Decision Support System. http://greenleaf.unl.edu. University of Nebraska–Lincoln.

Wells, N., S. Goddard, and M.J. Hayes, 2004: A self–calibrating Palmer Drought Severity Index. *Journal of Climate,* 17, 2335–2351.

Wilhite, D.A., 1993: Drought assessment, management and planning: *Theory and case studies.* Kluwer. Boston.

Wilhite D.A., and Glantz, M.H., 1985: Understanding the drought phenomenon: the role of definitions. *Water International,* 10, 111–120.

Wu, H., M.D. Svoboda, M.J. Hayes, D.A. Wilhite, F. Wen, 2007: Appropriate application of the Standardized Precipitation Index in arid locations and dry seasons. *International Journal of Climatology,* 27, 65–79.

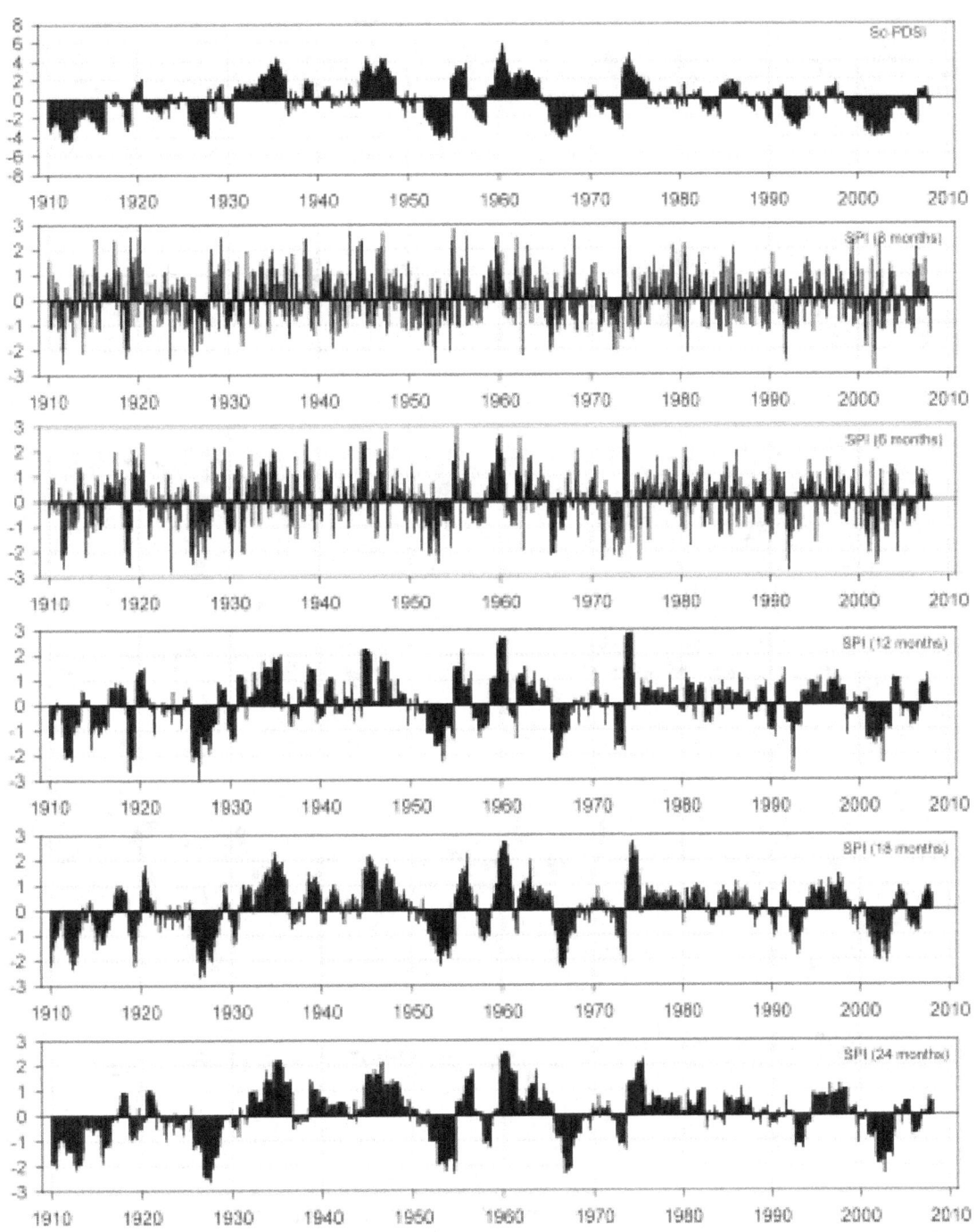

Figure 6.1 sc–PDSI and 3–, 6–, 12–, 18– and 24–month SPIs in Indore (India) (1910–2007)

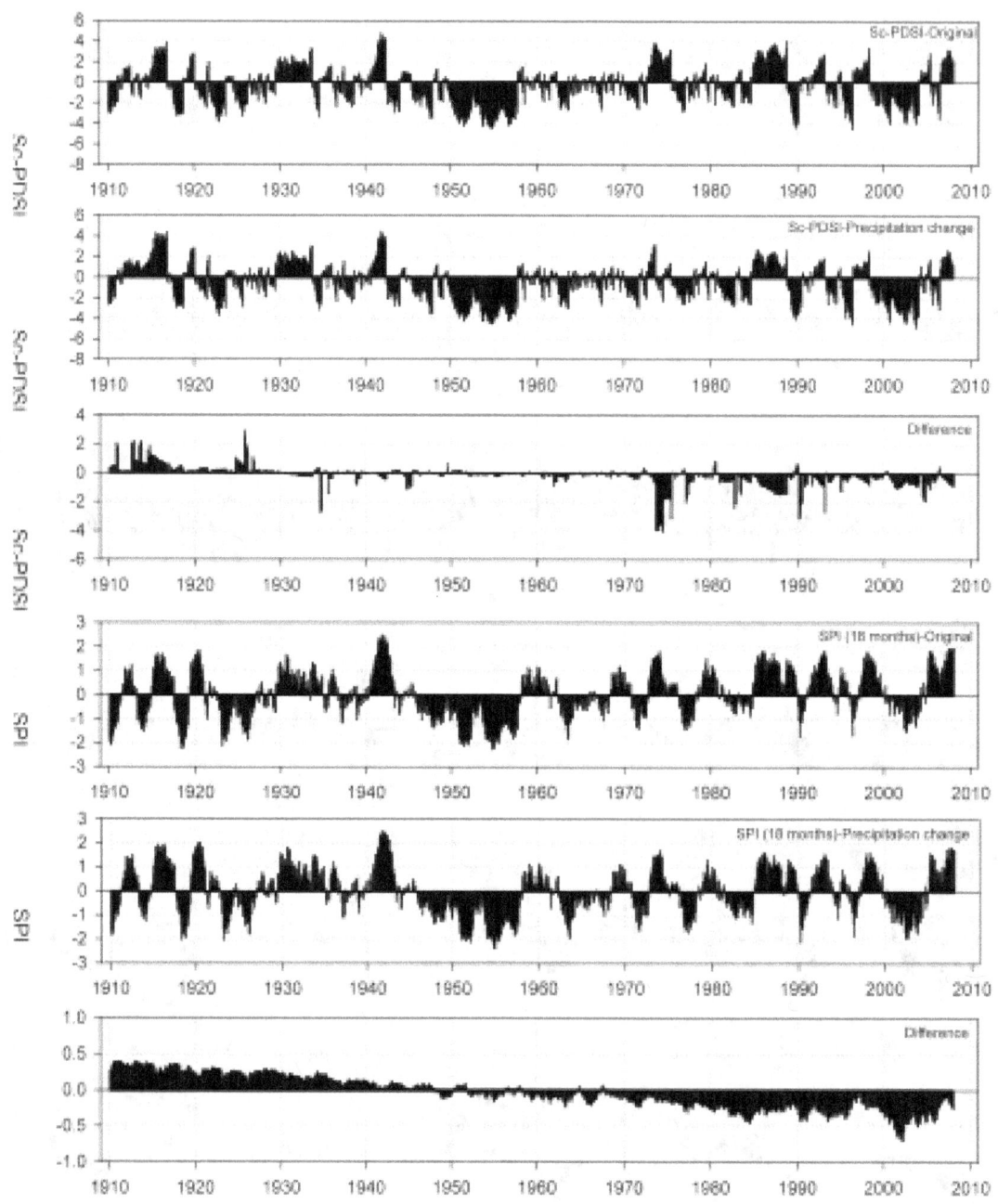

Figure 6.2 PDSI and 18–month SPI at the Albuquerque (New Mexico, USA) observatory (1910–2007). Both indices were calculated from precipitation series containing a progressive reduction of 15% between 1910 and 2007. The difference between the indices is also shown

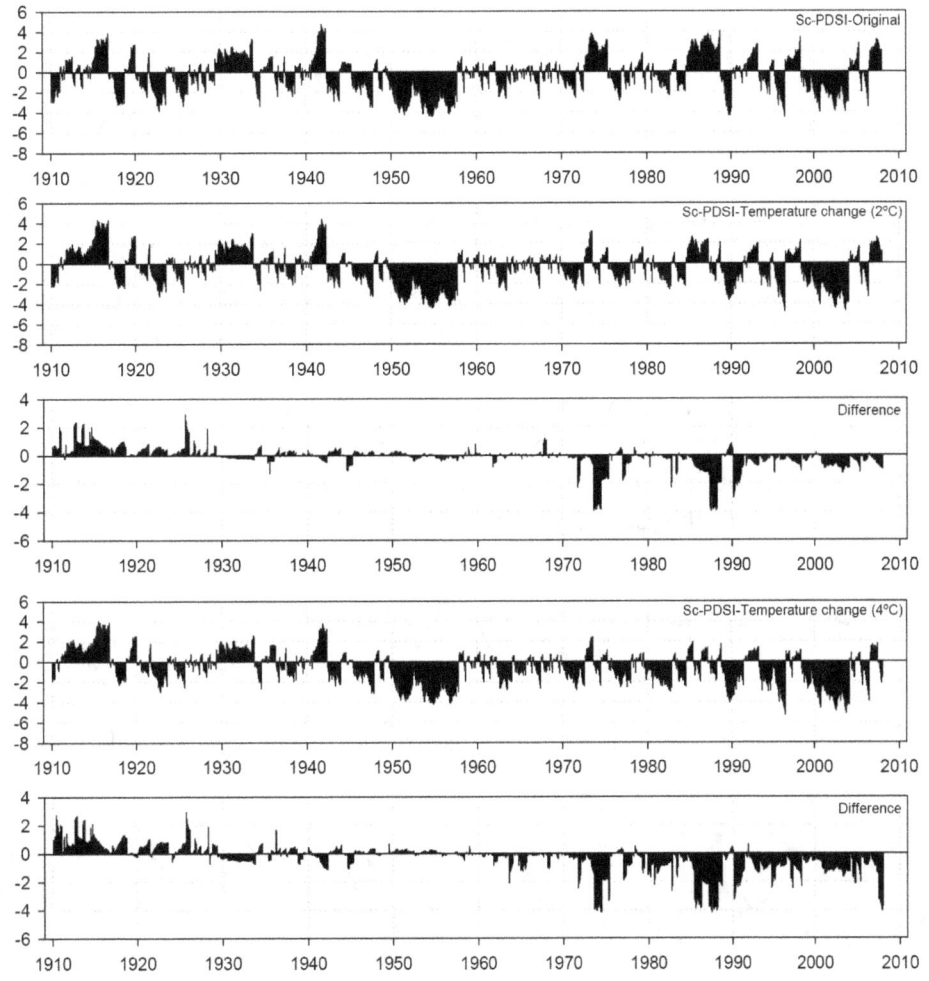

Figure 6.3 Evolution of the sc–PDSI at Albuquerque (New Mexico, USA) between 1910 and 2007, and under progressive temperature increase scenarios of 2°C and 4°C during the same period. The difference between the indices is also shown

Figure 6.4 Location of the 11 observatories used in the study

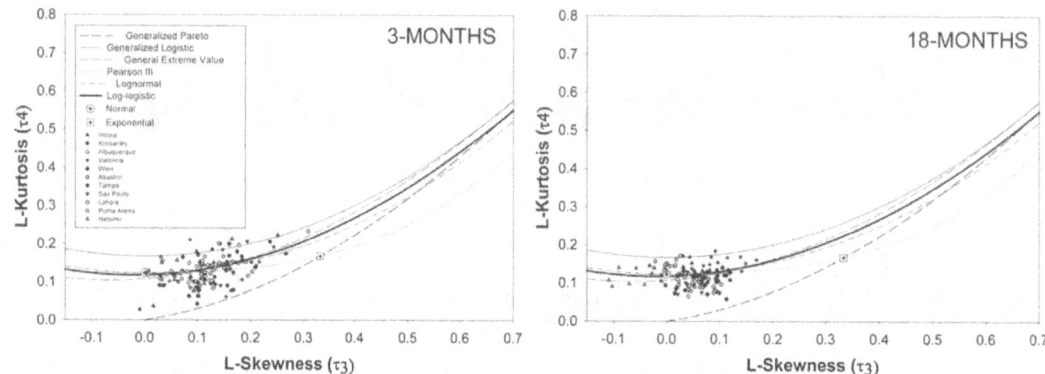

Figure 6.5 L–moment ratio diagrams for D series calculated at the time scales of 3– and 18–months. The theoretical L–moment ratios for different distributions are shown as are the empirical values obtained from the monthly series at each observatory

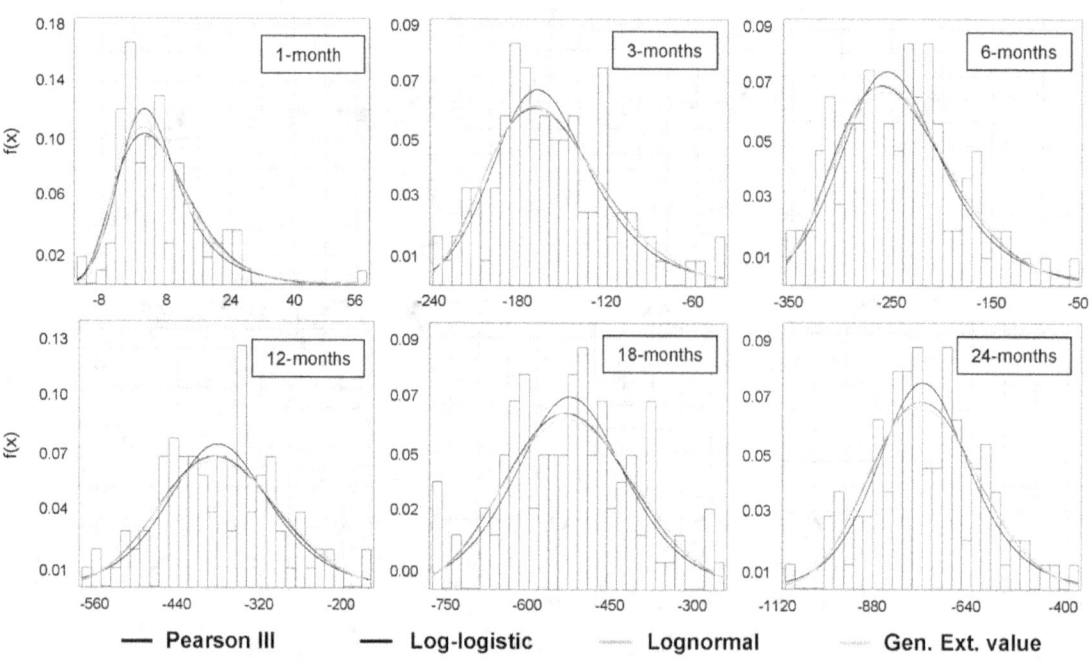

Figure 6.6 Empirical and modeled f(x) values using the Pearson III, Log–logistic, Lognormal and General Extreme Values distributions of the D series at the time scales of 1, 3, 6, 12, 18 and 24 months at the Albuquerque (New Mexico, USA) observatory

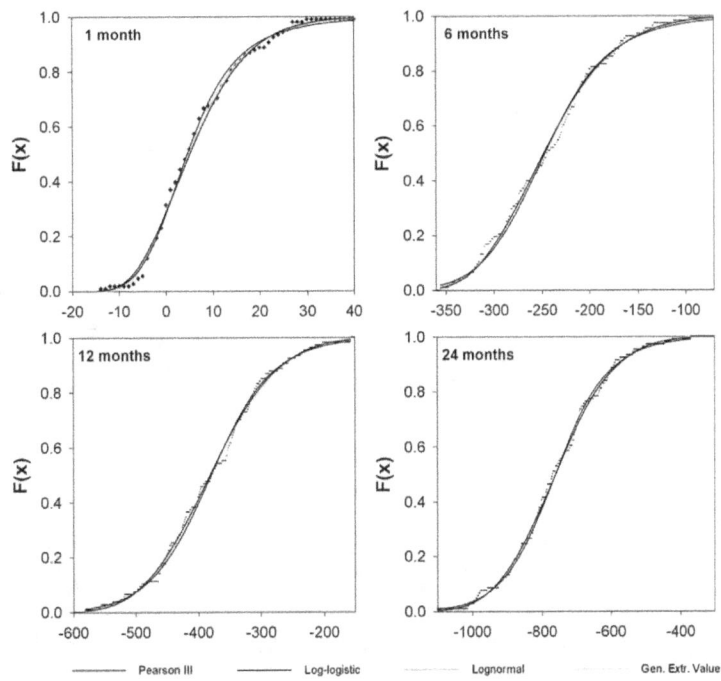

Figure 6.7 Empirical vs. modeled F(x) values from Pearson III, Log–logistic, Lognormal and General Extreme value distributions for D series at time scales of 1, 6, 12 and 24 months at the Albuquerque (New Mexico, USA) observatory

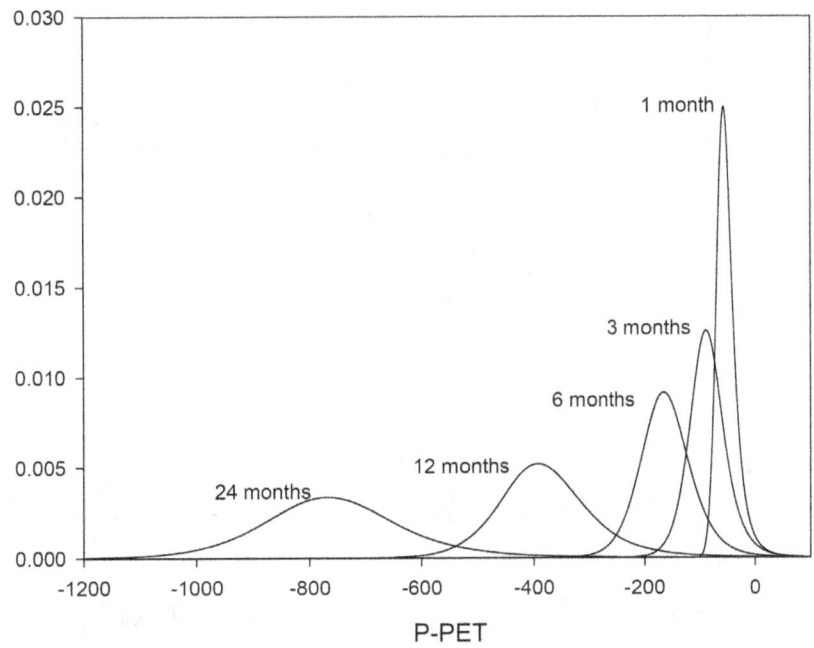

Figure 6.8 Probability density functions of the Log–logistic distribution for D series calculated at different time scales at the Albuquerque (New Mexico, USA) observatory

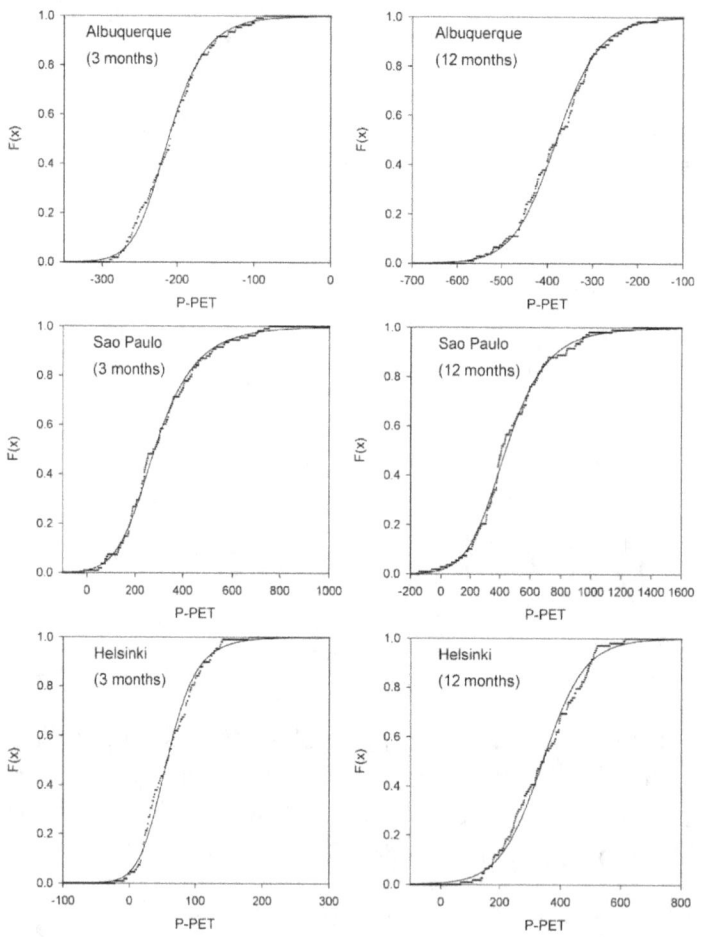

Figure 6.9 Theoretical according the Log–logistic distribution (black line) vs. empirical (dots) F(x) values for D series at time scales of 3 and 12 months for the observatories at Albuquerque, Sao Paulo and Helsinki

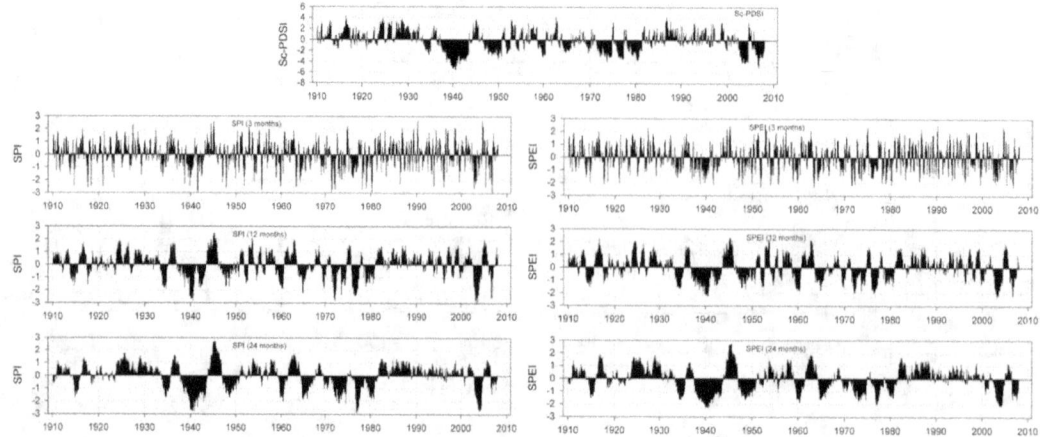

Figure 6.10   sc–PDSI, 3–, 12– and 24–month SPI and SPEI at Helsinki (1910–2007)

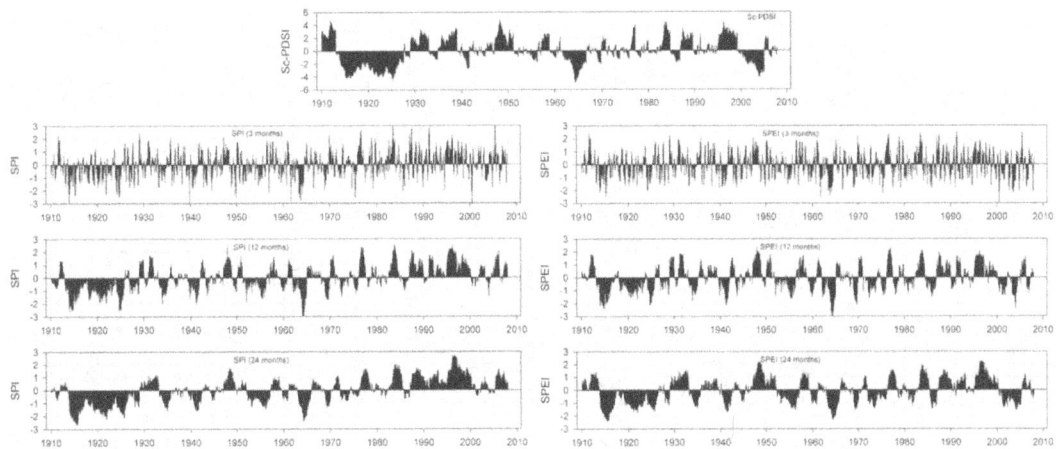

Figure 6.11   sc–PDSI, 3–, 12– and 24–month SPI and SPEI at Sao Paulo (1910–2007)

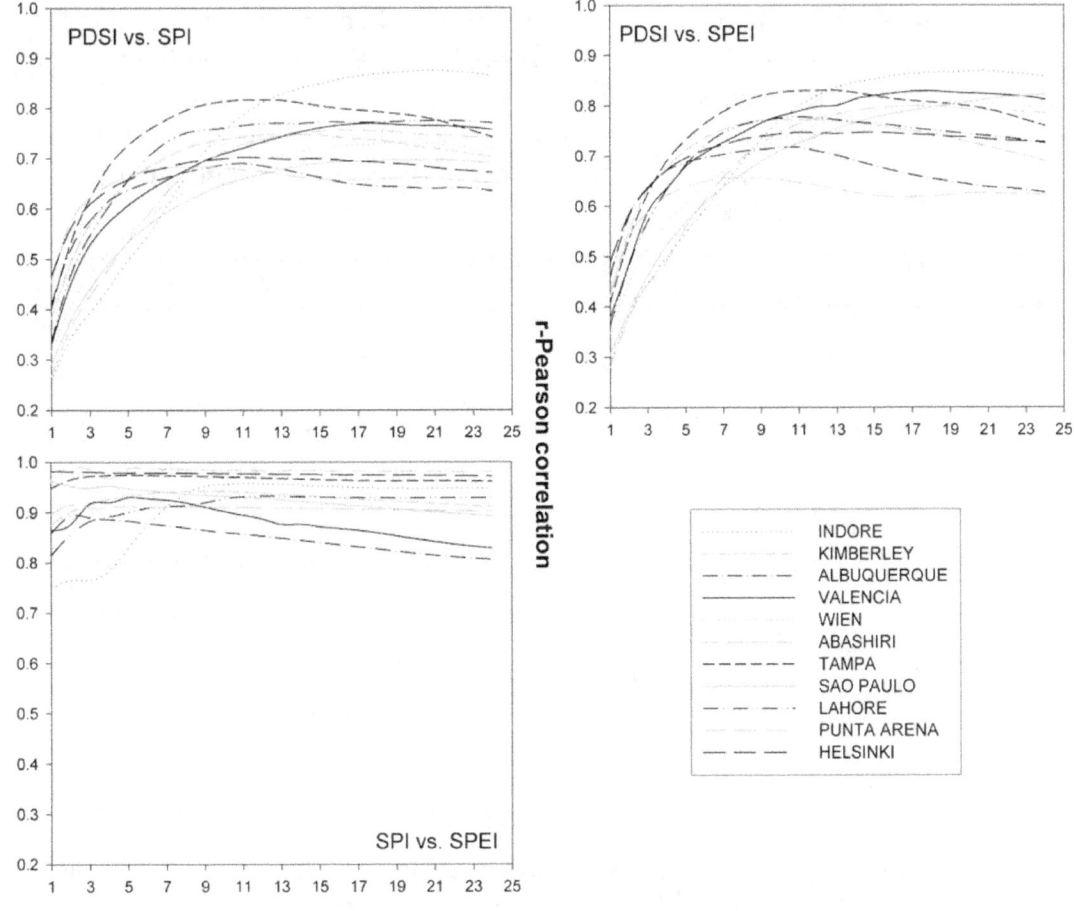

Figure 6.12   Correlation between the 1910–2007 series for the sc–PDSI, and 1–24–month SPI and SPEI at the 11 analyzed observatories

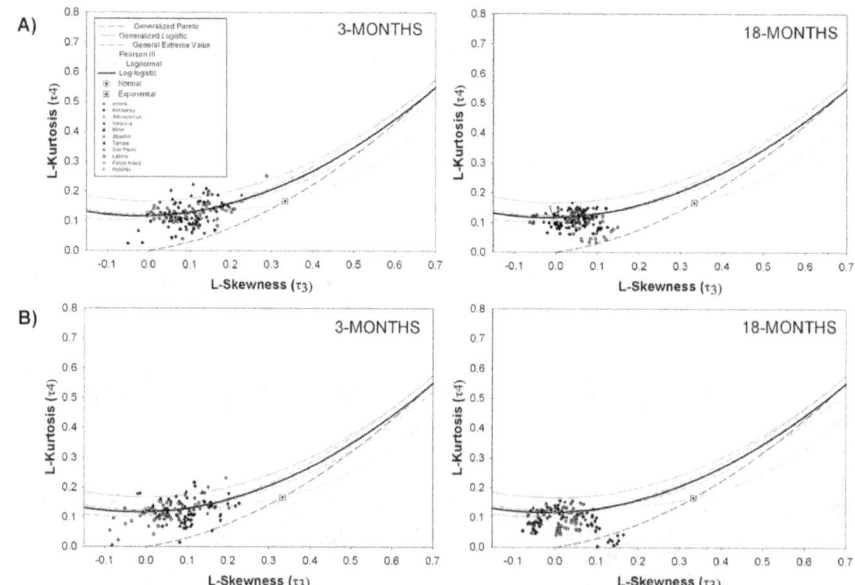

Figure 6.13   L–moment ratio diagrams for the D series calculated at the time scales of 3– and 18–months. A) Progressive temperature increase of 2°C. B) Progressive temperature increase of 4°C. The theoretical L–moment ratios for different distributions are shown as are the empirical values obtained from the monthly series at each observatory

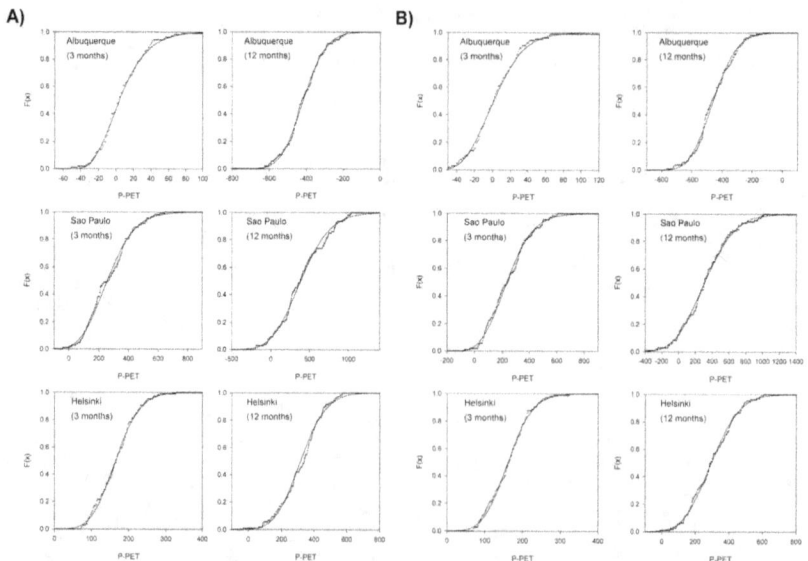

Figure 6.14   Theoretical according the log–logistic distribution (black line) vs. empirical (dots) F(x) values for D series at time scales of 3 and 12 months for the observatories at Albuquerque, Sao Paulo and Helsinki. (A) Temperature increase of 2°C. (B) Temperature increase of 4 °C

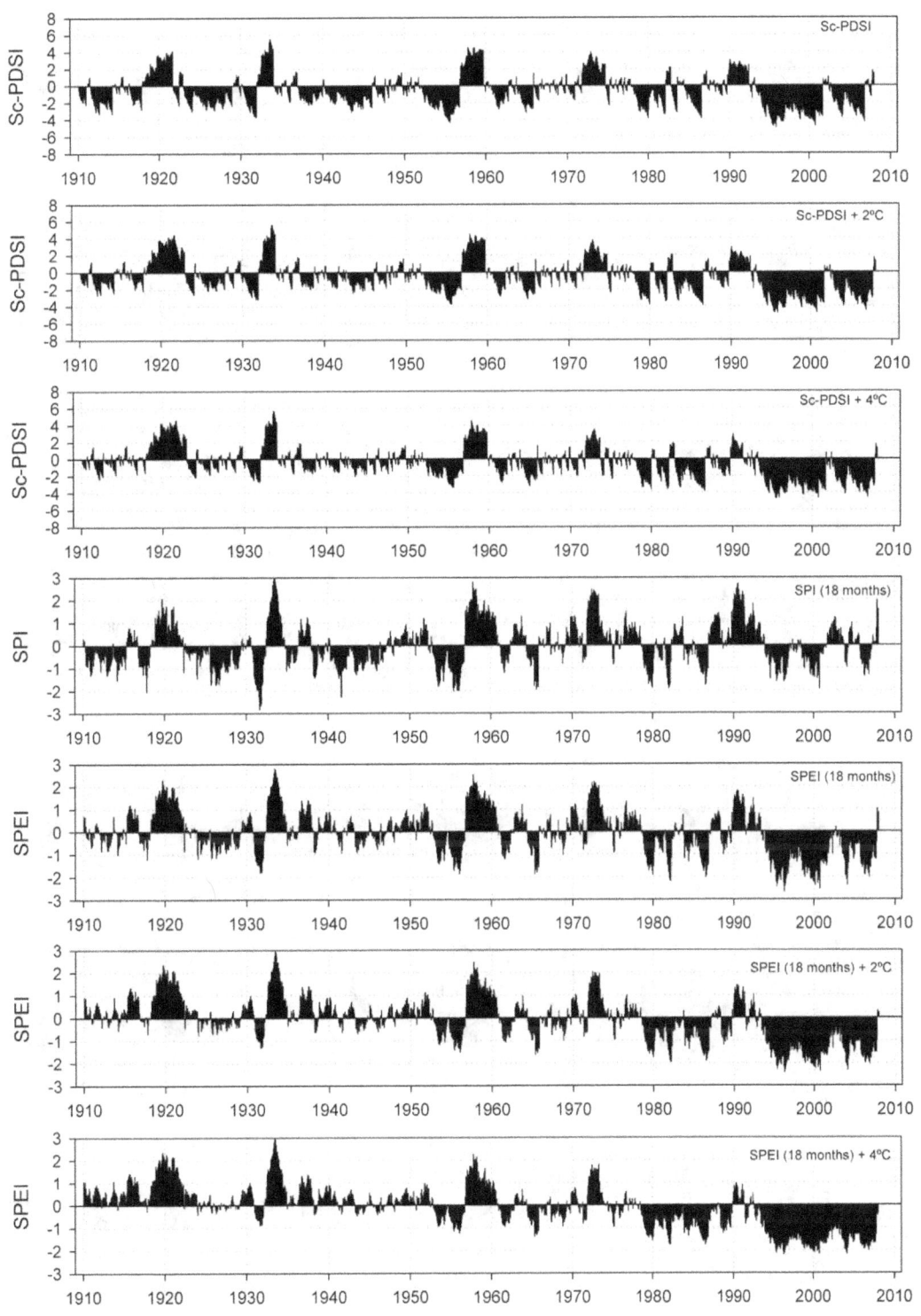

Figure 6.15 Evolution of the sc–PDSI, and 18–month SPI and SPEI in Valencia (Spain). The original series (1910–2007) and the sc–PDSI and SPEI were calculated for a temperature series with a progressive increase of 2°C and 4°C throughout the analyzed period

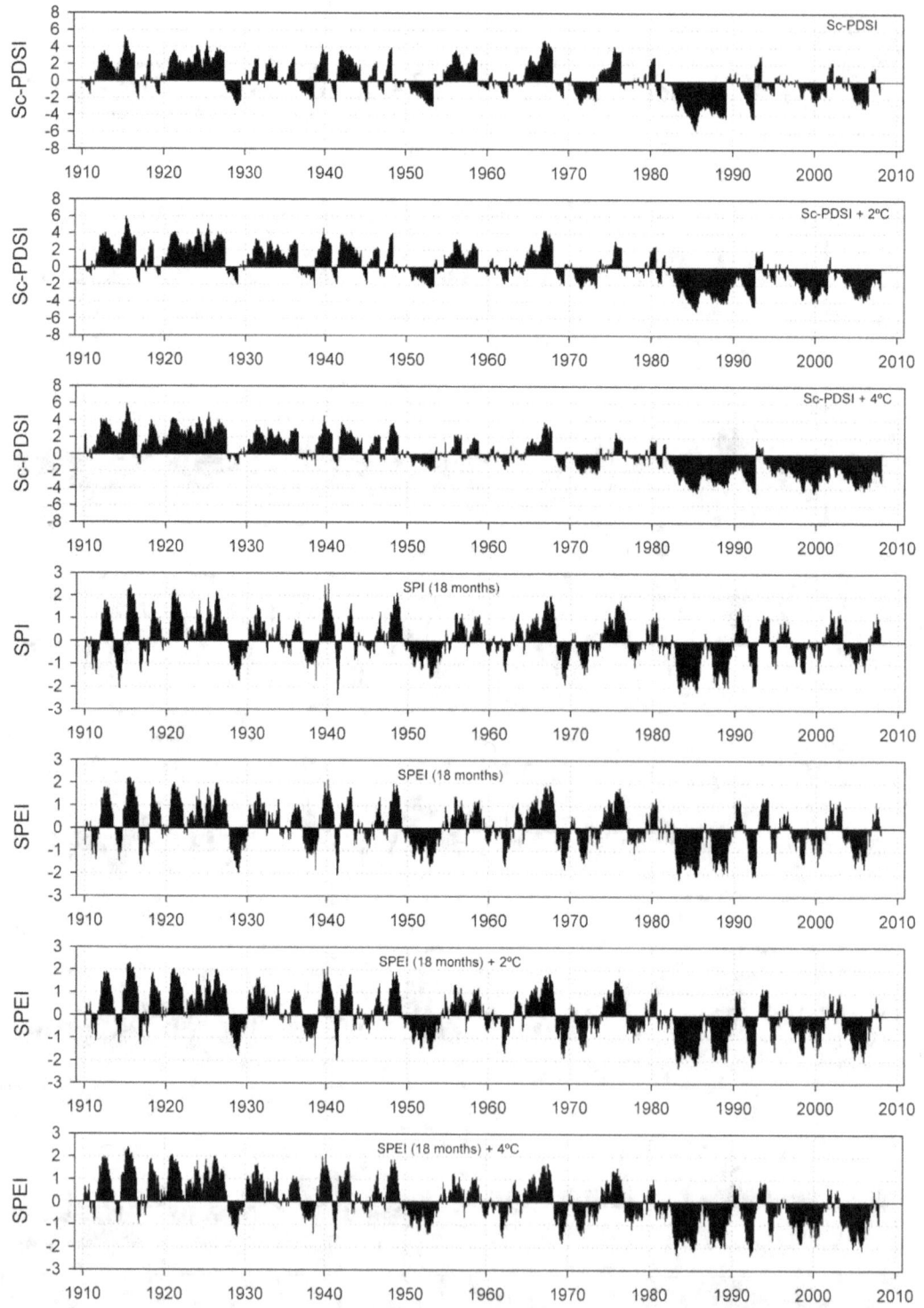

Figure 6.16   Evolution of the sc–PDSI, and 18–month SPI and SPEI at the Abashiri (Japan) observatory. The original series (1910–2007) and the sc–PDSI and SPEI were calculated from a temperature series with a progressive increase of 2°C and 4 °C throughout the analyzed period

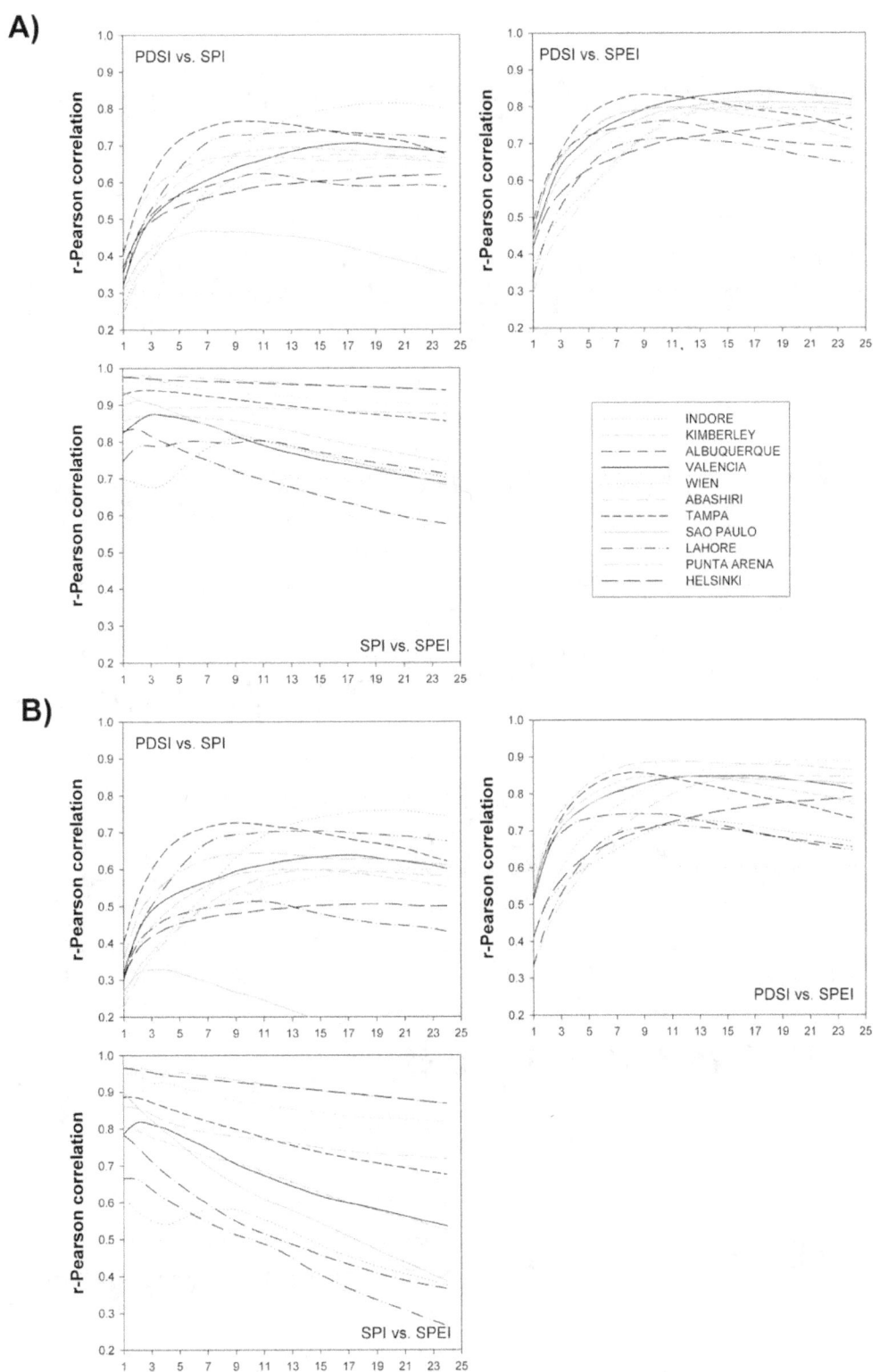

Figure 6.17   Correlation between the 1910–2007 series for the sc–PDSI, and 1–24–month SPI and SPEI in the 11 analyzed observatories. A) Temperature increase of 2°C. B) Temperature increase of 4°C

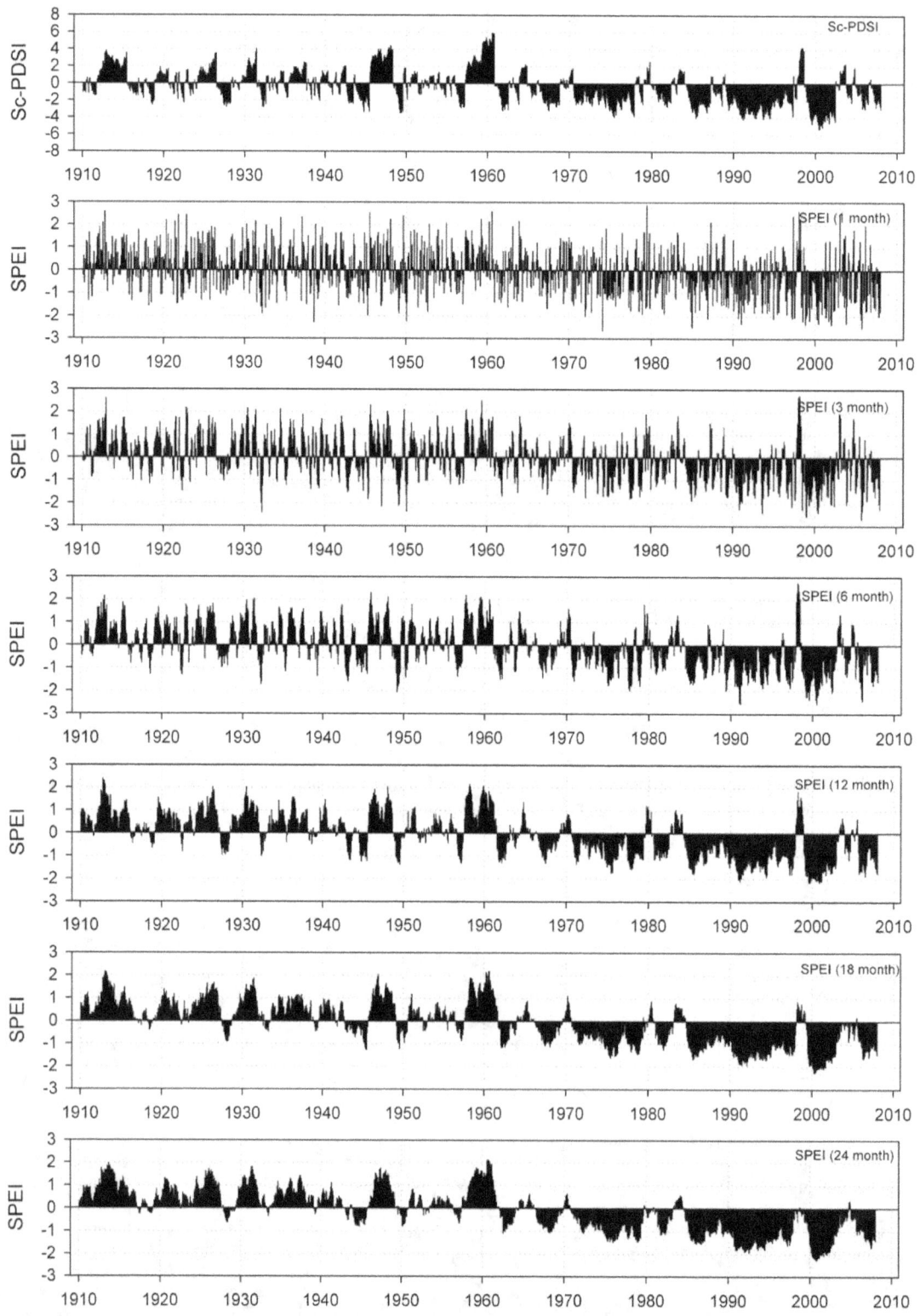

Figure 6.18 Evolution of the sc–PDSI, and 1–, 3–, 6–, 12–, 18– and 24– month SPEI at Tampa (Florida, USA) under a 4°C temperature increase scenario relative to the origin

PERFORMANCE OF DROUGHT INDICES FOR ECOLOGICAL, AGRICULTURAL AND HYDROLOGICAL APPLICATIONS

Sergio M. Vicente–Serrano[1,*], Santiago Beguería[2], Jorge Lorenzo–Lacruz[1], Jesús Julio Camarero[3], Juan I. López–Moreno[1], Cesar Azorin–Molina[1], Jesús Revuelto[1], Enrique Morán–Tejeda[1] and Arturo Sánchez–Lorenzo[4]

[1]*Instituto Pirenaico de Ecología, Consejo Superior de Investigaciones Científicas (IPE–CSIC), Campus de*

Aula Dei, P.O. Box 13034, E–50059, Zaragoza, Spain

[2]*Estación Experimental de Aula Dei, Consejo Superior de Investigaciones Científicas (EEAD–CSIC), Zaragoza, Spain*

[3]*ARAID–Instituto Pirenaico de Ecología, CSIC (Consejo Superior de Investigaciones Científicas), Campus de Aula Dei, P.O. Box 13034, E–50059, Zaragoza, Spain*

[4]*Institute for Atmospheric and Climate Science, ETH Zurich, Zurich, Switzerland*

*Corresponding author: svicen@ipe.csic.es

ABSTRACT

In this study we provide a global assessment of the performance of different drought indices for monitoring drought impacts on several hydrological, agricultural and ecological response variables. For this purpose, we compare the performance of several drought indices (the Standardized Precipitation Index, SPI; four versions of the Palmer Drought Severity Index, PDSI; and the Standardized Precipitation Evapotranspiration Index, SPEI) to predict changes in streamflow, soil moisture, forest growth and crop yield. We found a superior capability of the SPEI and the SPI drought indices, which are calculated on different time–scales, than the Palmer indices to capture the drought impacts on the aforementioned hydrological, agricultural and ecological variables. We detected small differences in the comparative performance of the SPI and the SPEI indices, but the SPEI was the drought index that best captured the responses of the assessed variables to drought in summer, the season in which more drought–related impacts are recorded and in which drought monitoring is critical. Hence, the SPEI index shows improved capability to identify drought impacts as compared with the SPI one. **In conclusion, it seems reasonable to recommend the use of the SPEI if the responses of the variables of interest to drought are not known a priori.**

Key–words: Drought index, drought vulnerability, agricultural droughts, dendrochronology, hydrological droughts, Standardized Precipitation Evapo-transpiration Index (SPEI), Standardized Precipitation Index (SPI), Palmer Drought Severity Index (PDSI).

INTRODUCTION

Drought is among the most complex climatic phenomena affecting society and the environment (Wilhite, 1993). The root of this complexity is related to the difficulty of quantifying drought severity since we identify a drought by its effects or impacts on different types of systems (agriculture, water resources, ecology, forestry, economy, etc.), but there is not a physical variable we can measure to quantify droughts. Thus, droughts are difficult to pinpoint in time and space since it is very complex to identify the moment when a drought starts and ends, and also to quantify its duration, magnitude and spatial extent (Burton *et al.* 1978; Wilhite, 2000).

These characteristics explain the vast scientific effort devoted to develop tools providing an objective and quantitative evaluation of drought severity. The quantification of drought impacts is commonly done by using the so–called drought indices, which are proxies based on climatic information and assumed to adequately quantify the degree of drought hazard exerted on sensitive systems. Many studies have shown strong relationships between the temporal variability of different drought indices and response variables of natural systems such as tree growth (e.g., Orwig and Abrams, 1997; Copenheaver *et al.* 2011; Pasho *et al.* 2011), river discharge (e.g., Vicente–Serrano and López–Moreno, 2005; Hannaford *et al.* 2011), groundwater level (Khan *et al.* 2008; Fiorillo and Guadagno, 2010), crop yields (e.g., Vicente–Serrano *et al.* 2006; Vergni and Todisco, 2011), vegetation activity (e.g., Lotsch *et al.* 2003; McAuliffe and Hamerlynck, 2010; Vicente–Serrano, 2007), the frequency of forest fires (Littell *et al.* 2009; Drobyshev *et al.* 2012), etc. Drought indices are currently used to monitor drought conditions in real time manner that is easily understood by end users (Svoboda *et al.* 2002; Shukla *et al.* 2011). Indeed, drought monitoring has been recognized as crucial for the implementation of drought plans (Wilhite, 1996; Wilhite *et al.* 2007).

Recent works have reviewed the development of drought indices and compared their advantages and drawbacks (Heim, 2002; Keyantash and Dracup, 2002; Mishra and Singh, 2010; Sivakumar *et al.* 2010). However, very few studies have performed robust statistical assessments by comparing different drought indices which may allow recommending the preferential use of one of them based on objective criteria (Guttman, 1998; Keyantash and Dracup, 2002; Steinemann, 2003; Paulo and Pereira, 2006; Quiring, 2009; Vicente–Serrano *et al.* 2010; Barua *et al.* 2011; Anderson *et al.* 2011). In addition, few researchers have compared the relative performance of different drought indices to identify drought impacts on several systems. In the case of drought impacts on hydrological systems, Vasiliades *et al.* (2011) compared five drought indices in Greece. Lorenzo – Lacruz *et al.* (2010) compared the performance of two drought indices to identify hydrological droughts in river discharges and reservoir storages in central Spain, and Zhai *et al.* (2010) compared the relationship between the Standardized Precipitation Index (SPI) and the Palmer Drought Severity Index (PDSI) and streamflow data in ten regions of China. Sims *et al.* (2002) compared the PDSI and the SPI to assess soil moisture variations in North Carolina, USA. In relation to vegetation activity and crop productivity, Potop (2011) compared different indices to assess drought impacts on corn yields in Moldava, and Mavromatis (2007) and Quiring and Papakryiakou (2003) followed a similar approach by quantifying wheat production in Greece and the Canadian prairies, respectively. Quiring and Ganesh (2010) compared drought indices to assess the responses of vegetation activity to drought severity in Texas (USA). Kempes *et al.* (2008) assessed tree–ring growth response to different drought indices in the southwestern USA. Recently, Drobyshev *et al.* (2012) analyzed the correlation between different drought indices and fire frequency in Sweden. The results of these studies are diverse, since the best drought index for detecting impacts changes as a function of the analyzed system and the performance of the drought indices varied spatially. As a result, at present there is high uncertainty among scientists, managers and end users of drought information when they aim to select one drought index for a specific purpose.

To the best of our knowledge, at present there is no global study analyzing and comparing to which degree the most widely used drought indices are able to identify drought impacts on vulnerable systems. This task is necessary in order to have solid and objective criteria for selecting a drought index to be used for specific tasks. In this study we provide the first global assessment of the performance of different drought indices for monitoring drought impacts on streamflows, soil moisture, forest growth and crop yields. For this purpose, we compare two of the most widely used drought indices, the Standardized Precipitation Index, SPI (McKee *et al.* 1993), and four versions of the Palmer Drought Severity Index, PDSI (Palmer, 1965). In addition, we also include in our comparison the recently developed Standardized Precipitation Evapotranspiration Index (SPEI), which has been claimed to outperform the two previous indices (Vicente–Serrano *et al.* 2010b).

DATASETS AND METHODS

Drought Indices

(a) *The Palmer Drought Indices*

The PDSI was a landmark in the development of drought indices. It enables measuring both wetness (positive value) and dryness (negative values), based on the supply and demand concepts of the water balance equation, and thus incorporates prior precipitation, moisture supply, runoff and evaporation demand at the surface level. Although the PDSI presents several deficiencies (Alley, 1984; Karl, 1986; Soulé, 1992; Akimremi *et al.* 1996; Weber and Nkemdirim, 1998; Vicente– Serrano *et al.* 2011), currently it is still one of the most widely used drought indices. The PDSI is calculated based on precipitation and temperature data, as well as the water content of the soil. All the basic terms of the water balance equation can be determined from those inputs, including evapotranspiration, soil recharge, runoff, and moisture loss from the surface layer. The complete calculation procedure of the PDSI can be consulted in many publications (e.g., Karl, 1983 and 1986; Alley, 1984).

The modified Palmer Drought Severity Index (WPLM) was proposed by the National Weather Service Climate Analysis Center for operational meteorological purposes (Heddinghaus and Sabol 1991), modifying the original rules of accumulation during wet and dry spells. The Palmer Hydrological Drought index (PHDI) was derived from the PDSI to quantify the long– term impact of drought on hydrological systems. Values of the PHDI tend to be negative for up to several months after PDSI have returned to normal levels, i.e. it usually returns to near–normal levels more gradually than the PDSI (Karl *et al.* 1987). Therefore, the PHDI is considered a measure of long–term hydrological drought since streamflows, reservoir storages and groundwater tend to stay below normal values for some time after a meteorological drought ends. Finally, the Palmer Z–Index is also derived from the Palmer model and it is much more responsive to short–term moisture deficiencies than the PDSI. The Palmer Z–Index shows how monthly moisture conditions depart from normal, and it is sensitive to unusual wet (and dry) months even in extended dry (or wet) spells. Therefore, the Palmer Z–index is usually used for the detection of short term droughts. One of the main problems of the Palmer indices is that the parameters necessary to calculate them were determined empirically and mainly tested in the USA, which restricts its use in other regions (see Akimremi *et al.* 1996) and limits the geographical comparisons based on the PDSI (Heim, 2002; Guttman *et al.* 1992). This problem was solved by developing of the self–calibrated Palmer indices (Wells *et al.* 2004), which are spatially comparable and report extreme wet and dry events at frequencies expected for rare conditions. Therefore, in this study we have used the self–calibrated versions of the four Palmer drought indices, which are more suitable for drought quantification and monitoring at a global scale than the corresponding Palmer indices.

(b) *The Standardized Precipitation Index (SPI)*
The Standardized Precipitation Index (SPI) was proposed by McKee *et al.* (1993) and it has been increasingly used during the two last decades because of its solid theoretical development, robustness and versatility in drought analyses (Redmond, 2002). The SPI is based on the conversion of the precipitation data to probabilities based on long–term precipitation records computed on different time scales. Probabilities are transformed to standardized series with an average of 0 and a standard deviation of 1. The main advantage of the SPI as compared with the Palmer indices is that the former allows analyzing drought impacts at different temporal scales while the latter does not (Edwards & McKee, 1997). Further, the SPI is able to identify different drought types since particular systems and regions can respond to drought conditions at very different time scales. In the case of water resources, the advantages of the SPI have been illustrated in several studies (Vicente–Serrano and López–Moreno, 2005; Szalai *et al.* 2000; Fiorillo and Guadagno, 2010; Lorenzo– Lacruz *et al.* 2010; Khan *et al.* 2008; Vicente–Serrano *et al.* 2011). In addition, several studies have also demonstrated variation in the

response of agricultural (Vicente–Serrano *et al.* 2006; Quiring and Ganesh, 2010) and ecological variables (Ji and Peters, 2003; Vicente–Serrano, 2007; Pasho *et al.* 2011) to different time scales of the SPI.

McKee *et al.* (1993) used the Gamma distribution to transform precipitation series to standardized units. Nevertheless, the frequency distributions of the precipitation series show significant changes that depended on the time scale (Vicente–Serrano, 2006). Among the different evaluated models, the Pearson III shows enhanced adaptability to precipitation series at different time scales (Guttman, 1999, Vicente–Serrano, 2006; Quiring, 2009). Therefore, here we use the algorithm described by Vicente–Serrano (2006) and López–Moreno and Vicente–Serrano (2008) to calculate 1– to 48–month SPI values based on the Pearson III distribution and the L–moments approach to obtain the distribution parameters.

(c) *The Standardized Precipitation Evapotranspiration Index (SPEI)*

The main criticism of the SPI is that its calculation is based only on precipitation data. The index does not consider other variables that can influence drought severity, since the SPI relies on two assumptions: i) the variability of precipitation is much higher than that of other variables, such as temperature and potential evapotranspiration (PET); and ii) the other variables are stationary (i.e. they have no temporal trend). The importance of variables other than precipitation is negligible in this framework, and droughts are assumed to be mainly controlled by the temporal variability of precipitation. Nevertheless, the role of warming–induced drought stress has been made evident in recent studies that analysed drought impacts on tree growth and mortality (e.g., Barber *et al.* 2000; Martínez–Villalta *et al.* 2008; Allen *et al.* 2010; Vicente–Serrano *et al.* 2010c; Carnicer *et al.* 2011; Camarero *et al.* 2011; Linares and Camarero, 2011) and on water resources (Cai and Cowan, 2008; Lespinas *et al.* 2010; Yulianti and Burn, 2006; Liang *et al.* 2010; Yang and Liu, 2011).

Therefore, the use of drought indices which include temperature data in their formulation, such as the PDSI, seems to be preferable than using indices without temperature information to identify warming–related drought impacts on different ecological, hydrological and agricultural systems. However, the PDSI lacks the multi–scalar character essential for assessing drought in relation to different hydrological systems, and also for differentiating among different drought types. The SPEI, based on precipitation and potential evapotranspiration, combines the sensitivity of PDSI to changes in evaporation demand, caused by temperature fluctuations and trends, with the simplicity of calculation and the multi–temporal nature of the SPI. The SPEI is based on a monthly climatic water balance (precipitation minus PET), which is adjusted using a 3–parameter log–logistic distribution. The values are accumulated at different time scales, following the same approach used in the SPI, and converted to standard deviations with respect to average values.

Datasets

The six drought indices here assessed (PDSI, PHDI, WPLM, Z–index, SPI and SPEI) were computed globally based on the CRU TS3.1 climate dataset (Mitchell and Jones, 2005; available online at http://badc.nerc.ac.uk/data/cru/), covering the period 1901–2009 at a spatial resolution of 0.5°. Given that the different hydrological, ecological and agricultural datasets used in this study contain temporal information available since 1948, and since the quality of the meteorological records in the CRU TS3.1 dataset is lower for the oldest records than for the most recent ones, we used only data for the period 1945–2009. Monthly precipitation and mean temperature were used to obtain the SPI and the SPEI at different time–scales. In addition, the different Palmer drought indices also required information on the water field capacity (Webb *et al.* 1993), which was obtained at a spatial resolution of 1° from http://daac.ornl.gov/SOILS/guides/Webb.html.

To determine the performance of the different drought indices to quantify the impact on the analyzed systems we used global data of four different variables with hydrological, agricultural and ecological implications. On the one hand, we used monthly streamflow data, recorded from 1945 to 004, in 925 gauges at the mouth of hydrological basins across the world (Dai *et al.* 2009). From the original dataset we selected 151 gauges in which a maximum of the 15% of the data gaps were filled. The drainage basins of each gauge were determined based on the GTOPO30 Digital Elevation Model (Figure 1a). Monthly streamflow records were used to obtain a streamflow drought index, the Standardized Streamflow Index (SSI) (Vicente–Serrano *et al.* 2012), which allows performing spatial and temporal comparison between streamflow data independently of the river regimes and streamflow magnitudes. The gauging data correspond in some cases with managed river and in other cases with unmanaged ones. This difference is not a problem for the analyses, and it is even interesting to assess how drought indices may be used and adapted to determine hydrological droughts both in managed and unmanaged basins.

Global soil moisture data was acquired from the International Soil Moisture Network (Robock *et al.* 2000; available online at http://www.ipf.tuwien.ac.at/ insitu/). Most of the series cover short periods or have data gaps, so we selected those series with a minimum of 10 years of data (Figure 1b). Some of the soil moisture stations provide daily or hourly data at different soil depths (commonly every 10 cm in depth from the top soil up to 1 or 1.5 m deep) whereas other stations provide monthly averages for the complete soil column from the top to 1 m deep. We homogenized all the existing information and converted the data to monthly averages of soil moisture for the soil column up to 1 m deep. Although the world soil moisture network uses different instruments and techniques (Dorigo *et al.* 2011) the measurements at the different sites are recorded in the same units (% of the water field capacity) and given that each sample was compared independently with the different drought indices, the techniques of soil moisture measurements did not affect the analyses. Most of the soil moisture stations do not provide soil moisture data for winter months as a consequence of soil freezing or soil saturation during this season. For this reason the analyses focused on the period from April to October, when data were available for all the stations. Concerning tree growth data, we compiled 1840 annual tree–ring width series or mean site chronologies encompassing the period 1945–2009 and archived by the National Climate Data Center (NCDC) in the International Tree–Ring Data Bank, ITRDB (Grissino–Mayer and Fritts, 1997; available online at: http://www.ncdc.noaa.gov/paleo/ treering.html) (Figure 1c). Each chronology represents the average annual radial growth series of several trees (typically more than ten) of the same species growing in the same site. The wood samples are taken following standard dendrocrhonological protocols which include sampling at least ten trees within a stand, taking usually two radial cores per tree at 1.3 m (Fritts, 1976; Cook *et al.* 1990). The selected sites corresponded to those chronologies listed in the ITRDB with at least ten trees sampled after 1940, which we regarded as an acceptable criterion for robust replication within each site. Raw tree–ring widths are detrended and standardized to remove long term biological growth trends, associated with tree ageing and increasing trunk diameter, and most of the first–order temporal autocorrelation, although this transformation preserves the inter–annual and inter–decadal variability.

Crop yield data of wheat cultivations was obtained from the Food and Agricultural Organization (FAO; available online at http://faostat.fao.org) for the period 1960–2009. Wheat crops were selected because they have a widespread distribution across the world and because it is mostly a non–irrigated crop, and hence it presents a higher vulnerability to drought than other crops such as rice or corn. Time series of annual crop productions in 173 countries were selected considering only those time series with a minimum of 15 years of records. Since the wheat productions show a large linear trend that is attributable to technological advances in cropping systems, the series were detrended assuming a linear model for each country series following Lobell *et al.* (2011).

Methods

The different drought indices were calculated using the monthly precipitation and mean temperature of the CRU TS 3.1 dataset. For the 151 basins we obtained the average precipitation and temperature for the entire basin from the same dataset. Therefore, we obtained one precipitation and one temperature series for each basin. In the case of the soil moisture and tree–ring width datasets we selected the 0.5° precipitation and mean temperature series that corresponded to the location of the sample. In the case of the national wheat crop data we calculated a weighted average series of monthly precipitation and mean temperature over each country using the percent area covered by wheat crops in that country as a weighting factor. The percent surface covered by wheat crops in each pixel was obtained from the Harvest Choice web site (http://harvestchoice.org) at a spatial resolution of 0.5° (see Figure 1d). The latitude necessary to obtain the SPEI and the Palmer indices, and the water field capacity used in the Palmer indices were also weighted for each country according to the percentage of surface cultivated by wheat.

Usually, the different hydrological, ecological and agricultural systems respond to different drought time scales due to the varied strategies of natural vegetation and crops to cope with water deficit (Chaves *et al.* 2003) or the different lithologic, land cover and/or water management regimes in the case of streamflow data (López–Moreno *et al.* 2012). **Therefore, the SPI and the SPEI were calculated at different time scales from 1 up to 48 months.** The multi–scalar character of these two correlations between SPI and SPEI were minor for most of the analysed months, but for the boreal summer months the correlations tended to be marginally higher for the SPEI than for the SPI.

Figure 4 shows the spatial distribution of correlations between the SSI series and four of the most widely used drought indices (SPI, SPEI, PDSI, Z–index) either considering continuous series (Figure 4a) or separately for January (Figure 4b) and July (Figure 4c) monthly series. Large differences existed between basins. In general, and independently of the drought index used, the strongest correlations between SSI and drought severity were found for the Atlantic basins of North America, the basins of central Europe and some basins of South America and Africa. On the contrary, poor correlations were found in the Asian basins, mainly those that drain to the Arctic Ocean. Nevertheless, in the latter basins, when monthly correlations were analysed separately, noticeable seasonal impacts were observed since correlations were much higher in July (Figure 4c) than in January (Figure 4b). In addition, in these zones it is clearly observed that differences between the SPI and SPEI correlations were important during the summer months, with the SPEI showing higher correlations than the SPI. The PDSI had higher correlations in the northern north hemisphere latitudes during the summer months than the SPI, although it did not outperform the SPEI. Figure 4d shows the drought index with the highest SSI–drought correlation for the annual continuous series and for the January and July series. The Palmer indices did not provide the best results with respect to streamflow data with a few exceptions. For the continuous SSI data the best correlation was found using SPEI in 44.4% of the basins, the 38.4% with the SPI and the remaining 17.3% with one of the four Palmer indices (Table 1). There were strong seasonal differences among areas since precipitation seems to be the main driver for the occurrence of streamflow droughts in the boreal winter when evapotranspiration rates are low, whereas in the boreal summer, when strong evapotranspiration rates are recorded in the Northern hemisphere, higher SSI–drought correlations were recorded when using the SPEI.

Soil Moisture

Figure 5 shows the box plots displaying the correlations between the different drought indices and the monthly soil moisture data obtained from April to October. Strong differences arise when comparing the SPI and the SPEI and the Palmer drought indices, with the first two indices outperforming the latter in all cases. It is interesting to note that correlations between soil moisture and drought indices were higher from July to October than for other months, being the former a period in which soils tend to be less saturated by water than in spring. The highest correlation between soil moisture and drought was found using the SPI or SPEI indices in a range of stations varying from 80% to 95% depending on the analyzed month, whereas in only 5 to 15% of the sites the highest correlation was found with the Palmer indices (Table 2). It was in the warmest months (July, August and September), in which evapo–transpiration rates are the highest, when a much higher percentage of sites showed higher correlations with the SPEI than with the SPI. Figure 6 shows the spatial distribution of correlations between the July soil moisture and the July series of SPI, SPEI, PDSI and Z–Index for the sites available in North America and also the drought index at which the maximum correlation is found. Higher correlations are found again with the SPI and the SPEI. **In addition, the SPEI shows the maximum correlation with soil moisture in most sites.**

Tree–Ring Width Series

Correlations between tree–ring width series and the drought indices are depicted in Figure 7. The median of the correlations oscillated between 0.44 for the SPI and 0.30 for the PHDI. The highest correlations were found during late spring and early summer, as it could be expected since most of the tree–ring series were located in the North hemisphere and most tree–ring growth occurs there during those seasons (Figure 8). There were very few differences in the magnitude of correlations between the SPI and the SPEI, but large differences were found between the Palmer drought indices. We show the spatial distribution of correlations for North America, in which the highest density of tree–ring width series was recorded. Figure 9a shows the maximum correlation found between the tree–ring width and several indices (SPI, SPEI, PDSI, Z–index). Higher growth–drought correlations were found in the central and southwestern areas of USA than elsewhere, being the former arid areas in which tree growth is highly driven by water availability. In humid sites of the East, North and North–West USA, where tree–ring growth is less constrained by drought, we obtain lower growth–drought correlations than elsewhere, independently of the selected drought index. Nevertheless, although the spatial pattern was quite similar considering the four drought indices, the magnitude of the correlations differed noticeably. In the areas with the highest growth–drought correlations of central and southwest USA, higher correlation values were found for the SPI and SPEI than for the Palmer drought indices. Figures 9b and 9c show correlations between annual tree growth and the series of the drought indices in January and July, respectively. Higher correlation coefficients were found in July than in January since higher growth activity is recorded in summer than in winter months. Again, higher growth–drought correlation values were also found for the SPI and the SPEI than for the PDSI and Z indices. With a few exceptions, the highest correlations in the different forests corresponded to the SPI or the SPEI (Figure 9d). **The SPEI showed higher correlation values than the other drought indices in almost 50% of all analyzed sites (Table 3).** The SPI showed the highest correlation in 37.9% of the forests. Only in the 13.7% of the forests the highest correlation corresponded to Palmer indices. Similar results were found at a monthly basis.

Wheat Crop Yields

A summary of the relationship between the global wheat yields and the six different drought indices is illustrated in the Figure 10, which records the maximum correlation between the annual wheat yields and the drought indices independently of the month of the year in which the highest correlation was found. This approach minimizes the impact of the different crop cycles and harvest dates in the different parts of the world. Stronger yield–drought correlations were obtained for the SPI and the SPEI than for the Palmer drought indices. However, important differences were found between the Palmer indices since the Z–index provided much better results than the other three indices. The median yield–drought correlation for the SPI was 0.33, for the SPEI it was 0.37 and for the Z–index it was 0.29. Figure 11 shows the maximum correlation between the annual wheat yields and the evaluated drought indices. Large differences in the influence of drought conditions on wheat crop productions are evident across the world. Thus, the highest worldwide correlations were found in those countries in which the surface cultivated by wheat corresponds to semi–arid lands, which is the case of Russia, Kazakhstan, Australia, Morocco or Spain, among others (Figure 1d), in which correlations were higher than 0.5. In other regions of the world, the prevailing humid conditions or the irrigation may reduce the vulnerability of wheat crops to drought.

Nevertheless, independently of the existing spatial differences, we found that the yield–drought correlations tended to be higher for the SPI and the SPEI than for the PDSI and the Z–index with very few exceptions such as Australia, India or Angola. In any case, when the countries were classified according to the drought index showing the highest yield–drought correlation, **we found that the wheat yields of most of the analyzed countries of the world were best correlated with the SPEI (49.5%) or with the SPI (34.3%).** The percentage of countries in which the highest correlation was found with one of the different Palmer indices was quite low (2.9% for the PDSI, 5.7% for the PHDI, 2.9% for the Z–index and 4.8% for the WPLM). Excepting Australia and Ethiopia, which showed the highest yield–drought correlation when considering the WPLM index, the national wheat yields tended to be more closely correlated to the SPEI than to the other drought indices.

DISCUSSION AND CONCLUSIONS

This study has provided the first global assessment of different indices to detect drought impacts on hydrological, ecological and agricultural systems. We must highlight the difficulty of developing this kind of studies based on empirical information given the existing methodological problems to quantify damages caused by water shortage on different systems that can be related to the severity of droughts. In addition, the global character of the study introduces other point of complexity given the varied sources of information and the need of an interdisciplinary approach.

We have used the two most widely drought indices worldwide. On the one hand, the Palmer drought indices that are currently implemented in drought monitoring systems, and the Standardized Precipitation Index (SPI), accepted by the World Meteorological Organization as the reference drought index for more effective drought monitoring and climate risk management (Hayes *et al.*, 2011). In addition, we also included the Standardized Precipitation Evapotranspiration Index (SPEI), which is similar to the SPI but considers the influence of potential evapotranspiration on drought severity, in our analyses.

Independently of the hydrological, agricultural or ecological system analyzed we have found a higher capability of the drought indices that are calculated on different time scales, i.e. the SPEI and the SPI, to correlate with the temporal variability of the different variables. The Palmer indices, which lack the flexibility of reflecting the intrinsic multi–scalar nature of droughts, performed systematically worse than the SPI and SPEI.

The response of a specific system to drought can be very complex, and according to the analyzed system and its spatial location it may have large differences in the cumulative period of water deficit required causing negative impacts on the considered system (Vicente–Serrano *et al.* 2011). Different studies showed that particular systems and regions respond to drought conditions at different time scales, including hydrological (e.g., Szalai *et al.* 2000; Vicente–Serrano and López–Moreno, 2005; Khan *et al.* 2008; Fiorillo and Guadagno, 2010; López–Moreno *et al.* 2012), agricultural (Quiring and Ganesh, 2010) and ecological variables (Ji and Peters, 2003; Vicente– Serrano, 2007; Pasho *et al.* 2011). Thus, it is commonly accepted that dry conditions occur only during part of the hydrological cycle and so it is not usual to find simultaneous water deficits in soil moisture, streamflows, reservoir storages and groundwater. The problem is even more complex when diverse hydrological, agricultural, environmental and socioeconomic systems affected by droughts are considered, since the response times to water deficits and the resistance or resilience (ability to recover after the drought) of each system to drought can vary substantially. Therefore, although different Palmer indices representing various time scales of drought have been included in this analysis (Karl, 1986), they are not sufficiently flexible to quantify the strong variability in the response to droughts that can be found across a particular region. This is the rationale behind the results obtained in this article, which demonstrate that multi–scalar indices such as the SPI or the SPEI outperform other indices and allow adapting a wide range of drought vulnerabilities. The magnitudes of the correlations between various hydrological, agricultural and ecological variables and the compared drought indices clearly show that the SPI and the SPEI are more capable to monitor drought conditions in different systems. Thus, the highest correlation between the response variable and the drought index was found from 70% up to 95% of the cases for the SPI or the SPEI indices, depending of the variable and the season of the year, whereas the Palmer drought indices commonly represented less than the 15% of the highest correlations.

However, this finding does not mean that the Palmer indices are not useful for some purposes. For example, Dai *et al.* (2004) and Dai (2011) showed good correlations between the PDSI and annual streamflows and soil moisture worldwide. When monthly temporal scales are used,, the capability of the Palmer indices diminishes. Further, several studies have also found significant correlations of streamflow data (Alley, 1985; Smith and Richman, 1993; Tang and Piechota, 2009; Zhai *et al.* 2010), tree–ring width series (Meko *et al.* 1993; Orwing and Abrams, 1997; Piovesan *et al.* 2008) and crop yields (Akinremi *et al.* 1996; Quiring and Papakryiakou, 2003; Scian, 2004; Mavromatis, 2007) with monthly Palmer indices (commonly the PDSI). Probably, in all these systems stronger correlations would have been found considering different time scales of the SPI or the SPEI, but we must also note that globally, in some of the analyzed sites, the best response between the temporal variability of the different variables is found with one of the four Palmer drought indices. This highlights the necessity of testing and comparing the local performance of different drought indices to select the most appropriate one according to the variable of interest.

There are small differences in the performance of the SPI and the SPEI for capturing the variability of the studied systems since the magnitude of the correlations is similar between the two indices in many of the analyzed variables. This result could suggest the better use of the SPI regarding the SPEI since SPI has less data requirements. Nevertheless, some differences found between both indices must be emphasized, which suggests a better performance of the SPEI as compared with the SPI :

i.   Independently of the variable of interest the SPEI renders higher correlations than the SPI. The SPEI recorded the highest percentage of cases showing the maximum variable–drought index correlations in all the analyzed variables. **The difference in the percentage of maximum correlations between SPI and SPEI is about 10% higher for the SPEI than for the SPI in the different analyzed systems.**

ii. The differences between the magnitude of correlations found for the SPI and the SPEI tend to be higher in the boreal summer, which represents the season in which soil moisture samples and forests are affected by drought stress in most of the analyzed sites, since most of them are located in the Northern hemisphere. Water demand by the atmosphere is higher in summer months than in other seasons due to higher incoming radiation and temperature. For this reason, in the season in which more drought–related impacts are recorded (water supply restrictions, decreased soil moisture, reduced tree growth, forest fires, etc.) and in which drought monitoring is more critical, the SPEI outperforms the SPI being the former index able to identify drought impacts better than the latter one.

These results clearly demonstrate that, although precipitation is the main driver of drought severity, the influence of the atmospheric evaporative demand cannot be neglected, mainly in the context of current global warming. Empirical studies have shown that temperature rise affects the severity of droughts. For example, Abramopoulos *et al.* (1988) used a general circulation model experiment to show that evaporation and transpiration can consume up to 80% of rainfall. The strong role of temperature as a major driver of drought severity was evident in the devastating 2003 central European heat wave, which drastically reduced tree growth and the Aboveground Net Primary Production (ANPP) across most of the continent (Ciais *et al.* 2005). **Thus, observational and empirical studies have demonstrated that higher temperature increases drought stress and enhances forest mortality under water shortage** (Adams *et al.* 2009; Allen *et al.* 2010). Warming processes are also involved in triggering the decline in world agricultural productions observed in the last years (Lobell *et al.* 2011). Zhao and Running (2010) have recently shown at a global scale that between 2000 and 2009 the annual ANPP decreased because of the combined effects of severe drought stress and high temperatures which induced high autotrophic respiration levels, indicating that ANPP decreases because of warming–associated drying trends.

Therefore, given the observed impacts of global warming processes on water availability and on related agricultural, ecological and hydrological systems, the expected future rise of temperatures (Solomon *et al.* 2007), and the results obtained in this study based on the objective comparison of different drought indices, **it seems reasonable to recommend the use of the SPEI if a priori we do not know the possible response to drought of the variable of interest.** Studies comparing the performance of several drought indices, like those evaluated here, would be preferable to determine the best drought index for identifying a certain drought type and its impacts on different systems. Nevertheless, this is sometimes expensive, time consuming and commonly there are not quantitative information and long time series of the variable of interest available to establish the comparisons. **Therefore, the low data requirements of the SPEI, the facility and flexibility of its calculation, and the consideration of the two main elements that determine drought severity, namely precipitation and atmospheric evaporative demand, are solid reasons to recommend its use over other drought indices.**

Drought indices is their major advantage as compared with other existing indices. Since the times of response to drought of the different systems is not known a priori, the Pearson correlation coefficients (r) between the time series of these variables and the 1– to 48–month SPI and SPEI series were computed, and the time–scale at which the strongest correlation was found was kept for further analyses. The different Palmer indices were also correlated with the time–series of SSI, tree– ring width, wheat yields and soil moisture. The monthly series of the different drought indices were detrended before calculating the Pearson coefficients between the drought indices and the annual tree–ring widths and wheat yields since the latter two series have been previously detrended to remove the respective effects of the tree ageing and technological advances on these variables.

RESULTS

Streamflow Data

Figure 2 shows a box plot illustrating the correlations obtained between the SSI series at 151 worldwide basins and the six assessed drought indices. Correlations were obtained for the continuous series between 1945 and 2004, independently of the month of the year, since the standardized character of the SSI allowed directly comparing with the drought indices at a monthly basis. In general, correlations tended to be higher for the SPI and the SPEI indices than for the Palmer ones (PDSI, PHDI, Z–index and WPLM). The median correlation coefficients for SPI and SPEI were 0.57 and 0.58, respectively, whereas for the PDSI it was 0.45, 0.39 for the PHDI, 0.42 for the Z–index, and 0.46 for the WPLM. **This shows that the SPI and SPEI tended to record better the occurrence of streamflow droughts than the Palmer indices.** Figure 3 shows the same analyses at a monthly basis, since streamflow response to climatic droughts may be very different as a function of the river regimes. Higher correlations were found again for the SPI and SPEI than for other indices, irrespective of the month. It is interesting to note that differences in the magnitude of indices. In addition, we must also stress that the SPEI formulation used in this study was based on PET estimates obtained by means of the Thornthwaite equation, which only requires data of mean temperature and has some deficiencies to obtain reliable estimates of the variable (Donohue *et al.* 2010). **Future improvements of the SPEI, including more reliable PET estimates based on the Hargreaves or Penman–Monteith equations, could reflect better the role played by PET on drought severity, and make the SPEI even more suitable to identify drought–related impacts across systems.**

ACKNOWLEDGEMENTS

We would like to thank the Climate Research Unit of the University of East Anglia (UK) for providing the gridded precipitation and temperature data used in this study, Dr. Aiguo Dai of the National Center for Atmospheric Research (NCAR) for providing the global streamflow data, the International Soil Moisture Network for providing soil moisture dataset, Contributors of the International Tree–Ring Data Bank, IGBP PAGES/World Data Center for Paleoclimatology, NOAA/NCDC Paleoclimatology Program (Boulder, Colorado, USA) for providing the tree–ring data and the Food and Agricultural organization (FAO) for providing the wheat productions. This work has been supported by the research projects CGL2008–01189/BTE, CGL2011–27574–CO2–02, CGL2011–24185, CGL2011–26654 and CGL2011–27536 financed by the Spanish Commission of Science and Technology and FEDER, EUROGEOSS (FP7–ENV–2008–1–226487) and ACQWA (FP7–ENV–2007–1– 212250) financed by the VII Framework Programme of the European Commission, "Efecto de los escenarios de cambio climático sobre la hidrología superficial y la gestión de embalses del Pirineo

Aragonés" financed by "Obra Social La Caixa" and the Aragón Government and "Influencia del cambio climático en el turismo de nieve" (CTTP01/10) financed by the Comisión de Trabajo de los Pirineos. JJC acknowledges the support of ARAID.

REFERENCES

Abramopoulos, F., C. Rosenzweig, and B. Choudhury, (1988): Improved ground hydrology calculations for global climate models (GCMs): Soil water movement and evapotranspiration. Journal of Climate, 1, 921–941.

Adams, H.D. *et al.* (2009): Temperature sensitivity of drought–induced tree mortality portends increased regional die–off under global–change–type drought. Proceedings of the National Academy of Sciences of the United States of America, 106: 7063–7066.

Allen, C.D., A.K. Macalady, H. Chenchouni, D. Bachelet, N. McDowell, M. Vennetier, T. Kizberger, A. Rigling, D.D. Breshears, E.H. Hogg, P. Gonzalez, R. Fensham, Z. Zhang, J. Castro, N. Demidova, J.H. Lim, G. Allard, S.W. Running, A. Semerci, and N. Cobb. (2010): A global overview of drought and heat–induced tree mortality reveals emerging climate change risks for forests. Forest Ecology and Management, 259, 660–684.

Alley, W.M., (1984): The Palmer drought severity index: limitations and applications. Journal of Applied Meteorology, 23, 1100–1109.

Alley, W.M., (1985): Palmer Drought Severity Index as a measure of hydrologica drought. Water Resources Bulletin, 21: 105–114

Anderson, M.C., Hain, C., Wardlow, B., Pimstein, A., Mecikalski, J.R., Kustas, W.P., (2011): Evaluation of drought indices based on Thermal remote sensing of evapotranspiration over the continental United States. Journal of Climate, 24: 2025–2044.

Akinremi, O.O., Mcginn, S.M., Barr, A.G., (1996): Evaluation of the Palmer Drought index on the Canadian prairies. Journal of Climate, 9: 897–905.

Barber, V.A., Juday, G.P., Finney, B.P., (2000): Reduced growth of Alaskan white spruce in the twentieth century from temperature–induced drought stress. Nature 405: 668–673.

Barua, S., Ng, A.W.M., Perera, B.J.C., (2011): Comparative Evaluation of Drought Indexes: Case Study on the Yarra River Catchment in Australia. Journal of Water Resources Planning and Management 137: 215–226.

Burton, I., R.W. Kates, and G.F. White, (1978): The environment as hazard. Oxford University Press. Nueva York, 240 pp.

Cai, W. and Cowan, T., (2008): Evidence of impacts from rising temperature on inflows to the Murray–Darling Basin. Geophysical Research Letters 35, art. no. L07701.

Camarero, J.J., Bigler, C., Linares, J.C. and Gil–Pelegrín, E. (2011): Synergistic effects of past historical logging and drought on the decline of Pyrenean silver fir forests. Forest Ecology and Management, 262, 759–769.

Carnicer, J., Coll, M., Ninyerola, M., Pons, X., Sánchez, G., Peñuelas, J., (2011): Widespread crown condition decline, food web disruption, and amplified tree mortality with increased climate change–type drought. Proceedings of the National Academy of Sciences of the United States of America, 108: 1474–1478.

Chaves, M.M., Maroco, J.P. and Pereira, J.S. (2003): Understanding plant responses to drought – from genes to the whole plant. Functional Plant Biology 30, 239–264.

Ciais, Ph. *et al.* (2005). Europe–wide reduction in primary productivity caused by the heat and drought in 2003. Nature 437, 529–533.

Cook, E.R., and Kairiukstis, L.A. (1990): Methods of Dendrochronology. Kluwer Academic Publishers, Dordrecht, The Netherlands.

Copenheaver, C.A., Crawford, C.J. and Fearer, T.M., (2011): Age–specific responses to climate identified in the growth of Quercus alba. Trees – Structure and Function 25: 647–653.

Dai, A., Trenberth, K.E., Qian, T. (2004): A global dataset of Palmer Drought Severity Index for 1870–2002: Relationship with soil moisture and effects of surface warming. Journal of Hydrometeorology, 5: 1117–1130.

Dai, A., T. Qian, K. E. Trenberth, and J. D Milliman, (2009): Changes in continental freshwater discharge from 1948–2004. J. Climate, 22: 2773–2791.

Dai, A. (2011), Characteristics and trends in various forms of the Palmer Drought Severity Index during 1900–2008, J. Geophys. Res., 116, D12115, doi:10.1029/2010JD015541.

Donohue, R.J., McVicar, T.R., Roderick, M.L., (2010): Assessing the ability of potential evaporation formulations to capture the dynamics in evaporative demand within a changing climate. Journal of Hydrology, 386: 186–197.

Dorigo, W. A., *et al.* (2011). The International Soil Moisture Network: a data hosting facility for global in situ soil moisture measurements, Hydrol. Earth Syst. Sci., 15: 1675–1698.

Drobyshev, I., Niklasson, M., Linderholm, H.W. (2012): Agricultural forest fire activity in Sweden: Climatic controls and geographical patterns in 20th century. Agricultural and Forest Meteorology, 154–155: 174–186

Edwards, D.C. and McKee, T.B., (1997): Characteristics of 20th century drought in the United States at multiple time scales. Atmospheric Science Paper No. 634.

Fiorillo, F., Guadagno, F.M. (2010): Karst spring discharges analysis in relation to drought periods, using the SPI Water Resources Management, 24: 1867–1884.

Fritts, H.C. (1976): Tree Rings and Climate. Blackburn Press, Caldwell, New Jersey.

Grissino–Mayer H.D., Fritts, H.C., (1997): The International Tree–Ring Data Bank: an enhanced global database serving the global scientific community. The Holocene, 7: 235–238.

Guttman, N.B., Wallis, J.R. and Hosking, J.R.M., (1992): Spatial comparability of the Palmer Drought Severity Index. Water Resources Bulletin. 28: 1111–1119.

Guttman, N.B., (1998): Comparing the Palmer drought index and the Standardized Precipitation Index. Journal of the American Water Resources Association, 34, 113–121.

Guttman, N.B., (1999): Accepting the standardized precipitation index: a calculation algorithm. Journal of the American Water Resources Association. 35: 311–322.

Hannaford, J., Lloyd–Hughes, B., Keef, C., Parry, S., Prudhomme, C. (2011): Examining the large– scale spatial coherence of European drought using regional indicators of precipitation and streamflow deficit Hydrological Processes 25: 1146–1162.

Hayes, M., M. Svoboda, N. Wall, and M. Widhalm (2011), The Lincoln Declaration on Drought Indices: Universal Meteorological Drought Index recommended, Bull. Am. Meteorol. Soc., 92, 485–488.

Heddinghaus, T. R. and P. Sabol, (1991). A review of the Palmer Drought Severity Index and where do we go from here? In: Proc. 7th Conf. on Applied Climatology, September 10–13,1991. American Meteorological Society, Boston, pp. 242–246.

Heim, R.R., (2002): A review of twentieth–century drought indices used in the United States. Bulletin of the American Meteorological Society, 83, 1149–1165.

Ji, L. and A.J. Peters, (2003): Assessing vegetation response to drought in the northern Great Plains using vegetation and drought indices. Remote Sensing of Environment, 87, 85–98.

Karl, T.R., (1986): The sensitivity of the Palmer Drought Severity Index and the Palmer z–Index to their calibration coefficients including potential evapo–transpiration. Journal of Climate and Applied Meteorology, 25, 77–86.

Karl, T.R., (1986): The sensitivity of the Palmer Drought Severity Index and the Palmer z–Index to their calibration coefficients including potential evapo–transpiration. Journal of Climate and Applied Meteorology, 25: 77–86.

Karl. T.R., Quinlan, F. and Ezell, D.D., (1987): Drought termination and amelioriation: its climatological probability. Journal of Climate and Applied Meteorology. 26: 1198–1209.

Kempes, C.P., O.B. Myers, D.D. Breshears, and J.J. Ebersole, (2008): Comparing response of Pinus edulis tree–ring growth to five alternate moisture indices using historic meteorological data. Journal of Arid Environments, 72, 350–357.

Keyantash, J. and J. Dracup., (2002): The quantification of drought: an evaluation of drought indices. Bulletin of the American Meteorological Society, 83, 1167–1180.

Khan, S., H.F. Gabriel, and T. Rana, (2008): Standard precipitation index to track drought and assess impact of rainfall on watertables in irrigation areas. Irrigation and Drainage Systems, 22, 159–177.

Lespinas, F., Ludwig, W. and Heussner, S., (2010): Impact of recent climate change on the hydrology of coastal mediterranean rivers in Southern France. Climatic Change, 99: 425–456.

Liang, S., Ge, S., Wan, L. and Zhang, J., (2010): Can climate change cause the Yellow River to dry up?. Water Resources Research, 46, W02505, doi:10.1029/2009WR007971.

Linares, J. C. and Camarero, J. J. (2011). From pattern to process: linking intrinsic water–use efficiency to drought–induced forest decline. Global Change Biology. doi: 10.1111/j.1365–2486.2011.02566.

Littell, J.S., Mckenzie, D., Peterson, D.L., Westerling, A.L. (2009): Climate and wildfire area burned in western U.S. ecoprovinces, 1916–2003 Ecological Applications 19: 1003–1021.

Lobell, D. A., Schlenker, W., Costa–Roberts. (2011): Trends and global crop production since 1980. Science, doi:10.1126/science.1204531

López–Moreno, J.I. and Vicente–Serrano, S.M., (2008) Extreme phases of the wintertime North Atlantic Oscillation and drought occurrence over Europe: a multi–temporal–scale approach. Journal of Climate 21, 1220–1243.

López–Moreno, J.I., S.M., Vicente–Serrano, J. Zabalza, S. Beguería, J. Lorenzo–Lacruz, C. Azorin– Molina, E. Morán–Tejeda. (2012) Hydrological response to climate variability at different time scales: a study in the Ebro basin. Journal of Hydrology, Under review.

Lorenzo–Lacruz, J., Vicente–Serrano, S.M., López–Moreno, J.I., Beguería, S., García–Ruiz, J.M., Cuadrat, J.M. (2010) The impact of droughts and water management on various hydrological systems in the headwaters of the Tagus River (central Spain). Journal of Hydrology, 386: 13–26.

Lotsch, A., Friedl, M.A., Anderson, B.T., Tucker, C.J. (2003): Coupled vegetation–precipitation variability observed from satellite and climate records. Geophysical Research Letters 30: CLM 8–1 – 8–4.

Martínez–Villalta, J., López, B.C., Adell, N., Badiella, L., Ninyerola, M., (2008): Twentieth century increase of Scots pine radial growth in NE Spain shows strong climate interactions. Global Change Biology 14, 2868–2881.

Mavromatis, T., (2007): Drought index evaluation for assessing future wheat production in Greece. International Journal of Climatology, 27, 911–924.

Meko, D., Cook, E.R., Stahle, D.W., Stockton, C.W., Hughes, M.K. (1993): Spatial patterns of tree– growth anomalies in the United States and southeastern Canada. Journal of Climate 6: 1773–1786.

McAuliffe, J.R., Hamerlynck, E.P. (2010): Perennial plant mortality in the Sonoran and Mojave deserts in response to severe, multi–year drought Journal of Arid Environments 74: 885–896.

McKee, T.B.N., J. Doesken, and J. Kleist, (1993): The relationship of drought frecuency and duration to time scales. Eight Conf. On Applied Climatology. Anaheim, CA, Amer. Meteor. Soc. 179–184.

Mishra, A. K., and V. P. Singh (2010), A review of drought concepts, J. Hydrol., 391, 202–216. Mitchell, T. D., and P. D. Jones (2005), An improved method of constructing a database of monthly climate observations and associated high resolution grids, Int. J. Climatol., 25, 693–712. Orwig, D.A., Abrams, M.D. (1997): Variation in radial growth responses to drought among species, site, and canopy strata Trees – Structure and Function 11: 474–484.

Palmer, W.C., (1965): Meteorological droughts. U.S. Department of Commerce Weather Bureau Research Paper 45, 58 pp.

Pasho, E., J. Julio Camarero, Martín de Luis and Vicente–Serrano, S.M. (2011) Impacts of drought at different time scales on forest growth across a wide climatic gradient in north–eastern Spain. Agricultural and Forest Meteorology. 151: 1800–1811.

Paulo, A.A., Pereira, L.S. (2006): Drought concepts and characterization: Comparing drought indices applied at local and regional scales Water International 31: 37–49.

Piovesan, G., Biondi, F., Di Filippo, A., Alessandrini, A., Maugeri, M. (2008): Drought–driven growth reduction in old beech (Fagus sylvatica L.) forests of the central Apennines, Italy Global Change Biology 14: 1265–1281.

Potop, V. (2011): Evolution of drought severity and its impact on corn in the Republic of Moldova Theoretical and Applied Climatology 105: 469–483.

Quiring, S.M., Papakryiakou, T.N. (2003): An evaluation of agricultural drought indices for the Canadian prairies Agricultural and Forest Meteorology 118: 49–62.

Quiring, S.M. (2009): Developing objective operational definitions for monitoring drought. Journal of Applied Meteorology and Climatology 48: 1217–1229.

Quiring, S.M., Ganesh, S. (2010): Evaluating the utility of the Vegetation Condition Index (VCI) for monitoring meteorological drought in Texas Agricultural and Forest Meteorology 150:330–339.

Redmond, K.T., (2002): The depiction of drought. Bulletin of the American Meteorological Society, 83, 1143–1147.

Robock, A., Vinnikov, K.Y., Srinivasan, G., Entin, J.K., Hollinger, S.E., Speranskaya, N.A., Liu, S., Namkhai, A. (2000): The Global Soil Moisture Data Bank Bulletin of the American Meteorological Society 81: 1281–1299.

Scian, B., Don, M. (1997): Retrospective analysis of the palmer drought severrity index in the semi– arid Pampas Region, Argentina International Journal of Climatology 17:. 313–322.

Shukla, S., Steinemann, A.C., Lettenmaier, D.P. (2011): Drought monitoring for Washington State: Indicators and applications Journal of Hydrometeorology 12: 66–83.

Sims, A.P., D.S. Nigoyi, and S. Raman, (2002): Adopting indices for estimating soil moisture: A North Carolina case study. Geophysical Research Letters, 29, 1183, doi:10.1029/2001GL013343.

Sivakumar, M. V. K., R. P. Motha, D. A. Wilhite, and D. A. Wood (Eds.) (2010): Agricultural Drought Indices: Proceedings of an Expert Meeting, 2–4 June 2010, Murcia, Spain, 219 pp., World Meteorol. Org., Geneva, Switzerland.

Smith, K., Richman, M.B. (1993): Recent hydroclimatic fluctuations and their effects on water resources in Illinois Climatic Change 24: 249–269.

Solomon, S. et al. (2007) Climate Change 2007: The Physical Science Basis. Cambridge University Press, Cambridge, United Kingdom and New York, NY, USA.

Soulé, P.T., (1992): Spatial patterns of drought frequency and duration in the contiguous USA based on multiple drought event definitions. International Journal of Climatology, 12, 11–24.

Svoboda, M., LeCompte, D., Hayes, M., Heim, R., Gleason, K., Angel, J., Rippey, B., Tinker, R., Palecki, M., Stooksbury, D., Miskus, D. and Stephens, S., (2002): The drought monitor. Bulletin of the American Meteorological Society. 83: 1181–1190.

Steinemann, A. (2003): Drought indicators and triggers: A stochastic approach to evaluation Journal of the American Water Resources Association 39: 1217–1233.

Szalai, S., C.S. Szinell, and J. Zoboki, (2000): Drought monitoring in Hungary. In Early warning systems for drought preparedness and drought management. World Meteorological Organization. Lisboa: 182–199.

Tang, C., Piechota, T.C. (2009): Spatial and temporal soil moisture and drought variability in the Upper Colorado River Basin Journal of Hydrology 379: 122–135.

Vasiliades, L., Loukas, A., Liberis, N. (2011): A Water Balance Derived Drought Index for Pinios River Basin, Greece Water Resources Management 25: 1087–1101.

Vergni, L., Todisco, F. (2011): Spatio–temporal variability of precipitation, temperature and agricultural drought indices in Central Italy. Agricultural and Forest Meteorology 151: 301–313.

Vicente–Serrano, S.M. and López–Moreno, J.I., (2005), Hydrological response to different time scales of climatological drought: an evaluation of the standardized precipitation index in a mountainous Mediterranean basin. Hydrology and Earth System Sciences 9: 523–533.

Vicente–Serrano, S.M., (2006), Differences in spatial patterns of drought on different time scales: an analysis of the Iberian Peninsula. Water Resources Management 20: 37–60.

Vicente–Serrano, S.M., Cuadrat, J.M. and Romo, A., (2006), Early prediction of crop productions using drought indices at different time scales and remote sensing data: application in the Ebro valley (North–east Spain). International Journal of Remote Sensing 27: 511–518.

Vicente–Serrano, S.M. (2007), Evaluating The Impact Of Drought Using Remote Sensing In A Mediterranean, Semi–Arid Region, Natural Hazards, 40: 173–208.

Vicente–Serrano, S.M., Beguería, S., López–Moreno, J.I., Angulo, M., El Kenawy, A. (2010a): A new global 0.5° gridded dataset (1901–2006) of a multiscalar drought index: comparison with current drought index datasets based on the Palmer Drought Severity Index. Journal of Hydrometeorology. 11: 1033–1043.

Vicente–Serrano S.M., Santiago Beguería, Juan I. López–Moreno, (2010b) A Multi–scalar drought index sensitive to global warming: The Standardized Precipitation Evapotranspiration Index – SPEI. Journal of Climate 23: 1696–1718.

Vicente–Serrano, S.M., Lasanta, T., Gracia, C., (2010c): Aridification determines changes in leaf activity in Pinus halepensis forests under semiarid Mediterranean climate conditions. Agricultural and Forest Meteorology 150, 614–628.

Vicente–Serrano, S.M., Beguería, S. and Juan I. López–Moreno (2011). Comment on "Characteristics and trends in various forms of the Palmer Drought Severity Index (PDSI) during 1900–2008" by A. Dai. Journal of Geophysical Research–Atmosphere. 116, D19112, doi:10.1029/2011JD016410.

Vicente–Serrano, S.M., Juan I. López–Moreno, Santiago Beguería, Jorge Lorenzo–Lacruz, Cesar Azorin–Molina and Enrique Morán–Tejeda, (2012): Accurate computation of a streamflow drought index. Journal of Hydrologic Engineering doi:10.1061/(ASCE)HE.1943–5584.0000433

Webb, R.S., C.E. Rosenzweig, and E.R. Levine, (1993): Specifying land surface characteristics in general circulation models: soil profile data set and derived water–holding capacities. Global Biogeochemical Cycles, 7, 97–108.

Weber, L., and L.C. Nkemdirim, (1998): The Palmer drought severity index revisited. Geografiska Annaler, 80A, 153–172.

Wells, N., S. Goddard, and M.J. Hayes, (2004): A self–calibrating Palmer Drought Severity Index. Journal of Climate, 17, 2335–2351.

Wilhite, D.A., (1993): Drought assessment, management and planning: Theory and case studies. Kluwer. Boston.

Wilhite, D.A., (1996): A methodology for drought preparedness. Natural Hazards. 13: 229–252. Wilhite, D.A., (2000): Drought as a natural hazard: concepts and definitions. In Drought: a global assessment. (D. Wilhite ed.). Vol 1: 3–18.

Wilhite, D. A., M. D. Svoboda, and M. J. Hayes (2007): Understanding the complex impacts of drought: A key to enhancing drought mitigation and preparedness, Water Resour. Manage., 21, 763–774, doi:10.1007/ s11269–006–9076–5.

Yang, Z and Liu, Q., (2011): Response of streamflow to climate changes in the yellow river basin, China. Journal of Climate, 12: 1113–1126.

Yulianti, J.S. y Burn, D.H., (2006): Investigating links between climatic warming and low streamflow in the Prairies region of Canada. Geophysical Research Letters, 33: L20403.

Zhai, J., Su, B., Krysanova, V., Vetter, T., Gao, C., Jiang, T. (2010): Spatial variation and trends in PDSI and SPI indices and their relation to streamflow in 10 large regions of china Journal of Climate 23: 649–663.

Zhao, M. and Running, S.W. (2009): Drought–induced reduction in global terrestrial net primary production from 2000 through 2009. Science 329, 940–943.

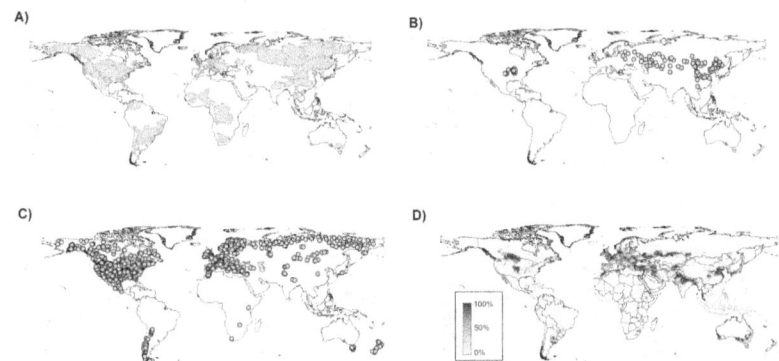

Figure 6.1 Response variables used in this study: A) Basins with monthly stream flow data for the period 1945–2004, B) soil moisture sample series with at least 10 years with data for the period 1950–2009, C) tree–ring width series with at least 25 years of data for the period 1945–2009, D) countries with series of wheat productions with at least 10 years of data for the period 1960–2009 (black outline). In D) the colors represent the percentage of lands cultivated by wheat, which was used as a weight to obtain the precipitation and temperature series, the water field capacity and the latitude used to obtain the time series of drought indices for each country.

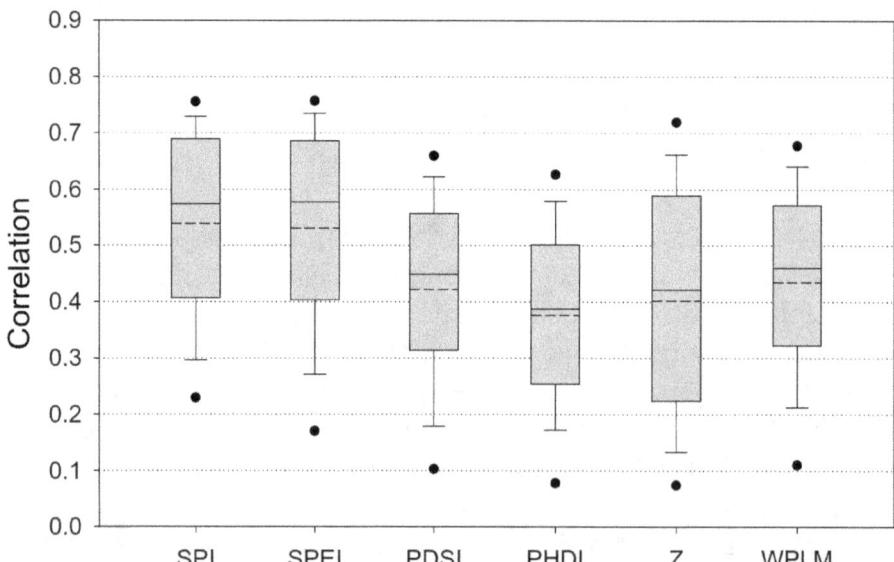

Figure 6.2 Box plots (the solid and dashed lines within the boxes correspond to the median and mean values, respectively) showing the correlations (Pearson coefficient) between the continuous series of the Standardized Streamflow Index (SSI) in 151 basins across the world and the six drought indices compared in this study.

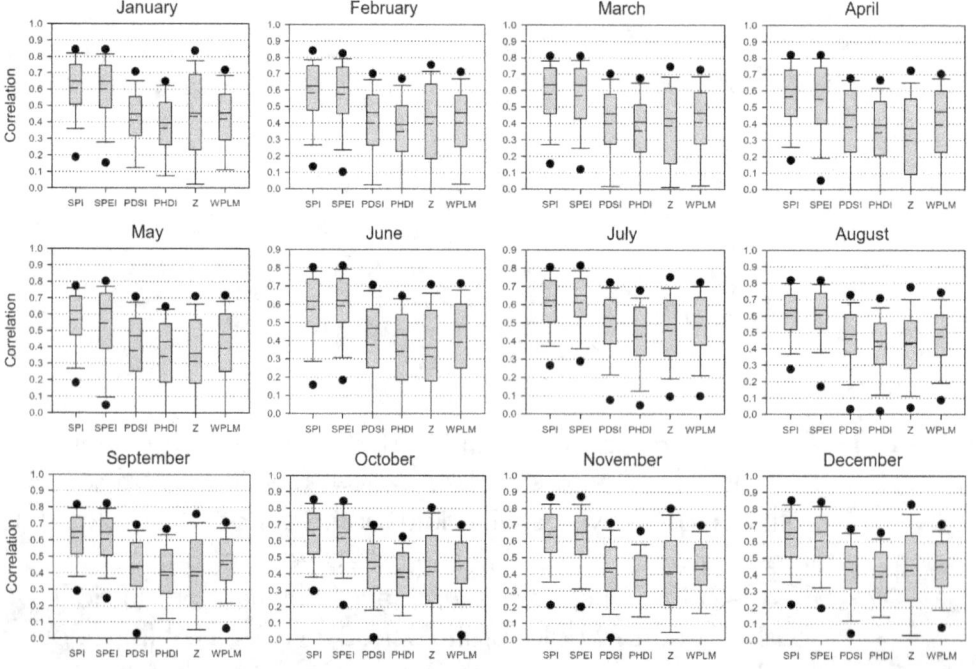

Figure 6.3 Box plots showing the correlations between the monthly series of the Standardized Streamflow Index (SSI) in 151 basins across the world and the six drought indices compared in this study.

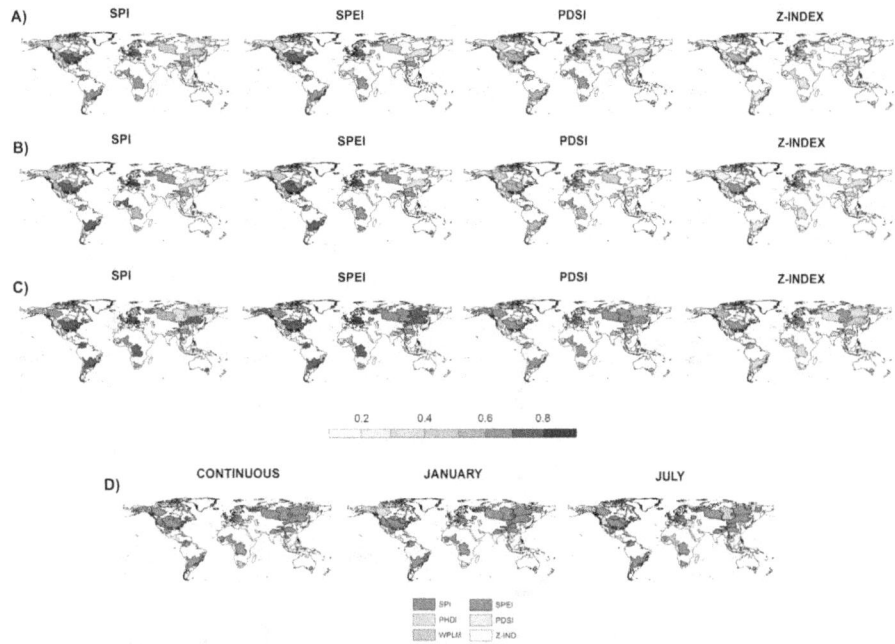

Figure 6.4 : Spatial distribution of correlation coefficients obtained between the Standardized Streamflow Index (SSI) and several drought indices (SPI, SPEI, PDSI and Z–index). A) Correlations considering continuous series from January 1948 up to September 2004, B) Correlations considering only the January series, C) Correlations considering only the July series, D) Drought index which presented the maximum correlation with the continuous (January 1948–September 2004), January and July SSI series.

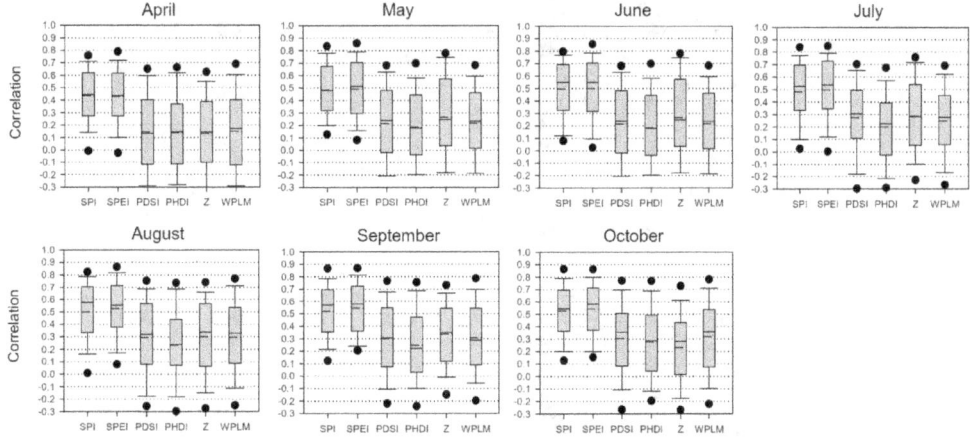

Figure 6.5   Box plots showing the correlation coefficients obtained between the monthly series of the Standardized Soil Moisture data in 117 sampling points across the world and the six drought indices assessed in this study.

Table 6.1 Percentage of the 151 analyzed worldwide basins at which the maximum correlation with the SSI series is found for any of the six drought indices compared. The percentage are given for the continuous SSI series and for each one of the monthly series

	Continuous	Jan	Feb	Mar	Apr	May	Jun	Jul	Aug	Sep	Oct	Nov	Dec
SPI	38.4	49.0	57.6	52.3	48.3	37.1	31.8	33.8	29.8	42.4	57.6	59.6	53.0
SPEI	44.4	33.1	31.1	37.7	40.4	52.3	54.3	47.0	53.0	43.0	31.8	32.5	30.5
PDSI	4.0	0.7	2.0	3.3	3.3	2.0	4.0	4.6	6.0	2.6	1.3	2.6	2.0
PHDI	0.0	2.0	1.3	1.3	2.0	1.3	2.0	2.0	2.0	2.6	0.7	1.3	2.6
Z–Index	7.3	13.9	4.0	4.0	5.3	4.6	5.3	6.0	4.0	5.3	5.3	2.6	10.6
WPLM	6.0	1.3	4.0	1.3	0.7	2.6	2.6	6.6	5.3	4.0	3.3	1.3	1.3

Table 6.2 Percentage of the 117 analysed sampling points at which the maximum correction with the standardized soil moisture series is found for any of the six drought indices compared. The percentages are given for the monthly series from April up to October

Drought Index	April	May	June	July	August	September	October
SPI	48.3	43.1	44.8	31	31.9	32.8	42.2
SPEI	44.0	46.6	44.8	56	51.7	49.1	44.0
PDSI	4.3	3.4	2.6	6.9	5.2	3.4	3.4
PHDI	0.9	2.6	3.4	0.9	3.4	3.4	2.6
Z	1.7	2.6	2.6	1.7	4.3	6.0	3.4
WPLM	0.9	1.7	1.7	3.4	3.4	5.2	4.3

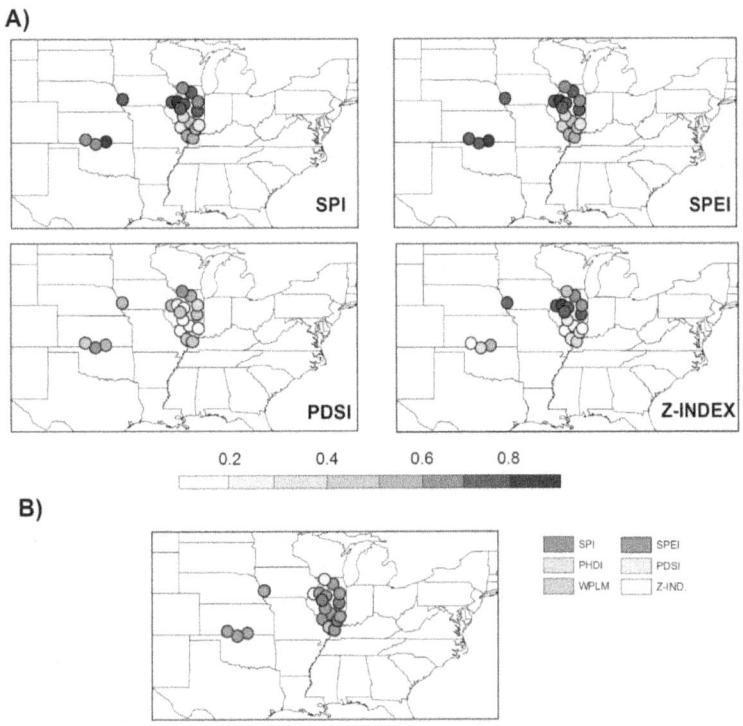

Figure 6.6 Spatial distribution of correlation coefficients obtained between the July soil moisture and several drought indices (SPI, SPEI, PDSI and Z–index) in North America. A) July correlations, B) Drought index which showed the maximum correlation with soil moisture in July

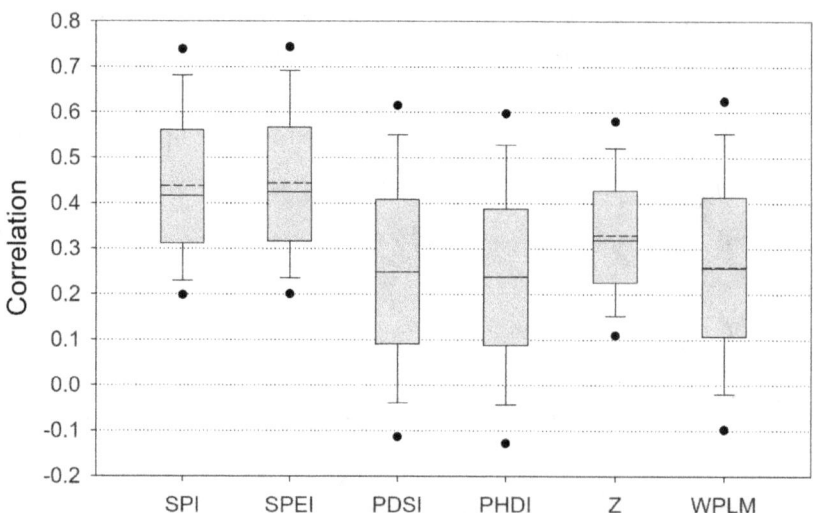

Figure 6.7   Box plots showing correlation coefficients obtained between the tree–ring width series in 1840 worldwide forests and the six drought indices compared in this study. The graph represents the maximum correlations found for any of the twelve monthly series.

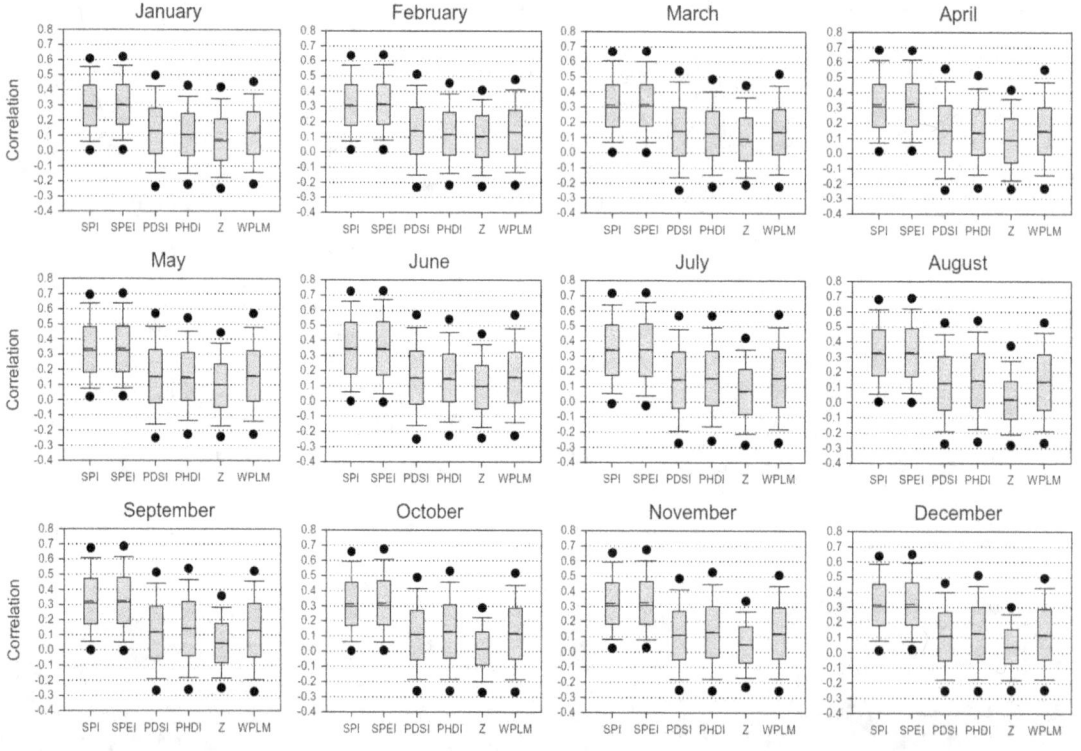

Figure 6.8   Box plots showing correlation coefficients obtained between the annual tree– ring width series in 1840 forests located across the world and the monthly series of the six drought indices compared in this study.

Table 6.3 Percentage of the 1840 analyzed tree-ring width showing the maximum correlations is found for any of the six drought indices compared. The percentages are given for the annual maximum correlations, independently of the month of the year in which they are found, and for each one of the monthly series

	Maximum	Jan	Feb	Mar	Apr	May	Jun	Jul	Aug	Sep	Oct	Nov	Dec
SPI	37.9	37.3	38.2	41.1	41.1	43.4	43.5	43.3	43.2	40.9	41.9	40.0	42.1
SPEI	48.5	48.4	49.3	46.3	46.8	45.2	46.6	47.3	48.2	49.5	49.9	49.9	49.8
PDSI	2.1	5.7	4.0	4.5	4.5	4.3	2.7	2.4	2.2	2.2	1.7	2.1	2.3
PHDI	2.2	4.5	3.4	3.5	3.4	3.6	3.4	3.9	3.3	3.9	3.5	2.8	3.0
2.8	7.0	3.0	3.9	3.3	2.2	2.1	2.2	1.8	1.5	2.1	1.6	4.0	1.4
WPLM	2.4	1.1	1.4	1.3	2.0	1.5	1.6	1.3	1.3	1.5	1.4	1.3	1.3

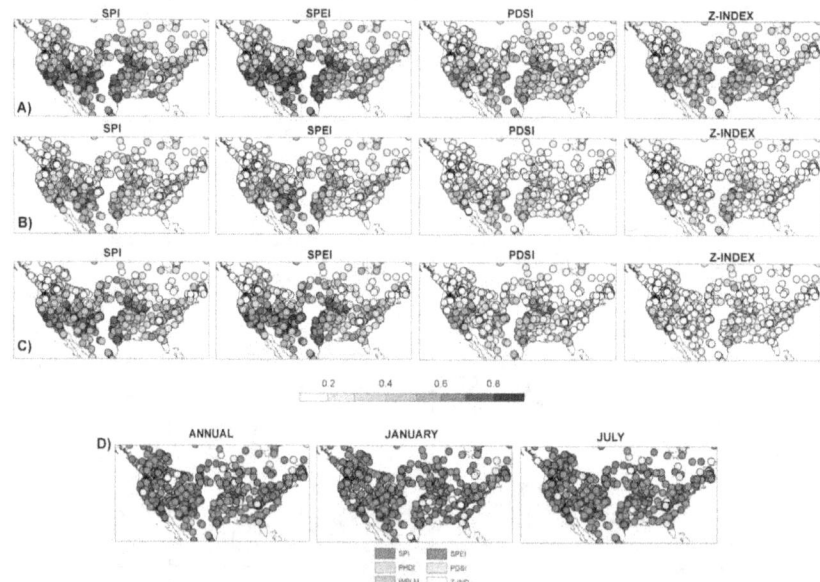

Figure 6.9   Spatial distribution of correlation coefficients obtained between the tree–ring width chronologies and several drought indices (SPI, SPEI, PDSI and Z–index) in the USA. A) Maximum annual correlations independently of the month of the year, B) Correlations considering only the January (B) and July (C) drought series, D) Drought index showing the maximum correlation with tree growth for the annual, January and July drought series.

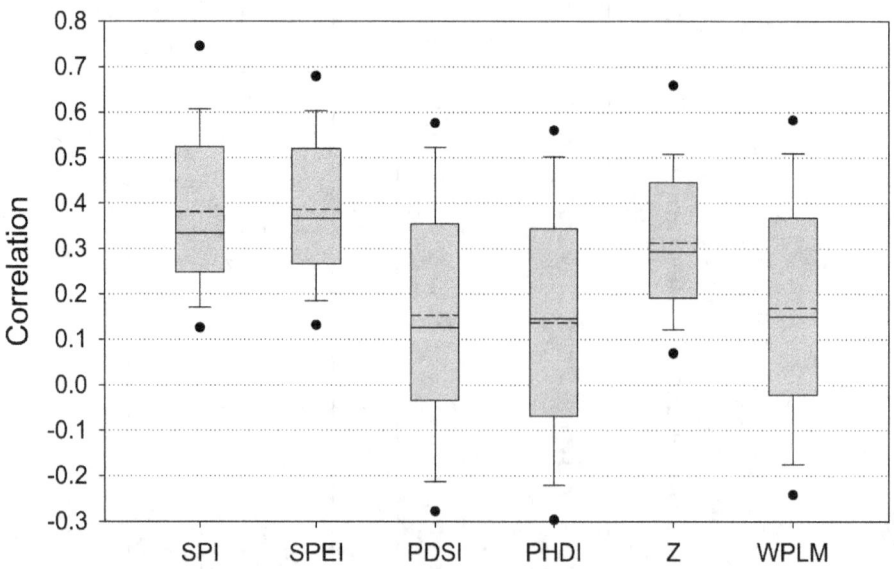

Figure 6.10   Box plots showing correlation coefficients between the wheat yields in 173 countries and the six drought indices compared in this study. The graph represents the maximum correlations found for any of the twelve monthly series.

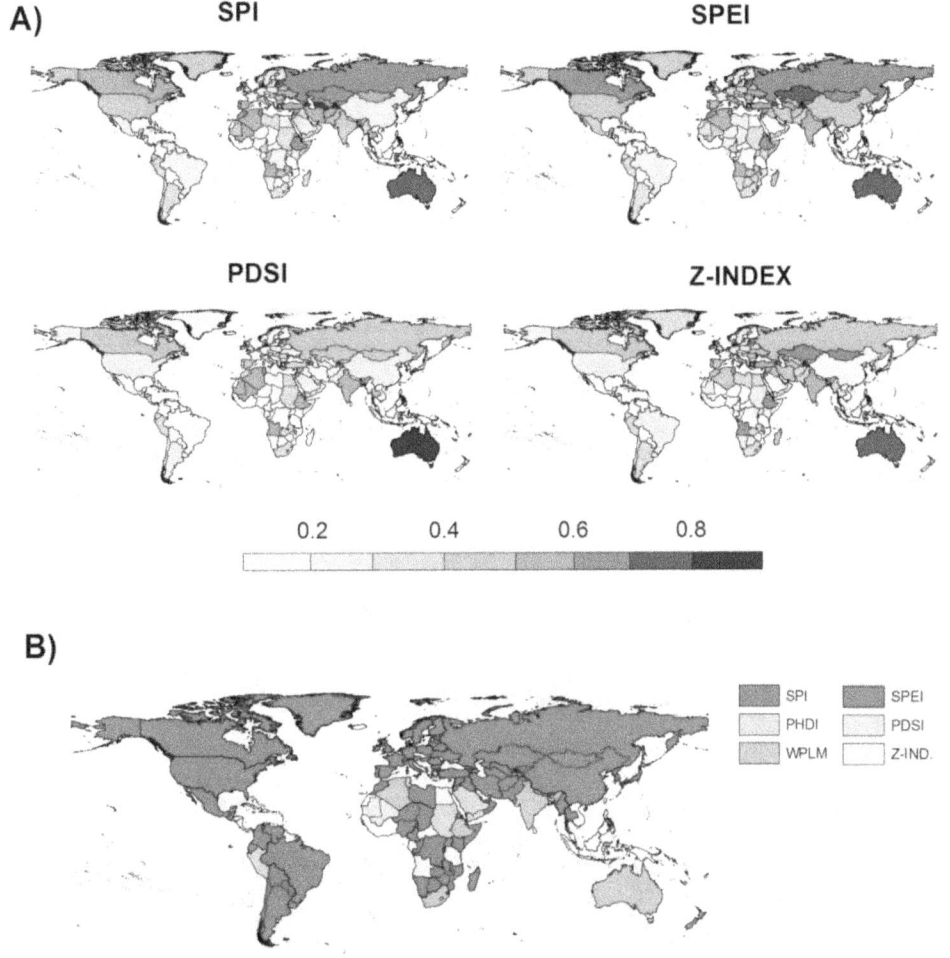

Figure 6.11 Spatial distribution of correlation coefficients obtained between the annual wheat yields and several drought indices (SPI, SPEI, PDSI and Z–index). A) Maximum annual correlations independently of the month of the year, B) Drought index showing the maximum correlation with crop yield for the different monthly series of the drought indices. The countries with white areas lack wheat cultivation or do not have available long time series of wheat yields.

7

AGRICULTURAL DROUGHT

📋 CHAPTER OUTLINE

- To identify Agricultural drought "climatological water balance" done for data of Anantapur Dist. A.P. Frequency analysis of Rainfall – Categorization of droughts – climate shifts – suitability of crops for any area are put forth with graphs.

Agricultural Drought Scenario in India and Need to Effective Monitoring of Agricultural Drought

Drought is a climatic anomaly, characterized by deficient supply of moisture resulting either from sub-normal rainfall, erratic rainfall distribution, higher water need or a combination of all the three factors.

About two thirds of the geographic area of India receives low rainfall (less than 1000 mm) which is also characterized by uneven and erratic distributions. Out of net sown area of 140 million hectares about 68% is reported to be vulnerable to drought conditions and about 50% of such vulnerable area is classified as 'severe' where frequency of drought is almost regular. Abnormally low rainfall in 1979 in India reported to have reduced the overall food grain by as much as 20%. The 1987 drought in India damaged 58.6 million hectares of cropped area affecting over 285 million people.

The 2002 drought had reduced the sown area to 112 million hectares from 124 million hectares and the food grain production to 174 million tons from 212 million tons. The total food grain production in India has to be stepped up from 212 million metric tons to 300 million metric tons by 2020 to meet the food demands of growing population. Therefore, there is a need for effective monitoring of agricultural drought, its onset progression and impact on crops to minimize the damages.

All the developing countries, being primarily agrarian, are very much dependent on the vagaries of seasonal rainfall and climatic conditions and hence more vulnerable to droughts. On an average severe drought occurs once in every five years in most of the tropical countries, though often they occur on successive years causing severe losses to agriculture and allied sectors. More than 500 million people live in the drought prone areas of the world and 30% of the entire continental surface is affected by droughts or desertification process. The water needs in agricultural sector are going to be very high, as several thousand tones of water is required to produce each metric ton of food grains.

Therefore, there is a need for effective monitoring of agricultural drought, its onset, progression and impact on crops to minimize the damages.

Source: NRSC (National Remote Sensing Centre) (Decision support centre) Agricultural drought.

Conventional mechanisms for agricultural drought monitoring Conventional agricultural drought conditions are characterized by ground observations on meteorological parameters such as rainfall, aridity and agricultural parameters such as sown crop area, crop condition and crop yield.

Meteorological observations The IMD prepares rainfall maps on sub-divisional basis every week throughout the year. These maps show the rainfall received during a week and corresponding departures from the normal. In addition, IMD also provides the information on weekly rainfall and its deviation from normal at district level for the entire country. This information is useful to indentity the districts with deficict / scanty rainfall and the prevailing meteorological drought.

IMD also monitors drought using water balance Technique which addresses agricultural drought. The aridity index is calculated using the formula.

$$\text{Aridity index} = \frac{\text{Water deficit}}{\text{Water need}}$$

$$\frac{(\text{Actual evapotranspiration} - \text{Potential evapotranspiration})}{\text{Potential evapotranspiration}}$$

The departure of aridity index from normal percentage terms is used to define the various categories of drought severity. Anomaly upto 25% is attributed to mild drought, 26 – 50% to moderate drought and > 50% to severe drought.

IMD has been bringing out weekly aridity anomaly charts from 1979 onwards, based on data from different observations, covering south west monsoon period.

These chart show the departures of actual aridity from normal aridity giving indication of severity of water deficit to water demand relationship on weekly basis. IMD also preparing detailed maps of rainfall, temp (max & min) cloud cover, relative humidity and analyse this information with prevailing crop conditions and an agreement advisory Bulletin is prepared and disseminated to users. Based on rainfall temp, soil moisture and evaporation, various indicators of meteorological drought indicators have been developed by researchers as shown in Table-2. Some of these indices like palmer index, standardized precipitation index, crop moisture index are being used operationally in some of the countries.

Rainfall as an agricultural drought indicator is limited by the sparse ground observations (especially in view of high spatial variability of tropical rainfall) as well as the lack of spatially and temporally unique relationship between incident rainfall and vegetation development. Though the daily reporting network of IMD is supplemented by the State Governments rain gauge network in each state, real – time information from the latter is normally limited to rain gauges located at Tahril head quarters. The possibility of observational errors also makes it necessary to process this data prior to its use leading to time delays. Many studies have proved that the spatial and temporal distribution of rainfall is more important than the total rainfall in a month or season. A review of past agricultural droughts in the country reveal the lack of unique relationship between incident ground measured rainfall (only a part of which replenishes soil moisture and thus available to vegetation) and the resulting vegetation development within and between seasons as well as across space. The "rainfall use – efficiency" varying over both time and space and the vegetation species dependence limits the use of rainfall as a sole or major agricultural drought indicator. Aridity anomaly data currently available is only representative of large areas such as meteorological sub-divisions. The aridity anomaly suffers from the same limitations as that of rainfall.

Table 7.1   Meteorological Drought Indicators Developed by Researchers

Year	Index
1916	Munger's index
1919	Kincer index
1930	Morkowitch index
1942	Blumenstock index
1954	Antecedent precipitation index
1957	Moisture adequacy index
1965	Palmers index (PDSi, PHDI)
1968	Keetch byseem drought index
1968	Crop moisture index
1981	Surface water supply index
1993	Standardised precipitation index

Agricultural Observations

Conventionally agricultural drought conditions are characterized by general observations on meteorological parameters such as rainfall aridity and agricultural parameters such as sown crop area, crop condition and crop yield.

Though the general observations of agricultural conditions made by the state departments of agriculture and revenue are exhaustive such a system involves a significant amount of subjective judgement at various stages. The periodicity and extent of ground observations also vary significantly between different states. The nature of sparse ground observations also make it difficult to assess, in near real-time, average drought conditions over the district. Such ground monitoring of both causative factors as well as impact of drought assessment suffer from various limitations such as sparse observations, subjective data etc.

Thus general monitoring of both causative factors as well as impact of drought suffer from various limitations such as sparse observations, subjective data etc. **There is a need for building up the capabilities by using innovative technologies and management measures for effective management of agricultural droughts in the country.** A system for national / regional and sub regional assessment and monitoring of agricultural droughts conditions through the cropping season to provide periodic information on the prevalence, severity level and persistence of agricultural drought is the utmost need of the hour. Such information provided in time affective manner, will help the resource managers in optimally allocating scarce financial and other resources to where and when they are most needed.

Source: National Remote Sensing Centre, Govt. of India recent update in Jan 2014.

Space technology for agricultural drought monitoring Unlike point observations of ground data, satellite sensors provide direct spatial information on vegetation stress caused by drought conditions. Satellite remote sensing technology is widely used for monitoring crops and agricultural drought assessment. Over the last 20 years, coarse resolution satellite sensors are being used routinely to monitor vegetation and detect the impact of moisture stress on vegetation. AVHRR on NOAA's polar orbiting satellites have been collecting coarse resolution imagery world wide with twice daily coverage and general survey view. The NOAA AVHRR NDVI has been extensively used for drought / vegetation monitoring, detection of drought and crop yield estimation. The drought monitoring of USA using NOAA – AVHRR data Global information and early warning system (GIEWS) and advanced real time environmental monitoring information system (ARTEMIS) of FAO using Meteosat and Spot – VGT data, international water management institute (IWMI)'s drought assessment in South West Asia using Modis data are proven examples for application of course resolution statellite images for operational drought assessment. **India's national agricultural drought assessment and monitoring system (NADAMS) Project stands as an example for operational use of both moderate resolution and coarse resolution satellite data for operational drought assessment at different spatial scales.**

Welcome to NRSC – Drought – DSC – National remote... www.dsc.nrsc.gov.ing DSC/drought/index VSP? include b.JSP...

Drought indices from satellite data   The crop / vegetation reflects high in the near infrared due to its canopy geometry, the health of the standing crops / vegetation and absorbs high in the red reflected radiance due to its biomass and accumulated photosynthesis.

Stressed vegetation has a higher reflectance than healthy vegetation in the visible (0.4 – 0.7 microns) region and lower reflectance in the near infrared (0.7 – 1.1 microns) region of the electromagnetic spectrum. Vegetation indices take the advantage. Of this differential response in the visible and near infra red regions of the spectrum.

Using these contrast characteristics of near infrared, red and middle infra red bands which indicate both the health and conditions of the crops / vegetation, different types of vegetation indices have been developed as shown below

1. Difference vegetation index
2. Ratio vegetation index
3. Infrared percent vegetation index
4. perpendicular vegetation index
5. Soil adjusted vegetation index
6. Weighted difference vegetation index
7. Greenness vegetation index
8. Atmospherically resistant vegetation index
9. Normalised difference vegetation index
10. Normalised difference wetness index
11. Enhanced vegetation index

Among the various vegetation indices that are now available, normalized difference vegetation index (NDVI) is an universally acceptable index for operational drought assessment because of its simplicity in calculation, easy to interpret and its ability to partially compensate for the effects of atmosphere, illumination geometry etc. NDVI is transformation of reflected radiation in the visible and near infrared bands of NOAA AVHRR and is a function of green leaf area and biomass.

NDVI is derived as under

$$NDVI = \frac{(NIR - Red)}{(NIR + Red)}$$

Where near infra red and red are the reflected radiations in these two spectral bands. Water, clouds and snow have higher reflectance in the visible region and consequently NDVI assumes negative values for these features. Base soil and rocks exhibit similar reflectance in both visible and near IR regions and the index values are near zero. The NDVI values for vegetation generally range from 0.1 to 0.6, the higher index values being associated with greater green leaf area and biomass.

In general growth and decay of crop canopy represents similarities in the temporal vegetation index profile during the crop growth. The peak of this profile corresponds to peak vegetation cover of the crop. Interpretation of vegetation index (VI) profile can be used to derive information on the crop stage.

Further VI level at peak vegetation stage or the time integration of VI profile is related with accumulated biomass in the crop or crop condition or crop yields. Lowering of VI values reflects moisture stress in vegetation, resulting from prolonged rainfall deficiency. Such a decrease in VI could also be caused by other stresses such as pest / disease attack, nutrient deficiency or geochemical effects. Comparing of the seasonal VI profile is thus reflective of vegetation dynamics and condition. Comparison of VI profile of the reporting year and a previous normal agriculture year provides assessment of drought impact in the scale of previous agricultural scenario. Agricultural situation is monitored from June to November every year.

Drought indices derived from NDVI and temperature Severity of drought situation is assessed by the extent of NDVI deviation from its long term mean. The concept of relative greenness (i.e.) the ratio of current NDVI to the historic mean NDVI for the same period is given by :

NDVI dev = NDVI i – NDVI m

NDVI i in the NDVI in the ith month

NDVI m = long term average of the same month

Maps produced using relative greenness are quite useful to assess drought situation and hence this indicator is being used widely.

Kogan (1990) developed vegetation condition index (VCI) using the range of NDVI and temp condition index (TCI) with brightness temperature data as under;

$$VCI = \frac{\left(NDVI - NDVI_{min}\right)}{\left(NDVI_{max} - NDVI_{min}\right)} \times 100$$

$$TCI = \frac{\left(BT_{max} - BT\right)}{\left(BT_{max} - BT_{min}\right)} \times 100$$

VTI = a × VCI + b × TCI

where NDVI $NDVI_{max}$ and $NDVI_{min}$ are smoothed weekly

NDVI absolute maximum and its minimum. BT, BT_{max} and BT_{min} are smoothed weekly brightness temperature absolute maximum and its minimum.

VCI is vegetative condition index

TCI is temperature condition index

VTI is vegetation health index

The VCI values between 50 to 100% indicate optimal and above normal conditions. At the VCI value of 100% and NDVI value for this month (or week) is equal to $NDVI_{max}$. Different degrees of a drought severity are indicated by VCI values below 50%.

Process based indicators By modeling the energy and matter transfer between atmosphere and surface a process based indicator known as evaporative fraction (EF) is derived, EF is defined as the fraction of available energy used for evapotranspiration. EF is derived as under :

$$Rn = G + H + IE$$

where   G = soil heat flux

H = sensible heat flux

Rn = net radiation

IE = latest heat flux

$$IE = \frac{(R_n - G) - CP \times rain \times I}{ra \times (Ts - Tx)}$$

where   CP = specific heat of air

rain = density

Ts = Surface temperature

ra = surface roughness

Tx = max temperature

$$EF = \frac{IE}{Rn - G}$$

where EF is evaporative fraction.

As long as moisture is available, energy will be used for its evapotranpiration and EF will be close to one (no water stress). In the absence of soil moisture all available energy will be directed into warming up surface and the ambient air and EF will approach zero indicating serious water stress.

Although there are many vegetation indices and derived indices, VCI and NDVI deviation from historic NDVI are being widely used on operational basis to assess drought situation.

Time series of NDVI permit monitoring of the dynamic nature of vegetation phenology. Variations or noise in radiometric data are caused by instrument calibration, scan angle, sun angle and atmospheric conditions. The effect of noise can be reduced in several ways (1) compositing via the greenest pixel method (2) using vegetation indices to compensate for variations of instrument calibration and cloud contamination (3) aggregating and averaging channel values over a defined geographic area and using these averages to calculate vegetation index and smoothing the profile with curve filling or filtering techniques.

National agricultural drought assessment and monitoring system (NADAMS) Agricultural drought assessment using space technology inputs has been operational in India since 1989, through a project "National Agricultural assessment and monitoring system" NADAMS provides near real time information on prevalence, severity level and persistence of agricultural drought at state / district / sub dist. level. Currently the project covers 13 states of India which are predominantly agricultural based and prone to drought situation.

Agricultural conditions are monitored at state / district level using daily observed coarse resolution (1.1 km) NOAA AVHRR DATA from advanced wide field sensor (AWIFS) sensor of resource sat I (IRS P$_6$) of 56 meters and wide field sensor (WIFS) of IRS IC and ID of 188 meters are being used for detailed assessment of agricultural drought at district and sub district level in four states namely Andhra Pradesh, Karnataka, Haryana and Maharashtra.

Satellite based assessment for India – 2015   The methodology being adopted in (NADAMS) essentially reflects the harmonization of satellite derived crop condition with ground collected rainfall and crop area progression to evolve decision rules on the prevalence, intensity and persistence of agricultural drought situation. The agricultural area of each district is monitored using time series NDVI with the support of ground data. The assessment of agricultural drought situation takes into consideration, the satellite derived information on

(a)   sensonal NDVI progression – (i.e.) transformation of NDVI from the beginning of the season

(b)   comparison of NDVI profile with previous normal years and

(c)   vegetation condition index, integrated with ground information on (a) weekly rainfall status compared to normal and (b) weekly progression of sown area.

During June to August, drought warning information is issued in terms of "watch, alert and normal" categories. In case of 'watch' external intervention is required, if similar drought like conditions persist during the successive month while 'alert' calls for immediate external intervention, in terms of crop contingency plans. During September and October, based on NDVI anomalies corroborated by ground situation, drought declaration is done in terms of mild, moderate and severe drought.

NDVI Image Gallery

Among the various vegetation indices that are now available, normalized difference vegetation index (NDVI) is an universally acceptable index for operational drought assessment because of its simplicity in calculation, easy to interpret and its ability to partially compensate for the effects of atmosphere, illumination geometry etc. NDVI is a transformation of reflected radiation in the visible and near infrared bands of NOAA AVHRR and is a function of green leaf area and biomass.

Agricultural Drought – Anantapur District, Andhra Pradesh, India

For analysis of this type of drought water balance analysis is made.

Availability of water in right quantity and in right time and its management with suitable agronomic practices are essential for good crop growth, development and yield.

Table 7.2 Agricultural Drought Warning and Declaration in NADAMS Project

Month	Assessment	Implications
June July August	Normal	Agricultural situation is normal.
	Watch	o Progress of agricultural situation is blow. o Ample scope to recovery. o No external intervention needed.
	Alert	Very slow progress of agriculture situation. Need for intervention. Develop and implement contingency plans to minimize loss.
September October	Mild Drought	Crop have suffered stress slightly.
	Moderate Drought	Considerable loss in production. Take measure to alleviate suffering.
	Severe	High risk significant reduction in crop yield. Management measures to provide relief.

To assess water availability to crops, soil moisture is to be taken into account and the net water balance through soil moisture can be estimated using the water balance technique.

The concepts of PET (potential evapotransporation) and water balance have been extensively applied to studies such as climate classification, aridity, humidity and drought.

The water balance study is made for Anantapur District of Andhra Pradesh, India as an example as this district is 2nd most drought prone district in India. The water balance parameters such as Index of aridity (Ia), Index of humidity (Ih), Index of moisture (Im), Index of moisture adequacy (Ima), AE (actual evapotranspiration) are found.

Agriculture Drought

Andhra Pradesh is basically an agricultural State, with 34% of its GDP contributed by agriculture. Agriculture is the major source of employment to the people. Droughts during the season affect the agricultural production and agricultural drought, it is necessary to measure the extent to which rainfall and soil moisture is falling short of the water requirement of crops during the cropping season. Moisture Adequacy Index (MAI) is a better measure for assessing the degree of adequacy of rainfall and soil moisture to meet the potential water requirement of crops. Hence, to identify the mandals that can be vulnerable to agricultural drought, weekly MAI was worked out for the mandals having more than 20 years of rainfall data.

Methodology for Calculating MAI

Moisture Adequacy Index (MAI) is the ratio of actual evapotranspiration (AET) to the potential evapotranspiration (PET). AET can be obtained as an output parameter from water balance calculations. Thornwait and Mather (1955) weekly water balance model was used for estimating water balance of mandals. Agricultural droughts during different seasons (years) were classified into four groups based on average MAI during the season.

Table 7.3 Drought Classification Based on MAI

Drought Severity	MAI
No drought	MAI > 0.75
Mild drought	MAI < 0.75 AND > 0.50
Moderate drought	MAI < 0.50 and > 0.25
Severe drought	MAI < 0.25

Rain Year 2012-2013

Agricultural status of mandals based on rainfall from June 1 to Sept. 30 map enclosed.

Climate and Water Balance of Anantapur District, Andhra Pradesh

Climatic water balance is a basic concept in modern climatology. It plays a key role in the rational classification of the climate of a place. It was originally developed by C.W. Thornthwaite. Its practical utility of agroclimatology and conservation of water resources especially a crop farming is of vital importance and it is much related to the water balance concept.

The term "water balance" refers to the balance between the income of water from rainfall and the outflow of water by evapotranspiration. It is a climatic balance since the quantities of rainfall and evapotranspiration are the active viable factors of the climate. Precipitation is the main source of water supply and evapotranspiration is the main cause of water loss in any region. Proper assessment of both water supply and water loss are of importance and play a key role in development of agriculture and water resources of the region.

The relative difference between precipitation (P) and potential evapotranspiration (PE) in a place determines whether a region is wet or dry. The net surplus or deficit of water supply on account of precipitation and evapotranspiration is expressed in terms of water balance in climatology.

The income of water through precipitation is measured directly from rain gauge stations but water loss is on difficult parameter to measure directly. Eminent meteologists have developed different formulate to determine the water loss through evapotranspiration and among them Thornthwaite and Mather (1955) Penman (1956) figure prominently. Indian scientists who have made adequate contribution in this field are Subrahmanyam (1956, 1957, 1958, 1963 and 1982), Subrahmanyam and Subrahmanian 1964), Subrahmanyam and Murthy (1968), Subrahmanyam and Sastry (1969), Subrahmanyam et al (1964, 1970, 1979), Rammohan (1978), Subrahmanian and Umadevi (1982).

Methodology

For the competation of water balance of any region, precipitation (P) is treated as income potential evapotranspiration (P.E) as expenditure and the soil moisture as a kind of reserve for use in times of low rainfall periods. The most important derivative elements of the water balance studies is actual evapotranspiration (E). It is the actual amount of water lost to the atmosphere by evaporation and transpiration. Water deficiency (W.D) and the water surplus (W.S) are the derivatives through which the other components such as aridity index (Ia) humidity index (Ih), moisture index (Im) and index of moisture adequacy (Ima) are determined.

Potential Evapotranspiration (PE)

To determine the potential evapotranspiration of any region Thornthwaite (1948) empirical formula is used. The formula for the computation of PE is

$$e = 1.6 \left(\frac{10tm}{I} \right)^a$$

where,

e = "unadjusted" P.E in cm

t_m = mean monthly temperature in centigrade

I = annual heat index (it is the sum of the 12 monthly heat indices)

$$i = \left(\frac{t}{5} \right)^{1.514}$$

a $= 6.75 \times 10^{-7} \times I^3 - 7.71 \times 10^{-5} \times I^2 + 1.792 \times 10^{-2}I + 0.49239$

(i.e.) $0.000000675\ I^3 - 0.0000771 I^2 + 0.01792\ I + 0.49239$

The formula gives unadjusted values of thermal efficiency since the number of days in a month ranges from 28 to 31 (nearly by 11%) and the number of hours of sunshine in the day between sunrise and sunset (when evapotranspiration principally takes place) varies with the latitude and season of the year, it becomes necessary to reduce or increase the unadjusted T.E by a factor that varies with the latitude and the month under question.

The monthly potential evapotranspiration for any month is then calculated from the equation

$$P.E = 1.6 \times b \left(\frac{10tm}{I} \right)^a \text{ cm / month}$$

where a is an exponent gives by the equation $a = 6.75 \times 10^{-7} I^3 - 7.71 \times 10^{-5} I^2 + 1.792 \cdot 10^2 I + 0.49239$ and b is a factor to correct for unequal day length between months. The value of b is obtained from the expression

$$b = \frac{\text{Possible sunshine hours for the particular month}}{12 \times 30 \, (\text{or } 31)}$$

The value of 'e' thus obtained is based on a 12 hrs day and 30 days month. Hence it has to be adjusted by taking into account the actual number of hours in a day as well as the variation in the number of days in the month from 28 to 31.

Using the conversion tables and monograms prepared by Thornwaite the PE of the area can be determined and it has to be adjusted for changing number of days in a month and the changing length of any day in different seasons.

Table 7.4   Meteorological Data of Anantapur Dist., A.P. India for the year 2011

Month	Temp. in 0°C		Mean tm	Monthly Heat Index $im = \left(\dfrac{tm}{5}\right)^{1.514}$
	Mean Maximum	Mean Minimum		
January	31.7	14.3	23	10.08
February	33.6	16.9	25.25	11.61
March	37.8	20.3	29.05	14.35
April	38.4	23.9	31.15	15.95
May	38.5	25.3	31.9	16.54
June	34.5	24.1	29.3	14.54
July	33.9	23.7	28.8	14.17
August	32.5	23.3	27.9	13.50
Sept	33.6	22.6	28.1	13.65
Oct	32.8	22.8	27.8	13.43
Nov	30.5	18.7	24.6	11.16
Dec	30.6	16.5	23.55	10.45
Annual Heat Index I = Σim				159.43

Potential evapotranspiration

$$PE = 1.6 \times b \times \left(\frac{10tm}{I}\right)^{a}$$

where a = $6.75 \times 10^{-7}I^{3} - 7.71 \times 10^{-5} I^{2} + 1.792 \times 10^{-2} I + 0.49239$

where I = 159.43 em / month

a = $2.74 - 1.96 \times 2.86 + 0.49239$

   = 4.13

where b is a factor to correct for unequal day length between months.

North latitude 13°14′ and 15°14′

Average = 14°14′

Table 7.5 Possible Sunshine Hours during the Month

Month	Monthly Percentage of Day Time Hours in the Year Latitude 14°14′	Possible Sunshine Hours During the Month		
January	7.95	$7.95/100 \times 365 \times 12$ $= \dfrac{348.21}{12 \times 31}$	=	0.94
February	7.34	$7.34/100 \times 365 \times 12$	=	0.89
March	8.34	$8.34/100 \times 365 \times 12$	=	0.98
April	8.47	$8.47/100 \times 365 \times 12$	=	0.99
May	8.81	$8.81/100 \times 365 \times 12$	=	1.04
June	8.84	$8.84/100 \times 365 \times 12$	=	1.04
July	9.01	$9.01/100 \times 365 \times 12$	=	1.06
August	8.83	$8.83/100 \times 365 \times 12$	=	1.04
Sept	8.30	$8.30/100 \times 365 \times 12$	=	0.98
Oct	8.25	$8.25/100 \times 365 \times 12$	=	0.97
Nov	8.02	$8.02/100 \times 365 \times 12$	=	0.94
Dec	8.30	$8.30/100 \times 365 \times 12$	=	0.98

Table 7.6 Potential Evapotranspiration of Anantapur District, A.P. India as per 2011 Year Temperature

Month	Details	PET
January	$1.6 \times 0.94 = \left[\dfrac{10 \times 23}{159.4}\right]^{4.13}$ $= 1.6 \times 0.94 \times (4.54)$	6.83 cm
February	$1.6 \times 0.89 \times 6.68$	9.51 cm
March	$1.6 \times 0.98 \times 11.92$	18.69 cm
April	$1.6 \times 0.99 \times 15.89$	25.17 cm
May	$1.6 \times 1.04 \times 17.54$	29.19 cm
June	$1.6 \times 1.04 \times 12.35$	20.55 cm
July	$1.6 \times 1.06 \times 11.49$	19.49 cm
August	$1.6 \times 1.04 \times 10.09$	16.79 cm
September	$1.6 \times 0.98 \times 10.39$	16.29 cm
October	$1.6 \times 0.97 \times 9.94$	15.43 cm
November	$1.6 \times 0.94 \times 5.99$	9.01 cm
December	$1.6 \times 0.98 \times 5.00$	7.84 cm

Note: Highest potential evapotranspiration during May being 29.19 cm and low in the month of December as 7.84 cm.

In order to compute water balance, it is essential to determine the field capacity or water holding capacity of the soil of the region under study which is an amount of water that can be stored in the root zone and varies with soil type, vegetation etc. Thornwaite and Mather (1957) have tabulated the field capacities of different combinations and soil and vegetation types. It enables to estimate the soil moisture storage of any region by referring to the standard tables provided the soil and vegetation conditions of a region are known

The soils of Anantapur District, Andhra Pradesh, India are classified as Red alfisol soils accounting for 78% while black soils are found in 20% of the total Geographical area (vide Swaminathan Foundation, Chennai report on Anantapur District). So the water holding capacity is considered as 200 mm as per the Thornwaite and Mather Table 1957.

Precipitation of Anantapur district, Andhra Pradesh, India   Anantapur district is located in the middle of the Peninsula renders it the driest part of the State of Andhra Pradesh. The Western ghats prevent the South West monsoon and being far away from the sea it could not derive full benefit of north east monsoon. Hence it is considered as the arid area. At Anantapur, the rainfall is not only low, but erratic in distribution. Hence the yields of groundnut major crop of the region are unstable. Even the chance for success of redgram is remote and district will get normal yields once in five years.

During the decade 1990 – 2000, the mean annual rainfall was 610.1 mm received in 35.6 rainy days. The ground nut crop failed in almost 50% of the years due to the dry spells occurring at different stage of the crop. During the decade 1990 - 2000 the annual rainfall ranged between 401.7 mm (1994) and 937.2 mm in (1998). Deficit or excess of moisture during pod development stage of the crop results in yield reduction.

For the present for water balance study of Anantapur District rainfall from 1946 to 1975 (30 years) has been considered, whose normal rainfall is 537.88 mm. As the details of rainfall particulars from 1876 to 2000 the district normal has been furnished as 544 mm.

Once P.E is calculated and since P is already known, the next step in water balance is to calculate P-PE. This may give a series of positive as well as negative values since the P and PE never coincide. Positive values indicate excess of water available for the soil storage and run off whereas negative values represent deficiency of water.

The next step is to compute the accumulated potential water loss. This accumulation starts from an estimated value of the potential water deficiency at the beginning of the first month when P – PE negative (estimated potential water loss can be determined by using the soil moisture retention tables of Thornthwaite and Mather 1957).

Using the water budget parameters Thornthwaite has derived the following indices so as to use them in the classification of climates being on thermal and moisture regions.

1. Index of aridity (Ia) is the percentage ratio between water deficiency (W.D) and potential evapotranspiration

$$Ia = \frac{W.D}{P.E} \times 100$$

2. Index of humidity (Ih) is the percentage ratio between water surplus (W.S) and potential evapotranspiration (PE)

$$Ih = \frac{W.S}{PE} \times 100$$

3. The moisture index is the index of moisture effectivity given by the equation
Im = Ih – 0.6 Ia.

4. Index of moisture adequacy (Ima) is defined as the percentage ratio of the actual evapotranspiration (A.E) to potential evapotranspiration (PE) on an annual basis. (Subrahmanyam et. al., 1965)

Water balance computation   The water balance parameters have been computed on the basis of book keeping procedure of Thornwaite and Mather (1955). Rainfall data of Anantapur District has been collected from the district administration and analysed to compute the water balance parameters. The field capacity or water holding capacity of the District is determined as 200 mm based on the soil texture and type of vegetation. The potential evapotranspiration (P.E) is calculated from the mean monthly temperature data collected. Using both precipitation and P.E data, the actual evapotranspiration (AE) water deficit (W.D) and water surplus (W.S) have been computed. The index of moisture Im, Index of Aridity (Ia) Index of humidity (Ih) and Index of moisture adequacy (Ima) have also been worked out to determine the climate of the Anantapur District

WD = Water Deficit (the amount by which the actual Evapotranspiration (AE) and the potential

Evapotranspiration (PE) differ in any month is the water deficit in that month

WS = Water surplus (After the soil moisture storage reaches the water holding capacity (i.e. in example 200 mm) any excess precipitation is counted as water surplus (WS) and is subject to run off.

PET = Potential evapotranspiration. The PET is calculated from the mean monthly temperature data collected from IMD and similar for monthly rainfall.

Computation of Indices like Aridity Index (Ia) Humidity Index (Ih) Moisture Index (Im) and Moisture Adequacy index (MAI)

The indices are output of water balance analysis. These are useful in climatic classification and to find climatic type of the particular place. Moisture adequacy index (MAI) provides a good indication of the moisture status of the soil in relation to the water need, high values of the index signifying good moisture availability and vice versa.

Components like water surplus water deficit and actual evapotranspiration (AET)   Water surplus (WS) and water deficit (WD) occur in different seasons at most places and both are significant in water balance study. The information about when the period of water surplus and deficit occurring in a season or year is helpful to find ideal period for starting of crop season and stages which may fall in deficit period. It also helps in flood and drought analysis. The analysis of drought made for Anantapur District, Andhra Pradesh, India and graphs have been drawn.

Beginning and end of growing period   As the start of rainy season, seed germination and initial crop growth depend on the amount and distribution of rainfall.

The beginning and end of growing period is identified based on IMA values. The growing season begins when the IMA is above 50% consecutively for at least three weeks. The end of season is identified when the IMA falls below 25% for four consecutive weeks.

Length of growing period analysis (LGP)   The length of growing period or the moisture availability period is an important parameter to assess the suitability of climate for agricultural production. It is a key parameter, which helps in assessing climatic suitability to go for particular cropping system. For example, mono cropping with short duration pulses can be adopted for areas with 75 days of LGP and mono cropping with short to medium duration crops for areas with 75-140 days LGP and double cropping can be adopted for areas with more than 180 days LGP.

Potential evapotranspiration (PET) is defined as the maximum quantity of water, which is transpired and evaporated by a uniform cover of short dense grass (reference crop). When the water supply is not limited (reference crop). A hypothetical reference crop with an assumed crop height of 0.12 m, a fixed surface resistance of 70 s/m and an albedo of 0.23.

Water Balance

It refers to the climatic balance obtained, by comparing the rainfall as income with evapotranspiration as loss or expenditure, soil being a medium for storing water during periods of excess rainfall and utilizing or releasing moisture during periods of deficit precipitation.

Water surplus: It is the excess amount of water remaining after the evaporation needs of the soil have been met (i.e. when the actual evapotranspiration equals potential evapotranspiration) and soil storage has been returned to the water holding capacity level.

Water deficit: It is the amount by which the available moisture fails to meet the demand for water and is computed by subtracting the potential evapotranspiration from the actual evapotranspiration for the period of interest.

Actual evpotranspiration: It is the actual amount of water lost to the atmosphere by evaporation and transpiration under existing conditions of moisture availability.

$$\text{Aridity Index (Ia) (\%)} = \frac{\text{Water deficit}}{\text{PET}} \times 100$$

$$\text{Humidity Index (Ih) (\%)} = \frac{\text{Water surplus}}{\text{PET}} \times 100$$

$$\text{Moisture Index Im (\%)} = (\text{Ih} - \text{Ia})$$

$$\text{Moisture Adequacy Index MAI (\%)} = \frac{\text{AET}}{\text{PET}} \times 100$$

Moderate (25.1 to 50%)

Agri favourable (50.1 to 75%)

Humid region (> 75%)

Cumulative moisture adequacy index map can be prepared based on above classification while calculating MAI for each mandal for the four conventional seasons of the year. The moisture that is necessary for the sustenance of a crop or a vegetation species can be best derived from the knowledge of the index of moisture adequacy.

In order to find out the moisture status in the different parts of Anantapur District it is to select a station in each mandal for which temperature and rainfall data are available. At these stations AE & PE to be found. The index of moisture adequacy for the four conventional seasons, these seasons are, Southwest monsoon (June – Sept), post monsoon (Oct – Nov), winter (Dec – Feb), hot weather (March – May). The AE & PE values are expressed as percentage ratios.

The use of climatological water balance in substitution to complete water balances directly measured in the field allows a more practical crop management, since the climatological water balances are based on data monitored as a routine.

ABBREVIATIONS

PE = Potential Evapo-transpiration

P = Precipitation

APWL = Accumulated potential water loss

St = Soil moisture storage

(St of Dec – St of Jan) i.e. 169.49 = 120 put it as ΔSt

Remove 27 from 49 = 22 put it ΔSt

ΔSt = Change in soil moisture

AE = Actual Evapo-transpiration

WD = Water deficit

Index of aridity $Ia = \dfrac{WD}{PE} \times 100 = \dfrac{157.6}{212.954} \times 100 = 74.00\%$

Index of humidity

$$Ih = \frac{WS}{PE} \times 100 = 0$$

Index of moisture Im

Im = Ih – 0.6 Ia = 0 – 0.6 x 74 = -43.51%

Index of moisture adequacy $Ima = \dfrac{AE}{PE} \times 100 = \dfrac{94.4}{212.954} \times 10 = 44.33\%$

Climatological Water balance – Computation for Anantapur District, Andhra Pradesh, India for the year 1946

Field capacity : 200 mm

All values are in mm

Line	Item	Jan	Feb	Mar	Apl	May	June	July	August	Sept	Oct	Nov	Dec	Annual
1.	PE	77.2	120.4	207.94	313.2	326.6	244.0	193.9	177.6	166.8	138.7	96.1	67.1	2129.54
2.	P	-	1.0	4.6	8.9	74.4	14.2	37.3	106.2	142.0	70.9	114.8	34	608.3
3.	P-PE	(-)77.2	(-)119.4	(-)203.34	(-)304.3	(-)252.2	(-)229.8	(-)156.6	(-)71.4	(-)24.8	(-)67.8	+18.7	(-)33.1	
4.	APWL	(-)279.0	(-)398	(-)601	(-)905	(-)1157	(-)1387	(-)1544	(-)1615	(-)1640	(-)1708	0	(-)33.1	
5.	St	49 →	27 →	9 →	2 →	1	1	1	1	1	1	19.7	169	
6.	ΔSt	120	22	18	7	1	0	0	0	0	0	18.7	149.3	
7.	AE	120	23	23	16	75	14	37	106	142	71	134	183	944
8.	WD	40	97	185	297	252	230	157	71	25	68	38	116	1576
9.	WS	0	0	0	0	0	0	0	0	0	0	0	0	0

WS = Water Surplus

Climatological Water Balance – Computation for Anantapur District, Andhra Pradesh, India for the year 1947

Field capacity : 200 mm

All values are in mm

Line	Item	Jan	Feb	Mar	Apl	May	June	July	August	Sept	Oct	Nov	Dec	Annual
1.	PE	77.2	120.4	207.94	313.2	326.6	244.0	193.9	177.6	166.8	138.7	96.1	67.1	2129.54
2.	P	1.80	1.5	-	6.4	11.5	66.8	58.2	94.5	190.5	59.2	7.4	6.10	503.9
3.	P-PE	75.4	118.9	207.94	306.8	315.1	177.2	135.7	83.1	23.7	79.5	88.7	61.0	
4.	APWL	136.4	255.3	463.24	770.04	1085.14	1262.34	1398.04	1481.14	0	79.5	168.2	229.2	
5.	St	102.0→	55→	19	4	1	1	1	1	24.7	134	84	62	
6.	ΔSt	40	47	36	15	3	0	0	0	23.7	23.7	0	0	
7.	AE	102.8	48.5	36	21.4	14.5	66.8	58.2	94.5	190.5	59.2	7.4	6.10	(-)705.9
8.	WD	25.6	71.9	171.94	291.8	312.1	177.2	135.7	83.1	23.6	79.5	88.7	61.0	1522.14
9.	WS	-	-	-	-	-	-	-	-	-	-	-	-	-

Ia = 71.5%   Ih = 0   Im = (-) 42.9%   Ima = 33.15%

APWL : P-PE of Dec (61) + Jan 75.4 = 136.4

ΔSt = Dec St (62) – Jan St (Put it as ΔSt) Remove 55 from 102 and put it in ΔSt as 47

AE = P + ΔSt

WD = AE ΔPE

again 55 – 19 = 36

19 – 4 = 15

4 – 1 = 3   1 – 1 = 0

Climatological Water Balance -Computation for Anantapur District, Andhra Pradesh, India for the year 1948

Field capacity : 200 mm

All values are in mm

Line	Item	Jan	Feb	Mar	Apl	May	June	July	August	Sept	Oct	Nov	Dec	Annual
1.	PE	77.2	120.4	207.94	313.2	326.6	244.0	193.9	177.6	166.8	138.7	96.1	67.1	2129.54
2.	P	0.8	-	0.30	36.6	54.9	21.1	46.7	90.7	84.8	58.1	89.7	-	483.7
3.	P-PE	(-)76.4	(-)120.4	(-)207.64	(-)277	(-)272	(-)223	(-)147	(-)87.0	(-)82.0	(-)81.0	(-)6.0	(-)67.0	
4.	APWL	143	263.0	471.0	748	1020	1243	1390	1477.0	1559.0	1640	1646.0	1713.0	
5.	St	97	53	18	5	1	1	1	1	1	1	1	1	
6.	ΔSt	96	43	35	13	4	0	0	0	0	0	0	0	
7.	AE	97	43	35.3	50	59	21.1	46.7	90.7	84.5	58.1	89.7	0	675.4
8.	WD	20	77	17.3	263	268	223	147	87	82	81	6	67	1494
9.	WS	0	-	-	-	-	-	-	-	-	-	-	-	

Ia = 70.16%   Ih = 0   Im = (-) 42.1%   Ima = 31.72%

Climatological Water Balance -Computation for Anantapur District, Andhra Pradesh, India for the year 1949

Field capacity : 200 mm

All values are in mm

Line	Item	Jan	Feb	Mar	Apl	May	June	July	August	Sept	Oct	Nov	Dec	Annual
1.	PE	77.2	120.4	207.90	313.2	326.6	244.0	193.9	177.6	166.8	138.7	96.1	67.1	2129.54
2.	P	-	-	-	4.1	72.4	43.4	65.5	127.7	161.0	114.3	11.2	-	599.6
3.	P-PE	(-)77.2	(-)120.4	(-)207.94	(-)309.1	(-)254.2	(-)200.6	(-)128.4	(-)49.9	(-)5.8	(-)24.4	(-)14.9	(-)67.1	
4.	APWL	144.3	264.7	472.64	781.74	1035.94	1236.54	1364.94	1414.84	1420.64	1445.04	1529.94	1597.44	
5.	St	96	52	18	4	1	1	1	1	1	1	1	1	
6.	ΔSt	95	44	34	14	3	0	0	0	0	0	0	0	
7.	AE	95	44	34	18.1	75.4	43.4	65.5	127.7	161.0	114.3	11.2	-	789.6
8.	WD	18	76	174	295	251	201	128.0	50	6	24	85	67.00	1375
9.	WS	0	-	-	-	-	-	-	-	-	-	-	-	-

Ia = 64.57%   Ih = 0   Im = (-) 38.74%   Ima = 37.08%

Climatological Water Balance – Computation for Anantapur District, Andhra Pradesh, India for the year 1950

Field capacity : 200 mm

All values are in mm

Line	Item	Jan	Feb	Mar	Apl	May	June	July	August	Sept	Oct	Nov	Dec	Annual
1.	PE	77.2	120.4	207.94	313.2	326.6	244.0	193.9	177.6	166.8	138.7	96.1	67.1	2129.54
2.	P	-	5.1	-	2.0	69.3	33.8	88.6	88.9	102.6	117.3	30.7	-	538.3
3.	P-PE	(-) 77.2	(-) 115.3	(-) 207.94	(-) 311.2	(-) 257.3	(-) 210.2	(-) 105.3	(-) 88.7	(-) 64.2	(-) 21.4	(-) 65.4	(-) 67.1	
4.	APWL	144.3	259.6	467.54	778.74	1036.04	1246.24	1351.54	1440.24	1504.44	1525.84	1591.24	1658.34	
5.	St	96	54	19	4	1	1	1	1	1	1	1	1	
6.	DSt	95	42	35	15	3	0	0	0	0	0	0	0	
7.	AE	95	47.1	35	17	72.3	33.8	88.6	88.9	102.6	117.3	30.7	-	728.3
8.	WD	17.8	73.3	172.94	296.2	254.3	210.2	1053	88.7	64.2	21.4	65.4	67.1	1436.84
9.	WS	0	-	-	-	-	-	-	-	-	-	-	-	-

Ia = 67.47%   Ih = 0   Im = (-) 40.48%   Ima = 34.20%

Climatological Water Balance – Computation for Anantapur District, Andhra Pradesh, India for the year 1951

Field capacity : 200 mm

All values are in mm

Line	Item	Jan	Feb	Mar	Apl	May	June	July	August	Sept	Oct	Nov	Dec	Annual
1.	PE	77.2	120.4	207.94	313.2	326.6	244.0	193.9	177.6	166.8	138.7	96.1	67.1	2129.54
2.	P	-	-	17.8	2.5	78.7	52.8	63.5	17.8	135.3	17.8	8.6	-	394.8
3.	P-PE	(-)77.2	(-)120.4	(-)190.14	(-)310.7	(-)247.9	(-)191.2	(-)130.40	(-)159.8	(-)31.5	(-)120.9	(-)87.5	(-)67.1	(-)1734.74
4.	APWL	144.3	264.7	454.84	765.54	1013.44	1204.64	1335.04	1494.84	1526.34	1647.24	1734.74	1801.84	
5.	St	96	52	20	5	1	1	1	1	1	1	1	1	
6.	DSt	95	44	32	15	4	0	0	0	0	0	0	0	
7.	AE	95	44	49.8	17.5	82.7	52.8	63.5	17.8	135.3	17.8	8.6	-	584.8
8.	WD	17.8	76.4	158.14	295.7	243.9	191.2	130.4	159.8	31.5	120.9	87.5	67.1	1580.34
9.	WS	0	-	-	-	-	-	-	-	-	-	-	-	-

I_a = 74.21% I_h = 0 I_m = (-) 44.53% I_{ma} = 27.46%

Climatological Water Balance – Computation for Anantapur District, Andhra Pradesh, India for the year 1952

Field capacity : 200 mm

All values are in mm

Line	Item	Jan	Feb	Mar	Apl	May	June	July	August	Sept	Oct	Nov	Dec	Annual
1.	PE	77.2	120.4	207.94	313.2	320.6	244.0	193.9	177.6	166.8	138.7	96.1	67.1	2129.54
2.	P	-	-	10.4	19.1	84.6	7.6	50.8	38.1	40.6	106.6	-	35.5	393.3
3.	P-PE	(-)77.2	(-)120.4	(-)197.54	(-)294.1	(-)236	(-)236.4	(-)143.1	(-)139.5	(-)126.2	(-)32.1	(-)96.1	(-)31.6	
4.	APWL	103.8	224.2	421.74	715.84	951.84	1188.24	1331.84	1470.84	1597.04	1629.14	1725.24	1756.84	
5.	St	118	64	24	5	2	1	1	1	1	1	1	1	
6.	DSt	117	54	40	19	3	1	0	0	0	0	0	0	
7.	AE	117	54	50.4	38.1	87.6	8.6	50.8	38.1	40.6	106.6	-	35.5	627.3
8.	WD	39.9	66.4	157.54	275.1	233	235.4	143.1	133.5	126.2	32.1	96.1	31.6	1569.94
9.	WS	-	-	-	-	-	-	-	-	-	-	-	-	-

Ia = 73.72%   Ih = 0   Im = (-) 44.23%   Ima = 29.46%

Climatological Water Balance – Computation for Anantapur District, Andhra Pradesh, India for the year 1953

Field capacity : 200 mm

All values are in mm

Line	Item	Jan	Feb	Mar	Apl	May	June	July	August	Sept	Oct	Nov	Dec	Annual
1.	PE	77.2	120.4	207.94	313.2	326.6	244.0	193.9	177.6	166.8	132.7	96.1	67.1	2129.54
2.	P	-	-	-	50.4	22.9	55.8	137.1	20.3	134.6	218.5	0.2	-	639.60
3.	P-PE	77.2	120.4	207.94	262.8	303.7	188.2	56.8	157.3	32.2	85.8	95.9	67.1	
4.	APWL	144.3	264.7	472.64	735.44	1039.14	1227.34	1284.14	1441.44	1473.64	0	95.9	163	
5.	St	96	52	18	5	1	1	1	1	1	86.8	123	88	
6.	ΔSt	46	44	34	13	4	0	0	0	0	85.8	36.2	35	
7.	AE	46	44	34	63.4	26.9	55.8	137.1	20.3	134.6	304.3	36.4	35	937.8
8.	WD	31.2	76.4	173.94	249.8	299.7	188.2	56.8	157.3	32.2	171.6	59.7	32	1528.94
9.	WS	0	-	-	-	-	-	-	-	-	-	-	-	-

Ia = 71.80%   Ih = 0   Im = (-) 43.08%   Ima = 44.04%

Agricultural Drought   231

Climatological Water Balance – Computation for Anantapur District, Andhra Pradesh, India for the year 1954

Field capacity : 200 mm

All values are in mm

Line	Item	Jan	Feb	Mar	Apl	May	June	July	August	Sept	Oct	Nov	Dec	Annual
1.	PE	77.2	120.4	207.94	313.2	326.6	244.0	193.9	177.6	166.8	132.7	96.1	67.1	2129.54
2.	P	-	-	-	10.2	40.6	30.4	109.2	63.5	35.5	124.4	-	-	413.8
3.	P-PE	(-)77.2	(-)120.4	(-)207.94	(-)303.0	(-)286.0	(-)213.6	(-)84.7	(-)114.1	(-)131.3	(-)83	(-)96.1	(-)67.1	
4.	APWL	144.3	264.7	472.64	775.64	1061.64	1275.24	1359.94	1477.04	1605.34	1613.64	1709.74	1776.89	
5.	St	96	52	18	4	1	1	1	1	1	1	1	1	
6.	DSt	95	44	34	14	3	0	0	0	0	0	0	0	
7.	AE	95	44	34	24.2	43.6	30.4	109.2	63.5	35.5	124.4	0	0	603.8
8.	WD	17.8	76.4	173.94	28.9	28.3	213.6	84.7	114.1	131.3	8.3	96.1	67.1	1300.64
9.	WS	0	-	-	-	-	-	-	-	-	-	-	-	-

Ia = 61.1%   Ih = 0   Im = (-) 36.66%   Ima = 28.35%

Climatological Water Balance – Computation for Anantapur District, Andhra Pradesh, India for the year 1955

Field capacity : 200 mm

All values are in mm

Line	Item	Jan	Feb	Mar	Apl	May	June	July	August	Sept	Oct	Nov	Dec	Annual
1.	PE	77.2	120.4	207.94	313.2	326.6	244.0	193.9	177.6	166.8	132.7	96.1	67.1	2129.54
2.	P	-	-	10.2	30.5	106.7	30.4	55.4	101.6	167.6	104.1	25.4	2.5	634.4
3.	P-PE	77.2	120.4	197.74	282.7	219.9	213.6	138.5	76.0	0.8	28.6	70.7	64.6	
4.	APWL	141.8	262.2	459.94	742.64	962.54	1176.14	1314.64	1390.62	0	28.6	99.3	163.9	
5.	St	97	53	20	5	2	1	1	1	1.8	173	121	87	
6.	DSt	10	44	33	15	3	1	0	0	0.8	172.2	52	34	
7.	AE	10	44	43.2	45.5	109.7	31.4	55.4	101.6	165.8	276.3	77.4	36.5	996.5
8.	WD	67.2	76.4	164.74	267.7	216.9	212.6	138.5	76	1	143.6	18.7	30.6	1413.94
9.	WS	-	-	-	-	-	-	-	-	-				

Ia = 66.4%   Ih = 0   Im = (-) 39.84%   Ima = 46.79%

Climatological Water Balance – Computation for Anantapur District, Andhra Pradesh, India for the year 1956

Field capacity : 200 mm

All values are in mm

Line	Item	Jan	Feb	Mar	Apl	May	June	July	August	Sept	Oct	Nov	Dec	Annual
1.	PE	77.2	120.4	207.94	313.2	326.6	244.0	193.9	177.6	166.8	132.71	96.1	67.1	2129.54
2.	P	12.7	1.0	-	50.9	47.7	56.9	99.2	28.4	170.0	222.2	70.1	80.	767.4
3.	P-PE	(-)64.5	(-)119.4	(-)207.94	(-)262.3	(-)278.9	(-)187.1	(-)194.7	(-)149.2	(+)3.2	(+)89.5	(-)26.0	(-)58.8	
4.	APWL	123.3	142.7	350.64	612.94	891.84	1078.94	1173.64	1322.84	0	0	26.0	84.8	
5.	St	109.0	97	34	9	2	1	1	1	4.2	93.7	175	130	
6.	ΔSt	21	12	63	25	7	1	0	0	3.2	89.5	93.7	45	
7.	AE	33.7	13	63	75.9	54.7	57.9	99.2	28.4	173.2	311.7	163.8	53.3	1127.2
8.	WD	43.5	107.4	144.94	237.3	271.9	186.1	94.7	149.2	6.4	179	67.7	13.8	150.194
9.	WS	0	-	-	-	-	-	-	-	-	-	-	-	-

Ia = 70.53%   Im = (-) 37.78%   Ih=0   Ima = 52.93%

Climatological Water Balance – Computation for Anantapur District, Andhra Pradesh, India for the year 1957

Field capacity :   **All values are in mm**

Line	Item	Jan	Feb	Mar	Apl	May	June	July	August	Sept	Oct	Nov	Dec	Annual
1.	PE	77.2	120.4	207.94	313.2	326.6	244.0	193.9	177.6	166.8	132.71	96.1	67.1	2129.54
2.	P	-	2.3	10.9	16.0	53.1	110.7	52.1	47.0	77.7	87.1	42.1	0.2	499.2
3.	P-PE	(-)77.2	(-)118.1	(-)197.04	(-)297.2	(-)273.5	(-)133.3	(-)141.8	(-)130.6	(-)89.1	(-)45.6	(-)34.0	(-)66.9	
4.	APWL	144.1	262.2	459.24	756.44	1029.94	1163.24	1305.04	1435.64	1524.74	1570.34	1604.34	1671.24	
5.	St	98	53	20	5	1	1	1	1	1	1	1	1	
6.	ΔSt	97	45	33	15	4	0	0	0	0	0	0	0	
7.	AE	97	47.3	43.9	31.0	57	111.00	52	47	78	87	42	0.2	693
8.	WD	20	73	164	282	270	133	142	131	89	46	54	67	1471
9.	WS	0	-	-	-	-	-	-	-	-	-	-	-	

Ia = 69.09%   Ih = 00   Im = (-) 41.5 %   Ima = 32.6%

AP WL = Dcc (P-PE) + Jan (P-PE)

Δ St = Dcc St - Jan St

$$AE = P + \Delta St$$
$$WD = AE \sim PE$$

Climatological Water Balance – Computation for Anantapur District, Andhra Pradesh, India for the year 1958

Field capacity : 200 mm   **All values are in mm**

Line	Item	Jan	Feb	Mar	Apl	May	June	July	August	Sept	Oct	Nov	Dec	Annual
1.	PE	77.2	120.4	207.94	313.2	326.6	244.0	193.9	177.6	166.8	132.7	96.1	67.1	2129.54
2.	P	-	-	-	25.4	91.4	30.00	85.5	106.7	121.9	111.7	35.5	-	608.6
3.	P-PE	(-)77.2	(-)120.4	(-)207.94	(-)287.8	(-)235.2	(-)213.5	(-)108.4	(-)70.9	(-)44.9	(-)21	(-)60.6	(-)67.1	
4.	APWL	144.3	264.7	473	761	996	1210	1318	1389	1434	1455	1516	1583	
5.	St	96	52	18	5	2	1	1	1	1	1	1	1	
6.	DSt	95	44	34	13	3	1	0	0	0	0	0	0	
7.	AE	95	44	34	38	94	32	86	107	122	112	36	0	800
8.	WD	18	76	174	275	233	212	108	71	45	21	60	67	1360
9.	WS	0	-	-	-	-	-	-	-	-	-	-	-	-

Ia = 63.86%   Ih = 0   Im = (-) 38.32%   Ima = 37.57%

Climatological Water Balance – Computation for Anantapur District, Andhra Pradesh, India for the year 1959
Field capacity : 200 mm
All values are in mm

Line	Item	Jan	Feb	Mar	Apl	May	June	July	August	Sept	Oct	Nov	Dec	Annual
1.	PE	77.2	120.4	207.94	313.2	326.6	244.0	193.9	177.6	166.8	132.7	96.1	67.1	2129.54
2.	P	-	-	-	12.7	48.8	106.6	81.3	63.5	130.2	25.4	-	-	419.75
3.	P-PE	(-) 77.2	(-) 120.4	(-) 207.94	(-) 300.5	(-) 277.8	(-) 137.4	(-) 112.6	(-) 114.1	(-) 36.6	(-) 107.3	(-) 96.16	(-) 67.1	
4.	APWL	144	264	472	773	1051	1188	1301	1415	1452	1559	1655	1722	
5.	St	96	52	18	4	1	1	1	1	1	1	1	1	
6.	DSt	95	44	34	14	3	0	0	0	0	0	0	0	
7.	AE	95	44	34	27	52	107	81	64	130	25	-	-	659
8.	WD	18	76	174	286	275	137	113	114	37	108	96	67	1501
9.	WS	0	-	-	-	-	-	-	-	-	-	-	-	

Ia = 70.5%   Ih = 0   Im = (-) 42.3%   Ima = 30.95%

Climatological Water Balance – Computation for Anantapur District, Andhra Pradesh, India for the year 1960

Field capacity : 200 mm

All values are in mm

Line	Item	Jan	Feb	Mar	Apl	May	June	July	August	Sept	Oct	Nov	Dec	Annual
1.	PE	77.2	120.4	207.94	313.2	326.6	244.0	193.9	177.6	166.8	132.7	96.1	67.1	2129.54
2.	P	-	-	-	12.7	50.8	25.4	51.8	12.7	231.1	35.5	55.9	-	475.9
3.	P-PE	(-)	(-)	(-)	(-)	(-)	(-)	(-)	(-)	(-)	(-)	(-)	(-)	
4.	APWL	144	264	472	773	1150	1369	1511	1676	1740	1837	1877	1944	
5.	St	96	52	18	4	1	1	1	1	1	1	1	1	
6.	DSt	95	44	34	14	3	0	0	0	0	0	0	0	
7.	AE	95	44	34	27	54	25	52	13	231	36	56	0	667
8.	WD	18	76	174	286	273	219	142	165	46	97	40	67	1603
9.	WS	-	-	-	-	-	-	-	-	-	-	-	-	

$Ia = 75.3\%$ $Ih = 0$ $Im = (-)\ 45.16\%$ $Ima = 31.32\%$

Climatological Water Balance – Computation for Anantapur District, Andhra Pradesh, India for the year 1961

Field capacity : 200 mm

All values are in mm

Line	Item	Jan	Feb	Mar	Apl	May	June	July	August	Sept	Oct	Nov	Dec	Annual
1.	PE	77.2	120.4	207.94	313.2	326.6	244.0	193.9	177.6	166.8	132.7	96.1	67.1	2129.54
2.	P	-	-	-	15.2	76	73.6	35.7	35.5	30.2	137.2	15.5	-	418.9
3.	P-PE	77.2 (-)	120.4 (-)	207.94 (-)	298 (-)	250.6 (-)	170.4 (-)	158.2 (-)	142.1 (-)	136.6 (-)	4.5 (+)	80.6 (-)	67.1 (-)	
4.	APWL	144.3	264.7	472.64	770.64	1021.24	1191.64	1349.84	1491.94	1628.54	0	80.6	147.7	
5.	St	96	52	18	4	1	1	1	1	1	5.5	133	95	
6.	DSt	0	44	34	14	3	0	0	0	0	4.5	127.5	38	
7.	AE	0	44	34	29.2	79	73.6	35.7	35.5	30.2	141.7	143.0	38	
8.	WD	77.2	76.4	173.94	284	247.6	170.4	158.2	142.1	136.6	9	46.9	29.1	1551.44
9.	WS	0	-	~	-	-	-	-	-	-	-	-	-	

Ia = 72.85%   Ih = 0   Im = 43.72%   Ima = 32.11%

APWL of Nov next to wet Month dry   -80.6

80.6 + 67.1 = 147.7   for Dec. APWL

$$ws = \left(p - {}^{st}_{(200)}\right)$$

Climatological Water Balance – Computation for Anantapur District, Andhra Pradesh, India for the year 1962

Field capacity : 200 mm

All values are in mm

Line	Item	Jan	Feb	Mar	Apl	May	June	July	August	Sept	Oct	Nov	Dec	Annual
1.	PE	77.2	120.4	207.94	313.2	326.6	244.0	193.9	177.6	166.8	132.7	96.1	67.1	2129.54
2.	P				38.1	45.7	43.2	85.5	124.4	83.8	149.8	10.1	68.4	649.0
3.	PPE	(-)	(-)	(-)	(-)	(-)	(-)	(-)	(-)	(-)	(+)	(-)	(+)	
4.	APWL	75.9	196.3	404.24	679.34	960.24	1161.04	1269.44	1322.88	1405.96	0	86	0	
5.	St	136	74	26	6	2	1	1	1	1	18.1	129	130.3	
6.	DSt	0	62	48	20	4	1	0	0	0	17.1	110.9	0	
7.	AE	0	62	48	58.1	49.7	44.2	85.5	124.4	83.8	166.9	121.0	68.4	912
8.	WD	77.2	58.4	159.94	255.1	276.9	199.8	108.4	53.2	83	34.2	24.9	1.3	1332.34
9.	WS	0				0				0				

Ia = 62.56%   Ih = 0   Im = (-) 37.54%   Ima = 42.83%

Climatological Water Balance – Computation for Anantapur District, Andhra Pradesh, India for the year 1963

Field capacity : 200 mm

All values are in mm

Line	Item	Jan	Feb	Mar	Apl	May	June	July	August	Sept	Oct	Nov	Dec	Annual
1.	PE	77.2	120.4	207.94	313.2	326.6	244.0	193.9	177.6	166.8	132.7	96.1	67.1	2129.54
2.	P	-	-	6.3	30.8	40.6	30.5	33	101.6	106.7	152.4	6.3	-	508.2
3.	PPE	77.2	120.4	201.64	282.4	286	213.5	160.9	76	60.1	19.7	89.8	67.6	
4.	APWL	144.3	264.7	466.34	748.74	1034.74	1248.24	1409.14	1505.14	1565.24	0	89.8	156.9	
5.	St	96	52	19	5	1	1	1	1	1	20.7	127	90	
6.	DSt	6	44	33	14	4	0	0	0	0	19.7	106.0	37	
7.	AE	6	44	39.3	44.8	44.6	30.5	33	101.6	106.7	172.1	112.3	37	771.9
8.	WD	71.0	76	169	268	282	214	161	76	60	39	16	30	1462
9.	WS	0	-	-	-	-	-	-	-	-	-	-	-	

Ia = 68.65%   Ih = 0   Im = (-) 41.19%   Ima = 36.25%

Climatological Water Balance – Computation for Anantapur District, Andhra Pradesh, India for the year 1964

Field capacity : 200 mm

All values are in mm

Line	Item	Jan	Feb	Mar	Apl	May	June	July	August	Sept	Oct	Nov	Dec	Annual
1.	PE	77.2	120.4	207.94	313.2	326.6	244.0	193.9	177.6	166.8	132.7	96.1	67.1	2129.54
2.	P	-	-	-	6.3	10.1	80.9	134.6	35.5	210.8	40.2	43.2	-	561.6
3.	P-PE	(-)	(-)	(-)	(-)	(-)	(-)	(-)	(-)	(+)	(-)	(-)	(-)	
		77.2	120.4	207.94	306.9	316.5	163.1	59.3	142.1	44.0	92.5	52.9	67.1	
4.	APWL	144.3	264.7	472.64	779.54	1096.9	1259.14	1318.44	1460.54	0	92.5	145.4	212.50	
5.	St	96	52	18	4	1	1	1	1	45	125	98	68	
6.	DSt	28	44	34	14	3	0	0	0	44	80	27	30	
7.	AE	28	44	34	20.3	13.1	80.9	134	35.5	254.8	120.2	70.2	30	865
8.	WD	49	76	174	293	314	163	60	142	88	13	26	37	1435
9.	WS	0	-	-	-	-	-	-	-	-	-	-	-	

$Ia = 67.39\%$ $Ih = 0$ $Im = (-)\ 40.43\%$ $Ima = 40.62\%$

Climatological Water Balance – Computation for Anantapur District, Andhra Pradesh, India for the year 1965

Field capacity : 200 mm

All values are in mm

Line	Item	Jan	Feb	Mar	Apl	May	June	July	August	Sept	Oct	Nov	Dec	Annual
1.	PE	77.2	120.4	207.94	313.2	326.6	244.0	193.9	177.6	166.8	132.7	96.1	67.1	
2.	P	-	-	-	20.3	10.1	43.2	22.8	132.0	86.3	-	5.1	33.0	
3.	P-PE	(-) 77.2	(-) 120.4	(-) 207.94	(-) 292.9	(-) 316.5	(-) 200.8	(-) 171.1	(-) 45.6	(-) 80.5	(-) 132.7	(-) 91.0	(-) 34.1	
4.	APWL	111.3	231.7	439.64	732.54	1049.04	1269.84	1420.94	1466.54	1547.04	1679.74	1770.74	1804.84	
5.	St	114	62	22	5	1	1	1	1	1	1	1	1	
6.	ΔSt	113	52	40	17	4	0	0	0	0	0	0	0	
7.	AE	113	52	40	37.3	14.1	43.2	22.8	132.0	86.3	-	5.1	33	578.8
8.	WD	36	68	168	276	313	201	171	46	81	133	91	34	1618
9.	WS	-	-	-	-	-	-	-	-	-	-	-	-	

Ia = 75.98%   Ih = 0   Im = (-) 45.59%   Ima = 27.18%

Climatological Water Balance – Computation for Anantapur District, Andhra Pradesh, India for the year 1966

Field capacity : 200 mm

All values are in mm

Line	Item	Jan	Feb	Mar	Apl	May	June	July	August	Sept	Oct	Nov	Dec	Annual
1.	PE	77.2	120.4	207.94	313.2	326.6	244.0	193.9	177.6	166.8	132.7	96.1	67.1	2129.54
2.	P	32.0	-	-	15.3	38.1	83.8	66.0	88.1	137.2	81.3	86.3	12.7	640.8
3.	P-PE	(-) 45.2	(-) 120.4	(-) 207.94	(-) 297.9	(-) 288.5	(-) 160.2	(-) 127.9	(-) 89.5	(-) 29.6	(-) 51.4	(-) 9.8	(-) 54.4	
4.	APWL	99.6	220	427.94	725.84	1014.34	1174.54	1302.44	1391.94	1421.54	1472.94	1482.74	1537.14	
5.	St	120	66	23	5	1	1	1	1	1	1	1	1	
6.	DSt	119	54	43	18	4	0	0	0	0	0	0	0	
7.	AE	151	54	43	33.3	42.1	84.8	66.0	88.1	137.2	81.3	86.3	12.7	879.8
8.	WD	74	66	165	280	28.5	159.0	128	90	30	51.0	10	54	1392
9.	WS	-	-	-	-	-	-	-	-	-	-	-	-	

I_a = 65.37% I_m = (-) 39.22% I_{ma} = 41.31%

Climatological Water Balance – Computation for Anantapur District, Andhra Pradesh, India for the year 1967

Field capacity : 200 mm

All values are in mm

Line	Item	Jan	Feb	Mar	Apl	May	June	July	August	Sept	Oct	Nov	Dec	Annual
1.	PE	77.2	120.4	207.94	313.2	326.6	244.0	193.9	177.6	166.8	138.7	96.1	67.1	2129.54
2.	P	-	-	10.2	10.2	65.9	43.2	116.2	17.8	121.9	119.8	12.1	10.2	527.50
3.	P-PE	(-) 77.2	(-) 120.4	(-) 197.74	(-) 303	(-) 260.7	(-) 200.8	(-) 77.7	(-) 159.8	(-) 44.9	(-) 18.9	(-) 84	(-) 56.9	
4.	APWL	134.1	254.5	452.24	755.24	1015.94	1216.74	1294.44	1454.24	1499.14	1518.04	1602.04	1658.94	
5.	St	102	55	20	5	1	1	1	1	1	1	1	1	
6.	DSt	101	47	33	15	4	0	0	0	0	0	0	0	
7.	AE	101	47	43.2	25.2	69.9	43.2	116.2	17.8	121.9	119.8	12.1	10.2	727.5
8.	WD	23.8	73.4	164.74	288	256.7	200.8	77.7	159.8	44.9	18.9	84	56.9	1449.64
9.	WS	-	-	-	-	-	-	-	-	-	-	-	-	-

Ia = 68.07%   Ih = 0   Im = (-) 40.84%   Ima = 34.16%

Climatological Water Balance – Computation for Anantapur District, Andhra Pradesh, India for the year 1968

Field capacity : 200 mm

All values are in mm

Line	Item	Jan	Feb	Mar	Apl	May	June	July	August	Sept	Oct	Nov	Dec	Annual
1.	PE	77.2	120.4	207.94	313.2	326.6	244.0	193.9	177.6	166.8	138.7	96.1	67.1	2129.54
2.	P	-	16.7	16.4	30.3	39.8	42	39.4	6.2	184.9	83.0	28.8	10.3	497.70
3.	P-PE	(-)	(-)	(-)	(-)	(-)	(+)	(-)	(-)	(+)	(-)	(-)	(-)	
4.	APWL	134.1	237.8	429.34	712.24	999.04	1201.04	1355.54	1526.94	0	55.7	123	179.9	
5.	St	102	60	23	5	1	1	1	1	19.1	151	107	80	
6.	DSt	22	42	37	18	4	0	0	0	18.1	131.9	44	27	
7.	AE	22	58.7	53.4	48.3	43.8	42.0	39.4	6.2	203	214.9	72.8	37.2	841.7
8.	WD	55.2	61.7	154.54	264.9	282.8	202.0	154.5	171.4	36.2	76.2	23.3	29.9	1457.44
9.	WS	-	-	-	-	-	-	-	-	-	-	-	-	-

Ia = 68.44%   Im = (-) 41.06%   Ima = 39.52%

Climatological Water Balance – Computation for Anantapur District, Andhra Pradesh, India for the year 1969

Field capacity : 200 mm

All values are in mm

Line	Item	Jan	Feb	Mar	Apl	May	June	July	August	Sept	Oct	Nov	Dec	Annual
1.	PE	77.2	120.4	207.94	313.2	326.6	244.0	193.9	177.6	166.8	132.7	96.1	67.1	2129.54
2.	P	-	-	-	5.4	91.2	21.9	23.7	163.3	27.4	142.9	8.5	11.2	495.5
3.	P-PE	(-) 77.2	(-) 120.4	(-) 207.94	(-) 307.8	(-) 235.4	(+) 222.1	(-) 170.2	(-) 14.3	(-) 139.4	(-) 10.2	(-) 87.6	(-) 55.9	
4.	APWL	133.1	253.5	461.44	769.24	1004.64	1226.74	1396.94	1411.24	1550.64	0	87.6	143.5	
5.	St	102	56	19	5	1	1	1	1	1	11.2	128	96	
6.	DSt	6	46	37	14	4	0	0	0	0	10.2	116.8	32	
7.	AE	6	46	37	19.4	95.2	21.9	23.7	163.3	27.4	153.1	125.3	43.2	761.5
8.	WD	71.2	82.4	170.94	293.8	231.4	222.1	170.2	14.3	139.4	20.4	29.2	23.9	1469.24
9.	WS	-	-	-	-	-	-	-	-	-	-	-	-	-

Ia = 68.99%   Im = (-) 41.40%   Ima = 35.76%   Ih = 0

Climatological Water Balance – Computation for Anantapur District, Andhra Pradesh, India for the year 1970

Field capacity : 200 mm

All values are in mm

Line	Item	Jan	Feb	Mar	Apl	May	June	July	August	Sept	Oct	Nov	Dec	Annual
1.	PE	77.2	120.4	207.94	313.2	326.6	244.0	193.9	177.6	166.8	132.7	96.1	67.1	2129.54
2.	P	0.6	1.1	-	11.3	108.4	18.7	25.0	69.7	147.5	117.3	3.4	-	503.00
3.	P-PE	(-)	(-)	(-)	(-)	(-)	(-)	(-)	(-)	(-)	(-)	(-)	(-)	
4.	APWL	143.7	263	470.94	772.94	991.14	1216.44	1385.34	1493.24	1512.54	1527.94	1620.64	1687.74	
5.	St	96	53	18	4	2	1	1	1	1	1	1	1	
6.	DSt	95	43	35	14	2	1	0	0	0	0	0	0	
7.	AE	95.6	44.1	35	25.3	110.4	19.7	25.0	69.7	147.5	117.3	3.4	-	693.00
8.	WD	18.4	76.3	172.94	287.9	216.2	224.3	168.9	107.9	19.3	15.4	92.7	67.1	1467.34
9.	WS	-	-	-	-	-	-	-	-	-	-	-	-	-

Ia = 60.90% Im = (-) 41.34% Ima = 32.54%

Climatological Water Balance – Computation for Anantapur District, Andhra Pradesh, India for the year 1971

Field capacity : 200 mm

All values are in mm

Line	Item	Jan	Feb	Mar	Apl	May	June	July	August	Sept	Oct	Nov	Dec	Annual
1.	PE	77.2	120.4	207.94	313.2	326.6	244.0	193.9	177.6	166.8	132.7	96.1	67.1	2129.54
2.	P	1.6	1.0	2.3	13.3	41.5	23.1	25.7	96.4	117.2	165.8	6.8	1.4	496.1
3.	P-PE	(-)	(-)	(-)	(-)	(-)	(-)	(-)	(-)	(-)	(+)	(-)	(-)	
4.	APWL	141.3	260.7	466.34	766.24	1051.34	1272.24	1440.44	1521.64	1571.24	0	89.3	65.7	
5.	St	98	54	19	5	1	1	1	1	1	34.1	127	91	
6.	DSt	7	44	35	14	4	0	0	0	0	33.1	92.9	36	
7.	AE	7.6	45	37.3	27.3	45.5	23.1	25.7	96.4	117.2	198.9	99.7	37.4	761.1
8.	WD	69.6	75.4	170.64	285.9	281.1	220.9	168.2	81.2	49.6	66.2	3.6	58.7	1531.04
9.	WS	-	-	-	-	-	-	-	-	-	-	-	-	-

Ia = 71.9%   Im = (-) 43.14%   Ima = 35.74%

Climatological Water Balance – Computation for Anantapur District, Andhra Pradesh, India for the year 1972

Field capacity : 200 mm

All values are in mm

Line	Item	Jan	Feb	Mar	Apl	May	June	July	August	Sept	Oct	Nov	Dec	Annual
1.	PE	77.2	120.4	207.94	313.2	326.6	244.0	193.9	177.6	166.8	132.7	96.1	67.1	2129.54
2.	P	-	-	-	8.3	113.9	43.8	13.3	6.5	140.2	147.4	42.0	8.4	523.8
3.	P-PE	(-) 77.2	(-) 120.4	(-) 207.94	(-) 304.9	(-) 212.7	(-) 200.2	(-) 180.6	(-) 171.1	(-) 26.6	(+) 14.7	(-) 54.1	(-) 58.7	
4.	APWL	135.9	256.3	464.24	769.14	981.84	1182.04	1362.64	1533.74	1560.34	0	54.1	112.8	
5.	St	100	55	19	5	2	1	1	1	1	15.7	152	113	
6.	DSt	13	45	36	14	3	1	0	0	0	14.7	136.8	39	
7.	AE	13	45	36	26.3	116.9	44.8	13.3	6.5	140.2	162.1	178.8	47.4	830.3
8.	WD	64.2	75.4	171.94	286.9	209.7	199.2	180.6	171.1	26.6	29.4	82.7	197	1517.44
9.	WS	-	-	-	-	-	-	-	-	-	-	-	-	-

I_a = 71.26% I_m = (-) 42.75% I_{ma} = 38.99%

Climatological Water Balance – Computation for Anantapur District, Andhra Pradesh, India for the year 1973

Field capacity : 200 mm

All values are in mm

Line	Item	Jan	Feb	Mar	Apl	May	June	July	August	Sept	Oct	Nov	Dec	Annual
1.	PE	77.2	120.4	207.94	313.2	326.6	244.0	193.9	177.6	166.8	132.7	96.1	67.1	2129.54
2.	P	-	-	-	2.9	17.2	66.9	26.1	91.5	193.8	177.3	21.5	2.1	599.3
3.	P-PE	(-)	(-)	(-)	(-)	(-)	(-)	(-)	(-)	(-)	(+)	(-)	(-)	
4.	APWL	142.2	262.6	470.54	780.84	1090.24	1267.34	1435.14	1521.24	0	0	74.6	1396	
5.	St	97	53	18	4	1	1	1	1	28	72.6	137.0	98	
6.	DSt	1	44	35	14	3	0	0	0	27	44.6	64.4	39	
7.	AE	1	44	35	16.9	20.2	66.9	26.1	91.5	220.8	221.9	85.9	41.1	871.3
8.	WD	76.2	76.4	172.94	296.3	320.4	177.1	167.8	86.1	54	89.2	10.2	26	1552.64
9.	WS	-	-	-	-	-	-	-	-	-	-	-	-	-

Ia = 72.91%   Im = (-) 43.75%   Ima = 40.91%

Climatological Water Balance – Computation for Anantapur District, Andhra Pradesh, India for the year 1974

Field capacity : 200 mm   **All values are in mm**

Line	Item	Jan	Feb	Mar	Apl	May	June	July	August	Sept	Oct	Nov	Dec	Annual
1.	PE	77.2	120.4	207.94	313.2	326.6	244.0	193.9	177.6	166.8	132.7	96.1	67.1	2129.54
2.	P	-	-	0.9	12.9	107.4	45.3	50.3	30.7	217.7	164.6	2.7	-	623.5
3.	P-PE	(-) 77.2	(-) 120.4	(-) 207.04	(-) 300.3	(-) 219.2	(-) 198.7	(-) 143.6	(-) 146.9	(+) 50.9	(+) 31.9	(-) 93.4	(-) 67.1	
4.	APWL	144.3	264.7	471.74	772.04	991.24	1189.94	1333.54	1480.44	0	0	93.4	160.5	
5.	St	96	52	18	4	2	1	1	1	51.9	83.8	125	88	
6.	DSt	8	44	34	14	2	1	0	0	50.9	31.9	41.2	37	
7.	AE	8	44	34.9	26.9	109.4	46.3	50.3	30.7	268.6	196.5	43.9	37	896.5
8.	WD	69.2	76.4	173.04	286.3	217.2	197.7	143.6	146.9	101.8	63.8	52.2	30.1	1558.24
9.	WS	-	-	-	-	-	-	-	-	-	-	-	-	-

$I_a = 73.17\%$ $I_m = I_h - 0.6\, I_a = (-)\, 43.9\%$ $I_{ma} = 42.1\%$

Climatological Water Balance – Computation for Anantapur District, Andhra Pradesh, India for the year 1975

Field capacity : 200 mm

All values are in mm

Line	Item	Jan	Feb	Mar	Apl	May	June	July	August	Sept	Oct	Nov	Dec	Annual
1.	PE	77.2	120.4	207.94	313.2	326.6	244.0	193.9	177.6	166.8	132.7	96.1	67.1	2129.54
2.	P	-	-	12.4	2.5	48.0	21.3	128.5	87.3	127.9	247.9	72.4	1.0	749.2
3.	P-PE	(-)	(-)	(-)	(-)	(-)	(-)	(-)	(-)	(-)	(+)	(-)	(-)	
		77.2	120.4	195.54	310.7	278.6	222.7	65.4	90.3	38.9	115.2	23.7	66.1	
4.	APWL	143.3	263.7	459.24	769.94	1048.54	1271.24	1336.64	1426.94	1465.84	0	23.7	89.8	
5.	St	97	52	20	4	1	1	1	1	1	116.2	177	127	
6.	DSt	30	45	32	16	3	0	0	0	0	115.2	60.8	50	
7.	AE	30	45	44.4	18.5	51.0	21.3	128.5	87.3	127.9	363.1	133.2	51.0	1101.2
8.	WD	47.2	75.4	163.54	294.7	275.6	222.7	65.4	90.3	38.9	230.4	37.1	16.1	1557.34
9.	WS	-	-	-	-	-	-	-	-	-	-	-	-	-

$$Ia = \frac{155.734}{212.954} \times 100 = 73.13\%$$

$$Ima = \frac{AE}{PE} \times 100 = \frac{110.12}{212.954} \times 100 = 51.71\%$$

$$Im = Ih - 0.6\ Ia = 0 - 43.88 = (-)\ 43.88\%$$

$$WD = AE \sim PE$$

$$AE = P + \Delta St$$

$$APWL = Dec\ (P\text{-}PE) + Jan\ (P\text{-}PE)$$

$$\Delta St = Dec\ (St) - Jan\ (St)$$

Agricultural Drought   253

Components and Indices derived from climatological water balance computation for Anantapur District, Andhra Pradesh, India for period 30 years (1946 – 1975)

(All are in mm)

Year	Components					Year	Indices			
	P	PE	AE	WD	WS		Ia%	Ih%	Im%	Ima%
1946	608.3	2129.54	944	1576	0	1946	74	0	(-) 43.51	44.33
1947	503.9	2129.54	905.9	1522.14	0	1947	71.5	0	(-) 42.9	33.15
1948	483.7	2129.54	675.4	1494	0	1948	70.16	0	(-) 42.1	31.72
1949	599.6	2129.54	789.6	1375	0	1949	64.57	0	(-) 38.74	37.08
1950	538.3	2129.54	728.3	1436.84	0	1950	67.47	0	(-) 40.48	34.20
1951	394.8	2129.54	584.8	1580.34	0	1951	74.21	0	(-) 44.53	27.46
1952	393.3	2129.54	627.3	1569.94	0	1952	73.72	0	(-) 44.25	29.46
1953	639.6	2129.54	937.8	1528.94	0	1953	71.80	0	(-) 43.05	44.04
1954	413.8	2129.54	603.8	1300.64	0	1954	**61.1**	0	**(-) 36.66**	28.35
1955	634.4	2129.54	996.5	1413.94	0	1955	66.4	0	(-) 39.84	46.79
1956	767.4	2129.54	1127.2	1501.94	0	1956	70.53	0	(-) 42.32	**52.93**
1957	499.2	2129.54	693.0	1471.00	0	1957	69.09	0	(-) 41.50	32.6
1958	608.2	2129.54	800.0	1360.00	0	1958	63.80	0	(-) 38.32	37.57
1959	419.75	2129.54	659.0	1501.0	0	1959	70.50	0	(-) 42.3	30.95
1960	475.90	2129.54	667.0	1603.0	0	1960	75.3	0	(-) 45.16	31.32
1961	418.90	2129.54	683.9	1551.44	0	1961	72.85	0	(-) 43.72	32.11

Components						Indices				
Year	P	PE	AE	WD	WS	Year	Ia%	Ih%	Im%	Ima%
1962	649.00	2129.54	912.0	1332.34	0	1962	62.56	0	(-)37.54	42.83
1963	508.20	2129.54	771.9	1462.0	0	1963	68.65	0	(-)41.19	36.25
1964	561.6	2129.54	865	1435	0	1964	67.39	0	(-)40.43	40.62
1965	352.8	2129.54	578.8	1618	0	1965	75.98	0	(-)45.99	27.18
1966	640.8	2129.54	879.8	1392	0	1966	65.37	0	(-)39.22	41.31
1967	527.5	2129.54	727.5	1449.64	0	1967	68.07	0	(-)40.84	34.16
1968	497.7	2129.54	841.7	1457.44	0	1968	68.44	0	(-)41.06	39.52
1969	495.5	2129.54	761.5	1469.24	0	1969	68.99	0	(-)41.40	35.76
1970	503.0	2129.54	693.0	1467.34	0	1970	60.90	0	(-)41.34	32.54
1971	496.1	2129.54	761.1	1531.04	0	1971	71.90	0	(-)43.14	35.74
1972	523.8	2129.54	830.3	1517.44	0	1972	71.26	0	(-)42.75	38.99
1973	599.3	2129.54	871.3	1552.64	0	1973	72.91	0	(-)43.75	40.91
1974	632.5	2129.54	896.5	1558.24	0	1974	73.17	0	(-)43.90	42.1
1975	749.2	2129.54	1101.20	1557.34	0	1975	73.13	0	(-)43.88	51.71
Total	16136.45 or 537.882 mm annual average	2129.54 mm (A.P)	23915.1 or 797.17 mm (A.A)	44585.86 or 1486.19 mm (A.A)	0		2085.72 i.e. 69.524%	0	1255.41 (-)41.847%	1113.68 37.123%

Note :

P = Precipitation

PE = Potential Evapo-transpiration

AE = Actual Evapo-transpiration

WD = Water deficit

WS = Water Surplus

Ia = Index of aridity

Ih = Index of humidity

Im = Index of moisture

Ima = Index of moisture adequacy

Drought Years in Anantapur District, Andhra Pradesh, India

Table 7.7 As Per Climatological Water Balance Method (1946-1975) (30 Years Period) Departure of Index of Aridity from the Median Percentage

S.No.	Classification of Drought	No.	Details of Year
1	Moderate	5	1957, 1963, 1967, 1968, 1969
2	Large	4	1966, 1964, 1955, 1950
3	Severe	6	1970, 1965, 1962, 1958, 1954, 1949
4	Disastrous	Nil	-

Table 7.8 As Per Chronological Chart of Rainfall (Annual Rainfall) (Period 1946-1975) - 30 years

Deficient Years	17 Nos	1947, 1948, 1951, 1952, 1954, 1957, 1959, 1960, 1961, 1963, 1965, 1967, 1968, 1969, 1970, 1971, 1972

Frequency analysis of annual rainfall (1946-75)

Normal Rainfall = 538 mm

75% dependability = 430 mm

50% dependability = 510 mm

Out of 30 years of study 15 years we can expect 510 mm and 24 years we can expect 430 mm. For 17 years out of 30 years rainfall is below normal rainfall 538 mm. Remaining 13 years surplus (i.e.) 56.67% of time is drought (i.e.) alternative every year is almost a drought year.

Table 7.9   Decennial Frequency of Drought Years

Category	From 1946 to 1955	1956 to 1965	1966 to 1975	Total
Moderate	Nil	2	3	5
Large	2	1	1	4
Severe	2	3	1	6
Disastrous	Nil	Nil	Nil	Nil

This type of analysis of drought intensity is helpful to evolve a better drought management system in the district.

For the analytical study of droughts several methods and indices have been proposed in the past (Subrahmanyam 1967). Most of them are based on the deficiency of rainfall on a yearly, monthly and weekly basis. A clear understanding and better appreciation of the problem of droughts and aridity became available after the water balance approach was put forth by Thornthwaite.

Conclusion for moisture indices   In Anantapur District from the indices arrived from climatological water balance computations the following conclusion is made

(a) The average Index of moisture (Im) being 41.847% and so the area lies in semi arid climate as per ("Moisture Index and climatic types" Revised scheme of Thornthwaite and Mather 1955) i.e., (-) 41.847% between (-66.5% to -33.4%).

(b) In India where droughts are as severe, as they are frequent. It is essential to know the water requirements of individual regions and the extent to which the normal water supply by precipitation is not able to meet those requirements. The ratio of AE to PE varies with available moisture in the soil. The ratio termed the "index of moisture adequacy". This provides a good indication of the moisture status of the soil in relation to the water need, low percentage values of the index signifying poor moisture availability (Subrahmanyam, Subba Rao and Subramaniam, 1963).

Anantapur district "Index of Moisture Adequacy" is 37.122% (annual average) indicating poor moisture availability. Thus the agriculture activity is normally restricted only to the rainy season. Unless otherwise raised as irrigated crop, as the water deficit is existing in the entire district and there is no water surplus.

Moisture adequacy over India   From the records of precipitation and computed data of potential evapotranspiration and the actual evapotranspiration were worked out from all the climatological stations in India according to the 1955 water balance scheme of Thornth Waite using the AE and PE values the indices of moisture adequacy were computed and geographical distribution of this index over India is shown in the Fig.

The highest values of this index are found in north – east India and south western portion of the Peninsula, there are the areas of fairly heavy and prolonged precipitation and consequently they rarely experience moisture deficiencies at least on an annual basis. The index which is higher than 80% in the north – east falls below 10% near Baluchistan in the west. Over the Peninsular high values of the index are generally found along the west coast while in the central Deccan, the values are less than 40% every where.

Since large values of moisture adequacy index are due to the availability of abundant quantities of moisture in relation to water need, in such regions intense agricultural operations throughout the year are either to be expected or may be recommended. On the other hand, in regions of low indices, agricultural activity is normally restricted only to the rainy season. A comparison of the moisture adequacy map with map of vegetation types of India as prepared by champion (1936) shows (Fig enclosed) certain striking features.

Wet evergreen and semi green forests in regions with moisture adequacy indices in excess of 80% and desertic vegetation below 20%. Interestingly the isopleth of 40% for the moisture adequacy index seems to bound semi arid climates with thorn forests both in south and north.

Next an attempt has been made to compare the moisture adequacy map with maps of distribution of some crops in India (Ranahawa 1958). About a third of the cultivated area in India is under rice which requires high temperatures and large amounts of water where irrigation facilities or either lacking or poor, it is therefore, grown mainly in the south west monsoon season when rainfall is usually abundant. Figure shows the areas of rice cultivation not seriously influenced by supplemental irrigation.

Intensive cultivation is found in the northeast and in the coastal belt of Peninsular India, it is significant that these regions are bounded by the isolines of 60% and 100% moisture adequacy index. In north and west India, however, its cultivation is limited either by the low thermal regime of climate or inadequate amounts of precipitation. While rice has high thermal and moisture needs, wheat (fig.) requires low temperatures and has lower water needs and is, therefore primarily grown in the north India. Though also raised over elevated regions in South India in the cool season, it is not a significant crop here. In the north, the small amounts of precipitation produced by the Western disturbance, in winter appear to be adequate for its cultivation. Comparison of fig with the moisture map reveals that wheat – growing areas are generally surrounded by 40% and 60% isolines of the index.

Jowar is an important millet crop which requires high temperatures and tolerates low amounts of moisture, excess water is considered to be even injurious to its cultivation. Its distribution presented in Fig. shows that it is grown mainly in the interior plains of Peninsular and Central India where temperatures are high enough and moisture in the soil is quite low. Even here as in the case of wheat, the bounding lines of moisture adequacy index are 40% and 60% but its high thermal requirements make it essentially a crop of South India.

An interesting conclusion that emerges from this study is that most of the agricultural crops in India do not seem to have favourable conditions of development below the 40% isoline of moisture adequacy index, unless otherwise raised as irrigated crops. It would appear that in the reclamation of vast stretches of land for agricultural production, this index would be very useful for determining the suitable kinds of crops that may be grown with least hazard of **damage from droughts.**

In the successful introduction of exotic tree plantations in silviculture a careful study of the seasonal varitation of this parameter may be a practical utility.

For, it is only when the changing water demands of the plantation, through its vegetation cycle are matched by the seasonal fluctuations of moisture adequacy, that maximum development of the species may be expected.

Thus when systematically studied on a weekly or monthly basis, this index should provide valuable information on the liability to drought and therefore, the climatic suitability of a region for agricultural or forest development.

The highest aridity index of 75.98 percent is noticed in the year 1965 indicating the severity of drought in the district. The lowest aridity of about 61.1 per cent is noticed in the year 1954 indicating a less intensive drought in Anantapur District, A.P.

Another important derivative element determined from the water surplus is an index of humidity (Ih). It is the percentage ratio of water surplus to the water need. This parameter helps to determine the climate of the region as 'moist or dry'. The computed figures have shown that the index of humidity is zero for the entire district. It indicates that the district enjoys a dry climate.

Water balance and droughts   The agriculture potentialities of any region are dependent on the availability of water during different crop growth periods. The availability of water during different crop growth periods. The availability of water for plant growth both in its amount and period can be accurately estimated through the water balance studies of the region. The water balance is not only a handy technique for determining the moisture status of the soil in a particular region, but also helps in assessing the drought conditions and their severities. The climatic water balance on an annual basis is made use of to determine the drought years and their severities in the district area.

Among the water balance indices, aridity index (Ia) is the most suitable parameter for the analytical study of drought conditions with special reference to their frequency. Analysis of droughts and their intensities are made from the percentage departures of aridity index (Ia) from the median value (vide Table). The study shows the nature and intensities of drought conditions experienced by the region during the study period. Categorisation of droughts is made by following the scheme of standard deviation technique employed by Subrahmanyam et.al (1965).

Table 7.10 Categorization of Droughts

Departure of the Ia From the Median Value	Drought Intensity
Less $< \frac{1}{2}\,\sigma$	Moderate
$\frac{1}{2}\,\sigma$ to σ	Large
σ to $2\,\sigma$	Severe
Greater $> 2\,\sigma$	Disastrous

By applying the above taxonomic system the drought years of Anantapur district during the study period of 30 years + (1946 – 1975) are classified (Figure). The percentage departure of aridity index from the median value shows considerable variation in drought conditions prevailing in the district.

During the study period, the district experienced the following Decennial (ten years period) frequency of drought years in Anantapur District, Andhra Pradesh.

From 1946 to 1955

Moderate Droughts = Nil

Large = 2

Severe = 2

Disastrous = Nil

From 1956 to 1965

Moderate Droughts = 2

Large = 1

Severe = 3

Disastrous = Nil

From 1966 to 1975

Moderate Droughts = 3

Large = 1

Severe = 1

Disastrous = Nil

15 drought years are as follows out of 30 years of study period.

Moderate Droughts = 5 Nos (1957, 1963, 1967, 1968, 1969)

Large = 4 (1966, 1964, 1955, 1950)

Severe = 6 (1970, 1965, 1962, 1958, 1954, 1949)

Disastrous = Nil

This type of analysis of drought intensity is helpful to evolve a better drought management system in the district.

However the present study in expected to serve as an initial step in opening a new field of pursuit in drought climatology from the water balance angle.

While knowledge of aridity is essential from the point of view of large scale planning study of droughts is very important for planning short term operations especially in connection with agricultural development.

For the analytical study of droughts several methods and indices have been proposed in the past (Subrahmanyam, 1967). Most of them are based on the deficiency of rainfall on yearly monthly and weekly basis. A clear understanding and better appreciation of the problems of droughts and aridity became available after the water balance approach was put forth by Thornthwaite and Mather in 1955. According to this concept, drought is a situation in which the amount of water needed for maximum evapotranspriration exceeds the amount obtainable by precipitation and from the soil.

Droughts are permanent features of arid climates as conditions in Rajastan attest. In semiarid as Anantapur district subhumid climates as Cuddalore drought is a seasonal phenomenon alternating characteristically with period of adequate or surplus moisture.

The humid and prehumid climates of which Colombo and Mercara are good examples have prevailing water surplus conditions but droughts occur here inter mittently or in opportune as, for example, when the expected rainfall does not came due to an interruption of the monsoonal circulation.

Thus, three kinds of droughts have been recognized in drought climatology.

Permanent droughts   These are characteristic of arid climates in which sparse hardy types of vegetation are fully adopted to the severe water shortages and agriculture is impossible in them without artificial irrigation.

The second are the "**Seasonal droughts**" found in climates with well-recognized rainy and dry seasons. Most of India which has a typical monsoonal climatic regime, suffers from this latter kind of droughts whose incidence duration as well as cessation are usually well-known and established by ages of observation experience and record. Agricultural is successfully practiced in these zones by adjusting the dates of planting and harvesting and also by choosing crop varieties of proper seasonality duration and drought-tolerance the third kind of droughts are called the "**Contingent droughts**" (Tharnth Waite 1947) which result from the rainfall being irregular and variable and which may occur in any season almost anywhere but are typical of and more frequent in the sub humid climates, the so called "borderline", climates, with moist and dry types an either side.

One of the challenging problems of droughts climatology is the rational understanding and effective combating of these contingent droughts whose incidence, duration severity as well as frequency are quite unpredictable.

For the analytical study of droughts with special reference to their intensity and frequency, the aridity index Ia of Tharnthwaite has been found to be a very useful parameter (Subrahmanyan and Subramaniam, 1965) years of drought may be easily identified and their intensities assessed from a study of the departures of the Ia from the median value for the corresponding station, as now been done for Anantapur District of Andhra Pradesh, India. Decennial frequency of droughts (number of drought years in each successive decade interval) is another aspect which could also be studied as was now done for Anantapur District. In the figures these aspects are illustrated by Reference to Anantapur which has a semiarid types of climate possessing highly variable water balance.

From the diagrams it can be readily seen that Anantapur experienced diverse climatic conditions of moisture regime during the period under study (1946-1975). However, the fluctuations of Ia show a greater tendency (at least in respect of frequency) of the climate to lean towards less dry conditions.

Categorization of droughts may be done based on a procedure similar to that employed by Ramdas (1950), but rather than taking the mean deviation as was done by him using the standard deviation as detailed above.

Water balance and climatic shifts   The water balance of any region is not stable over a period of time. Study of climatic shifts due to changes in the moisture regime of climate and their periodicities in another important aspect of drought climatology. Hence it is necessary to study the variation in the index of moisture for certain consecutive years to understand the shifts in the climate to wetter or 'drier' condition. This aspect is of great practical value in long term economic planning. This aspect has been studied for Anantapur District. For the period (1946-1975) for 30 years. The average index of moisture percentage (Im%) for the district was (-) 41.847%. The index values of 30 years are plotted for individual years to understand years to understand the climatic shifts of the District. The study has revealed that the climate of the District is semiarid to Dry semi humid (D to C1) i.e., -66.5% to -33.4%. Though climatic shifts are temporary such study helps to understand the extreme conditions of climate which may result in severe floods and droughts depending upon the intensity and duration of water surplus or deficiency.

The duration of drought spell again is as important a study as its intensity in as much as both of them together determine the ultimate effect of water shortage on economic situation (Subrahmanyam and Sastry (1969).

Water balance and agriculture   The seasonal moisture regime and the length of the day in relation to the march of thermal regime are the major determinants of plant development. Social factors are generally of secondary importance though they play a decisive role in controlling the distribution of species in the primary zones mainly through variations in ground moisture.

In agriculture planning, detailed information on the soil moisture status (i.e) absolute moisture content, its seasonal variation, periods durations and magnitudes of moisture deficiency and surplus etc are very essential.

Agricultural planning of any region especially in low rainfall areas requires a sound knowledge of the local climatology as climate determines to a great extent of its agricultural potential. The suitability of a region for agricultural development can be clearly assessed with "Index of moisture adequacy" which is another important derivative element in water balance study.

Index of moisture adequacy (Ima) is the percentage ratio of the actual evapotranspiration (AE) to potential evapotranspiration (PE). Based on the percentage of moisture adequacy values

The suitability of crops that can be grown successfully in the absence of supplemental irrigation are given by Subrahmanyam et.al (1963) in the following table.

Table 7.11   Moisture adequacy and suitability of crops

Index of Moisture Adequacy Percentage Ima	Crop Suitability
80 – 100	Paddy (high yields)
60 – 80	Paddy (low yields)
Less < 60	Paddy (uneconomical)
40 to 60	Millets
20 to 40	Drought resistant crops
< 20	Unsuitable for irrigation

Index of adequacy (Ima) values of Anantapur district vary from 27.18 to 52.93% (1965 and 1956 respectively). The Ima values range from 27% to 37% in 16 years, from 37% to 47% in 12 years and above 47 in two years. Hence the area (in an average) can be classified as below 40% range which is drought resistant crops such as Jowar, ragi, bajra etc., to be grown. Most of the Ima values of the district are less than 40% (20 years out of thirty years) and remaining ten years also falls in the category of 40 – 60 range which area is also favourable for cultivation of millets.

But in the villages of district the following cropping pattern is being adopted. Kharif is the major crop season in Anantapur District of the 9.75 lakh hectares of gross cropped area in the district in 2006 – 2007 - 7.94 lakh hectares, i.e., 81% of gross cropped area gets cultivated during the kharif season.

Traditionally, food grains dominated the cropping pattern of the district. Minor millets such as samai, varagu, korra and major millets such as sorghum, various pulses and paddy were the major food grains in the district.

An analysis of cropping pattern in Anantapur District over 1960 – 1961 to 2006 – 2007 shows that during early 1960 more than two thirds of gross cropped area was cultivated with food grains predominantly millets, with some amount of pulses and paddy (G.O. AP 2006 GO AP various years). Among non-food crops cotton and groundnut were important.

The area under groundnut has increased fourfold from slightly less than 2 lakh hectare in the early 1960's to 8 lakh hectare by 2005-06. In 1961-62 the area cultivated with millets was much greater than the total ground nut area, while by 2005-06 millets area was a mere 38,000 hectares. Minor millets have more or less disappeared for cultivation while the area under major millets has reduced by 90 percent.

As groundnut is largely cultivated with pulses as an intercrop, the importance of pulses has remained more or less stable over the years. In Anantapur ground nut is essentially a Kharif crop with 98% of total groundnut area being cultivated during Kharif season.

The groundnut (rainfed) is sowed <u>during July</u> and for irrigated groundnut sowing in June. Harvesting is done in the first <u>week of December.</u>

Those farmers who have irrigation facilities and plan to take up irrigated groundnut in Kharif sowing is done in <u>last week of May</u> and harvested in the <u>month of October.</u>

Irrigation groundnut crop is usually followed by sunflower cultivation during the period <u>October to January.</u>

Sunflower crop is followed with groundnut in Rabi season <u>January to August.</u>

Alternatively on irrigated fields that over not suitable for groundnut cultivation three rounds of crop, ragi, paddy and paddy in that order is grown. Ragi is taken from June to August, followed by paddy from Sept. to Dec. end, followed by paddy again during the months January to March.

As per irrigation statistics, under 5353 tanks and 87000 wells, gross irrigated area is 1,54,000 ha and net irrigated area is 1,25,000 ha in the district.

Out of net irrigated 31% is from surface irrigation and 69% is from ground water. Out of 9.75 lakh ha of gross cropped area, rainfed area is 8.21 lakh ha 84.2% and 1.54 lakh ha 15.79% irrigated under tanks and canals and wells.

The district population as 2001 year is 36,40,478 and 70% of the population are dependent on agriculture (i.e.) 25,48,334 nos.

The District Gazetter published in 1905 notes that natural conditions in Anantapur district are extremely unfavourable for agricultural growth. To quote "The nature conditions of Anantapur could scarcely be more inimical to agricultural prosperity than they are. The soil is, most of it, wretchedly infortile the rainfall is light and uncertain, fuel and fodder are scarce, irrigation facilities are few, the indigeneous cattle are bad manure is difficult to get and people are few in number" GoAP 1993 page 67. These conditions remain true even to-day in Anantapur.

South west Monsoon of Anantapur District (Period 1946 to 1975)   South West Monsoon period of rainfall in the district is from June 1st to September 30th. During south west monsoon, the quantity of rainfall is lower than potential evapotranspiration, thus having serious implications for crop growth.

South west monsoon period rainfall from June to Sept.	Potential evapotranspiration during the period June to Sept.
308.94 mm	782.3 mm

Comparison of monthly rainfall to PET

	PET
Rainfall in the month of June 1946 to 1975 $\}$ $\dfrac{1367.8}{30} = 45.59$ mm	244.00 mm
- do - July $\dfrac{1908.1}{30} = 63.60$ mm	193.9 mm
- do - August $\dfrac{2094.1}{30} = 69.80$ mm	177.6 mm
- do - September $\dfrac{3898.3}{30} = 129.94$ mm	166.8 mm

Table 7.12   Rainfall Deficit with Respect to PET

Month	Rainfall mm	PET mm	Deficit mm
June	45.59	244.0	81.32%
July	63.60	193.9	67.20%
August	69.80	177.6	61.00%
September	129.94	166.8	22.00%

Due to this the soil moisture stress condition under different stages of crop growth would result in inadequate plant population, high percentage of flower drop, poor seed scatting etc., and there by implications for crop yields.

Coefficient of variation of rainfall for the period 1946 – 1975

$$C.V. = \frac{\text{Standard deviation}}{\text{Mean}} \%$$

$$\sigma = \sqrt{\frac{\sum\left(x - \overline{x}\right)^2}{n-1}}$$

Total rainfall from 1946 – 1975 (30 years) = 16136.45 mm

$$\overline{x} = \frac{\sum x}{n} \quad \text{mean} \quad \frac{16136.45}{30} = 537.88 \text{ say } 538 \text{ mm}$$

$$\frac{\left(x - \overline{x}\right)^2}{n-1} = \frac{292032.17}{29} = 10070.07$$

n - 1 = 29 years

Standard deviation $= \sqrt{10070.07} = 100.35$

$$C.V. = \frac{100.35}{538} = 18.35\%$$

Coefficient of variation of rainfall varies from place to place and its value ranges from 15 to 70 C.V. is least in regions of high rainfall and largest in regions of scanty rainfall. The area being in zone of scanty rainfall.

Frequency analysis of Rainfall in Anantapur District

The return period (or) recurrence interval (or) simply the frequency is denoted by Weibul's formula $Tr = \frac{n+1}{m}$.

If a plot is made between the value of the random variable as abscissa and the corresponding exceedence probability as ordinate, the resulting graph is called a distribution graph or probability plot (or) an empirical distribution.

The return period indicates the average number of years within which a given event will be equaled or exceeded.

When the frequency analysis is carried out on annual rainfall information such as 50% dependable rainfall or 75% dependable rainfall can be obtained.

From the analysis the following results are obtained (1946 to 1975 – Anantapur District)

Normal rainfall = 538 mm

75% dependability = 430 mm

50% dependability = 510 mm

Out of 30 years of study 15 years we can expect 510 mm and 24 years, we can expect 430 mm. For 17 years out of 30 years rainfall is below normal rainfall i.e. 538 mm. Remaining 13 years surplus years (i.e.) 56.66% of time is drought alternatively every alternate year is almost a drought year. Now the graph is drawn between the annual rainfall of column (3) as abscissa and the corresponding exceedence probability of column (6) as the ordinate. The graph is shown in Page No.276. From this figure the rainfall (annual) corresponding to the exceedence probabilities of 0.75 and 0.5 are read as 430 mm and 510 mm respectively. Therefore the 75% and 50% dependable rainfalls are 430 and 510 mm. In other wards at a given place we can expect a rainfalls of 430 mm or more in 75 years out of 100 years and 510 mm or more in 50 years out of 100 years.

Percentage departure of yearly aridity index Ia and moisture adequacyIma from median of Anantapur District A.P., India

Year	"Ia" median value = 70.33			"Ima" median value = 36.01		
	Ia%	Actual	Percentage	Ima%	Actual	Percentage
1946	74	(-) 3.67	(-) 5.22	44.33	8.23	22.85
1947	71.5	(-) 1.17	(-) 1.66	33.15	2.95	8.19
1948	70.16	(+) 0.17	(-) 0.24	31.72	4.29	11.91
1949	64.57	(+) 5.76	(+) 8.19	37.08	1.07	2.97
1950	67.47	(+) 2.86	(+) 4.07	34.20	1.81	5.03
1951	74.21	(-) 3.88	(-) 5.52	27.46	8.55	23.74
1952	73.72	(-) 3.39	(-) 4.02	29.46	6.55	18.19
1953	71.80	(-) 1.47	(-) 2.09	44.04	8.03	22.30
1954	61.1	(+) 9.23	(+) 13.12	28.35	7.66	21.27
1955	66.4	(+) 3.93	(+) 5.59	46.79	10.78	29.94
1956	70.53	(-) 0.20	(-) 0.28	52.93	16.92	46.99
1957	69.09	(+) 1.24	(+) 1.76	32.6	3.5	9.72
1958	63.80	(+) 6.53	(+) 9.28	37.57	1.56	4.33
1959	70.50	(-) 0.17	(-) 0.24	30.95	5.06	14.05
1960	75.30	(-) 4.97	(-) 7.07	31.32	4.69	13.02

Year	"Ia" median value = 70.33			"Ima" median value = 36.01		
	Ia%	Actual	Percentage	Ima%	Actual	Percentage
1961	72.85	(-) 2.52	(-) 3.58	32.11	3.90	10.83
1962	62.56	(+) 7.77	(+) 11.05	42.83	6.82	18.94
1963	68.65	(+) 1.68	(+) 2.39	36.25	0.24	0.67
1964	67.39	(+) 2.94	(+) 4.18	40.62	4.61	12.80
1965	75.98	(-) 5.65	(-) 8.03	27.18	8.83	24.52
1966	65.37	(+) 4.96	(+) 7.05	41.31	21.7	60.26
1967	68.07	(+) 2.26	(+) 3.21	34.16	1.94	5.39
1968	68.44	(+) 1.89	(+) 2.69	39.52	3.15	8.75
1969	68.99	(+) 1.34	(+) 1.91	35.76	0.25	0.69
1970	60.90	(+) 9.43	(+) 13.41	32.54	3.43	9.64
1971	71.90	(-) 1.57	(-) 2.23	35.74	0.27	0.75
1972	71.26	(-) 0.93	(-) 1.32	38.99	2.98	8.28
1973	72.91	(-) 2.58	(-) 3.67	40.91	4.81	13.36
1974	73.17	(-) 2.84	(-) 4.04	42.1	6.09	16.91
1975	73.13	(-) 2.80	(-) 3.98	51.71	15.70	43.60

Frequency analysis of annual Rainfall (1946-1975) 30 years period of Anantapur District, Andhra Pradesh, India

Year	Annual Rainfall in mm	Annual rainfall in descending order x_m	Rank m	Return Period $T_M = \dfrac{n+1}{m} = \dfrac{31}{m}$	Exceedence Probability $P(x_m) = \dfrac{1}{T_-} = \dfrac{m}{31}$
1946	608.3	767.4	1	31.00	0.03
1947	503.9	749.2	2	15.5	0.06
1948	483.7	649.0	3	10.3	0.09
1949	599.6	640.8	4	7.8	0.13
1950	538.3	639.6	5	6.2	0.16
1951	394.8	634.4	6	5.2	0.19
1952	393.3	632.5	7	4.4	0.22
1953	639.6	608.6	8	3.9	0.25
1954	413.8	608.3	9	3.4	0.29
1955	634.4	599.6	10	3.1	0.32
1956	767.4	599.3	11	2.8	0.35
1957	499.2	561.61	12	2.6	0.39
1958	608.6	538.3	13	2.4	0.42
1959	419.75	527.5	14	2.2	0.45
1960	475.90	523.8	15	2.1	0.48
1961	418.90	508.2	16	1.9	0.52

Year	Annual Rainfall in mm	Annual rainfall in descending order x_m	Rank m	Return Period $T_M = \dfrac{n+1}{m} = \dfrac{31}{m}$	Exceedence Probability $P(x_m) = \dfrac{1}{T_r} = \dfrac{m}{31}$
1962	649.0	503.9	17	1.8	0.55
1963	508.2	503.0	18	1.7	0.58
1964	561.6	499.2	19	1.6	0.61
1965	352.8	497.7	20	1.6	0.65
1966	640.8	496.1	21	1.5	0.68
1967	527.5	495.5	22	1.4	0.71
1968	497.7	483.7	23	1.3	0.74
1969	495.5	475.90	24	1.3	0.77
1970	503.0	419.75	25	1.2	0.81
1971	491.1	418.9	26	1.2	0.84
1972	523.8	413.8	27	1.1	0.87
1973	599.3	394.8	28	1.1	0.90
1974	632.5	393.3	29	1.1	0.94
1975	749.2	352.8	30	1.0	0.96
Total	$\dfrac{16136.45}{30} = 538$				

Standard deviation of (Ia) Index of aridity for Anantapur District, Andhra Pradesh, India for the period (1946-1975) – 30 years to analyse the classification of droughts following the scheme of standard deviation technique employed by Subrahmanyam et al., (1965)

Table 7.13   Categorization of Droughts

Departure of the Ia from Median Value	Drought Intensity
Less $< \dfrac{1}{2}\sigma$	Moderate
$\dfrac{1}{2}\sigma$ to σ	Large
σ to 2σ	Severe
Grater $> 2\sigma$	Disastrous

1. Standard Deviation of (Ia) (σ)

$$\sigma = \sqrt{\dfrac{\sum\left(x-\overline{x}\right)^2}{n-1}}$$

Where $\overline{x} = \dfrac{\sum x}{n}$

$\overline{x}$ = Mean Strength

n = Number of Samples

X = Particular value of observation

$$\sum\left(x-\overline{x}\right)^2 = 1810.71 \qquad x - 1 = 30 - 1 = 29$$

$$\sigma = \sqrt{\dfrac{1810.71}{29}} = 7.9$$

2. For median value of (Ima) putting Ima in order and finding the middle number resulting a pair of middle numbers since even numbers are 30. So adding and dividing 35.76 and 36.25 the median value is obtained as 36.01.

3. Similarly for median value of Ia the middle numbers are 70.16 and 70.50. So adding and dividing the median value obtained is 70.33 chronological chart of Rainfall for the Period 1946-1975 – 30 years of Anantapur District, Andhra Pradesh, India is presented in Fig............ In this presentation the annual rainfall is plotted as the ordinate against the year in which it is recoded as the

abscissa. The points are then joined by straight lines in between as shown in the Fig……….. above. The normal annual rainfall which was 538 mm for the above period is also plotted on the graph, which indicates that 17 years are deficient years as they fell below normal and 13 years are surplus years.

Agricultural drought according to National Commission on Agriculture 1976, Drought is an occasion when the rainfall for a week is half of the normal or less, when the normal weekly rainfall is 5 mm or more.

Agricultural drought is a period of four such consecutive weeks in the period from middle of October or Six consecutive weeks during rest of the year. Seasonal drought occurs when the achal seasonal rainfall is deficient by more than twice the mean deviation.

When soil moisture and rainfall are inadequate during the crop growing season to support healthy crop growth to maturity which situation causes extreme crop stress and wilting. Agricultural drought in defined as a period of four consecutive weeks (of severe meteorological drought) with a rainfall deficiency of more than 50% of the LTA or with a weekly rainfall of 5 cm or less during the period from mid-May to Mid October (the Kharif season) when 80% of the countrys total crop is Planted or six such consecutive weeks during the rest of the year.

DROUGHTS IN ANANTAPUR DISTRICT (1946-1975) 30 YAERS

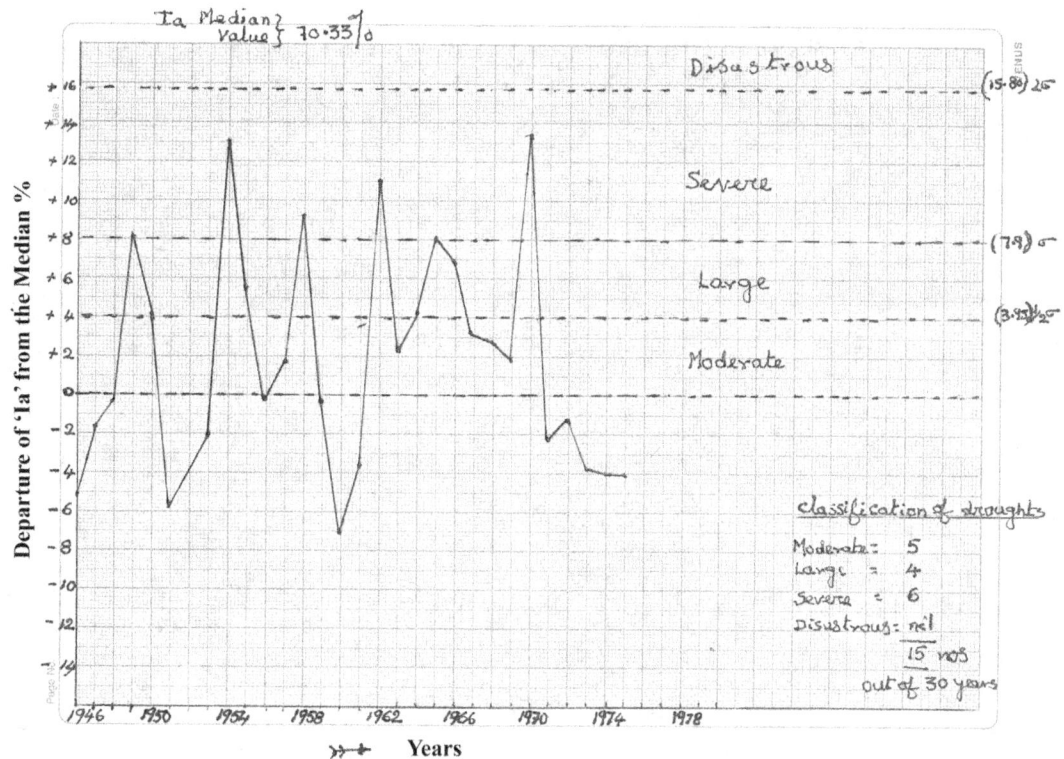

CLIMATIC SHIFTS AT ANANTAPUR DISTRICT (CLIMATE TYPES)

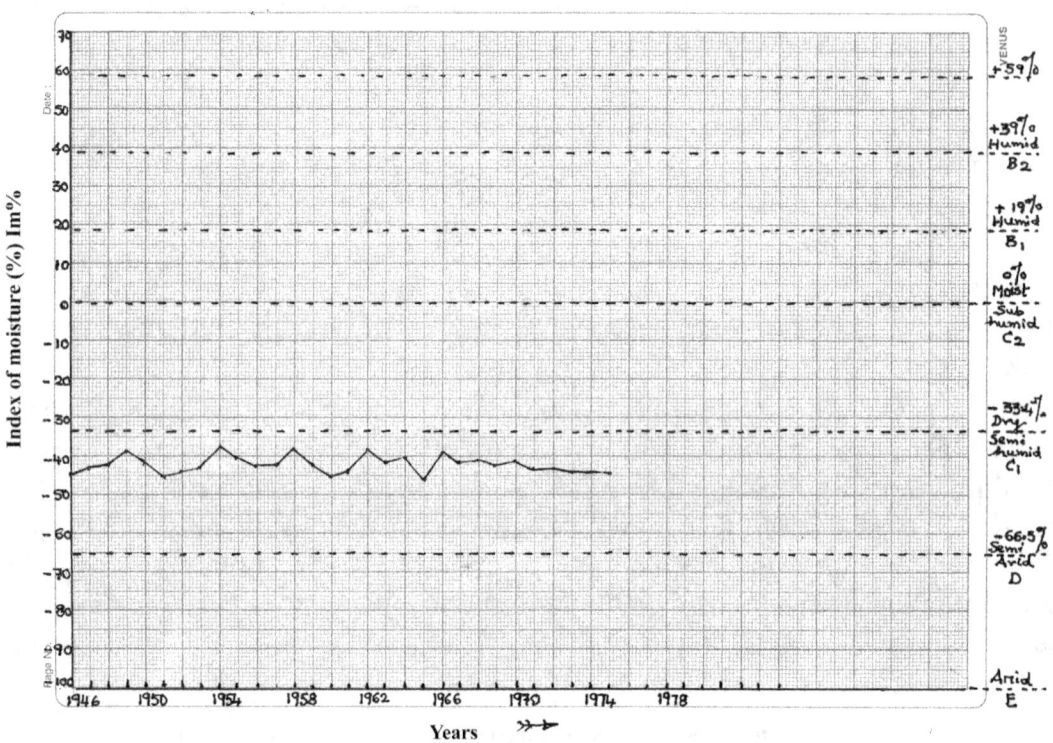

DECENNIAL FREQUENCE OF DROUGHT YAERS - ANANTAPUR DISTRICT (1946-75)

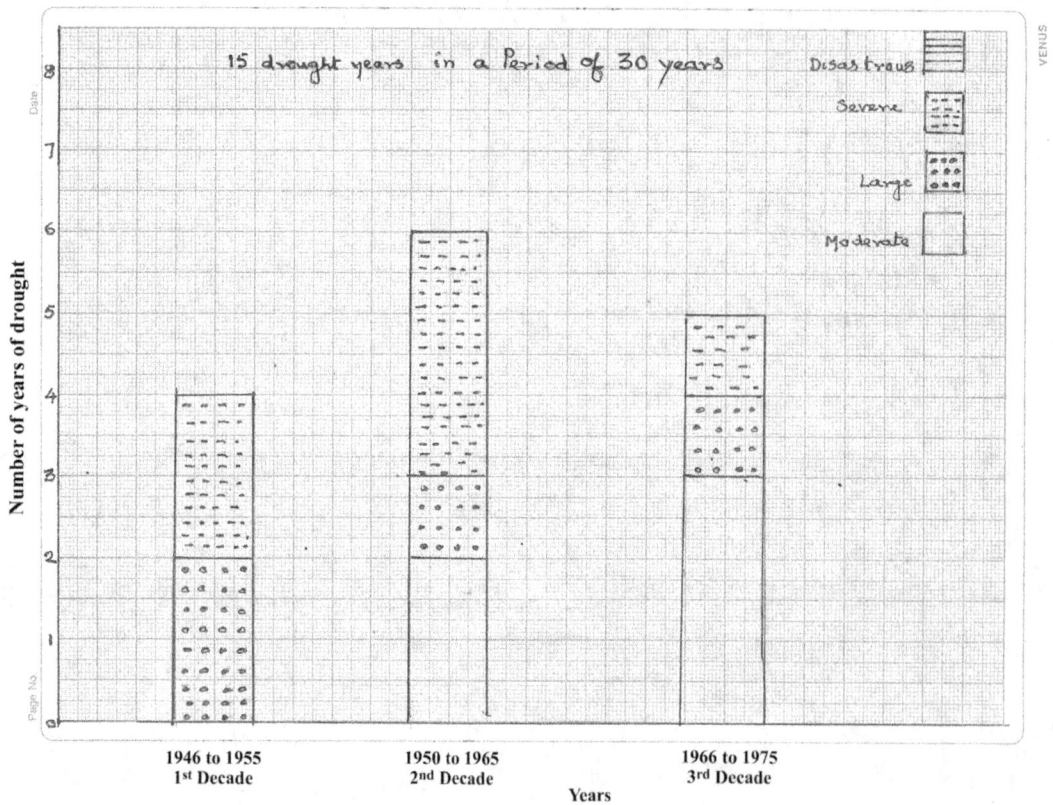

CHRONOLOGICAL CHART OF RAINFALL (ANNUAL RAINFALL) 1946-75

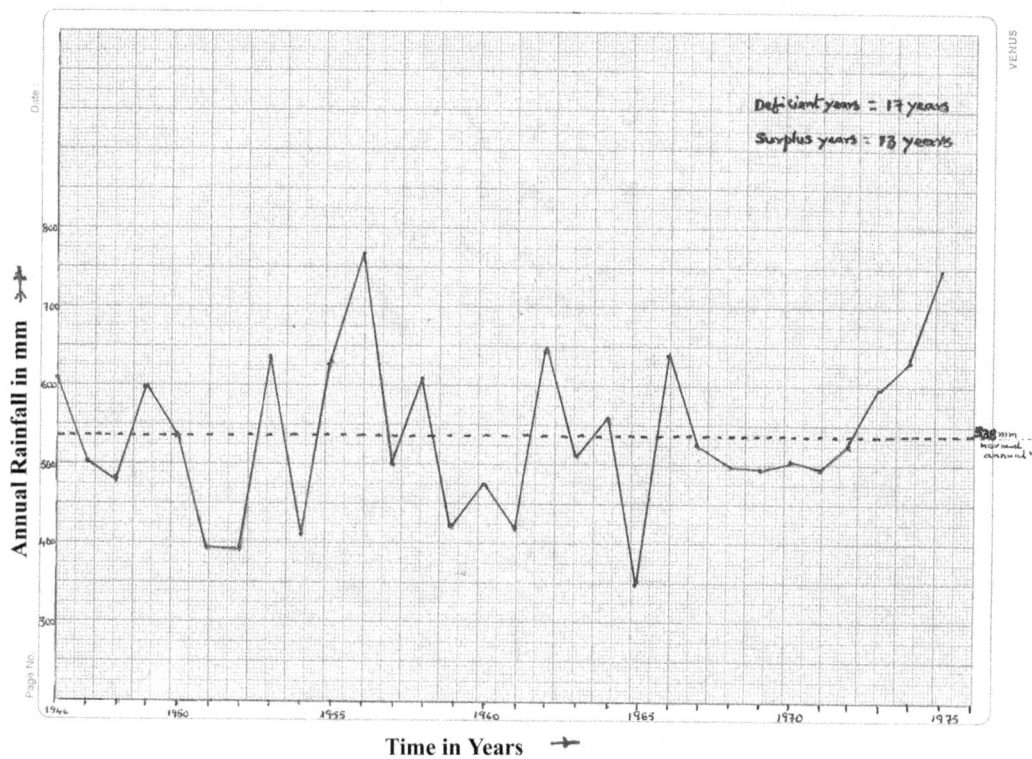

Time in Years

FREQUENCY ANALYSIS OF ANNAUL RAINFAL (1946-75)

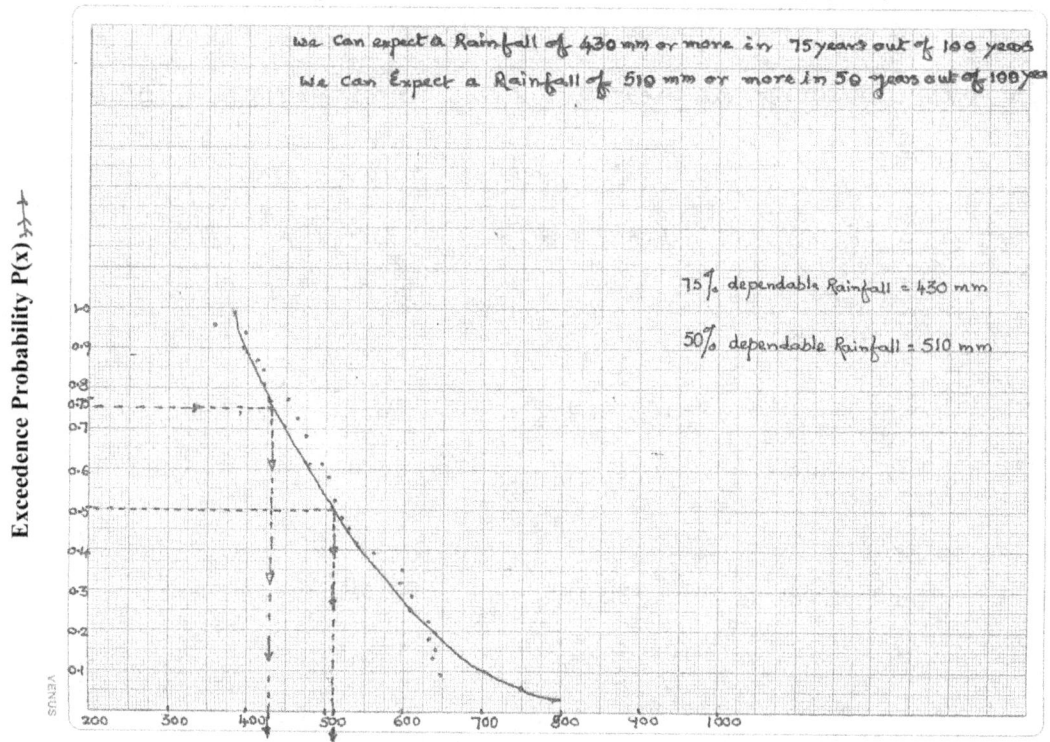

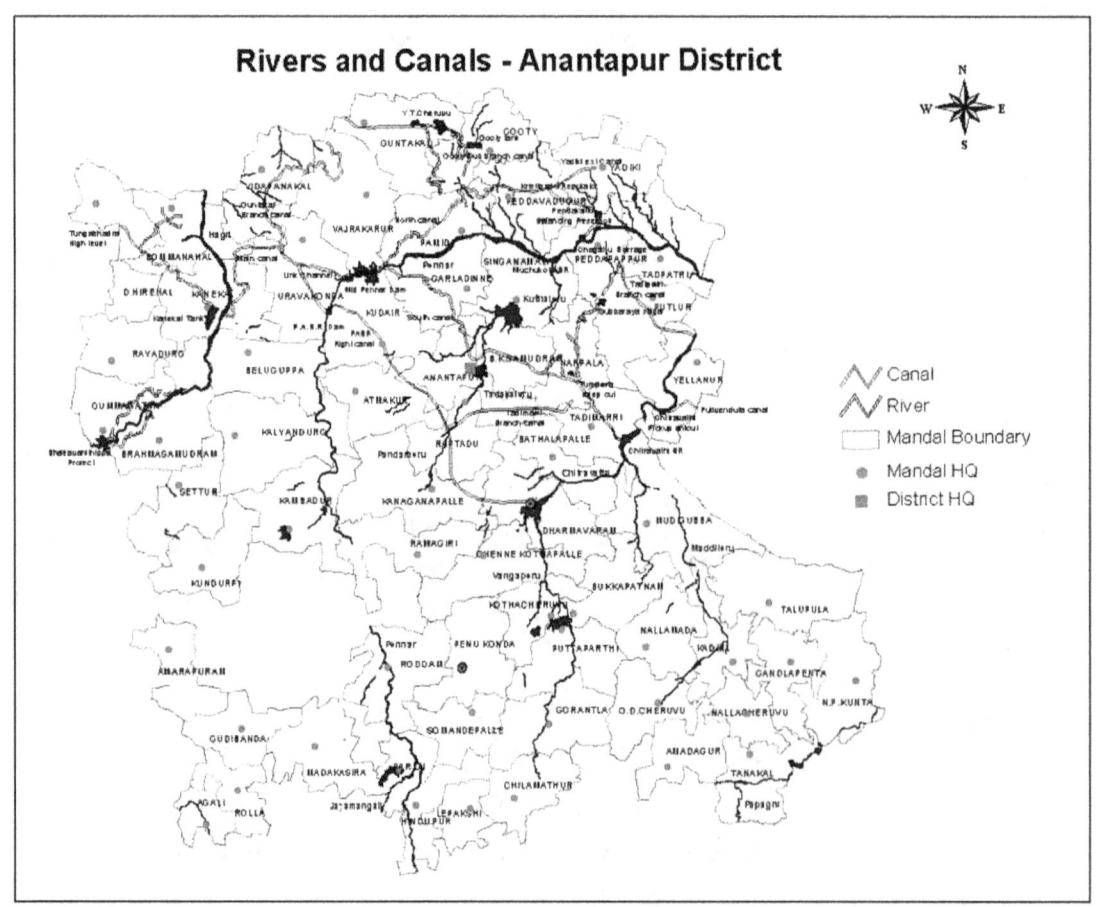

Raingauge Stations in Anantapur District, Andhra Pradesh, India

68	ANDHRA	ANANTAPUR	MADAKASIRA	REV
69	ANDHRA	ANANTAPUR	HINDUPUR	REV
70	ANDHRA	ANANTAPUR	GORANTLA	
71	ANDHRA	ANANTAPUR	BEVINAPALLI	
72	ANDHRA	ANANTAPUR	TANAKAL % TANAKALL	REV
73	ANDHRA	ANANTAPUR	C.G.PROJECT	PWD
74	ANDHRA	ANANTAPUR	PENUKONDA	
75	ANDHRA	ANANTAPUR	BUKKAPATNAM	REV
76	ANDHRA	ANANTAPUR	DHARMAVARAM	REV
77	ANDHRA	ANANTAPUR	KALYANDRUG	REV
78	ANDHRA	ANANTAPUR	ANANTAPUR	OBSY

79	ANDHRA	ANANTAPUR	PERUR	PWD
80	ANDHRA	ANANTAPUR	GARLADINNE	PWD
81	ANDHRA	ANANTAPUR	MIDPENAR REGULATOR	PWD
82	ANDHRA	ANANTAPUR	NARAPALA	PWD
83	ANDHRA	ANANTAPUR	ANATPUR	REV
84	ANDHRA	ANANTAPUR	THOGARAKANTA	PWD
85	ANDHRA	ANANTAPUR	KANAGANAPALLI	REV
86	ANDHRA	ANANTAPUR	KUDERU	REV
87	ANDHRA	ANANTAPUR	M.C.PALLI	
88	ANDHRA	ANANTAPUR	ENUMALADODDI	REV
89	ANDHRA	ANANTAPUR	KADAVAKOLLU	REV
90	ANDHRA	ANANTAPUR	RODDAM	REV
91	ANDHRA	ANANTAPUR	KADIRI	REV
92	ANDHRA	ANANTAPUR	TADPATRI	REV
93	ANDHRA	ANANTAPUR	YELLANDUR&ELDANU	PWD
94	ANDHRA	ANANTAPUR	DAHAYANA CHOU	PWD
95	ANDHRA	ANANTAPUR	THALUPULA	PWD
96	ANDHRA	ANANTAPUR	GOOTY	REV
97	ANDHRA	ANANTAPUR	YADKI	REV
98	ANDHRA	ANANTAPUR	GUNTAKAL	REV
99	ANDHRA	ANANTAPUR	KUNEKONDLA	REV
100	ANDHRA	ANANTAPUR	PEDDAWADAGURU	REV
101	ANDHRA	ANANTAPUR	RAYADURG	REV
102	ANDHRA	ANANTAPUR	AMARAPURAM	REV
103	ANDHRA	ANANTAPUR	B.T.PROJECT	PWD
104	ANDHRA	ANANTAPUR	URVAKONDA	REV

105	ANDHRA	ANANTAPUR	KANEKAL	REV
106	ANDHRA	ANANTAPUR	KANEKAL I.B	PWD
107	ANDHRA	ANANTAPUR	BELUGUPPA	REV
108	ANDHRA	ANANTAPUR	C.K.PALLI	REV
109	ANDHRA	ANANTAPUR	CHILAMALUR	REV
110	ANDHRA	ANANTAPUR	KANBADUR	REV
111	ANDHRA	ANANTAPUR	MOTANCHINTPALLI	REV
112	ANDHRA	ANANTAPUR	MANAGAKALLUMARI	REV
113	ANDHRA	ANANTAPUR	MOTAN CHINTLAPAL	
114	ANDHRA	ANANTAPUR	NUTUMADUGU	REV
115	ANDHRA	ANANTAPUR	PADMIDI	REV
116	ANDHRA	ANANTAPUR	PEDDAVADUGURU	PWD
117	ANDHRA	ANANTAPUR	PONDIPARTHY	REV
118	ANDHRA	ANANTAPUR	POTHAPALLI	REV
119	ANDHRA	ANANTAPUR	SRIGANAMALA	REV
120	ANDHRA	ANANTAPUR	U.P. PROJECT / PERU	PWD
121	ANDHRA	ANANTAPUR	A.KONDAPURAM	PWD
122	ANDHRA	ANANTAPUR	AGALI	REV
123	ANDHRA	ANANTAPUR	AMADAGURU	REV
124	ANDHRA	ANANTAPUR	ATMAKUR	REV
125	ANDHRA	ANANTAPUR	B.K.SAMUDRAM	REV
126	ANDHRA	ANANTAPUR	BATTALAPALLI	REV
127	ANDHRA	ANANTAPUR	BOMMANHAL	REV
128	ANDHRA	ANANTAPUR	BOMMANHALLI HLC	PWD
129	ANDHRA	ANANTAPUR	BRAHMA SAMUDRAM	REV
130	ANDHRA	ANANTAPUR	D.HIREHAL	REV

131	ANDHRA	ANANTAPUR	EGUVAPALLI	PWD
132	ANDHRA	ANANTAPUR	GANDLAPENTA	REV
133	ANDHRA	ANANTAPUR	GARLADINNE	REV
134	ANDHRA	ANANTAPUR	GONEGANDLA	PWD
135	ANDHRA	ANANTAPUR	GORANTLA	REV
136	ANDHRA	ANANTAPUR	GUDIBANDA	REV
137	ANDHRA	ANANTAPUR	GUMMAGATTA	REV
138	ANDHRA	ANANTAPUR	KALLUMARRI	REV
139	ANDHRA	ANANTAPUR	KOTHACHERUVU	REV
140	ANDHRA	ANANTAPUR	KUNDURPI	REV
141	ANDHRA	ANANTAPUR	LEPAKSHI	REV
142	ANDHRA	ANANTAPUR	M.P. DAMSITE	PWD
143	ANDHRA	ANANTAPUR	MUDIGUBBA	REV
144	ANDHRA	ANANTAPUR	N.P. KUNTA	REV
145	ANDHRA	ANANTAPUR	NAGASAMUDRAM	REV
146	ANDHRA	ANANTAPUR	NALLACHERUVU	REV
147	ANDHRA	ANANTAPUR	NALLAMADA	REV
148	ANDHRA	ANANTAPUR	O.D.CHERUVU	REV
149	ANDHRA	ANANTAPUR	PALASAMUDRAM	REV
150	ANDHRA	ANANTAPUR	PANDIPARTHI	REV
151	ANDHRA	ANANTAPUR	PARIGI	REV
152	ANDHRA	ANANTAPUR	PARLAPALLI	PWD
153	ANDHRA	ANANTAPUR	PEDDAPAPPUR	REV
154	ANDHRA	ANANTAPUR	PUTLURU	REV
155	ANDHRA	ANANTAPUR	PUTTLUR	PWD

156	ANDHRA	ANANTAPUR	RAMAGIRI	REV
157	ANDHRA	ANANTAPUR	RAPTHADU	REV
158	ANDHRA	ANANTAPUR	ROLLA	REV
159	ANDHRA	ANANTAPUR	SETTURU	REV
160	ANDHRA	ANANTAPUR	SOMANDEPALLI	REV
161	ANDHRA	ANANTAPUR	TADIMARRI	REV
162	ANDHRA	ANANTAPUR	TALUPULA	REV
163	ANDHRA	ANANTAPUR	VAJRAKARUR	REV
164	ANDHRA	ANANTAPUR	VIDAPANAKAL	REV
165	ANDHRA	ANANTAPUR	YENUMULADODDI	REV
166	ANDHRA	ANANTAPUR	NARPALA	
167	ANDHRA	ANANTAPUR	YELLANUR	
168	ANDHRA	ANANTAPUR	PEDDAVANDBGURU	

8

HYDROLOGICAL DROUGHT

⌨ CHAPTER OUTLINE

- Characterization of Hydrological drought. Application of drought indices and explanation about various indices in India and Asia – SPATSIM package.

Drought indices may be used for drought warning and lead time assessment. Drought indices summarize different data on rainfall snow pack stream flow and other water supply indicators into a comprehensible big picture. Drought indices are quantitative indicators, used to identify, characterize and analysis drought events with relative accuracy and objectivity. There are several indices that measure how much precipitation for a given period of time has deviated from historically established norms like PDSI / PHDI, SPI, CMI, SWSI, ADI. The various drought indices are usually discipline specific and have their own pros and cons, some indices are better suited than others for certain uses. Most water supply planners find it useful to consult one or more indices before making a decision.

In general different studies have indicated the usefulness of the spi to quantify different drought types. The long time scales (over 6 months) are considered as hydrological drought indicators (river discharges or reservoir storages).

To identify the main dry period it is necessary to analysis the time scales and larger than 6 months because the high frequency of spi values at the shorter time scales hide the most important dry periods. Time scales shorter than 6 months show non-significant autocorrelations considering lags shorter than 4 months, whereas considering the spi at time scales larger than 6 months the autocorrelations with lags of 4 of more months increase noticeably.

Therefore, with time scales shorter than 6 months, it is difficult to identify periods of consecutive 4 months with day conditions. Although the spi is widely used, there are not many empirical studies that provide evidence about the usefulness of the different time scales for drought monitoring in surface water resources.

In Hungary analysed the relationships between time scales of spi river discharges and reservoir storages showing important spatial differences.

NORMALIZED DIFFERENCE VEGETATION INDEX

Outline

- Hydrological drought
- Monitoring and evaluation of hydrological drought
- Different hydrological drought indices
- Case studies

Hydrological Drought

Figure 8.1   Hydrological Drought Condition

Figure 8.2   Praying for Rain -Drought Period

Figure 8.3   Going in Search of Drinking Water during Drought.

HYDROLOGICAL DROUGHT

Hydrological drought is defines as a significant decrease in the availability of water in all its forms appearing in the land phase of the hydrological cycle (Nalbantis, 2009).

Forms of water

- Surface water
- Stream flow (snowmelt and spring flow)
- Lake and reservoir level
- Ground water
- Ground water level

HYDROLOGICAL DROUGHT

Hydrological drought is described as a sustained and regionally extensive occurrence of below average natural water availability (Tallaksen and van Lanen, 2004)

Hydrological drought as period of time below the average water content in streams, reservoirs, groundwater aquifers, lakes and soils. The period is associated effects of precipitation (including snowfall) shortfall on surface and subsurface water supply, rather than with direct shortfall in precipitation (Yevjevich *et al.*, 1977).

Hydrological drought may be the result of long term meteorological droughts that results in the drying up of reservoirs, lakes, streams, rivers and a decline in groundwater levels (Rathore, 2004).

HYDROLOGICAL DROUGHT

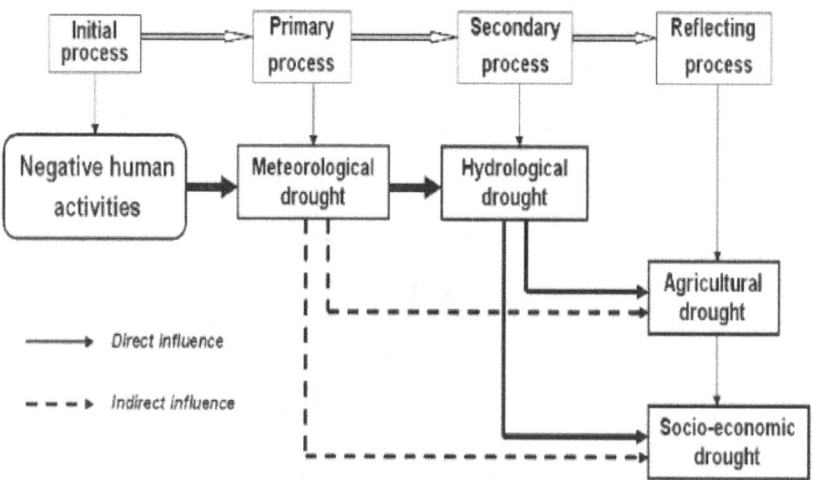

Source: Shaban, 2009.

HYDROLOGICAL DROUGHT

Characterization

o   Its severity expresses by a drought index
o   Its time of onset and its duration,
o   Its areal extent and
o   Its frequency of occurrence.

HYDROLOGICAL DROUGHT INDICES

Operational Requirements

o   Easily understood
o   Carrying physical meaning
o   Sensitive to wide range of drought conditions
o   Independent of area of application
o   Reveal drought with short lag after its occurrence and
o   Based on the data which are readily available

HYDROLOGICAL DROUGHT INDICES

Data and Instrumentation

o   Rainfall gauges in several representative sites
o   Hydrographs and flow meters on rivers and springs water courses, outlets of watershed
o   Scale levels to measure water levels in lakes, reservoirs, ponds etc
o   Remote sensing data (RADARSAT, NOAA, MODIS etc.) to monitor snow coverage and areal extent of lakes and wetlands
o   Piezometers to measure groundwater levels
o   Available equipments and laboratory for soil moisture

HYDROLOGICAL DROUGHT INDICES

Indices Based on Precipitation

Standardized Precipitation Index (SPI) used in South Asia. To quantify the precipitation deficit in the monsoon and non-monsoon periods

$$SPI = (X_{ij} - X_{im}) / \sigma$$

Where X_{ij} is the seasonal precipitation at the ith rain gauge station and j^{th} observation, X_m is the long term seasonal mean and σ is standard deviation.

HYDROLOGICAL DROUGHT INDICES

Indices Based on Precipitation

Effective Drought Index (EDI)   It is an intensive measure which considers daily water accumulation with weighting function for time passage.

Advantages

- It calculates daily drought severity.
- Rapid detection and precise measurement of short term drought.
- Indicates current level of available water resources.
- It is able to diagnose prolonged droughts that continue for several years (because: It calculates the total precipitation period)

HYDROLOGICAL DROUGHT INDICES

Indices Based on Precipitation

Effective Drought Index (EDI)

Steps to calculate EDI   Calculate the daily EP

$$EP_i - \sum_{i}^{n=1}\left[\left(\sum_{n}^{m=1} p_m\right)\Big/n\right]$$

Where i = 365 is the period over which precipitation is summed. P_m denotes precipitation m days ago.

- Calculate 30 year mean EP (MEP) for each calendar year
- Calculate the DEP, which is the difference between EP and MEP
- When DEP is negative, it signifies dryer than average, add the days of prolonged dryness to the existing period (i = 365) and
- Recalculate MEP and DEP
- Divide the DEP for each calendar day by standard deviation of DEP over the past 30 years. This will results in EDI.

HYDROLOGICAL DROUGHT INDICES

Indices Based on Precipitation

Effective Drought Index (EDI)

Definition of states of drought with EDI

Drought Classes	Criterion
Extreme drought	$EDI \leq 2.0$
Severe drought	$-2.0 \leq EDI \leq -1.5$
Moderate drought	$-1.5 \leq EDI \leq -1.0$
Near normal	$-1.0 \leq EDI \leq 1.0$

HYDROLOGICAL DROUGHT INDICES

Indices Based on Streamflow

Streamflow Drought Index (SDI)

This index $SDI_{i,k}$ requires streamflow volume values $Q_{i,j}$ where i denote the hydrological year and j^{th} month within a hydrological year. We can obtain $V_{i,k}$ cumulative streamflow volume for the i-th hydrological year and k-th reference period

$$V_{j,k} = \sum_{j=1}^{8k} Q_{i,j} \quad i = 1, 2, \ldots\ldots, \qquad j = 1, 2, \ldots\ldots 12, \qquad k = 1, 2, 3, 4$$

$$SDI_{j,k} = \frac{V_{j,k} - V_k}{S_k}$$

Where V_k and s_k are respectively the mean and standard deviation of the cumulative streamflow volumes for the k-th reference period.

HYDROLOGICAL DROUGHT INDICES

Indices Based on Streamflow

Streamflow Drought Index (SDI)

Definition of states of drought with SDI

Description of State	Criterion
Non drought	SDI ≥ 0.0
Mild drought	−1.0 ≤ SDI < 0.0
Moderate drought	−1.5 ≤ SDI < −1.0
Severe drought	−2.0 ≤ SDI < −1.5
Extreme drought	SDI < −2.0

HYDROLOGICAL DROUGHT INDICES

Indices Based on Streamflow

Surface Water Supply index (SWSI)

This index integrate reservoir storage, streamflow and two precipitation types (snow and rain) at high elevations into a single index. The SWSI is given by

$$SWSI = \frac{aP_{snow} + bP_{prec} + cP_{stream} + dP_{resv} - 50}{12}$$

Where a, b, c, d = weights for snow, rain, streamflow and reservoir storage respectively (a + b + c + d = 1) and P = the probability (%) of non-exceedence for each of these four water balance components. The estimation is carried out with a monthly time step. In winter months, SWSI is computed using snowpeck, precipitation and reservoir storage. In summer, streamflow, precipitation and reservoir storage data are used.

The range of SWSI is similar to PDSI as −4.2 to +4.2.

HYDROLOGICAL DROUGHT INDICES

Indices Based on Low Flows

o   (WMO 1974) defines low flow as a flow of water in a stream during prolonged dry weather.

o   Droughts include low-flow periods, but a continuous seasonal low-flow event does not necessarily constitute a drought, although many researchers refer to a continuous low-flow period in one year as an Annual drought.

HYDROLOGICAL DROUGHT INDICES

Indices based on Low Flows

- A number of consecutive time intervals where the selected flow variable (a discharge or flow volume) has lower values than a reference flow level indicate the duration of a drought event.
- For each such event, the sum of deviations of a flow variable from the reference level represents the cumulative flow-deficit amount (drought severity).
- This deficit divided by the duration is the measure of drought intensity.

HYDROLOGICAL DROUGHT INDICES

Indices Based on Low Flows

Definition of Water Shortages

- A deep shortage–when annual runoff is lower than the mean, by at least one standard deviation.
- A continuous shortage–when annual volumes are lower than the mean, during at least 4 consecutive years.
- An extended shortage–when a deep or continuous shortage extends over the entire region under consideration.

Drought indices and definitions based solely on low flow or reservoir storage are normally designed for reservoir operation and are seldom (if at all) used as triggers for drought relief, or for drought monitoring over vast territories.

HYDROLOGICAL DROUGHT INDICES

Index Based on Runoff

Steps for defining the Drought Index

- Normalization of runoff

  Runoff data should be fit in to follow normal distribution or other type of distribution.
- Normalizing runoff would convert the probability density function of Pearson type III distribution into the standard normal distribution as function of Z.
- Define the runoff anomaly percentage. For e.g. the categories of runoff are separated into 5 according to their percentage anomalies.

HYDROLOGICAL DROUGHT INDICES

Index Based on Runoff

Definition of states of drought index based on runoff

Runoff Anomaly Percent, Δ	Corresponding Category	Runoff Status
Δ < –30%	–2	Low
–30% ≤ Δ ≤ –10%	–1	Lower
–10% ≤ Δ < 10%	0	Normal
10% ≤ Δ < 30%	1	Higher
Δ > 30%	2	High

HYDROLOGICAL DROUGHT INDICES

Index Based on Runoff

Definition of states of drought index based on runoff

- Define the runoff denoted drought index and its categories.
- Develop a set of standards for classifying runoff levels (water deficiency or abundance) in the rivers to indicate the associated drought/flood categories.
- Based on the Z value calculated for normal distribution, drought/flood categories could be defined.

Category	AF	Z value	D/F	TFD
1	> 95%	$Z > 1.6448$	Flood	5%
2	70%–95%	$0.5244 < Z \leq 1.6448$	Light flood	25%
3	30%–70%	$-0.5244 \leq Z \leq 0.5244$	Normal	40%
4	5%–30%	$-1.6448 \leq Z < -0.5244$	Light drought	25%
5	< 5%	$Z < -1.6448$	Drought	5%

AF = accumulation frequency,
D/F = drought or flood, and
TFD = theoretical frequency distribution.

HYDROLOGICAL DROUGHT INDICES

Indices based on Groundwater Levels

Standardized Water Level Index (SWI)

Standardized Water Level Index has been developed by to scale ground water recharge deficit. The SWI expression is given by

$$SWI = (W_{ij} - W_{im}) / \sigma$$

W_{ij} is the seasonal water level for the ith well and j^{th} observation.

W_{im} is the long term seasonal mean and σ is standard deviation.

HYDROLOGICAL DROUGHT INDICES

Indices Based on Groundwater Levels

Standardized Water Level Index (SWI)

Definition of states of drought with SWI

Drought Classes	Criterion
Extreme drought	SWI > 2.0
Severe drought	SWI > 1.5
Moderate drought	SWI > 1.0
Mild drought	SWI > 0.0
Non drought	SWI < 0.0

Positive anomalies correspond to drought and negative anomalies correspond to no-drought or normal condition.

HYDROLOGICAL DROUGHT INDICES

Indices based on Groundwater Levels

Groundwater Resource Index (GRI)

A groundwater resource index has been developed by Mendicino *et al.* (2008) to quantify groundwater detention for the assessment of drought condition. The index is given by

$$GRI_{y,m} = \frac{D_{ym} - \mu_{Dm}}{\sigma_{Dm}}$$

$GRI_{y,m}$ and $D_{y,m}$ are respectively the values of the index and of the groundwater detention for the year y and month m. $\mu_{D,m}$ and $\sigma_{D,m}$ are respectively the mean and standard deviation of groundwater detention for the month m in a defined number of years.

HYDROLOGICAL DROUGHT INDICES

Indices Based on Water Balance

Palmer Drought Severity Index (PDSI)
Palmer (1965) developed soil moisture algorithm which uses precipitation, temperature data and available water content of the soil. This model relates regional soil moisture conditions to the normal using a water balance model. PDSI indicates standardized moisture conditions and allows comparisons to be made between locations and between months. PDSI values are normally calculated on a monthly basis.

The major problem associated with using PDSI is that its computation is complex and requires substantial input of meteorological data. Its application in Asia, where observational networks are scarce, is therefore limited.

Applications of Hydrological Drought Indices in Asia
Drought dynamics in Aravali regions of Rajasthan, India

A study carried out by Bhuiyan *et al.* (2006) analysed the seasonal drought dynamics in the Aravali region of Rajasthan State of India. The study identified the spatio-temporal patterns in Hydrological drought.

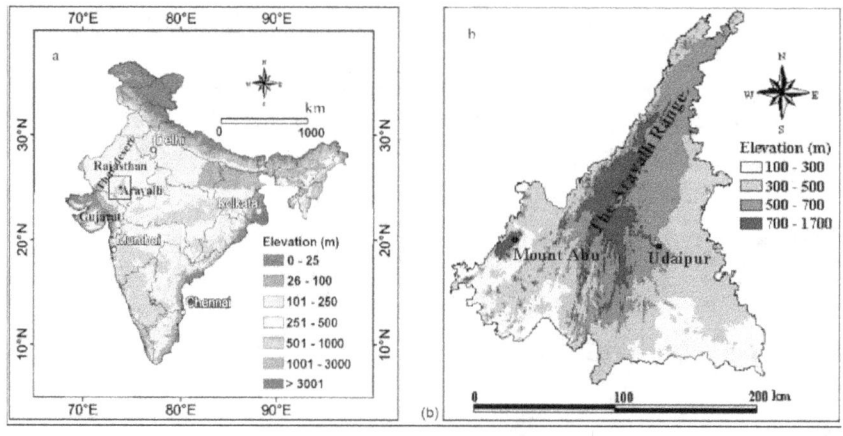

Figure 8.5 Drought Dynamics in Aravali Regions of Rajasthan, India

The study focused on drought during the monsoon and non monsoon periods. For hydrological drought analysis, Standardized Water Level Index (SWI) was used. The data used were of groundwater levels of 541 wells of the region. SWI was calculated using the mean seasonal water levels of 20 years (1984–2003). SWI values of the wells were interpolated using spline interpolation technique in a GIS environment to generate SWI maps of the region.

Drought dynamics in Aravali regions of Rajasthan, India

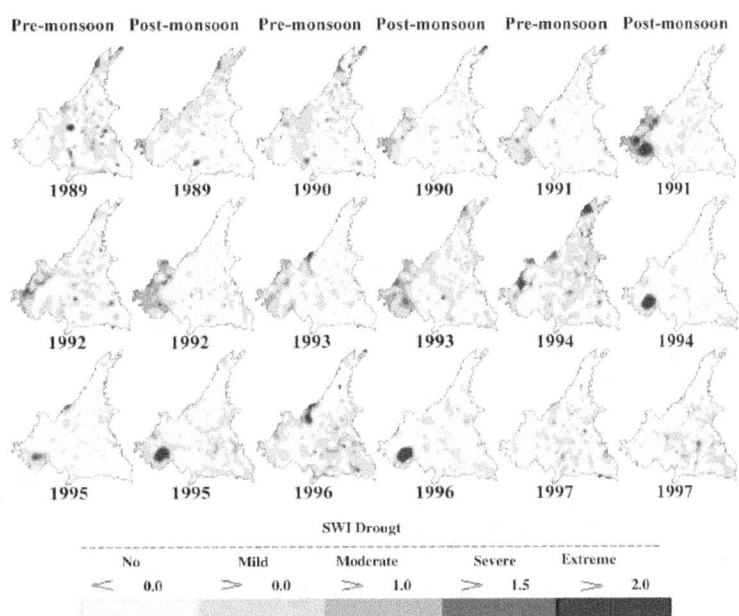

Figure 8.6 Water Table Depletion Zones

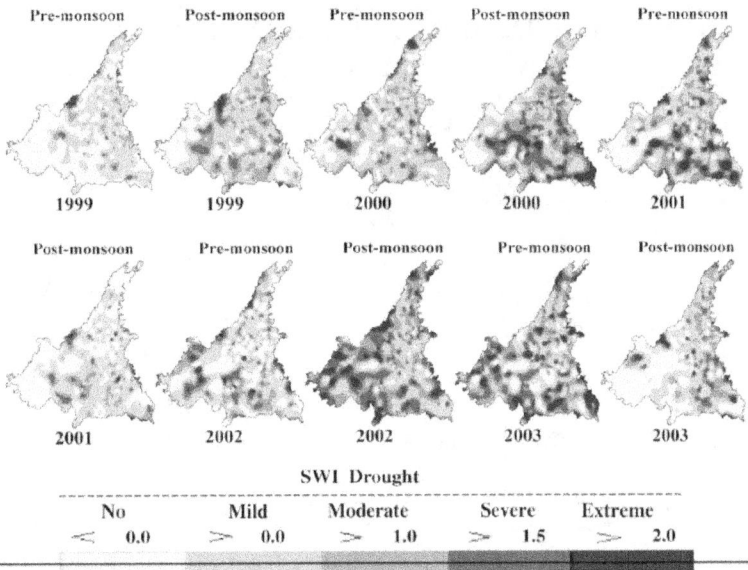

Figure 8.7 Years of Continuous Hydrological Drought

Drought dynamics in Aravali regions of Rajasthan, India

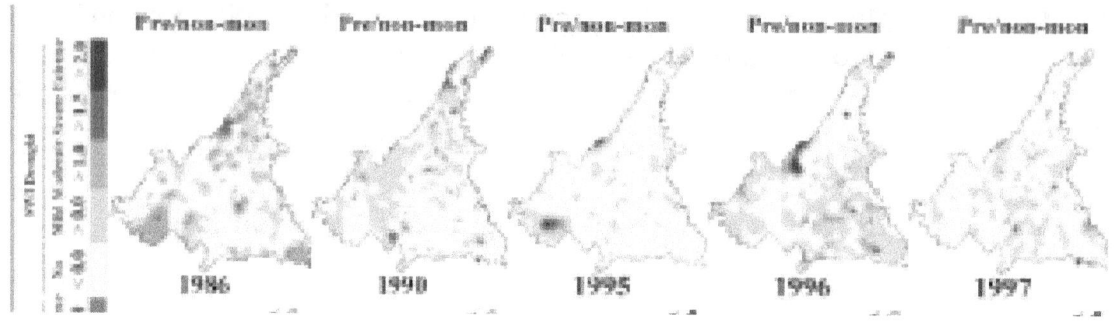

Figure 8.8 Spatio – Temporal Shift of Hydrological Drought

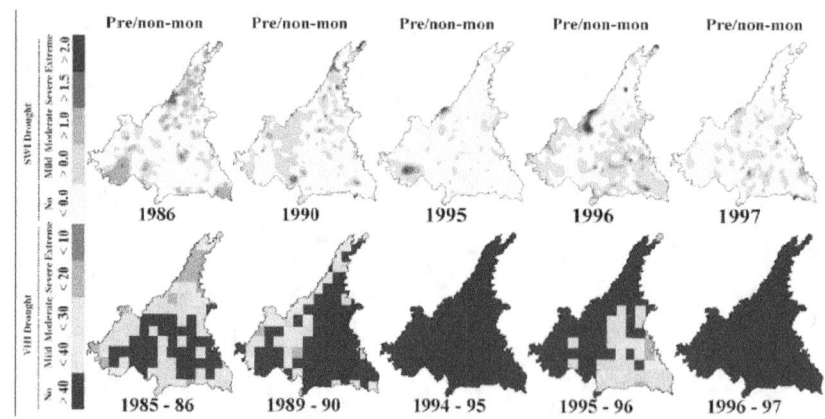

Figure 8.9 Correlation Between SWI and VHI During Non-monsoon Periods

Application of Effective Drought Index (EDI) for Drought Assessment in Seoul, Korea

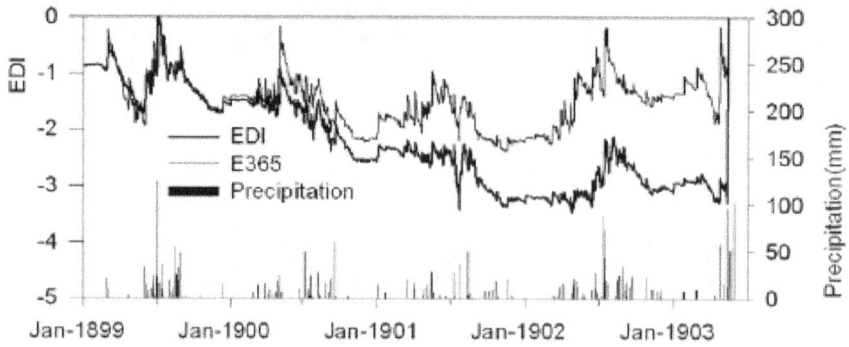

Figure 8.10 EDI and E365 for Long Term Drought

E365: EDI calculated based on precipitation without considering any continued dry period Kim *et al.* (2009)

Application of Effective Drought Index (EDI) for drought assessment in Seoul, Korea

Type	Explanation	Characteristics	Use
EDI	○ Intensive measure that considers the daily water accumulation with the weighting function of time passage ○ Although the calculation period for total precipitation is set at 365 days, when a day period begins, the total days of the dry period is added to the 365 days.	○ Measures the drought severity on a daily basis ○ Measures the drought severity without time-scale limitations	○ Monitoring ongoing drought severity ○ Defining the onset and secession of drought.
CEDI	○ In case of heavy rain, it considers the factor of large outflow over a short period of time	○ Corrects the lowered description of the drought severity arisen by the intermittent heavy rainfall.	○ As EDI but at regions that experience occasional periods of heavy rainfall
AEDI	○ Summation of negative EDI values during the consecutive dry period	○ Approximation of accumulated drought severity during the case ○ Provides effective scale in comparing different drought cases	○ Comparative analysis of severity between drought events ○ Estimation of the accumulated damage
YAEDI$_{365}$	○ Sum the negative EDI values for yearly unit and then divide it by 365	○ Quantitative expression of the dryness of each year	○ Comparative analysis on the annual drought severity
YAEDI$_{ND}$	○ As in YAEDI365 but divide it by the number of days that EDI was negative during that year.	○ Shows how intensive the drought was during the year	○ Utilized as supplementary material in measuring annual drought severity
AWRI	○ Estimates current available water resources accumulated for 365 days.	○ Reflects the actual amount of available water resources ○ An absolute scale that is independent of seasonality	○ Utilized as supplementary material in determining drought status

Specifications of EDI and its derivative indexes

Runoff derived drought index for the arid area of Hexi corridor, Northwest China

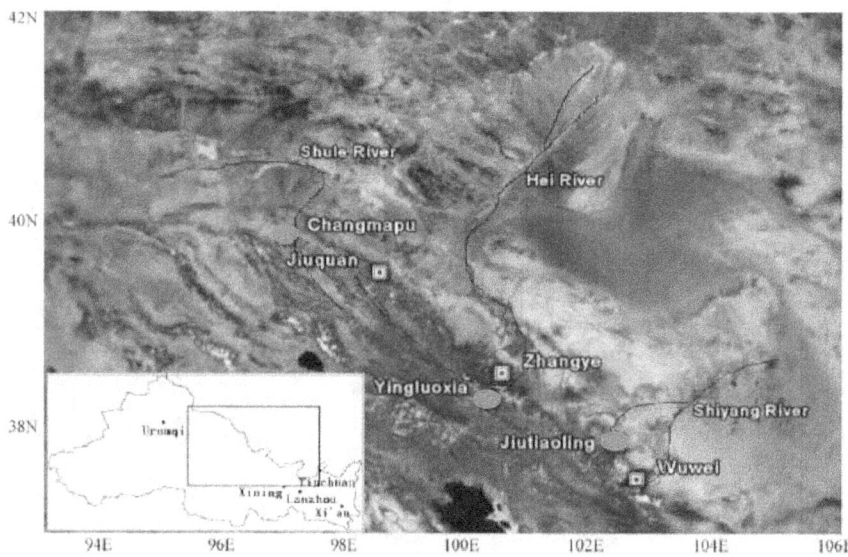

Figure 8.11   Location Map of Shule, Hei and Shiyang river and their Hydrological Monitoring Stations Wang *et al.* (2009)

Runoff derived drought index for the arid area of Hexi corridor, Northwest China

Runoff Anomaly Percent, Δ	Corresponding Category	Runoff Status	Percent of Different Categories at Three Rivers (%)			Mean Percent of Different Categories of Three Rivers (%)
			Shule	**Hei**	**Shiyang**	
Δ < -30%	−2	Low	13.0	15.2	15.2	14.5
−30% ≤ Δ ≤ −10%	−1	Lower	21.7	19.6	26.1	22.5
−10% ≤ Δ < 10%	0	Normal	41.3	36.9	34.7	37.7
10% ≤ Δ < 30%	1	Higher	15.2	17.4	15.2	15.9
Δ > 30%	2	High	8.7	10.9	8.7	9.4

Table 8.1: Runoff Anomaly Categories of Shule, Hei and Shiyang Rivers (1959 – 2004)

Runoff derived drought index for the arid area of Hexi Corridor, North West China

Category	D/F	Events			MFD
		Shule	**Hei**	**Shiyang**	
1	Flood	4	2	3	6.5%
2	Light Flood	10	13	12	25.4%
3	Normal	19	19	17	39.9%
4	Light drought	11	10	12	23.9%
5	Drought	2	2	2	4.3%

* D/F = Drought or flood, and MFD = Mean frequency distribution.

Drought / flood events in Shule, Hei and Shiyang rivers (1959 – 2004)

Table 8.1:   Runoff Derived Drought Index for the Arid Area of Hexi Corridor, Northwest China

Category	Z value	Related Drought Severity and Irrigation
1	$Z \geq -0.5244$	For "normal", and no operation needed
2	$-1.0846 < Z < -0.5244$	For "slightly light drought", and delaying the time of first irrigation, use of reservoir water
3	$-1.6448 \leq Z \leq -1.0846$	For "heavier drought", and delaying the time of first irrigation, use of reservoir water and appropriate extraction of underground water
4	$Z < -1.6488$	For "drought", and necessity of utilizing underground water

Z of this Table is denoted as Z_{ir} to indicate the levels of normal, light drought, heavier dryness and real drought for categories 1–4, respectively.

Table 8.2   Runoff Index Drought Severity and Irrigation

Application of SPI and using stochastic models and neural network for drought forecasting for Kansabati River Basin, West Bengal India

Station No.	Rain Gauge Station	Area (km²)	Elevation (m) (a.m.s.l.)	Geographic Coordinates	
				Latitude	**Longitude**
1	Simulia	1279.5	220.97	23°10'	86°22'
2	Rangagora	1151.55	222.92	23°4'	86°24'
3	Tusuma	554.45	158.6	23°08'	86°43'
4	Kharidwar	682.4	135.96	23°00'	86°38'
5	Phulberia	597.1	144.32	22°55'	86°37'

Rain gauge stations in the river basin

Mishra and Desai (2006)

Application of SPI and using stochastic models and neural network for drought forecasting for Kansabati River Basin, West Bengal India

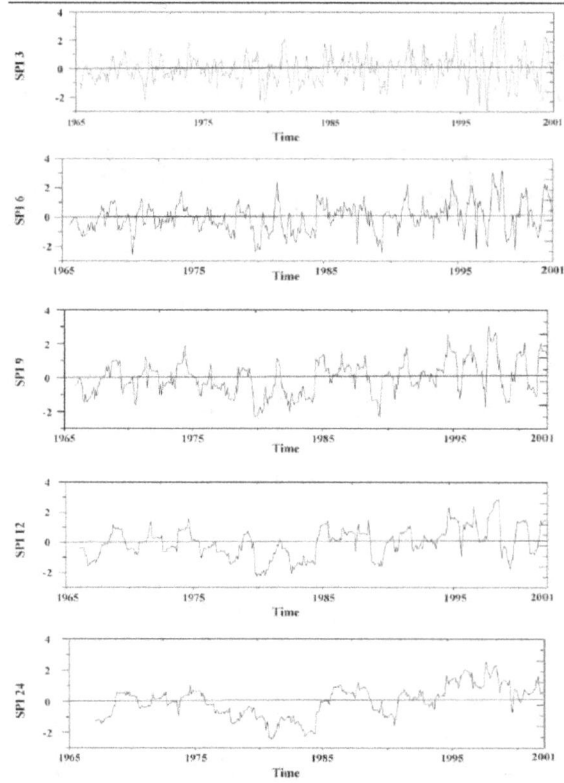

Figure 8.12   SPI Series over Different Time Scale

Application of SPI and using stochastic models and neural network for drought forecasting for Kansabati River Basin, West Bengal India

The neural network models were useful for forecasting of drought which could help local administration and water resource planners to take precautions considering severity of drought known in advance.

Identification of drought venerable areas using remote sensing data

Jain *et al*. (2009)

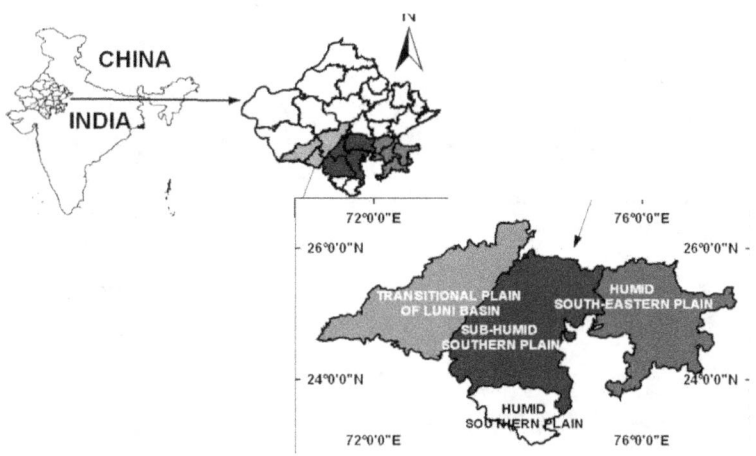

Figure 8.13 Remote Sensing Data were used for Drought Assessment using NDVI and Water Applying Vegetation Index (WSVI) along with SPI for Southern Rajasthan, India

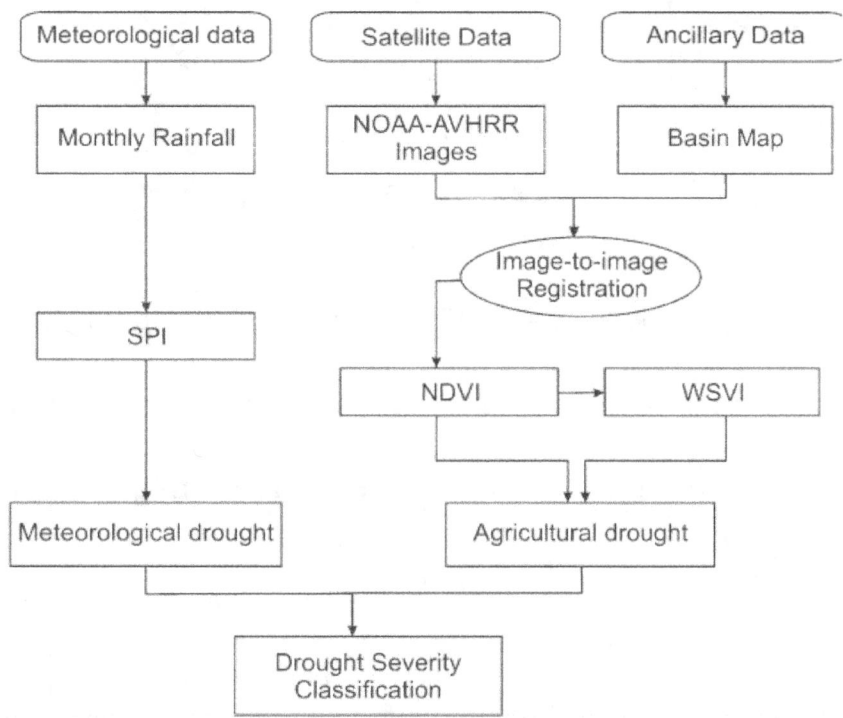

Figure 8.14 Identification of Drought Venerable Areas using Remote Sensing Data

Remote sensing data were used for drought assessment using NDVI and Water applying Vegetation Index (WSVI) along with SPI for Southern Rajasthan, India

Identification of drought venerable areas using remote sensing data

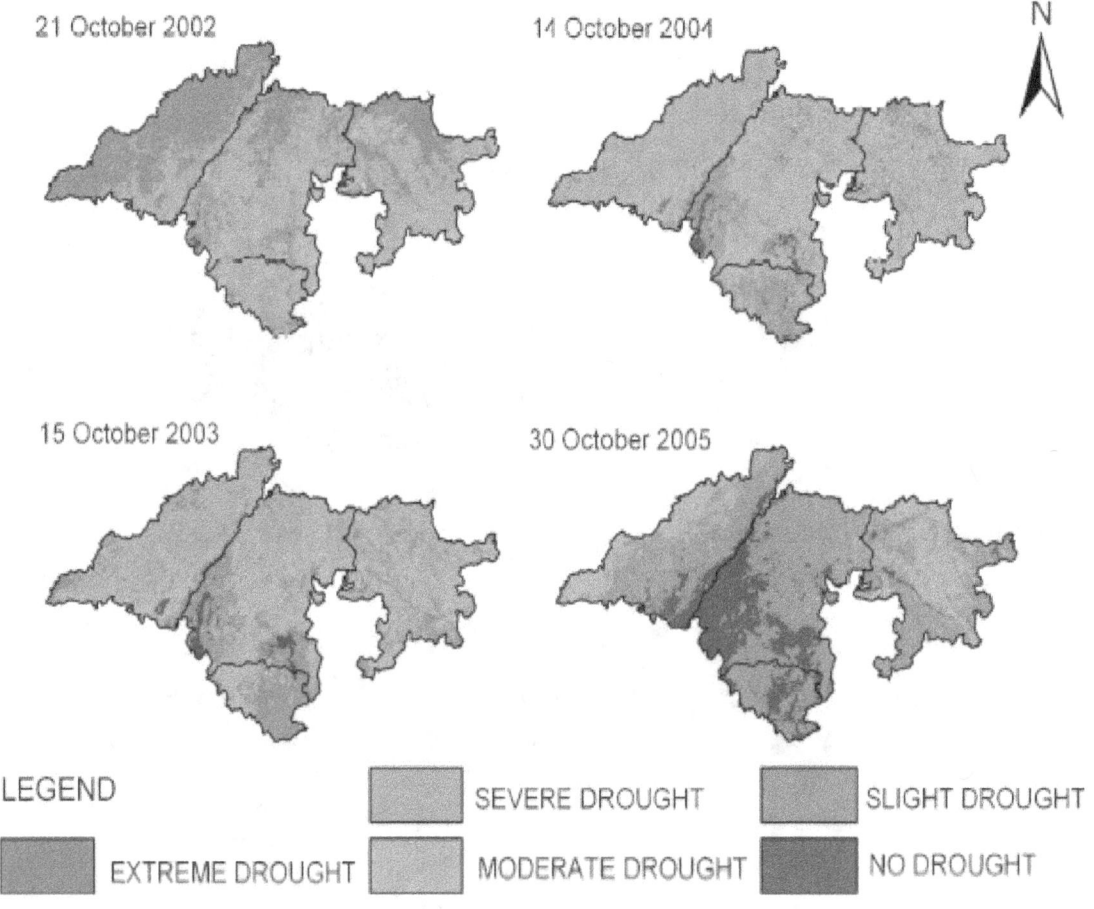

Figure 8.15 Classified Post – NDVI Maps of the Study Areas in Different Years

Identification of drought venerable areas using remote sensing data

Year	Extreme	Severe	Moderate	Slight	Normal
2002	33.50	33.80	25.30	7.00	0.40
2003	2.80	34.50	37.30	22.70	2.70
2004	4.70	38.40	40.30	14.70	1.90
2005	0.19	0.56	21.20	56.49	21.57

Percentage of areas under different category of drought using NDVI

Year	Extreme	Severe	Moderate	Slight	Normal
2002	36.6	35.1	22.4	5.8	0.6
2003	3.7	36.6	36.8	15.9	7.1
2004	3.2	35.6	48.0	19.7	0.6
2005	0.1	2.4	20.1	61.1	16.2

Percentage of areas under different category of drought using WSVI

Drought Software

Drought software is developed jointly by IWMI and IWR.

Part of the SPATSIM package

Spatial and Time Series Information Modeling

SPATISM is developed by the Institute for Water Research (IWR), South Africa. It is permanently expanding to include more options for various water resources analyses.

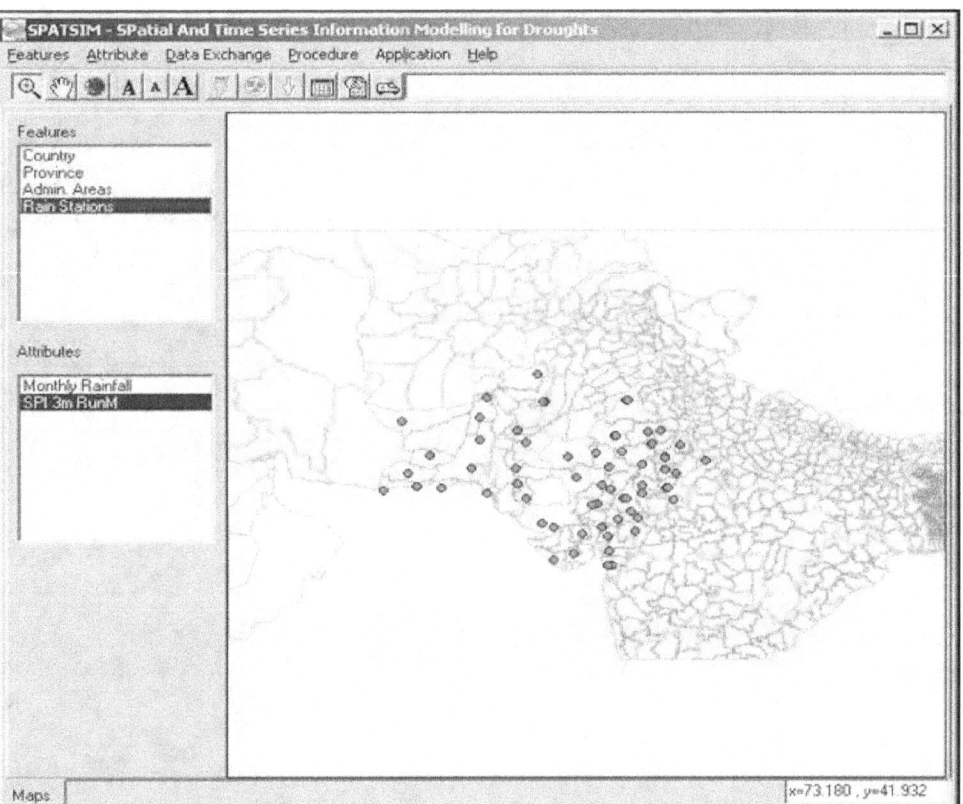

Main SPATSIM screen showing a coverage of SW Asia and rainfall stations' locations

It calculates, displays, spatially plots, exports/imports areal rainfall and variety of drought indices from rainfall time series data.

Strategies to enhance adaptive capacity to climate change in vulnerable regions

Lead institution

Indian Agricultural Research Institute (IARI), New Delhi

In collaboration with

CMFRI, Mumbai, OUAT, Bhubneshwar, CRRI, Cuttack

- Identification of current and future risks to livelihoods due to climatic variability

- Development of drought indices to facilitate Early Warning System (EWS) for Drought & promoting it's use in adaptation by farmers and other stakeholders

- Develop community based sustainable rural livelihoods strategies to minimize adverse climatic impact in droughts as well as floods prone vulnerable districts

- Capacity building of the stakeholders on strategies for alternate livelihoods strategies in future climate change.

CONCLUSIONS

- SPI is most commonly used index for hydrological drought assessment in conjunction with other indices.

- Other hydrological drought indices based on stream flow are hindered by the data availability.

- Indices based on groundwater levels in conjunction with other index are used for hydrological drought assessment for non monsoon periods.

- Lack of coordination between data monitoring agencies.

REFERENCE

Manoj Khanna, Water Technology Centre, Indian Agricultural Research Institute, New Delhi, INDIA

SOCIOECONOMIC DROUGHT

Socio–Economic Drought

This condition is when some supply of some goods and services such as energy food and drinking water are reduced or threatened by changes in meteorological and hydrological conditions.

Socio–economic drought occurs when the demand for an economic goods exceeds supply as a result of a weather related short fall in water supply.

When human activities are affected by reduced precipitation and related Water availability, this form of droughts associates human activities, with elements of meteorological agricultural and hydrological droughts.

Socioeconomic definitions of drought associate the supply and demand of same economic goods with elements of meteorological, hydrological and agricultural drought. It differs from the afore mentioned types of drought because its occurrence depends on the time and space processes of supply and demand to identify or classify droughts. The supply of many economic goods such as water forage, food grains, fish and hydroelectric power, depends on weather.

Because of natural variability of climate water supply is ample in some years but unable to meet human and environmental needs in other years. Socioeconomic droughts occurs when the demand for an economic good exceeds supply as a result of a weather related shortfall in water supply.

For example in URUGUAY in 1988–89 drought resulted in significantly reduced hydroelectric power production because power plants were dependent on streams flow rather than storage for power generation. Reducing hydro electric power production requiring the Government to convert to more expensive (Imported) Petroleum and implement stringent energy conservation measures to meet the nation's power needs.

In most instances the demand for economic goods is increasing as a result in increasing population and per capita consumption. Supply may also increase because of improved production efficiency, technology or the construction of reservoirs that increase surface water storage capacity. If both supply and demand are increasing the critical factor is the relative rate of change is demand increasing more rapidly than supply? If so vulnerability and the incidence of drought may increase in the future as supply and demand trends converge. All droughts originate from a deficiency of precipitation or meteorological drought but other types of drought and impacts cascade from this deficiency.

Source: National drought mitigation centre university of Nebraska–Lincoln USA.

Threats to water and food security   Water is life and climate change is threatening this precious resource. Nearly every U.S. region is facing some increased risk of seasonal drought.

Climate change will significantly affect the sustainability of water supplies in the coming decades. As hart of the country gets drier the amount of water availability and its quality will likely to decrease – impacting peoples health and food supplies.

As temperatures rise and precipitation decreases, water quality can be Jeepardized. Shrinking amounts of water can concentrate contaminants such as heavy metals, industrial chemicals and pesticides and sediments and salts.

During drought drinking water supplies are susceptible to harmful algal blooms and other micro organisms of course, drought means more than not having access to clean drinking water, changes in precipitation and water availability could have serious consequences for commercial agricultural crops, yield less and food security suffers. Drought conditions can also help fuel out of control wildfires.

Local communities across the country can prepare for drought by learning to conserve water and improving drinking water safe guards. Nine states in USA and several local Governments have developed preparedness measures to address the drought impacts associated with climate change. The most common recommendation is improving general preparedness measurers for drought and ensuring an adequate water supply.

California's plan includes measures that focus on ensuring adequate water availability during times of drought.

The city of Berkeleys strategy includes a city focused vulnerability assessment that will include assessing water resources and Los angles has a measure to prepare for increased drought conditions.

Climate change will worsen smog and causes plants to produce more pollution increasing respiratory health threats particularly for people with allergies and asthma.

Climate Change Health Threads in California

1. Air pollution–smog smoke and pollen
2. Extreme heat: more intensive hot days and Heat waves.
3. Infections diseases: Dengue fever, West Nile virus and Lyme disease
4. Drought–Threats to water and food security
5. Flooding–Devasting floods and heavy rains
6. Extreme Weather–Record Breaking events in 2011 year

California's Changing Climate

1. The frequency of extreme heat and prolonged drought are already increasing
2. In the future, with climate change, average temperatures in the state could rise anywhere from 4.7–10.5°F (2.2 – 5.8°C) by last century.
3. Californians will experience greater exposure to public health, threats such as heat-related sickness air pollution and water scarcity.
4. California has a strategy to prepare for the health impacts of climate change

The city of Berkeley, Los Angeles, San Fracisco and San Rafael have also identified local climate change – related health threats. Maryland and New York also follow the above socio economic impact is the impact that development on society and the economy. Some examples of this include cars, cell phones, laws & stocks

Types of drought impacts   Drought affects all parts of our environment and our communities. The many different drought impacts are often grouped as "Economic" "Environmental" and "Social" impacts. All of these impacts must be considered in planning and for and responding to drought conditions.

Economic impacts   Economic impacts are those impacts of drought that cost people (or businesses) money. The few examples of economic impacts are as follows:

1. Farmer may lose money if a drought destroys their crops.
2. If a farmer's water supply is too low, the farmer may have to spend more money on irrigation and to drill new wells.
3. Ranchers may have to spend more money on feed and water for their animals.
4. Businesses that depends on farming like companies that make tractors and food may loose business when drought damages crops or live stock.

5. People who work in the timber industry may be affected when wild fires destroy stands of timber.

6. Business that sell boats and fishing equipment may not be able to sell some of their goods because drought has dried up lakes and other water resources.

7. Power companies that normally rely on hydroelectric lower may have to spend more money on other fuel source if drought dries up too much of the water supply. The power companies customers would also have to pay more.

8. Water companies may have to spend money on new or additional water supplies.

9. People might have to pay more for food.

10. Barges and ships may have difficulty to navigating in streams, rivers and canals because of low water levels which would also affect businesses that depend on water transportation for receiving and sending goods and materials.

Environmental impacts   Drought also effects the environment in many different ways. Plants and animals depend on water just like people. When a drought occurs, their food supply can shink and their habitat can be damaged. Sometimes the damage is only temporary and their habitate and food supply return to normal when the drought is over. But sometimes droughts impacts on environment can last a long time, may be forever. Examples of environmental impacts are as follows:

1. Losses on destruction of fish and wild life habitite due to lack of food and drinking water for wild animals.

2. Increase in disease will occur among wild animals because of reduced food and water supplies & migration will take place in wild life.

3. Increase stress on endangered species or even extinction.

4. Due to low water levels in reservoirs lakes or ponds loss of wet lands occur.

5. More wild fires will also occur.

6. Wind & Water erosion of soil and poor soil quality will happen.

Social impacts   Social impacts of drought are ways that drought affects peoples health and safety. Social impacts include public safety health conflicts between people when there is not enough water to go around and changes in life style.

Examples of social impacts are as follows:

1. Anxiety or depression about economic losses caused by drought.

2. Health problems related to low water flows and poor water Quality. Health problems related to dust lost of human life.

3. Threat to public safety from an increased number of forest and range fires.

4. Reduced incomes.

5. People may have to move from farms into cities or from one city to another.

6. Fewer recreational activities.

What are the effects of drought Drought can have serious health social economic and political impacts with far reaching consequences. Water is the one of the most essential commodities for human survival, second only to breathable air.

So when there is drought, which by definition means having too little water, to meet current demands conditions can become different or dangerous very quickly. The consequences of drought may include:

1. Hunger and famine: Drought conditions often provide to little water to support food crops, through natural precipitation or irrigation using reserve water supplies. The same problem affects grass and grain used to food live stock and poultry. When drought under mines or destroys food sources, people go hungry, when the drought is severe and continues over a long period famine may occur.

Thirst: All living things must have water to survive. People can live for weeks without food, but only a few days without water.

Disease: Drought often creates a lack of clean water for drinking. Public Sanitation and personal hygiene which can lead to a wide range of life threatening diseases.

Wild fires: The low moisture and precipitation that often characterize droughts can quickly create hazardous conditions in forests and across range lands, setting the stage for wildfires that may cause injuries and deaths as well as extensive damage to property and already shrinking food supplies.

Social conflicts and water When a precious commodity like water is on short supply due to drought and the lack of water creates a corresponding lack of food, people will complete and eventually fight and kill to secure enough water to survive.

Common Causes for Drought in India

Meteorology

1. In adequate monsoon rainfall
2. High temperature evaporation, wind speed
3. Unseasonal rains and fog snowfall

Water resources

1. In adequate water availability
2. High water loss in storage and distribution utilities
3. Over exploitation of surface & ground water.

Agriculture crop-yield

1. Shift in agricultural practices (low to moderate) water demand crops to high crops
2. Crop damage due to rain and snow pest

Population

1. High greater rate of human and animals
2. Location of high water consuming miles stones at semiarid / arid regions

Drought impact is also felt due to:

1. Deficit ground water recharge.
2. Non availability of quality seeds.
3. Reduced drought power for agricultural operations due to distress sale of cattle.
4. Land degradation.
5. Fall in investment capacity of farmers rise in prices reduced grain trade and power supply.

Social impact of droughts   The social implication of droughts are perhaps the most felt, as they directly involve us and our families. Some people (especially those from developed countries) have never experienced what is it to live without adequate water, it is nightmare.

Health has a direct link to the water supply of any settlement. Clean water for Drinking and water for cleaning and sanitation help society prevent and manage diseased.

Hunger, malnutrition, anemia and mortality impacts of droughts are indirect in nature, a droughts cause low food production (crops & livestock) and particularly in poor regions, people have less to eat. Food nutrition also is a problem and that leads to vulnerability diseases / illness and deaths. This is particularly so in remote communities of poorer countries, where communication and accessibility is usually poor.

Fresh water levels and water discharge during droughts are low, resulting in less dilution in ecosystem waters. This means that the concentration of chemicals, nutrients and solid particles increases and dissolved oxygen decreases.

People migrate to other places in search of better living conditions this makes a region in drought vulnerable, as many of its young and working population are forced to leave. Farm families suffer more when family members migrate. Droughts in more rural areas of the world causes strain on family lives. There is more pressure on woman to work outside farms to help provide for family.

Anxiety stress and the general low and drained feeling of not knowing when things will improve can have a negative effect on people. People are unhappy and depressed because all the things that they used to do is no longer available and they have to deal with a difficulty that has no end in sight. Community net works are broken and social interaction decreases. This results in low esteem and feeling of social isolation.

People feel unsafe and threatened by loss of forest and wild fires, as well as loss of human life.

There are many things an individual, community or Government can do to minimize the impact of drought if they occur.

Here are a Few

Education: It is important that each of us learn about how droughts occur and how they affect us. This empowers us to think of solutions and other things. We can do if we find ourselves in a drought. The Government also need to educate the public periodically about their environment, climate, weather and some natural disasters that can happen. The Government also needs to understand the terrain of the region and the livelihood of a drought, so that there are no surprises if they happen.

Stop Pollution

Taking measures to stop all forms of water pollution is important because as the onset of drought, humans resort to surface water such as streams and lakes and the like. If these are in great condition humans can depend on them for drinking and irrigation until things improve. If they are all polluted or contaminated and unsafe for any kind of use it makes the problem even more distressing.

Water Conservation and Storage

Water is precious and a scarce commodity every where in the world and humans need to use warm wisely as such. Even if these is water available, it is important because the practice makes us cope better when there is a shortage. Aslo preserving water leaves enough to be stored in dams, river water and even turned into ponds.

California Drought 2014 USA

On 15th July 2014, the state water resources control board announced that residents who waste water on out door lawns and landscaping would be fined $500 a day. This emergency regulation is in place to enforce water conservation efforts by the stage of California, as they fight a state – wide brought, believed to be the state's driest year on record.

Source: https://www.swrcb.ca.gov/press_room/press_releases/2014/pm071514.pdf

People must learn to conserve water, even if there is no drought. With the right attitude towards water, we are better prepared to face the impact in case there is a shortage.

Sometimes (especially in developed nations) canals and pipe lines are built to connect places with abundant water to places with less water. Projects like that can be very expensive, but they ensure that during droughts, there can be some water flow until traditional water supply sources improve.

Preparing for Drought

Try as we might we cannot control the weather. Thus we cannot prevent drought that are caused strictly by a lack of rainfall or abundance of heat. But we can manage our water resources to better handle these conditions so that a drought does not occur during short dry spells.

Ecologists can also use various tools to predict and assess drought conditions around, the world. In the U.S. the U.S. drought monitor provides a day by day visual of the drought conditions around the country. And the U.S seasonal drought over look predicts drought trends that may occur based on statistical and actual weather forecasts. Another program, the drought, impact reporter collects data from the media and other weather observers about the impact of drought in a given area.

Using the information from these tools ecologists can predict when and where a drought might occur, assess the damages caused by a drought and help an area recovery more quickly after a drought occurs.

REFERENCES

1. Drought basics, the national drought mitigation centre. https://drought.unl.edu./drought basics.aspx

2. Questions and answers about droughts. https://water.usgs.gov/edu/qadroughts.html

3. Drought management, water Encyclopedia https://www.waterencyclopedia.com/Da-En/Drought-Management.html.

4. About droughts, Australian Emergency management institute https://schools.aemi.edu.au/content/privacy - statement

5. Natural Hazards / Heat & drought .Natural disastors association. http://www.n-d-a.org/heat – drought.php.

6. Too little water – drought – BBC bite size http://www.bbc.co.uk/schools/g-csebite size/ geography / water_rivers / drought_mevl.shtm

DROUGHT MONITORING AND EARLY WARNING SYSTEMS

⬚ CHAPTER OUTLINE

- Drought disaster challenges and mitigation in India – strategic appraisal – Drought risk management – Key indicators for declaration of drought in India – Short comings of current early warning systems in the entire world – are dealt – Developing composite index and Indices used for it are developed.

There is an urgent need to develop better drought monitoring and early warning systems. Drought is an insidious natural hazard that results from a deficiency of precipitation from a expected or normal that, when extended over a season or longer period of time, is insufficient to meet the demands of human activities and the environment.

Drought must be considered a relative rather than absolute condition. A critical component of national drought strategies should be a comprehensive drought monitoring system that can provide early warning of droughts onset and end, determine its severity and deliver that information to a broad group of users in a timely manner. With this information the impacts of drought can be reduced or avoided in many cases.

One factor that distinguishes drought from other natural hazards in the absence of a precise and universally accepted definition. There are hundreds of definitions adding to the confusion about whether or not a drought exists and its degree of severity. Definitions also need to be application specific because drought impacts will vary between sectors.

Drought means different things to different users such as a water manager, an agricultural producer a hydroelectric power plant operator and a wildlife biologist.

Droughts are commonly classified by type as meteorological, agricultural and hydrological and droughts differ from one another in three essential characteristics, intensity duration and spatial coverage.

There are numerous natural indicators of drought that should be monitored routinely to determine the onset, ending and spatial characteristics of drought. Severity must also be evaluated on frequent time steps.

Although all types of droughts originate from a deficiency of precipitation, it is insufficient to rely solely on this climate element to assess severity and resultant impacts because of factors identified previously.

Effective drought early warning systems must integrate precipitation and other climatic parameters with water information such as stream flow snow pack, ground water levels, reservoir and lake levels and soil moisture into a comprehensive assessment of current and future drought and water supply conditions.

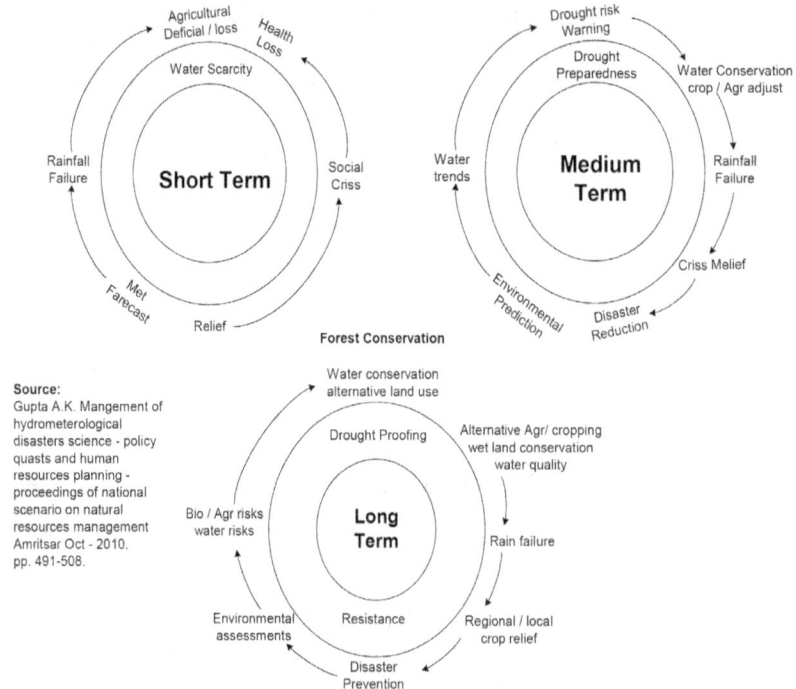

Figure 10.1 Drought Risk Management (Depicts Important Concerns of drought Management Cycles in the Short Medium Large Context)

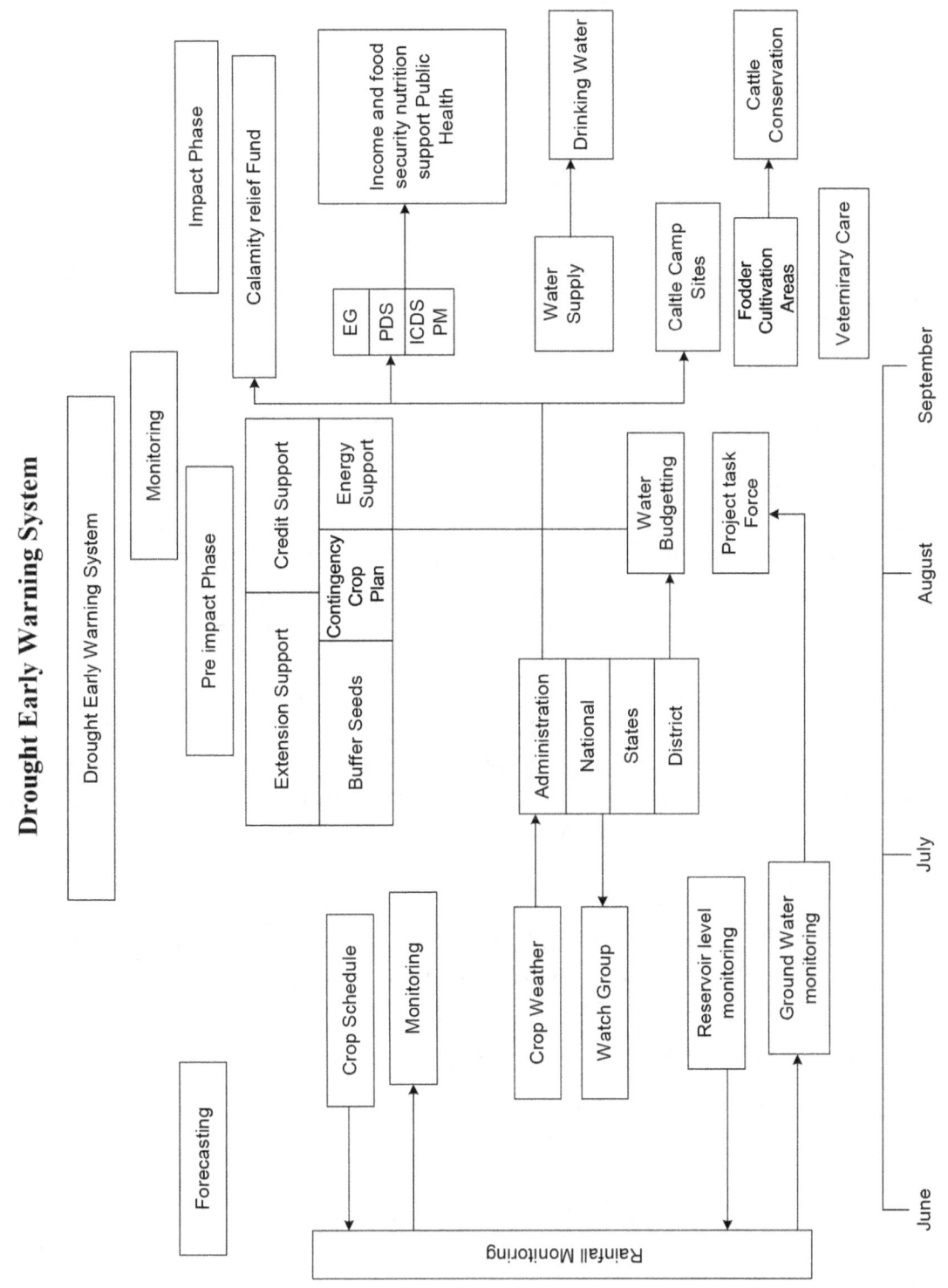

Figure 10.2 Drought Early Warning System

Source: Drought Disaster Challenges and Mitigation in India. Strategic Appraisal Anil K. Gupta, Vinar K. Sehgal, Pallavee Tyagi. Current Science. Vol. 100. No. 12, 25 June 2011.

Declaration of drought in India   The practice existing in India for declaration of drought is as follows :

For the states which receive rainfall during South - West monsoon (July to Sept), the authorities concerned will declare drought in the month of October.

The monsoon is over by this month and figures for total rainfall are available in this month similarly a final picture regarding the crop conditions as well as the reservoir storage is available by end of October.

For the states that receive rains from the north – east monsoon drought declaration should be done in January.

The key indicators on which focus is made in India for declaration of drought are as noted below

Key Index -1 : Rainfall Deficiency

There are two ways to be considered.

(a) Rainfall received during the months June and July is less than 50% of the average rainfall for these two months.

(b) The total rainfall for the entire duration of the rainy season of the state from June to Sept. (South – West monsoon) and (or) from Dec. to March (North – East monsoon) is less than 75% of the average rainfall for the season.

Key Index - 2 : Area Under Sowing

The area under bowing provides reliable information on the availability of water for agricultural operations. Drought conditions could be said to exist if the total sowing area of Kharif crops is less than 50% of the total cultivable area by the end of July/ August depending upon the schedule of sowing in individual states. Then Govt. should therefore declare a drought if along with other indicators.

In case of Rabi crops, the declaration of drought could be linked to the area of sowing being less than 50% of the total cultivable area by the end of November/ December along with the other indicators.

Key Index - 3 : Normalised Difference Vegetation Index

The National Agricultural drought assessment and monitoring system (NADAMS) instituted by the National Remote Sensing Centre (NRSC) issues a bi-weekly drought bulletin and monthly reports on detailed crop and seasonal condition during the Kharif season. The present the normalized difference vegetation index (NDVI) and normalized difference wetness index (NDWI) from the data obtained from National Oceanic and Atmospheric Administration – Advanced very High Resolution Radiometer (NOAA – AVHRR) and Indian Remote Sensing (IRS) satellite wide field sensor (WIFS) data. These reports provide quantitative information on sowing, surface water spread and District / Tahsil / Block level crop condition assessment along with spatial variation in terms of maps.

At present, 11 agriculturally important and drought – vulnerable states of Andhra Pradesh, Bihar, Gujarat, Hariyana, Karnataka, Maharastra Madhya Pradesh, Orissa,

Rajasthan, Tamilnadu and Uttar Pradesh are covered through these reports.

The values obtained for a given NDVI always ranges from −1 to +1. A negative number or a number close to zero means no vegetation and a number close to +1 (0.8 − 0.7) represents luxurious vegetation.

For declaring the drought, states need to obtain NDVI values through NDAMS. All the above − mentioned states receive NADAMS reports on a regular basis. Those states which do not receive the reports could approach the NRSC for receiving the information.

"It is necessary that the states declare drought only when the deviation of NDVI value from the normal is 0.4 or less. However, the NDVI value needs to be applied in conjunction with other indicators and values. The NDVI must not be invoked for the Declaration of drought in isolation from the other two key indicators.

Key Index - 4 : Moisture Adequacy Index

Moisture Adequacy Index (MAI) which is based on a calculation of weekly water balance, is equal to the ratio (expressed as percentage) of actual evapo-transpiration (AEI) to the potential evapo-transpiration (PET) following a soil − water balancing approach during a cropping season.

MAI is obtained by using the following equation.

$$MAI = \frac{AE}{PE}$$

Where, AE is actual evapotranspiration and PE is potential − evapotranspiration (in %) during different phonological stages of a crop.

Water balance calculation takes into account the soil characteristic, crop growth period and water requirements of major crops. Drought is specified crop-wise on a real-time basis. MAI values are shown in table below −

Table 10.1 Agricultural Drought Code

AE/PE (%) During Different Phenophases	Drought Intensity	Phenophase G wise Code		
		Seeding (S)	Vegetative (V)	Reproductive (R)
76 to 100	No drought	S_0	V_0	R_0
51 to 75	Mild drought	S_1	V_1	R_1
26 to 50	Moderate drought	S_2	V_2	R_2
25 and less	Severe drought	S_3	V_3	R_3

(Based on Moisture Adequacy Index) GAZRI

Table 10.2 Intensity of Agricultural Drought

S.No.	No Drought	Mild	Moderate	Severe
1	$S_0V_0R_0$	$S_0V_0R_2$	$S_0V_0R_3$	$S_0V_2R_3$
2	$S_0V_0R_1$	$S_0V_1R_2$	$S_0V_1R_1$	$S_0V_3R_2$
3	$S_0V_1R_0$	$S_0V_2R_0$	$S_0V_2R_2$	$S_0V_3R_3$
4	$S_0V_0R_0$	$S_0V_2R_1$	$S_0V_3R_0$	$S_1V_2R_3$
5	$S_1V_0R_1$	$S_1V_0R_2$	$S_0V_3R_1$	$S_1V_3R_3$
6	$S_1V_1R_0$	$S_1V_1R_1$	$S_1V_0R_3$	$S_0V_2R_3$
7	$S_0V_1R_1$	$S_1V_1R_2$	$S_1V_1R_1$	$S_2V_0R_3$

Source: Vittal, Kar and Rao, CAZRI

MAI values are critical to ascertain agricultural drought. The state agriculture Dept needs to calculate the MAI values on the basis of data available to it and provide it to the Dept. of Relief and Disaster management, which would ascertain that MAI values conform to the intensity of moderate drought as shown in the table above before drought is declared.

As mentioned above, MAI values need to be applied in conjunction with other indicators such as rainfall figures, area under sowing and NDVI values. In light of all the discussions and conclusions made during several workshops to build a comprehensive and flexible a drought early warning system (DEWS) as possible it is important to monitor drought across the many sectors mentioned earlier.

The use of a single index will rarely work for all places at all the times and for all types of droughts. In cases such as those it is important to consolidation of indices and indicators into one comprehensive "composite index". A composite index approach allows for the most robust way of detecting and determining the magnitude (duration + intensity) of droughts as they occur.

Considering the complexity of drought and the many indices and indicators necessary to assess its severity and likely impacts the most successfully approach to data (drought.unl.edu/dm) is the U.S. drought monitor. It incorporates multiple indices and indicators of drought including impacts into the assessment process fully.

Although many countries do not have the range of data available to replicate this process fully, any approach that incorporates information beyond precipitation and perhaps temperature data is going to provide a more accurate picture of drought severity.

Effective drought early warning systems (DEWS) are an integral part of efforts worldwide to improve drought preparedness, timely and reliable data and information must be the corner stone of effective drought polices and plans. Monitoring drought presents some unique challenges because of drought distinctive characteristics.

An expert group meeting on early warning systems for drought preparedness, sponsored by the world meteorological organisation (WMO) and others recently examined the status, short comings and needs of (DEWS) drought early warning systems and made recommendations on how these systems can help in achieving a greater level of drought preparedness (Wilhite et al., 2000b).

This meeting was organized as part of WMO's contribution to the UNCCD. The proceedings of this meeting documented recent efforts in drought early warning systems (DEWS) in countries such as Brazil, China, Hungary, India, Nigeria, South Africa, and United States, but also noted the activities of regional drought monitoring centers in West Asia and North Africa.

The shortcomings of current drought early warning systems (DEWS) were noted in the following areas

1. **Data networks:** Inadequate density and data quantity of meteorological and hydrological networks on all major climate and water supply parameters.

2. **Data sharing:** Inadequate data sharing between Government agencies and high cost of data, limit the application of data in drought preparedness mitigation and response.

3. **Early warning system products:** Data and information products are often not user friendly and users are often not trained in the application of this information to decision making.

4. **Drought forecasts:** Unreliable seasonal forecasts and lack of specificity of information provided by forecasts limit the use of this information by farmers and others.

5. **Drought monitoring tool:** Inadequate indices for detecting the early onset and end of drought, although the standardized precipitation index (spi) was cited as an important new monitoring tool to detect the early emergence of drought.

6. **Integrated drought/climate monitoring:** Drought monitoring systems should be integrated and based on multiple indicators to fully understand drought magnitude, spatial extent and impacts.

7. **Drought impact assessment methodology:** Lack of impact assessment methodology hinders impact estimates and the activation of mitigation and response programs.

8. **Delivery systems:** Data and information on emerging drought conditions, seasonal forecasts and other products are often not delivered to users in a timely manner.

9. **Global drought early warning system:** No historical drought data base exists and there is no global drought assessment product that is based on one

or two key indicators, which could be helpful to international organisations, non-Governmental organisations (NGO's) and others.

With the above short comings duly rectifying and creating a "composite index" in India will be a greater help to farmers and feed security and enhancement of drinking water. The early onset, duration and end along with quantifying severity of drought will be an imperative task laying on the authorities.

Developing a composite index and releasing drought information weekly duly consulting and incorporating review from a group of climatologists extension agents and others across the nation, will lead the country economically forward.

The following indicators and indices are suggested to be a part of "composite index" to be called as "Indian drought monitor".

1. **Standardized Precipitation Index (SPI)**

   To characterize meteorological drought with onset and end of drought and severity, duration, (on short term scatus). Longterm scales over 6 months are considered as hydrological drought indicators and hydrological drought using stochastic models and neural network for drought forecasting.

   > **Example** Carried for Kansabli River Basin, West Bengal, India and also in South Asia to quantify precipitation deficit in the monsoon and non-monsoon.

2. **Standardized Precipitation Evapotranspiration Index (SPEI)**

   This new index particularly suited to detecting monitoring and exposing the consequences of global warming on drought conditions.

3. **Aridity Anomaly Index (AAI)**

   To monitor the agricultural drought during the rainy Kharif and past rainy (Rabi) seasons, based on Thornthwaite (1948) Water Balance Techniques on weekly or fortnightly basis).

4. **Moisture Adequate/Availability Index (MAI)**

   It is obtained from weekly water balance developed by central arid zone research institute (CAZRI) Jodhpur for monitoring the Indian arid regions.

5. **Standardized Water level Index (SWI)**

   It is developed to scale ground water recharge deficit for hydrological drought analysis, used to analysis drought dynamics in aravali region of Rajastan India used mean seasonal water levels for 20 years from 541 wells (1984-2003). SWI values of the wells were interpolated using spline interpolation technique in a GIS environment to generate SWI maps of the region (for non-monsoon period hydrological drought).

6. **Normalised Difference Vegetation Index (NDVI)**

   It is based on the concept that vegetation vigour is an indication of water availability or lack thereof. (To characterize agriculture drought along with MAI).

7. **Normalised Difference Witness Index (NDWI)**

It is based on the use of short wave infra red (SWIR) band which is sensitive to moisture availability in soil as well as in the crop canopy. In the beginning of crop season, soil background is dominant, which makes the (SWIR) sensitive to soil moisture in top 1-2 cms. As the crop progresses SWIR becomes sensitive to leaf moisture content.

NDWI using SWIR can complement NDVI for drought assessment particularly in the beginning of the cropping season. Higher values of NDWI signify more surface wetness. Used to assessment of agricultural drought.

Note:

(a) The values of NDVI will be received to states through (NADAMS) National agricultural drought assessment and monitoring systems, instituted by National remote sensing center (NRSC) issues a biweekly drought bulletin and monthly reports on detailed crop and seasonal condition during the Kharif season.

(b) A negative number (NDVI) or a number close to zero means no vegetation and a number close to +1 (0.8 to 0.7) represents luxurious vegetation. The values obtained from a given (NDVI) always ranges from (–) 1.0 to (+) 1.0. It is necessary that the states declare drought only when the deviation of NDVI value from the normal in 0.4 or less. However the NDVI values needs to be applied in conjunction with other indicators and values.

8. **Vegetation Condition Index (VCI)**

This index was first suggested by Kogan 1995 & 1997. It shows effectively how close the current months NDVI is to the minimum NDVI calculated from the long term record of remote sensing images. This is also used to identify agricultural drought.

9. **Water Balance indices: Agro meteorological drought indices** (for agricultural drought)

Source: FAO irrigation and drainage paper – 25 – 1978

(a) Required irrigation water

I_i = Rei – Etci

I_i = Required irrigation water in month (i) mm

R_{ei} = Effective precipitation during month (i) mm that calculated according to USDA

ET_{ci} = is the monthly crop evapotranspirator mm

Important Note

The Indian Meteorological Department (IMD) collects its rainfall data using a network of 2800 rain gauge stations distributed across 36 meteorological subdivisions of the country. It categorizes rainfall in each subdivision as excess normal, deficient or scanty, on an average each meteorological subdivision has more than a dozen districts. It is very likely that within a meteorological subdivision certain districts are effected by drought while others are not. Therefore it is very essential to have a drought early warning system for each village having one rain gauge station. If onset end duration and severity of drought along with magnitude at village level is known, preparedness for drought and mitigation can be of ease for farmers and authorities. So the spi, SPEI and water balance parameters arrived along with remote sensing indices are to be worked out and information to villages and farmers to be given to mitigate drought conditions in India.

10. **Surface Water Supply Index (SWSI) (Hydrological drought)**

This index integrate reservoir storage stream flow and two precipitation types (snow and rain) at high elevations into single index.

It represents water supply conditions unique to each basin or water management requirement of each basin.

$$SWSI = \frac{ap_{snow} + bp_{pree} + cp_{stream} + dp_{reservoir} - 50}{12}$$

where a b c d = weights of snow, rain, stream flow and reservoir storage respectively (a + b + c + d = 1)

P = the probability (%) of non-exceedence for each of these four water balance components.

The estimation is carried out with a monthly time step. In winter months SWSI is computed using precipitation, snowpack and reservoir storage.

In summer, stream flow, precipitation and reservoir storage data are used.

The range of SWSI is similar to palmer drought severity index (PDSI) as (−) 4.2 to (+) 4.2

Remote sensing data were used to hydrological drought assessment using NDVI and water supporting vegetation index (WSVI) along with standardized precipitation index (spi) for Southern Rajastan India.

Table 10.3 Developing a "Composite Index" called as "Indian Drought Monitor"

1.	Standardized precipitation index (spi)	Meteorological drought (short term scale) Hydrological drought at longer scales.
2.	Standardised precipitation evapo-transpiration index (SPEI)	Meteorological drought with global warming on drought conditions

3.	Aridity anomaly index (AAI)	Agricultural drought during Kharif and Rabi season using Water Balance Technique
4.	Moisture adequacy index MAI	Obtained from water balance for agricultural drought to monitor arid regions in India (CAZRI)
5.	Normalised difference vegetation index NDVI	Agricultural drought along with MAI
6.	Normalised difference wetness index (NDWI)	Agricultural drought
7.	Vegetation condition index (VCI)	Agricultural drought
8.	Water Balance indices	Agricultural drought
9.	Standardised water level index (SWI)	For non-monsoon period hydrological drought
10.	Surface water supply index (SWSI)	Hydrological drought (unique to basin)

DROUGHT MITIGATION

Mitigation means actions that can be taken before or at the beginning of drought to help reduce the incidence or impacts of drought. These measures are important for adapting to climate change, restoring ecological balance and bringing development benefits to the people.

Most of these measures are related to integrated soil, water and forest management and form part of soil conservation, water shed development and forest programmes.

Drought mitigation programmes are never stand-alone interventions to be taken in the wake of a drought, but are very much a part of development planning. The measurers need to be implemented by Government of India and State Governments as part of their regular development programmes.

While drought mitigation programmes have been in existence for a long time, they have not been effective due to many reasons such as allocation being as thinly scattered over large areas. The peoples involvement is very megre. The programmes are fragmented in their implementation. The accountability mechanisms are very weak and all programme experience serious lacunae.

In India during 2009 integrated wasteland development programme (IWDP). Drought prone area programme (DPAP) and Desert Development Programme (DDP) have been consolidated into a single programme named as integrated watershed management programme (IWMP) in place of all the above mentioned three Area development programmes. The cost norms will be Rs.12,000/ha for plains and Rs.15000/ha for the hilly and different areas. The cost will be shared in the ratio of 90:10 between centre and state.

The objectives of mitigation measures are to reduce soil erosion, augment soil moisture retard the drainage of rain water and improve the efficiency of water use. It involves a wide range of soil and water conservation measures and farm practices.

Rain water harvesting and conservation   Water harvesting and conservation & refer to processes and structures of rainfall and run off collection from large catchment areas and channeling them for human consumption.

In India, these processes and structures have been in existence since antiquity, but the increasing frequency and severity of droughts and population growth have focused on the revival of these practices and structures. Every house hold's minimum water requirements can be easily met by collecting rainwater locally from village / community ponds / large manmade containers by diverting and storing water from local streams / springs and by tapping sub-surface water below river/stream buds.

There are two methods for water conservation

1.   Artificial recharge of ground water

2.   Traditional methods

While the artificial recharge of ground water is used extensively in all the water shed development programmes being implemented. Traditional methods of water collection and harvesting through ponds, tanks are even more important for assuring continuous and reliable access to water. Both methods include measures which are low-cost, community – oriented and environment – friendly.

These methods are considered very useful for ground water recharge both when rainfall is deficient and when there are flash floods. Harvesting and conservation of flood water to rejuvenate depleted high-capacity aquifers by adopting integrated ground water techniques, such as dams, tanks, anicuts, percolation tanks could improve water availability and create a water buffer for dealing with successive droughts.

Artificial Recharge of Ground Water

A typical water shed development programme has several components depending on the topography, nature and depth soil cover, type of rocks and their Pattern of formation and layout, water absorbing capacity of land, rainfall intensity and land use. These include the following.

Contour pounding, contour trenching, contour cultivation Bench Terracing, graded Bunding, Gully plugging check dams / Nalla bunding construction, Gabion structures, stream bank protection, farm ponds percolation tanks, spread basin, anicuts, sub-surface barriers, injection wells, Dug well recharge, village pond) Tank, Tankas / kunds / kundis, khadins, var, vardi baoli, bavadi, jhalara, Hill scope collection spring water harvesting, rain water harvesting in urban areas, Rooftops paved and unpaved areas, water bodies.

Water Saving Technologies

Drip and sprinkler irrigation system

The technologies are recommended for achieving higher irrigation efficiencies and could be used for very small – sized holdings.

While sprinklers require energized pump sets, micro tube drips can work under a very low – pressure head, with as little as a bucket full of water. Sprinklers tend to irrigate more uniformly than gravity systems and therefore efficiencies typically average about 70%. But in windy and dry areas much water can be lost due to evaporation in this system.

The sprinkler system is particularly effective in sandy undulating terrain. For fruits, vegetables and orchard crops, drip irrigation (also known as trickle irrigation) is more suitable.

These systems require much less maintenance when compared with the conventional pressurized irrigation systems.

Soil and water conservation   Conservation practices minimize the disruption of soil structure, composition and natural biodiversity there by reducing erosion and soil degradation, surface run off and water pollution. The following are established practices of soil and water conservation.

(1) Crop rotation (2) Contoured row crops (3) Terracing (4) Tillage practices (5) Erosion control structures (6) water detention and retension structures (7) Wind breaks and shelter belts (8) Litter management (9) Reclamation of salt affected soil. Soil and water conservation can be approached through agronomic and engineering measures.

Agronomic measures include contour forming, off season tillage, deep tillage mulching and providing vegetative barriers on the contour. These measures prevent soil erosion and increase soil moisture.

Engineering measures differ with location, slope, the land, soil type and amount and intensity of rainfall. Measures commonly used are exhibited in the earlier pages [artificial recharge of ground water]

Water supply projects can also be implemented for drought mitigation with a view to strengthen the drought preparedness. Activities such as water – use planning, rain water harvesting, runoff collection using surface and underground structure.

Improved management of channels and wells, exploration of additional water resources through drilling and dam construction, are implemented as part of drought – mitigation plan. To increase the moisture availability the following in situ moisture conservation practices can be adopted.

1. For agriculture crops – measure include ridges and furrows, basins and water spreading.

2. For tree crops – measure include saucer basins, semi circular bunds, crescent shaped bunds, catch pits and deep pitting

3. Rain water harvesting collects rainfall and moisture for immediate or eventual use in irrigation and domestic supplies. Part of rain water collected from roofs can be stored in a cistern or tank for later use.

4. Land scape contouring is used to direct runoff into areas planned with trees, shrubs and turf.

Farmers can prepare for drought by developing plans which cover all aspects of farm management and take into account variable climate conditions. Sustainable strategies include appropriate fencing to control overgrazing pest control measures.

Planting drought resistant crops and pasture stabilization of eroded soils, pruning plants to reduce leaf area removing weak plants and thinning dense bunds to reduce competition of and the protection of native plant species.

Improved water saving farm practices   It is necessary to adopt farm practices which can progressively reduce the water requirement of existing crops and improve primary productivity of the cultivated land. Such practices are particularly important for semi-arid regions which have already taken to intensive farming with irrigation water, both from canals and aquifers.

The practices include the increased use of organic manure with the gradual reduction of chemical fertilizers, vermiculture and agronomic practices, such as mulching, crop rotation and use of biopest control measures.

Organic manure can help regain structure and texture of soils and enhance their moisture retension capacity along with improving soil nutrients. Use of farm management practices, such as mulching can reduce evaporation from the soil surface, there by increasing the efficiency of irrigation water utilization.

Long Term Irrigation Management

A long term strategy is required for managing water resources through irrigation projects in India. It consists of several measures which would expand the area under irrigation and reduce the incidence of drought. All the state Governments need to develop policies and procedures for utilization of irrigation resources. The important elements of these polices and management polices are listed below.

(1) Monitoring reservoirs, (2) Setting up of water users associations, (3) Conjunctive use of surface and ground water, (4) Prevention of evaporation losses from reservoir, (5) Increasing storages through expeditious completion of irrigation projects, 6) Integrating small reservoir with major reserves, 7) Integrating basin planning 8) Inter – basin transfer of water.

Afforestation

It is well known that development of forests in areas, which are susceptible to periodic recurrence of drought, is indeed a very effective drought resistant measure.

Areas which are devoid of tree growth suffer serious erosion and need to be covered with vegetation in the shortest possible time with a view to mitigate drought conditions.

Drought affected areas have vast expanses devoid of vegetation, depleted of tree grow and exposed parent racks and boulders. The accelerated run off in these areas is so large that all the surrounding agricultural land cannot even support marginal or subsistence agriculture.

To remedy this, vegetation on hill-slopes catchments and other vulnerable areas need to be under taken particularly where rainfall is low.

Trees and vegetation not only protect the soil improve its water holding capacity minimum run off regulate drainage (Both surface and under ground) but also preserve and improve the productive capacity of the soil and fertility of the agricultural land in the vicinity. The tree foliage or foliage of any effective vegetation, whether it is shrubs bushes or even well-pastured grass, forms a sheltering shield or canopy which breaks down the intensity of torrential rain and thus reduces its erosive action on the soil below. Further more, when this water with reduced velocity reaches down it does not flow down to the rivers and on wards to sea but is absorbed at the surface due to the vegetation. This recharges ground and surface waters resulting in water reservoirs with perennial rather than seasonal storage.

Before the afforestation programme is taken up, a thorough inspection and classification of the areas needs to be conducted. The land identified for after expectation should be divided into three categories. (1) Areas with adequate depth of soil to make afforestation feasible. (2) Areas with shallow soils fit for supporting grass and shrub growth but not fit for tree growth. (3) Badly degraded and eroded areas unfit for tree growth and shrubs and where only soil and moisture conservation operations should be carried out.

In drought prone areas, planting of drought resistant varieties of trees should be considered. Fruit trees, such as sitafal (annona squamosa) and drought resistant fodder species, may not only be useful as an afforestation measure, but also enable the supply of fodder to the cattle. Different species of bushes and shrubs should be planted which not only prevent soil erosion but also provide on leaf-hedge against cattle and barrier against fire when planted like a boundary or fence.

Afforestation should be taken up on a large scale with participation of Panchayat Raj institutions through water shed development programme.

Community participation in drought mitigation   Community participation is an essential feature of drought mitigation programmed. As a local water management and rainwater harvesting hold the key to drought mitigation. Community – based institutions such as WUAS can play important roles in managing water resources at the micro level. This can be done by empowering Panchayati Raj institutions and strengthening women's self help groups and organizing community based consultation through gram sabha.

Climate variability and adoption   Climate variability refers to the climate parameter of a region varying from its long term mean every year in a specific time period the climate of a location is different. Some years have below average rainfall, source have average or above average rainfall.

Due to the phenomenon of climate charge affecting India, such variability would have an impact an agriculture. As a result of variability, the hydrological cycle is likely to be altered and the severity of drought and intensity of floods in various parts of India is likely to increase. Further, a general reduction in the quantity of available runoff is predicted. Simulations using dynamic crop models indicate a decrease in yield of crops as temperature increases in different part of India.

Incidence of pests and diseases may increase with climate variability and climate change. With long dry spells and move intense rainfall, the resulting decline in water quality will lead to greater risk of water borne diseases. Changing temperatures and rainfall in drought prone areas are likely to shift populations of insect pests and other vectors and change the incidence of existing vector borne diseases in both humans and imps.

Climate variability would require farmers to adopt. Adoptation is a process by which strategies to moderate, cope with or take advantage of the consequences of climatic events are enhanced, development and implemented.

Short term cropping, inter cropping, small scale fodder cultivation, small – scale fish cultivation in mini-ponds, home stead gardens and farm ponds for rainwater harvesting are some of the examples of adaptation practices that can be adopted at the local level. These livelihood practices which improve the adaptive capacity of farmers is likely to be a regular feature of drought management program. Drought can be mitigated by two kinds of measures either by adopting preventive measures or by developing a preparedness plant. The preventive measures and preparedness plans have been shown below.

1. **Preventive measures**

    (1) Dams / reservoirs and wetlands to store water, (2) water shed management (3) water rationing, (4) cattle management, (5) proper selection of crop for

drought affected areas, (6) levelling soil conservation Techniques, (7) reducing deforestation and fire wood cutting in the affected areas, (8) alternative land – use models for water sustainability (9) checking of migration and providing alternate employment, (10) education and training to the people, (11) participatory community programmes.

2. **Preparedness plans**

   (a) Improvement in agriculture through modifying cropping patterns and introducing drought resistance varieties of crops.

   (b) Management of range land with improvement of grazing patterns. Introduction of feed and protection of shrubs and trees.

   (c) Development of water resources system with improved irrigation. Development of improved storage facilities protection of surface water from evaporation and introduction of drip irrigation system.

   (d) Animal husbandry activities can help in mitigation with use of improved and scientific methods.

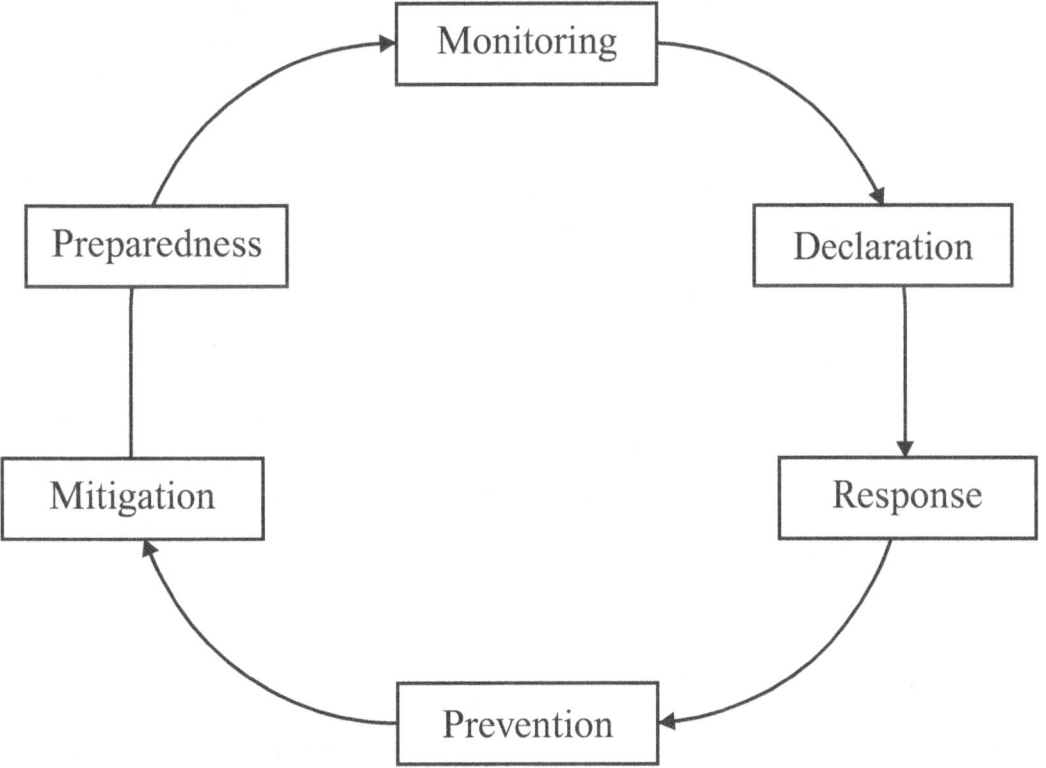

Figure 11.1 A Schematic Description of Drought Management Cycle in India.

Source: Current Science Vol. 100. No. 12. 25 June 2011

Table 11.1   Soil and Water Conservation Measures for Different Rainfall Regions

Rainfall (mm)	In Situ Soil Conservation Measures	Rainwater Harvesting/ Management Measures
Arid < 500 mm	Conservation furrows, contour farming, sowing across slope contour cultivation, mulching, Deep ploughing, inter row water conservation system, small basins, Dead furrows.	Inter plot water harvesting, small field ponds/tanks revival of old, neglected and abandoned traditional water harvesting system khadins
Semiarid 500 – 1000 mm	Conservation furrows, contour farming, compartmental bunding Run off strips, Tied Ridges, Graded Ridges, mulching Live hedges, Ridge and furrow system, off-season and pageon conserved soil moisture, Broad beds and furrows graded border strips.	On farms reservoirs, pond/ tank-cum well system, polythen – lined rain water structure Doba/Jalkund revival of old and abandoned traditional water harvesting system.
Sub-humid >1000 mm	Field bunds, graded, bunds vegetative bunds, level/ graded terraces, contour trenches, chos, inter plot water harvesting, Raising Bed and sunken system	Dug – wells / dug-cum- bore wells, ground water recharge measures. Recharge of old and abandoned wells check dams / gully plugs, Water harvesting structures with spill ways Efficient conveyance system supplemental irrigation through cooperative pumping system with flexible PVC pipes. Micro-irrigation (sprinkler and drip system)

		Water saving pratices Micro catchments (conservation benches) Terraces, checked dams seepage pits, sub-surface water collection, water harvesting structures with spillways, efficient conveyance system, Measures to reduce storage losses, low cost small irrigation devices like Drum/ Bucket Kit, drip irrigation etc. water management, water saving techniques.

Interlinking of Rivers in India.

Source: Eenadu daily newspaper April 9, 2012.

There is inequality of availability of water among the Indian states. Every year 68,969 TMC of surface water is available. Out of the above only 13% (i.e., 8814 TMC) of water is being utilized leaving 87% (i.e., 60115 TMC) water going as waste to running to sea. As per the experts version if one TMC of water is saved and utilized for agriculture, we can irrigate Rice and Pulses earning an amount of Rs.32.5 crores. That means to say that we are losing Rs.20 lakh crores worth by allowing 60115 TMC of water to go to sea as a waste.

The interlinking of Indian rivers is a great boon to the people of India, which will mitigate frequent drought occurring in the country. There is a possibility of irrigation of 3.4 crores Hectors of land apart from solving drinking water problem, catering water to industries and generating hydro-electricity.

The population of India may raise to 180 crores by 2050. At present approximately 14 crores hectors of land is under irrigation and it has to be increased to 16 crores hectors of land to be irrigated to meet the food security of the population. It is now contemplated by the Govt. of India to interlink 37 rivers of India at 30 places (i.e., Himalayan rivers 14 nos penensilar rivers 16 nos).

The following are the links proposed to Himalayan rivers :

(1) Kosi – Mechi (2) Kosi – Gagra (3) Gandak – Ganga (4) Gagra – Yamuna (5) Sarada – Yamuna (6) Yamuna – Rajasthan (7) Rajasthan – Sabarmati (8) Chunar – Son Barrage (9) Son dam – Distributari of South Ganga (10) Masas – Sankosh – Theesta – Ganga (11) Jogigopa – Theasta – Farakka (12) Farakka – Sunder bans (13) Ganga (Farakka) – Damodhar – Suverna Rekka (14) Suverna Rekka – Mahanadi

The following are the places where in the South Indian (peninsular) rivers are proposed to be inter linked)

1. Mahanadi (Mani badra)   –   Godavari (Dowleswaram)
2. Godavari (Echempalli)   –   Krishna (Nagarjuna sagar)
3. Godavari (Echempalli)   –   Krishna (Pulichentala)
4. Godavari (Polavaram)   –   Krishna (Vijayawada)
5. Kriushna (Almatti)   –   Pennar
6. Krishna (Sreesailam)   –   Pennar
7. Krishna (Nagarjuna sagar)   –   Pennar (Somasila)
8. Pennar (Somasila)   –   Palar – Kaveri
9. Kaveri (Kattalai)   –   Vaigee – Gundar
10. Ken   –   Betva
11. Parbathi   –   Kalisindh – Chenbal
12. Vaar – Thapi   –   Narmada
13. Daman ganga   –   Pinjal
14. Bendthi   –   Vardha
15. Netravathi   –   Hemavathi
16. Panbha   –   Achankovil – Vaippar

The cost of the project as per estimates of 2002 year is Rs.5.6 lakh crores. The Supreme Court of India during 2002 year has asked the Government of India to implement the product by 2010 and now it is extended to year 2016. If this project is completed drought recurrence in India will be more or less will be mitigated.

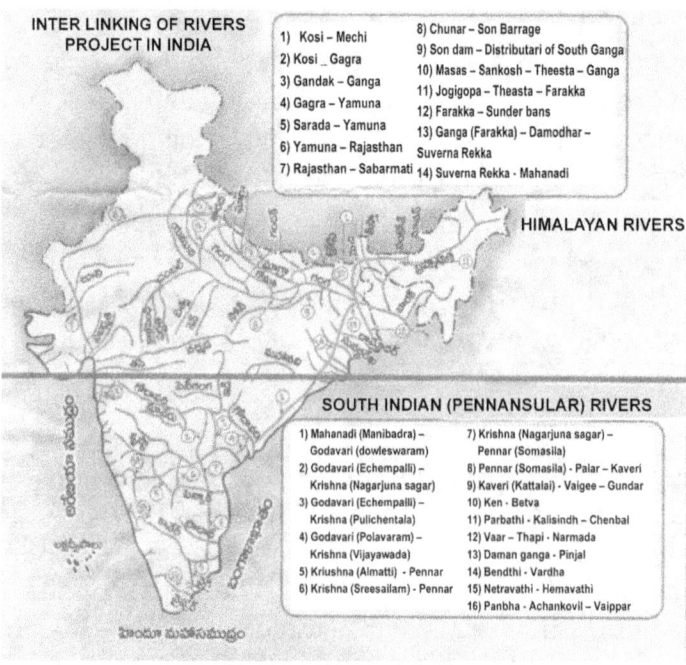

Figure 11.2 Interlinking of Rivers in India

REFERENCES

ADPC and FAO. 2007, Climate Variability and change. Adaptation to Drought in Bangladesh. A Resource Book and Training Guide. Rome.

Central Ground Water Board, ministry of Water Resources 2000. Guide on Artificial Recharge to Ground Water, New Delhi.

Department of Agriculture and Cooperation, Government of India, 2004. Drought 2002: A Report, Department of Agriculture and Cooperation, Ministry of Agriculture, New Delhi.

Department of Agriculture and Cooperation, Government of India, 2009. Crisis Management Plan, 2009.

Department of Drinking Water Supply, Ministry of Rural Development, Government of India, 2000. Guidelines for Implementation of Schemes and Projects on Sustainability under Accelerated Rural Water Supply Programme (ARWSP) & Prime Minister's Gramodaya Yojana (PMGY) – Rural Drinking Water.

Government of India. Year unspecified. Drought Management. Draft.

Kumar, M. Dinesh. 2003. Food Security and Sustainable Agriculture in India: The Water Management Challenge. Working Paper 60. International Water Management Institute. Colombo.

Mall R.K., Gupta Akhilesh, Sing R. and L.S. Rathore. 2006. Water Resources and Climate Change: An Indian Perspective. Current Science, Vol. 90 No.12, 1610-1626.

Ministry of Rural Development, Government of India, 2006. From Hariyali to Neeranchal, Report of the Technical Committee on Watershed Programmes in India.

Narain, P., M.A. Khan and G. Singh. 2005, Potential for Water Conservation and Harvesting against Drought in Rajasthan, India. Working Paper 104. Drought Series Paper 7. International Water Management Institute. Colombo.

National Remote Sensing Centre, ISRO / DOS, Government of India. http://dsc.nrcsc.gov.in. Accessed March 2009.

Public Health Engineering Department, Government of Meghalaya. Rainwater Harvesting Manual. http://megphed.gov.in/knowledge/RWHmanual.htm. Accessed on February 25, 2008.

Rathore, M.S., 2005. State Level Analysis of Drought Policies and Impacts in Rajasthan, India. Working paper 93. Drought Series Paper No. 6. International Water Management Institute, Colombo.

Sakhtivadivel, Ramaswamy. 2007. The Groundwater Recharge Movement in India. in. M. Giordano and K.G. Villholth. The Agriculture Ground water Revolution: Opportunities and Threats to Development. CAB International.

Samra. J.S. 2004. Review and Analysis of Drought Monitoring, Declaration and Management in India. Working Paper 84. Drought Series Paper 2. International Water Management Institute, Colombo.

Sathaye, J., P.R. Shukla and N.H. Ravindranath. 2006, Climate Change, Sustainable Development and India. Current Science, Vol.90 No. 3, 10 February.

Shah, M. 2006. Towards Reforms, Overhauling Watershed Programme. Economic and Political Weekly. July 8-15.

Sharma, A. 2003. Rethinking Tanks: Opportunities for Revitalizing Irrigation Tanks, Empirical Findings from Anantapur District, Andhra Pradesh. India, Working Paper 62. International Water Management Institute Colombo.

Sharma, B.R. and V.U. Smakhtin, 2004. Potential of Water Harvesting as a Strategic Tool for Drought Mitigation, International Water Management Institute. Colombo.

Sinha Ray. K.C. (Year Unspecified). Role of Drought Early Warning Systems for Sustainable Agriculral Research in India. Posted on http://drought.unl.edu/monitor/EWS/ch10_Sinha-Ray.pdf.Accessed 7 March 2009.

Steinmann, A, M.J.Hayes, and L. Cavalcanti. Drought Indicators and Triggers. In Donald Wilhite (ed.) Drought and Water Crises: Science, Technology, and Management Issues. CRC Press, Boca Raton, FL. Also available on http://water.washington.edu/Reseach/Reports/droughtindicatorsand striggerschapter.pdf.

Upadhyay,Bhawana.2004. Gender Roles and Multiple Uses of Water in North Gujarat. Working Paper 70. International Water Management Institute. Colombo.

Wilhite, D.A. M.J. Hayes, and C.L. Knutson, 2005. Drought Preparedness Planning: Building Institutional Capacity. pp. 93-135. In D.A. Wilhite (ed) Drought and Water Crises: Science, Technology and Management Issues. CRC Press, Boca Raton, FL.

Vaidyanathan, A. Watershed Development 2001. Reflections on Recent Developments. Discussion Paper No. 31. Kerala Research Programme on Local Development Centre for Development Studies.

12

SUCCESS STORIES

1. **Build on Micro-level experiences**

   The villages of Sukhomajri in Haryana and about 100 communities in ALWAR have improved their socio-economic conditions through communities used traditional water harvesting structures, such as village tanks and Johads, which increased the ground water table in the area, resulting in increased water storage and substantial increase in crop production and resultant income.

   Ralegan Siddhi and Hirve bazar from Ahmednagar District are the other successful examples of community based initiatives in water resources management. These micro level success stories need to be spread to other parts of the country for other communities to replicate.

   (**Source:** From catch water newsletter of April 2000 "Centre for Science and Environment, New Delhi).

2. **Water harvesting at office premises of Centre for Science and Environment at Tughlakabad Institutional area New Delhi during 1999 - Source:** CSE Catch water newsletter of April 2000)

3. **Village Raj – Samadiyala in Rajkot District, Gujarat State**
   (**Source :** Catch water newsletter of CSE, New Delhi June 2000 and daily Enadu May, 2000).

   During the recent drought in Gujarat these were many villages that used to wait for their supply of water through trains or tankers. However, residents of villages such as Raj – Samadiyala in Rajkot district considered it to be a "shame" to get Government water, nor have they felt the need. A sound water management system has ensured perennial water in their wells even as the rest of Gujarat reeled under a severe water crisis. Water harvesting work accomplished by the people under leadership of Sarpanch Hiardevsingh Jadeja. The Sarpanch consulted the India space department and with the aid of satellite map, recognized the water sources availability and less resources areas. In less resource areas, he at 45 places initiated artificial recharge measures. The extent of land in the village is 250 acres being irrigated apart controlled drinking water supply system 24 hours sweet drinking water through taps to houses is available. The farmers are having irrigating land three times in a year and in this drought year also the village earned Rs. two and half crores. Due to development of reservoir, sufficient water is available in the wells, resulting irrigation of wheat, sugar cane, pulses, cotton and mirchi. They have earned Rs.10 lakhs only through sale of vegetables.

4. Using the excavated silty soils from irrigation tanks - results – Enadu May 20, 2000[1].

   In combined Andhra Pradesh an extent of 65% of lands are red soils. Among them large extent of lands are light, and depth of soil is less. In these lands clay content and so these lands retain less the water. In order to enhance this character the soils that are being excavated from tank (under Jalarakshana program) to be transported to land and spread on the land to enhance the water retaining capacity and yield of crop. The clayee silt taken from tank to be spread at the rate of 40 tons / acre to enhance clay content of 2% in the soil of land. It has to be ploughed uniformly mixing the same. This activity will also save the manures laid which retain in the upper layers of land.

   In accordance to procedure said above, water has been utilized by farmers of Maheswaram (V) in Ranga Reddy dist. A.P. during the month May 2000. The results obtained by them using tank bed soil to their lands is as follows :

---

[3] In a democracy unless "Water becomes every body's business" – it will remain a source of conflict, Hindu, Oct. 13, 2002 by Dr. M.S. Swaminathan, Agrl. Scientist.

Details	Land With Clayee Silt Spreaded	Land With Out Clayee Silt Spread	Increased Yield Percentage
Soil density gr/cm^3	1.58	1.65	
Water infiltration cm/hour	4.25	7.07	
Soil moisture in land %	11.20	9.07	
Rainfed crops yield	Tons / hector		
Ladies finger	11.78	10.66	105
Tamoto	11.28	10.19	10.8

5. **Fluoride – hit turn to rain water harvesting for relief**
   Source: Hindu Wednesday Oct. 8th, 2003
   After decades of suffering, the people of fluoride affected Nalgonda Dist. A.P. finally seem to have found an alternative for the high fluoride water to serve their cooking and drinking needs.
   The use of roof top rain water harvesting structures is slowly gaining momentum in at least one mandal of the district – Narkatpalli. Several roof top water harvesting structures are fast coming up in the seven villages of the mandal as increasing number of residents preferring them to the conventional ground water sources.
   While the ground water in several parts of the district. Particularly Narkatpalli mandal has a fluoride content of more than 4.5 mg per litre, one of the highest in the country, tests proved that the rain water tapped from the roof top structures contained only 0.2 mg per litre in conformity with the prescribed standard of less than 1 mg per litre.
   "Unlike in the past when we used to suffer problems like pains in the muscles and bones, we are feeling much better after opting for consumption of rain water" says Narsi Reddy who plumped for a rain water harvesting structure about five months ago.
   Mr. Sisodia, Nalgonda dist. Collector inspected a few water harvesting structures, assured the people, that he would extend help and subsidise to those who come forward for constructing the rain water harvesting structures.

6. **A breached check dam restored by villagers with the support of NGO with contributions of various countries in Nambi (V) near Jaipur, Rajastan.**

(**Source:** India abroad news service, Enadu daily, Friday 27th April, 2001).
The breached check dam (which was damaged one hundred years ago) restored by villages of Nambi (V) Jaipur Rajastan with help of Tarun Bharat Sangh. NGO and people and contributions from abroad (as per Kanshiram, farmer)

The results obtained are:

By spending five lakhs for restoration of check dam the produce of worth forty lakhs is being received for villagers. 2) Migration to other areas stopped. 3) Alcohol drinking has been stopped. 4) Tree plantation is in progress 5) Two months time dam was restored. 6) Sufficient water obtained from reservoir and all wells are having full of water 7) Several people from other areas are coming here for employment.

National water commission chairman D.K. Chadda has inspected the village and said it should be a model to other villages in the country. Central Irrigation Minister Mr. Arun Sethi sent a message to emulate the water harvesting an other parts of India on seeing Nambi (V) of Rajasthan.

7. **A small tank breached 20 years back restored as percollation Tank in Vadigalavandlapalle B. Kotha Kota Mandal of Chittoor Dt. A.P.**
   Source: Enadu daily Sunday 14th May 2000.
   The breach for the tank in Vadigalavandlapalle (V) restored and revitted with a cost of 0.87 lakhs converted as percolation tank. Now the tank catering drinking water for five villages in B. Kothakota Mandal. Even in mid summer it possess 4 feet of water in the tank. It is catering sufficient water to cattle also in Vadigalavandlapalle (V) Gudlavariapalle (V) Bavagaripalle (V) Abbemma Kota Kandrapalli, Dennemedapalle. It is a model for the entire Chittoor Dist. to emulate the efforts made by villagers, watershed committee and NGO, Rogers.

8. **Rain water harvesting in Japan country Sumida city**
   Source: 7th June 2000 Vartha News Daily in A.P., India.
   For Sumida city in Japan water supply is being done from Ton river at a distance of 150 kms from an anicut constructed across the river. Water scarcity is being felt now the then due to non availability of water. Due to this the people of city selected water harvesting as a permanent solution. Another reason for their option to rain water harvesting is the occurrence of drought now the then. Concrete paving every where not allowing rainwater to seap into ground. For a little intensity of rain surroundings flooding with water, resulting sewage system damaged. As such rain water with all dirty and dabrise the other contaminants, reaching river is daily phenomino. Sumida river contamination is another reason. In order to circumvent this position they have selected rain water harvesting as a programme. The rain water

flowing on roads were made to sink into ground resulting rise of ground water levels, and tree plantation taken, has resulted in green belt.

In Sumida city under ground tanks were built under multistoried houses. The biggest example is city hall roof top area 5 lakh square meters, which collected in underground tank below city hall with capacity one thousand cubic meters rain water collection tank. The water being used for blushing toilets is being used. 1997 the water collected below city hall building tank were used for flushing toilets about 43.7% of tank capacity.

Another example is highest building is Sumeda Resselliy stadium "Kokugikan" having 8400 sqm roof with basement tank one thousand cubic meters capacity to rainwater to store. The water being used for flushing and supply for cooling towers.

For Fire extinguishing rain water is being used at Sumeda city's speciality. At street turnings they have established "Rogisan" (under ground tank) in five nos in the city. The rain water from nearby houses roof tops collected is being stored in "Rogisan" and in times of need they draw water with hand pump. In normal days the water is being used for plantation and exigencies, for fire extinging is being used. Still water storage is felt this water will be used for drinking also. Japan has been collecting rain water from all big building tops and storing water in under ground basement tanks and the same is being used for plantation flushing and cooling towers and other puposes. For all newly being built buildings the Govt. of Japan has made mandatory to have rain water collection procedure.

Water Harvesting at CSE

In 1999, CSE installed a water harvesting system in its office premises at Tughlakabad Institutional Area, New Delhi. The system was designed in a way to ensure that drainage of rainwater from the premises is practically prevented, and maximum amount of rainwater is effectively harvested.

For recharging the water, 13 soakaways have been constructed around the building. Each soakaway is 9 m deep and is filled with brickbats to act as a filter media. The mouth of each borewell is covered with an inverted earthen pot with a small hole to prevent the entry of debris which could clog the borewell. Further, the pot is surrounded by brickbats (broken brick pieces) to filter the runoff entering the well, and the entire assembly is covered with a *jali* – a perforated cover. Water that falls on the unpaved areas surrounding the building enters these soakaways and percolates into the ground.

An 45 m deep abandoned borewell in the premises has been converted into a recharge borewell by connecting the rooftop drains to it. Runoff from the terrace is passed through a grating to filter it before passing into the borewell. Water from the some portion of the terrace falls into an open pond in the front of the building, from where it overflows to an underground water tank of 8500 litres capacity.

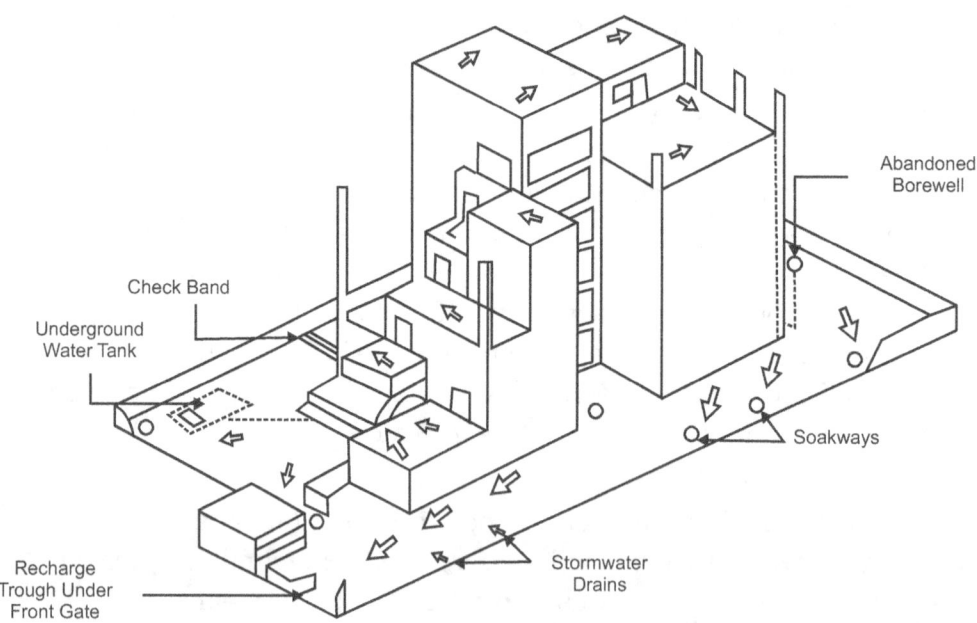

Figure 12.1   A View of the Entire System

Openings of the municipal storm water drains within the area have been raised slightly above the ground level, so that rainwater does not drain away, except in case of a heavy downpour (see figure 3). To prevent water from flowing out of the campus through the gate, three soakaways have been constructed in a trough under the ront gate.

Because of the water harvesting system, only in case of an extraordinary downpour would water flow out of the CSE campus and it is ensured that all the rainwater falling over the building area is recharged or stored.

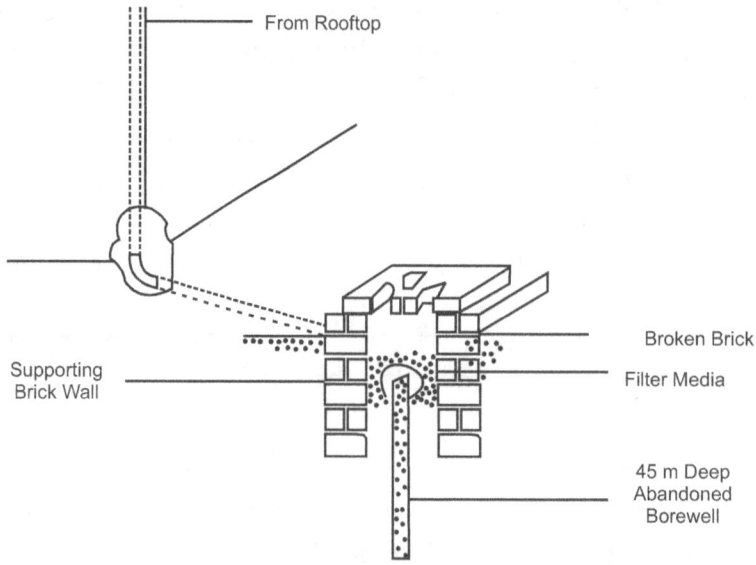

Figure 12.2   Detail of Recharge Borewell

GOVERNMENT HAVE PAID LITTLE ATTENTION TO EFFORTS NEEDED TO DROUGHT-PROOF THE COUNTRY

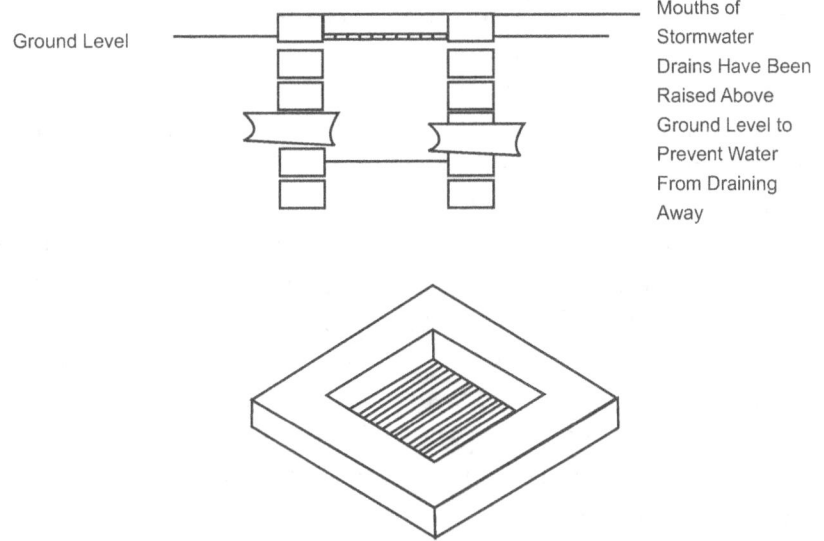

Ground Level

Mouths of Stormwater Drains Have Been Raised Above Ground Level to Prevent Water From Draining Away

Figure 12.3 Storm Drains Raised to Prevent Water from Draining

A rain gauge and water meters have been installed to monitor the rainfall and water-use in the building. On 13 March 2000, a short rainstorm of 7.5 mm occurring over 32 minutes was recorded, which effectively recharged about 4500 litres of water over the CSE area of 1000 sqm. Although this quantity is not very large, it is interesting to note that this small shower contributed an amount of water equal to about 6 days of drinking water requirement of the entire 110-strong CSE staff at 10 litres per person per day.

"State Center Prize Paper" by Institution of Engineers India, Hyderabad - 2006

REJUVENATION OF "SWARNAMUKHI RIVER BASIN" IN CHITTOOR DIST., A.P. INDIA

Analysis of Water Table Monitored Near Sub-Surface Dams Constructed (9 Nos. in Series)

Er. P. Munirathnam, B.E., C.E., MIE*

ABSTRACT

Swarnamukhi river lies in Chittoor district and drains an area of 3387 sq.km flows and for a length of 155 kms before it empties in Bay of Bengal. The basin is a semi-arid tract paddy 60% and groundnut 25% are the principal crops covered in basin (i.e.) 85%. The geology of the area is granites and granite gneisses. East – West trending dolerite dykes are found to intrude into the granitic country rocks. They are great in number and stand out as prominent wall like structures. The granitic rocks are weathered jointed and fractured. The depth of weathering ranges from 10m to 20m over which 3m to 7.5 m of valley fill is existing. The alluvial thickness covers with filter points and weathered thickness is penetrated by dug wells.

The river said to have been flowing two decades ago and there used to be natural springs which have now dried up and the farmers deepens their dug wells every year. The farmers are suffering for want for irrigation water for their Rabi irrigated groundnut crop during April and for Kharif paddy during July, August. In order to save the standing crops and to circumvent this situation it has become imperative for A.P. Govt. and CAPART (Corporation of Government of India) to take two steps to recharge the unconfined aquifer consisting of fairly permeable alluvial sediments by artificial method through construction of series of subsurfaces dams with cheaper material like clay and LDPE film (9 Nos) and arrested the flow in the natural aquifer, which caused to impound 48.28 mcft (under 7 Nos of S.S. dams with which this paper covers) in sediments. The impounded water has served to stabilize 1400 acres of Rabi irrigated groundnut and 700 acres of early Kharif paddy during the year 2002. The water levels in piezometers erected on U/s and D/s of S.S. dams have been monitored from October, 2001 (soon after construction of dams) to December, 2002. The well hydrograph along with rainfall have been prepared and enclosed for 7 Nos. of S.S. dams. The analysis of water table revealed that there is a rise of 1.43 m (average) in the basin (i.e. 24 kms length during January i.e. postmonsoon).

Further, during month of June (pre-monsoon) before construction of S.S. dams the fluctuation in water level is 0.65 m to 0.25 m, whereas after construction of dams the fluctuation is 0.3m to 3m. A net benefit of Rs.5,000/- per acre by irrigating groundnut or paddy crops is observed to have been realized on an average due to stabilization of crops with impounded water. The work has been carried out by nine numbers of voluntary organizations forming as consortium led by "Rastriya Seva Samithi" a leading organisation in the state duly participated by beneficiaries. The cost of construction of these sub-surface dams 7 Nos is Rs.113.91 lakhs. Cost benefit analysis is made. The cost in return for year is found to be 21.040% i.e. Rs.24,37,674/-. In about a span of 5 years the entire cost of investment is likely to be recovered. The cost benefit ratio between annual, equivalent of benefit and annual equivalent of capital cost taking an interest rate of 8% and 2% as depreciation per year worked out as 2.1 which is more than 1 which indicates that it is economical to invest on this project. The primary aim of the project is to harvest the excess run-off going as waste to the sea and to recharge the depleted unconfined aquifer (valley fill) to revive their full potential not only to protect existing agriculture but also to increase ground water resource potential to meet future water demands. To generate more base flows in the catchment for river to become perennial, the watershed development programme in the basin is to be carried out with prime precaution to undertake all the structures on scientific and technical lines lest they breach and entire concept of rejuvenation gets delayed. To access the performance of watershed programme the water level in 2 nos of Piezometers and other 8 Nos. observation wells at each site have to be monitored during the period January to June every year taking monthly water level readings for a period of four years.

It is very essential that financial institutions should come forward to assist the farmers on co-operative basis to construct such S.S. dams. The state and central government should give all the technical assistance for locating such sites and carrying out surveys and designing them on the lines, suggested in this paper. There is a pressing need to transfer the existing intensive agricultural pattern to a more light irrigated crop farming in view of limited availability of water in the basin.

Key Words: Rejuvenation – River Basin Development – Base Flow – Alluvium – Ground Water Resource – Subsurface Dam- Irrigation.

INTRODUCTION

The river Swarnamukhi at Sankampalli in Pakala Mandal, Chittoor District. It is one of the most important rivers in Chittoor district. After flowing 155 kms in Chittoor and Nellore districts of Andhra Pradesh, it joins Bay of Bengal directly. The drainage basin has an area of 3387 sq.kms. Geometrically, the catchment area lies between 13°28'N and 14° TN latitudes and 75°10E and 80° TE longitudes. The Swarnamukhi basin is essentially a semi-arid tract as per the moisture index of Thornthwait. The moisture index values in the basin ranges between (-) 53.9% and (-) 59.3% indicating the region as semi-arid type of climate.

About 67% of its area has a slop below 3% useful for agriculture. Only 24.3%, of the total area is actually under agriculture and another 12.8% is classified as cultivable waste. The cropping pattern in the area is paddy 60% cash crop groundnut 25%, other food crops like jowar, bajra, ragi and pulses 5%, sugarcane 3%, chillies 3%, garden crops 4%. There is pressing need to transfer the existing irrigation intensive agricultural pattern to a more light irrigated crop farming in view of limited availability of water in the basin.

PRECIPITATION

The average annual rainfall in the basin is 100 cm which spatially varies from 80.7 cm in the interior west to 125.9 cm in the coastal region to the east.

The rainfall distribution during South-West Monsoon shows decreasing trend eastwards, while during the North-East Monsoon, the decreasing trend is towards West.

The period between January to May being the main dry season accounts for only 10 to 15% of the annual rainfall. The bulk of annual rainfall occurs usually in a period of 3 to 4 months and hence it is heavily lost a surface runoff. The dominant aspect of the climate of this region is general aridity but for the rain during the monsoon for a short period of the year imparting humidity to the atmosphere and moisture to the soil. The low to moderate rainfall determines the pattern of agricultural operations.

The arrival and departure of the monsoon, the duration for which it stays, the intensity and spatial distribution are highly variables.

The water supply for agriculture critically depends on the monsoon which is unpredicted making the agriculture activity a gamble.

Evaporation & Potential Evapotraspiration

The temperature in the month of May is the highest of the year. With the onset of monsoon in June, the temperature slowly decreases until it registers the lowest temperature in the month of January.

The mean annual evaporation recorded through pan evaporation is about 860.2 mm. The total area of water bodies in the basin is 155.5 sq.k. It results in a total evaporation of 130.3 mcmt. From the water bodies (i.e.) 17% of the total surface water resources of the basin.

It is obvious from the above, the need to store water in underground than at surface.

The potential evapotranspirtaion calculated using thornthwaites formula is 2049 mm for a normal year. Except for the months of October and November which experiences moisture surplus and all the remaining months show moisture deficiency. Normally, the maximum deficit occurs during April to July.

Geology and Soils of the Area

About 62% of the basin is occupied by granites and granitic gneisses. The granitic rocks are weathered, jointed and fractured. The depth of weathering ranges from 10m to 20m veins of quartz and pegmatites and dolerite dykes are found to intrude into the granitic country rocks. Of these, the dolerite dykes are very important as they occur in great numbers and stand out as prominent wall-like structures. The trend being east-west.

The soils of the region are (1) sandy soils (red sandy soils), 2. loamy sands (alluvium near courses). There is an excess yield of 2.078 TMC in the basin as per statistics of Irrigation Department.

In light of the above facts, the CAPART has realized the need to arrest the run off in the sufficient depth of alluvium available and weathered zone existing, To increase storage of ground water potential in the basin.

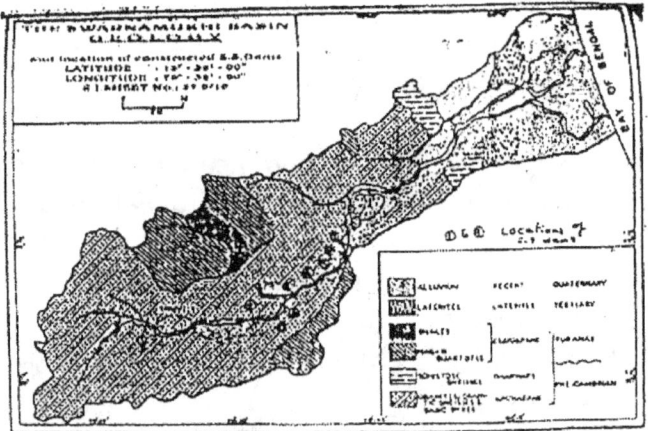

Figure 12.4 Swarnamukhi River Basin – Geology

Existing Agricultural Activity in the Basin

The climate from January to February is characterized by low temperature and scanty rainfall, below 2% of annual rainfall (rainfall records from 1962 to 2001). In the first half of .the January, sometimes moderate rains are recorded due to spill over effects of the North East Monsoon. February is practically a dry month. The monthly potential evapotranspiration during January and February are about 102.6mm and 122.4mm.

The above facts clearly reveal that despite low temperatures, this season experiences water deficit for want of rainfall and the crops need regular irrigation.

Sowing of groundnut irrigation starts in the second half of December through the first half of January. Normally, no rainfall occurs during March to May except certain occasional years during which some rainfall occurs. The average rainfall during the period is about 55.3mm. The highest amount of water deficiency occurs during this period of year. The potential evapotranspiration of March to April and May will be about 168.8 mm, 162.4mm and 167.5mm as against 55.39mm of rainfall.

Irrigation interval is to be at 7 day and if summer is severe at 5-day intervals irrigation is to be made. Harvesting of paddy and groundnut starts in March and continues till the middle of April.

Sugarcane planting takes place through out this season. For groundnut crop, the scarcity is felt during March and April (i.e.) effective flowering period which is most sensitive to water deficit followed by pod formation. Except some green patches of sugarcane and very limited paddy crop, the countryside presents a barren look.

Agricultural activity practically ceases by the middle of April except where irrigation is available to a limited extent. With the "mango showers" occurring in the second half of May agricultural activity starts with the preparation of fields for un-irrigated groundnut and millet.

COMMUNITY BASED RAINWATER HARVESTING HAS THE POTENTIAL TO DROUGHT – PROOF THE ENTIRE COUNTRY

South-West Monsoon (June to September) (Kharif Season)

South-West monsoon which is usually sets in the first half of June lasts till September. Sometimes, the onset of the monsoon is delayed. With the onset of, monsoon agricultural activities begin in June-July. The P .E. during June, July, August, September are 149.5mm, 131mm, 139.1 mm and 125.1 mm against rainfall of 59.35mm, 198.45mm, 90.04mm and 107.13mm. During this period, un-irrigated. groundnut gingerly millets are sown. Paddy is planted.

Though groundnut is sown as un-irrigated, at the end of crop season, water is insufficient and unless water is supplemented for one or two wettings, the entire crop will get dried or the expected yields will get denied. By September, the un-irrigated crops are already harvested or ready for harvest.

North-East Monsoon (October to December)

North-east monsoon or retreating monsoon begins in October and ends in December. It brings the second peak of rainfall to the interior part of the basin and the main peak of the rainfall to the lower basin. The season witnesses frequent occurrences of cyclones originated in the Bay of Bengal.

The potential evapotranspiration during October, November and December is 10.14 cm. 9.11 cm and 8.87 cm as against the rainfall of 23.12 cm, 32.85 cm and 14.53 cm resulting in water surplus throughout this season. The annual evapotranspiration is 1556.2 mm.

The season starts with harvesting of unirrigated groundnut and some varieties of paddy crop and the sowing of the next crop especially the irrigated groundnut millet and paddy.

Necessity for taking up of the construction of sub-surface dams across Swarnamukhi River

1. Sand quarrying decreased the thickness of sandy alluvium, the main sub-strata for the storage of pellicular water. This made to tap all the water from the aquifers on either side stored as a bank storage and 'as such almost all filter points became dry.

2. Due to over exploitation of ground water more than annual replenishment has I caused continuous decline .of water levels, decline of well yields of shallow wells and increase in the cost of energy required to lift water from greater depths.

3. In order to conserve surface runoff resulting during North-East and South-West Monsoons and to store water in the valley fill to augment the water table levels in the adjacent wells, it was decided to check the ground water flow in the river by construction of series of subsurface dams.

4. To restore the earlier status of river (i.e.) making perennial by taking basin development activities.

5. The farmers resort to irrigate their irrigated groundnut and paddy during Rabi season (December 15 to April) through dug wells and borewells. A precarious situation is confronted by them witnessing shortfall of water in their bores during months of May and April, They adopt devices to reduce the delivery pipe diameter to have a continuous flow of water and non-stopping of pumps. Even this arrangement meets to the requirement of certain extent and other extent being kept as fallow.

Similar situation is experienced during early Kharif season also in some parts of basin (i.e.) May to August. During the months of July and August water scarcity is felt for the planted paddy and also for un-irrigated groundnut crop and millets. In order to circumvent this situation by rendering adequate recharge to all the dug wells and borewells in the influencing zone, the movement of ground water has been thought of checking and get it retained in the basin to utilise it for stabilisation of crops during Rabi and Kharif season scarcity months.

Investigation carried out and Design adopted during Construction of 9.0 Dams   Ground water is dynamic in nature and its movement is controlled by gravity and flows continuously from recharge areas to discharge areas through the porous and fractured rock media like surface run off, ground water also leaves the basin if it is not harnessed. Hence, the movement of ground water has been checked and retained in the basin to utilise it for stabilisation of Rabi ground nut crop, paddy crop and early Kharif paddy crop by construction of 9.0 dams across Swarnamukhi River at appropriate places. The thickness of the valley fill ranges from 3m to 1.5m. The weathered granite gneiss ranges from 9m. to 16m where the dug wells are pierced. The average gradient being 1/450 in a length of 48 kms. The weathered rock along the line of dam is existing at 4m to 9.5m.

The catchment receives sufficient rainfall to recharge and also to flush out the excess of salt accumulated in the top soil due to recycling of water. The dam wall occupies very little space without any Buttresses. Suitable sites have been selected by studying the topographic map. The site map is prepared and proper alignment of dyke line is selected depending on the physical conditions and geological features .and economical viability.

The dyke line thus selected has been sounded with resistivity survey and ascertained the depth to rock and possibilities of fractures along the alignment. A few auger holes have been drilled to corroborate the results of the resistivity survey besides getting the soil samples for the determination of the physical properties and water bearing properties of the valley fill along the dyke and in the influencing zone. Well inventory at each of the constructed dam sites carried out and well section, nature of the aquifer tapped, its hydraulic properties determined. The specific yield of the valley fill has been arrived as 12%.

The existing cropping pattern in the area, the sources of water and quantity available and short fall periods have been studied and-found that May and June are the vulnerable months in the early Kharif period and months of March and April in the Rabi. Period during which water scarcity exists and crops will wither for want of stabilisation from external sources as the borewells and dug well will not yield sufficient water. In order to circumvent this problem and to save the crops, it was decided by State Government and CAPART to construct 9.0 dams to check the movement of ground water flowing in valley fill and retain sufficient water at the respective sites to stabilise the irrigated ground nut crop and paddy crop during Rabi (December 15 to April) and paddy crop in early Kharif season (May to August).

After ascertaining the suitability of the area and also the proposed alignment to construct the dyke, the dam has been designed and the cost of the construction has been worked out taking into consideration nature of the materials to be used for construction which depends on the area being benefited and profit to be accrued. It has been decided to construct the dam with' clay as abundant material is available in nearby irrigation tanks. The B.C. ratio has been worked out and found to be more than one and hence it was practicable to construct the darms.

The clay has been tested in Engineering College Laboratory, Tirupati its analysis shows as on average gravel 1%, sand 13% silty clay being 87%. The plastic limit is 25.5% and porosity of clay being 0.45%.

Based on the results of, investigations made it was decided to construct 9 Nos. of S.S. dams in a span of 48 kms of Swarnamukhi river. The construction of S.S. dams is one of the components envisaged for rejuvenation of swarnamukhi River (dried for the past 25 years). The watershed development has been taken up to increase the base flow into river for longer - periods.

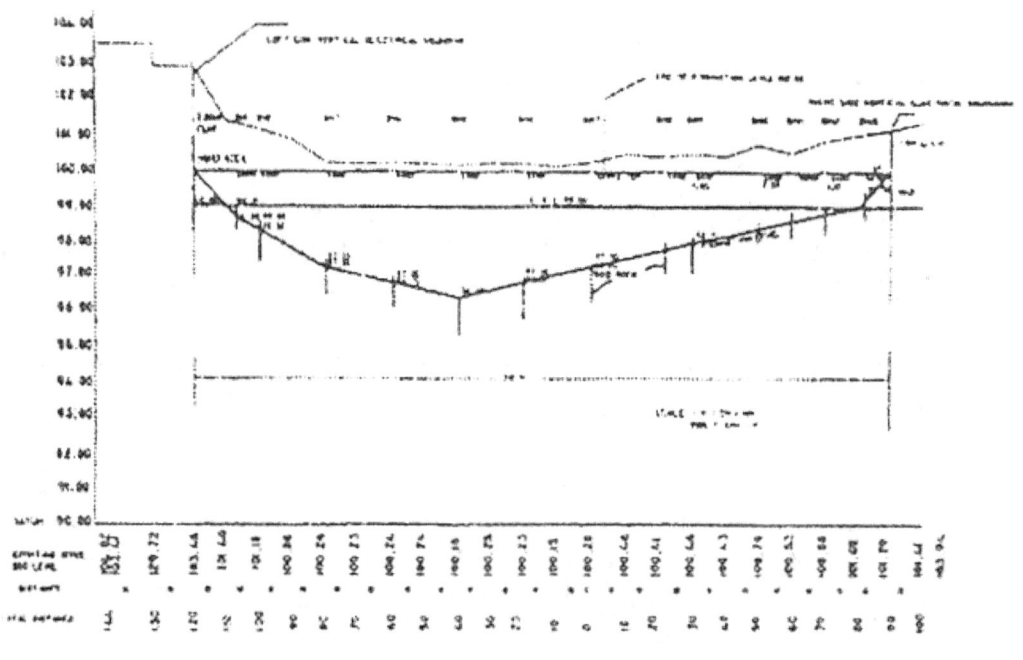

Figure 12.5 Longitudinal Section of Sub-Surface Dam on Swarnamukhi River near Munagalapalem (V), Chittoor Dist.

The S.S. dams were constructed with puddle clay cut off with O.9m thick which has been keyed into hard end connection on either side of river bank. The top of 9.0 dam is limited to 1 m below river bed level to allow excess flows to D/s and as well as to flush the salts accumulated and the riparian rights of the farmers D/s are not infringed. The dam is keyed in the bottom 0.5m into weathered zone. The clay wall was compacted with wooden rammers in order to obtain proctors density. The clay wall was constructed imbibing due optimum moisture content while preparing clay balls. In order to retain this moisture content for longer period 200 microns low density polythene ethylene film (The life of LDPE film is 50 years as established by publication of United States Department of the Interior Bureau of Reclamation captioned "Evaluation of the durability of Polythene Films" by VD. Glebov) was covered on U/s and D/s and top of dam. Increase in film thickness prolongs its life as the process of oxidization is considerably slowed down. This protects from cracks yielding in the clay wall which will be in due contact with sand in summer season thereby no stored water is lost from U/s to D/s of dam. As the seepage lesser in the impervious clay loam being 0.9 to 1.2 cumec million square meters of wetted area, in order to save this quantity also the LDPE is provided over clay wall. Experiments carried out at Irrigation and Power Research Institute (IPRI), Amritsar with LDEP Film with a head of 3.65 m permeability is found as nil for 100 microns LDPE even after 12 years. The black colour LDPE is used for protection against ultra violet light degradation of film.

The imperviousness of the construction of the dam has been proved by the study of U/s and to D/s water levels monitored in the constructed piezometers erected at 20m away from dam alignment as on U/s and D/s there is a difference in water level. During construction of dam as and when dam rises 1 m sand is deposited adjacent to the LDPE film and similarly entire excavated area is refilled with excavated sand.

On U/s. and D/s of the constructed dam piezometers have been erected, and water levels monitored at fortnightly intervals. In order to establish the water level fluctuations trend vis-avis rainfall and discharge, a well hydrograph is drawn for each dam, covering the water level reading monitored before construction from the observation wells and with piezometers water level readings after construction of 7.0 dam. The water levels have been observed for 9 months in January, 2001 to September in observation well before construction and after construction of SS. dam from October, 2001 to December, 2002. Piezometers U/s and D/s have been monitored.

Water level Fluctuations

Before construction of S.S. dam water table levels have been measured in observation wells from 2001, January to September, 2001 and shown in the well hydrograph. S.S. dams have been completed during June 5th to end of September and piezometers were erected on U/s and D/s have been measured from October, 2001 to December, 2002 and water table reading the U/s and D/s piezometers have been plotted. The well hydrograph enclosed has been incorporated with monthly rainfalls of 2001 and 2002 years measured from rain gauges of Tirupati (Mallimadugu Project); and Srikalahasti (Swarnamukhi anicut). The well hydrograph on U/s piezometers and D/s piezometers have been drawn and it shows that the dams are constructed perfectly as there is difference in U/s and D/s water levels in piezometers.

The water levels from observation wells 4 on left U/s and 4 numbers on right on U/s at each S.S. dam have also been fixed and water levels measured from 3/ 2002 to 12/2002.

The average rise of water table due to construction of S.S. dams is 1.43m during post monsoon season (January). Statement enclosed shows the quantity of water impounded during starting the Rabi.

25% of the total estimated quantity of water is taken as quantity of return of applied irrigation water. The stabilisation of 2100 acres (or) 854 ha under 7 dams have been achieved by construction of S.S. dams.

The transmissibility and specific capacity of weathered granitic gneiss as calculated from the yields by Nooka Raju (1974), G.S.I are 181.59 m³/day for 1 m and 7m³/day for 1 mt. draw down.

Stored Water on U/S of 7.0 Dams utilisation for Scarcity period in Rabi and early Kharif Seasons

Rabi Season: Groundnut crop-irrigated.

Period: December 15th to April 15th (120 days)

Total water required for crop period: 488mm (8 Nos. of irrigation)

Each wetting requires: 61 mm.

Quantity of water required to stabilize each wetting: 30.5mm. It was reported that the crop at the last month i.e., March 15th to April 15th is suffering for want of sufficient water due to rise of temperature beyond 33°C which leads to reduction of yields. Farmers are restricting the diameter of delivery pipe of pump fitted to well and irrigating a part of already sowed groundnut crop.

To circumvent this situation and to save the crop sown, the water is being supplemented from the stored water on the U/s of constructed subsurface dam by way of indirect irrigation (i.e.) artificial recharge to existing wells. The quantity of water required to stabilise the crop 122 mm/ha during four wettings i.e., 0.122 x 10000 = 1220 cum/ha (i.e.) 43114.8 cft/ha (or) 17526 cft/acre (or) 0.01753 mcft/acre. i.e., 58 acres/mcft stabilisation.

Early Kharif Paddy: April 15 to Aug., 15

The total requirement of water for early Kharif paddy is 1200mm. It is reported that paddy crops is being withered for want of adequate water in the borewells at the end of crop period i.e., July 15 to August, 15. This is the grain formation state. The water requirements for 30 days is 8 mm/day = 240 mm/ha i.e., 0.24 x 10000 = 2400 cum/ha (or) 0.03448 mcft/acres = 29 acres/mcft.

Total quantity of water retained behind the sub-surface dams (7 Nos.) during December = 48.28 mcft.

The groundnut crop during Rabi season for one month and during early Kharif season for one month, the crop requires stabilization with water available at 7.0 dams. 50% of water is being utilized : for stabilization of Rabi irrigated groundnut = 24.14 mcft.

The area that is being stabilized = 24.14 x 58 = 1400.12 acres, say 1400 acres.

For stabilization of early Kharif paddy water set apart is 24.14 .mcft.

The area that can be stabilized during Kharif = 24.14 x 29 = 700.06, say 700 acres.

Total area stabilized

Rabi season = 1400 acres

Kharif season = 700 acres

  Total = 2100 acres

For creating new ayacut in absence of stabilization with irrigated groundnut during Rabi season water requirement is 0.070 mcft/acre (or) 14.29 acres/mcft.

Total area with 24.14 mcft = 347 acres

For Kharif paddy with 24.14 mcft, the acreage that can be created is 0.1724 mcft/acre (or) 5.8 acres/ mcft = 140 acres.

Total area of ayacut stabilised during 2002 year i.e. 21.00 acres.

Total new ayacut = 347 + 140 = 487 acres

(a) Benefit due to groundnut crop Rs. 5000/acre

    347 × 5000   = 17,35,000 = 00

(b) Benefit due to paddy crop at Rs. 5000/acre =

    140 × 5000   = 7,00,000 = 00

                       Total = 24,35,000 = 00

Cost in return per year

$$\frac{24,35,000}{1,13,91,000} \times 100 = 21.40\% \text{ (or) Rs} = 2437674 - 00$$

No. of years required to get the cost of 7.0 dams to be recovered.

$$\frac{1,13,91,000}{24,37,674} = 4.6 \text{ (or say) 5 years}$$

Cost Benefit Ratio

Assuming an interest rate of 8%, the annual equivalent of capital cost is calculated by using the formula $G = il + P + D$.

G = Annual equivalent of capital cost

i = rate of interest;

I = Capital

P = Cost of maintenance / D = Depreciation (2% per year)

G = 8/100 × 11391000 + 0 + 2/100

   = 9,11,280 + 0 + 2,27,820 = Rs. 11,39,100/-

$$\text{B.C.Ratio} = \frac{\text{Annual Equivalent of Benefit}}{\text{Annual equivalent of captial cost}}$$

$$\frac{24,35,000}{11,39,100} = 2.1$$

As mentioned earlier if the B.C. ratio is more than one, it is economical to invest

RECOMMENDATION FOR FUTURE ACTION

The watershed management of forest and non-forest area of the basin is being carried out to convert the runoff into base flow. This storage of base flow will be released to river continuously making the storage capacity at 7 dams to improve and ultimately the river to rejuvenate and become perennial one.

The prime precaution to be taken is to construct the ponds, check dams etc., in catchment area to standards on technical lines lest they breach and entire concept of rejuvenation gets delayed. The need of the day is to conserve every drop of water and recharge the depleted aquifers.

Now, the performance of 7 dams have been established. In future care to be taken for monitoring of water levels pre-monsoon to postmonsoon for - a period of four years for accessing the quantity of base flows generated and to find out at what accuracy the watershed development has been carried out.

Ahmedabad, India, Jan., pp. 12C-l to 21.

Raju, KC.B. (1997). Water Conservation in Hard Rock Areas: A Necessity. Proceedings of the National Meet on 5 & T Inputs for Water Resources Management, organized by Dept. of Science and Technology, Govt. of India, April 18, 1997, New Delhi.

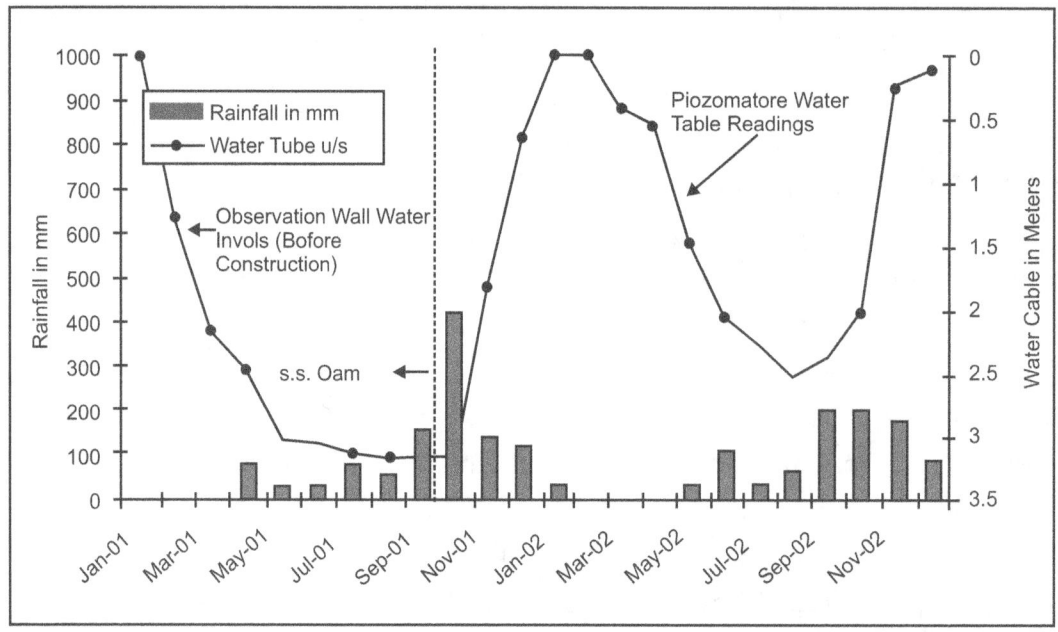

Figure 12.6 Rainfall, Water Table and Piozometer Water Table at S.S. dam Bandarupalle.

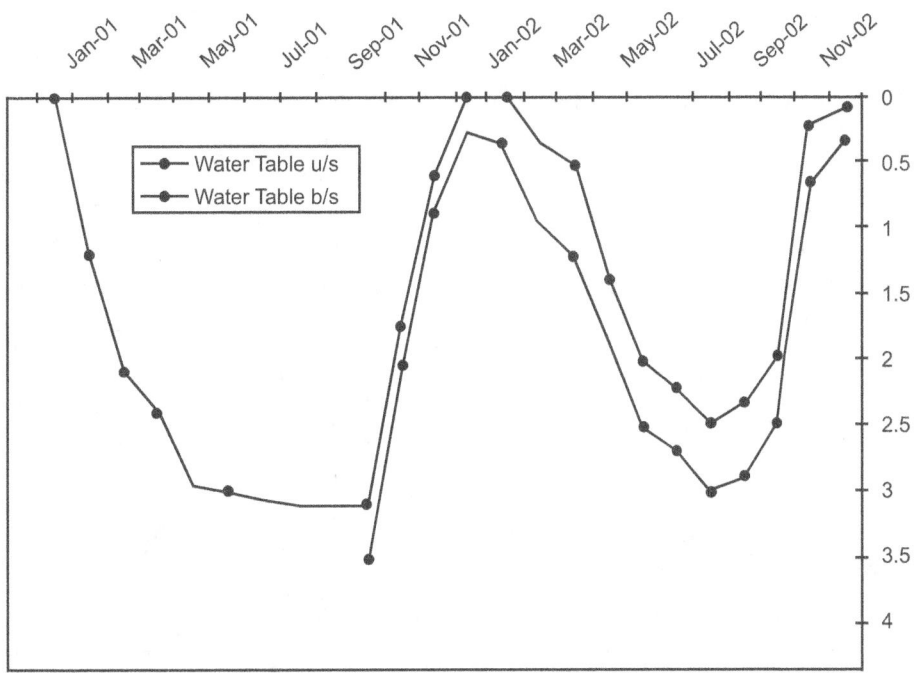

Construction of Subsurface Dam at Bandanupatle Scross Swarnamukhi River

Figure 12.7 Water levels u/s & d/s of SS dam

CONCLUSION

From the foregoing it could be seen that the sub-surface dams are feasible and these dams can be constructed by using cheaper material like clay and LDPE film. Thus, making it possible for a farmer to construct such structure to increase the productivity of the land holdings. It is very essential that financial institutions should come forward to assist the farmers on co-operative basis to construct such dams. The state and central government should give all the technical assistance for locating such sites and carrying out surveys and designing them on the above lines. Beneficiaries participation is a must to make the scheme of artificial recharge a success as has been done in this project. There is a pressing need' to transfer the existing irrigation intensive agricultural pattern to a more light irrigated crop farming in view of limited availability of water in the basin. They must adopt water conservation methods including drip and sprinklers.

REFERENCES

Central Ground Board, 1994. Manual on Artificial Recharge -of Ground Water.

Raju, K.C.B., 1998. Importance of Recharging Depleted Aquifer - State of the Art of Artificial Recharge in India - Journal 'Geological society of India, 51, 429-454.

Gupta, S.K.& Sarma, P. (1996). Rejuvenating our Rivers - The Akshayadhara Concept Current Science, V 70, No.8, p.694-696.

Narasimha Reddy and Prakasam, P. (1982). Evaluation of Artificial Recharge Projects in Granitic Terains of Andhra Pradesh, Workshop on Artificial Recharge of Ground Water in granitic terrain – CGwb – MOWR – Gol.10-11, Oct. pp.194-210.

Raju, K.C.B. (1985). Artificial Recharge through Percolation Tanks and sub-surface Dykes Proceedings of the International Seminar on Artificial Recharge of Ground Water.

CHERRAPUNJI HAS WATER SCARCITY FOR NINE MONTHS IN A YEAR DESPITE HAVING 11,000 MM OF ANNUAL RAINFALL. THIS SHOULD BE AN EYE OPENER IF YOU DON'T HARVEST THE RAIN THERE WILL NEVER BE ENOUGH WATER

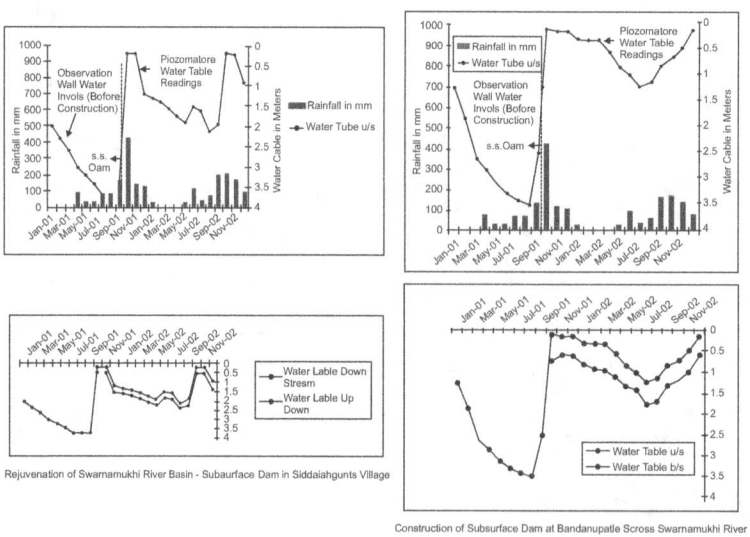

Figure 12.8   Water Table u/s & D/s and Rainfall and Water Table near S.S. dams, Siddigunta (V) and Bandarupalle (V), A.P.

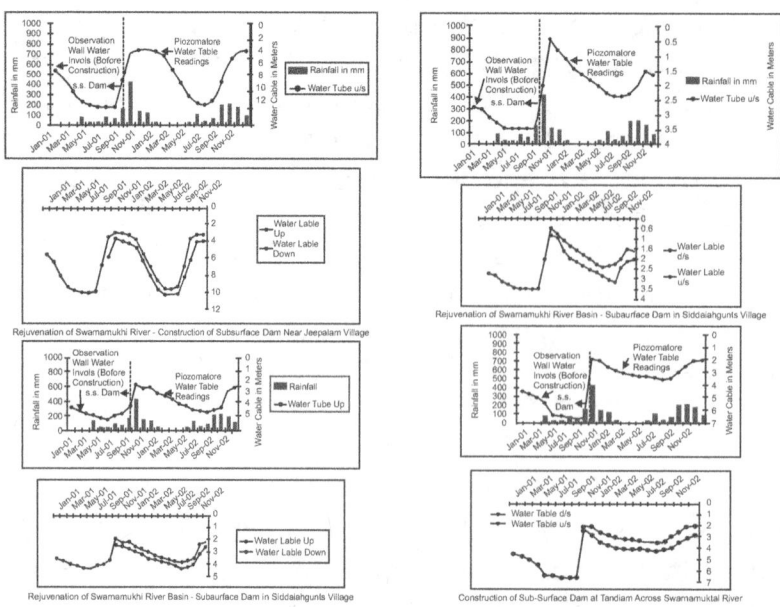

Figure 12.9   Water Table u/s & D/s, Rainfall at Jeepalem, Siddaigunta, Chandragiri and Tandlam (V), A.P.

ANALYSIS OF WATER TABLE MONITORED IN PIEZOMETERS AFTER CONSTRUCTION OF S.S. DAMS FROM OCTOBER, 2001 TO DECEMBER 2002 AND WATER TABLES LEVEL IN OBSERVATION WELLS FROM JANUARY, 2001 TO DECEMBER, 2002.

S.No.	Name of the Sub-surface Dam constructed	Observation well water colum (before construction of S.S Dams) January (Post-Monsoon) Meters	June (Pre-Monsoon) Meters	Water Column in U/S Piezometers after Construction January (Post-Monsoon) Meters	June (Pre-Monson) Meters	Rise in Water column in Curresponding Month after construction of S.S. Dam January (Post-Monsoon) Meters	June (Pre-Monsoon) Meters	Quantity of water stored at beginning of Rabi i.e., Dec. 15 mcft.	Quantity of water utilized for Rabi season for stabilization mcft.	Area of Ground nut Crop stabilized arcres	Quantity of water utilized for early Kharif Paddy mcft	Area of Paddy crop stabilized acres	Total area served with impounded water for stabilisation ares	Remarks
1	2	3	4	5	6	7	8	9	10	11	12	13	14	15
1	Bandarpulle (V) Srikalahasti (M)	1.95	0.60	3.10	2.35	1.15	1.75	2.69+0.67=3.36	1.68	97.44	1.68	48.72	146.16	With the impounded water of 48.28 mcft if new area is developed for irrigated at a later stage instead of stabilization of existing ayacut 347 acres of rabi irrigation groundnut crop and 140 acres of early Kharif paddy can be irrigated Based on this only C.B. ratio has been worked out.
2	Siddaiahgunta (V) Yerpedu (M)	1.85	0.65	2.50	2.00	0.65	1.35	2.37+0.59=2.96	1.48	85.84	1.48	42.92	128.76	
3	Munagalapalem (V) Yerpedu (M)	2.30	0.25	3.20	2.50	0.90	2.25	9.22+2.3=11.52	5.76	334.08	5.76	167.04	501.12	
4	Jeepalem (V) Renigunta (M)	4.25	0.45	6.75	0.75	2.50	0.30	14.832+3.71=18.54	9.27	537.66	9.27	268.83	806.49	
5	Nillsanipeta (V) Renigunta (M)	1.20	0.30	2.50	2.25	1.30	1.95	3+0.5+0.763=3.81	1.875	110.49	1.875	55.245	165.135	
6	Gajulamandyam(V) Renigunga (M)	1.00	0.25	2.85	1.65	1.85	1.40	3+0.75=3.75	2.17	108.75	2.17	54.375	163.125	
7	Tandlam (V) Renigunta (M)	2.25	0.30	3.90	3.30	1.65	3.00	3.47+0.87=4.34	2.17	125.86	2.17	62.93	188.79	
				10 m average 1.43 m				48.28	24.14	1400.12 or 1400 acres	24.14	700.06 say 700 acres	2100.18 (or) 2100 acres i.e, 854 hec	

SWARNAMUKHI RIVER REJUVENATION PROJECT

Chittoor District, Andhra Pradesh

Sl. No.	Location of Sub-Surface Dam	Length (m)	Max. Depth to Weathered Rock (m)	Average Depth of Valley Fill (m)	Slope. in the vicinity	Length of Effect (KM)	Thickness Weathered Mantle (Granite) (m)	Cost (Rupees in lakh)	Impounding Capacity mcft. No.of Ag. Wells Benefited nos.
1.	Thandlam	300.00	6.00	2.60	1 in 500	1.30	15	18.00	13.02 cft. 285 nos
2.	Gajulamandyam	280.00	8.00	4.00	1 in 400	1.60	12	15.44	1.90 mcft 81 nos
3.	Neslan Pet	148.50	6.67	3.60	1 in 360	1.29	10	13.50	4.73 mcft 100 nos
4.	Jeepalem	340.00	7.30	3.50	1 in 360	1.26	12	37.00	16.53 mcft 382 nos
5.	Munagalapalem	210.00	4.00	2.10	1 in 450	0.945	10	9.50	8.67 mcft 642 nos
6.	Siddaiahgunta	156.00	5.00	3.7	1 in 530	1.961	12	5.80	2.4 mcft 53 nos
7.	Bandarupalli	193.00	4.00	2.36	1 in 900	2.12	14	14.67	3.50 mcft 105 nos 50.75 mcft 1648 nos

CROP WATER REQUIREMENT BY MODIFIED PENMEN METHOD

Crop Period : 120 days   Rabi   I.D. Groundnut

I.No.	Description of Item	December	January	February	March	Total
1.	E.T. value in mm	88.7	102.6	122.4	168.8	482.50
2.	KC crop coefficient value	0.4	0.7	0.95	0.75	2.80
3.	Monthly water requirement (1×2)	35.48	71.82	116.28	126.60	350.18
4.	Add for initial watering	10.00	-	-	-	10.00
5.	Gross monthly water requirement	45.48	71.82	116.28	126.60	360.18

I.No.	Description of Item	December	January	February	March	Total
6.	Monthly rainfall at 50% P.L.	142.58	27.92	16.70	3.24	190.44
7.	Effective rainfall @ 50% of 50% P.L.	71.29	13.96	8.35	1.62	95.22
8.	Net irrigation requirement	Nil	57.24	107.93	124.98	290.15
9.	Requirement with 70% field efficiency	Nil	81.77	154.19	178.54	414.50
10.	Monthly requirement at canal head with 85% conveyance efficiency	Nil	96.20	181.40	210.05	487.65
11.	Total requirement in mm	Nil	96 mm	181 mm	210 mm	488.00 mm
12.	Total requirement per hectare	Nil	960 cum	1810 cum	2100 cum	4880 cum (or) 14.33 acre / mcft

C.C. DALAL Award – Paper by Institution of Engineers India –
Hyderabad, 2005

IMPROVEMENT OF QUALITY AND QUANTITYOF DRINKING WATER USED BY COLONIES OF DEPARTMENT OF SPACE FROM KALANGI RIVER BASIN

P.MUNIRATNAM, B.E., M.I.E., C.E.[2]

SYNOPSIS

In the coastal margins of potable ground water basin, the lowering of water level, beyond the allowable level can result in intrusion of sea water and as such the water quality in the basin is affected subject to the condition that permeable soil formations are well within the hydraulic connection with sea water either directly on the ocean floor or along with river having connection with estuary or bay which contains sea water. This paper presents the control measures taken to over come the problem of sea water intrusion into the coastal aquifer along Kalangi having connection with rive estuary (i.e) Pulicat lake which contains sea water and also to improve the quantity of potable drinking water supply to Sullurpet town and D.O.S. Housing colonies.

INTRODUCTION

Over-exploitation of ground water in and around Sullurpet town, Nellore District has led to landward movement of sea water-fresh water interface resulting in contamination of fresh water aquifers. The ingress of sea water into coastal aquifer along Kalangi river through estuary (@) Pulicat lake has affected the agricultural economy of Sullurpet area and also drinking water availability. The contamination of fresh water by the tidal water inflow is stopped by reconstructing latarite rough stone groyne which is badly damaged across river Kalangi. In addition now the sea water intrusion into ground water is also stopped by providing a wall of Timber piles. At a distance of 10.2 km from the groyne up stream to conserve the rain water runoff during monsoon and base flows during summer one water harvesting structure is constructed (by name subsurface dam). The storing fresh water will dilute the sea water intruded before construction of groyne. The river Kalangi has received water from its upper catchment and water has reached up to the site of constructed S.S. dam on 22.7.05 and water has been impounded at site hence 0.66 meters raise of S.S. dam wall is made with Bed pitching and revetment duly grouting the joints with cement mortar. There are back waters up to 5 km length as on 22.07.05. This occurrence has resulted in diluting the sea water already ingressed into river bed. As per water sample collected fromDOS-4 well on 27.07.05 and analysis made revealed that the intensity of has been considerably reduced reaching to the limits of drinking water.

---

[2] Formerly Technical Consultant to Govt. of A.P. – APERP – World Bank aided Programme. Facilitator cum Evaluator CAPART (New Delhi) and Executive Engineer (Irrigation) Rtd.

Due to construction of groyne sea water intrusion has been stopped improving quality of drinking water. By construction of S.S. dam the quantity of drinking water available has been increased by 41 lpcd for shar colonies and 18 lpcd for Sullurpet Town during summer period.

The construction of groyne & S.S. dam has been started during April 2005 & completed during 2nd week of July 05.

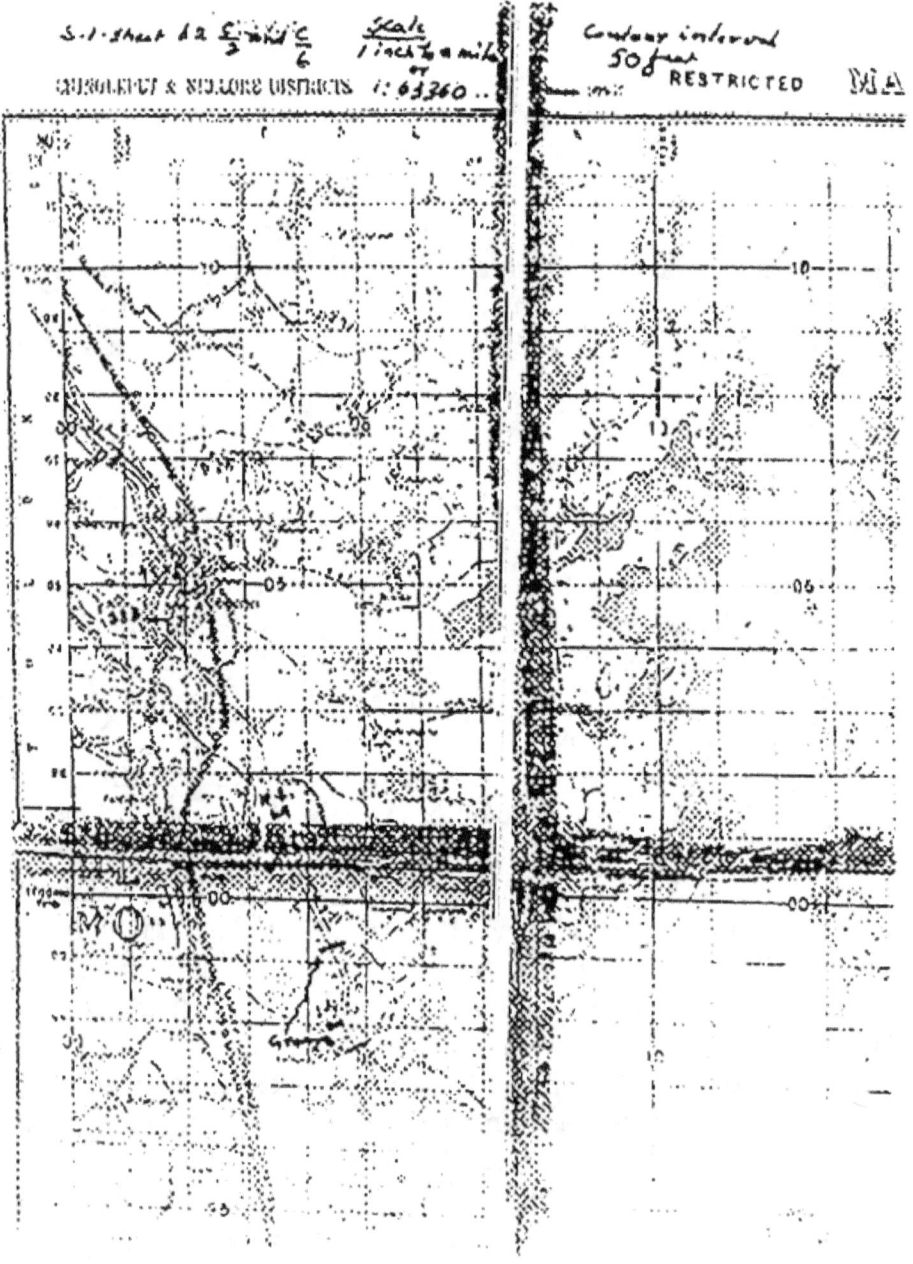

Figure 12.10   Survey of India Sheet Showing Area of Kalangi River

In order to avoid mistaken diagnosis of sea water intrusion as evidenced by temporary increases of total dissolved salts, criteria from recognition of sea water in ground waters, it is recommended that chloride and sum of carbonate and bicarbonate ratio is the basis. So the water analysis is made on these lines and found as 10.8 on 8/2004 before construction of S.S. Dam and 7.91 during construction and 1.845 on 27.07.05 after construction and receipt of water in Kalangi River.

Conclusion about Quality   Chloride is the dominant anion of ocean water which has gradually reduced bicarbonate is abundant anion in groundwater and it gradually increased. Ultimately the groundwater in DOS well is gradually falling within limits due to fresh water mixing already available sea water at DOS No.4.

Consequences caused due to this problem

1. Infiltration wells constructed along Kalangi Rive by "SHAR" Sriharikota for drinking water supply to their Housing colonies located at Sullurpet Town and around yielded saline water. The water table in the well at DOS No.4 being 3 meters below Kalangi river Bed level resulting short supply of drinking water and searching for alternative ways for the needs.

2. On the Bed of Kalangi river sea water was available to a length of 10.2 kms along the river made the adjacent farmers to stop pumping of surface water to their fields an extent of 500 acres on both sides of river.

3. After some time farmers also noticed salinity from the ground water from the filter points and dug wells located in their fields. The ingress of sea water into the coastal aquifer has effected the agricultural economy of this area and also adequate drinking water availability.

How the problem identified   Water sample were collected from 60 No. s of wells located 2km away from coast line in the coastal area of Nellore District and a rapid determination of total dissolved solids made by measuring specific electrical conductance of ground water samples and a tentative Saline Zone is marked which is 13 km from sea coast. But Sullurpet town is still 9.5km away from this tantive saline zone. This has directed to investigate the way from which entry of sea water towards Sullurpet Town has resulted. Then the confluence of Kalangi river with Pulicat lake was inspected and found from the villager of Vattembedu that a groyne constructed 1.4 km u/s of confluence of river has been badly damaged and Pulicat waters are entering into Kalangi river when there is no flow in the river at high tide conditions.

The surface water samples were collected all along river and up to Sullurpet town (10.2km) and salinity is noticed and ground water sample from the infiltration well No.4 of "SHAR" which is supplying drinking water to their Housing colonies is analysed and the results of the samples revealed that the ratio of chlorides and sum of carbonate bicarbonate is of the order of 10.47 against one.

Factors for causing the problem

1. Badly damaged rough stone laterite groyne allowing surface sea water to enter into river Kalangi.

2. River bed level from Sullurpet Town (10.2km length) to damaged groyne is below high tide depth of 0.8 meters at confluence causing inundation of entire length of 10.2 km by sea water.

3. The Kalangi river flows from the month of September to February and there is no surface sea water entry due to thrust of flowing water. But during the period from March to August the depth of surface water in river is minimum (i.e.) below H.T.L. and as such the sea water enters the river and heads up to Sullurpet Town.

4. During periods when there is no surface flow sand mining has taken place and a reverse gradient developed enhancing the velocity and depth of water to flow up stream towards Sullupet Town.

Efforts made to solve the problem   At the site where old groyne existed with number of breaches is a feasible site as R.B.L is high and width is 140meters only. So drilling has been taken up across the river and at a depth of 2.6 meters clay bed is encountered in all the bores drilled on D/s of groyne (i.e.) at Bed level of Pulicat lake. The river B.L. on u/s of groyne is 1.3 meters above Pulicat B.L. Therefore 3.8 meters long pile is selected to drive to the alluvium.

Figure 12.11   Applying Bentanite Clay Slurry into Bore Hole

Applying Bentanite clay slurry into bore hole to retain sides of bore at groyne site.

A drawing to stop sea water entry to river is finalized. The laterite rough stone groyne is constructed whose height is kept 1.2 meters above W.L of Pulicat lake and timber piles of size 0.3meters X 0.15 meters X 3.8 meters with shoe at Bottom & Tongue and grove arrangement on both side of pile is made. Bentonite clay slurry is poured into hole during augur drilling during to make the sides of hole to retain. When the clay bed is encountered the pile is placed in position duly inserting the same in the tongue & groove arrangement made on side of previous pile and hammered with over head mechanism till it intrudes into clay bed and the pile is kept 1.2m above Pulicat lake B.L. This procedure of drilling holes with augur and using Bentonite clay slurry to retain the sides and also to seal the gap between piles at tongue & groove arrangement has resulted in stopping the entry of sea water into ground water.

Figure 12.12   Lowering of Timber Piles.

Lowering of Timber pile into the bore hole drilled adjacent to already drilled lowered pile at groynes site

A change is seen on u/s of groyne to a stretch of 10.2 km by decrease of depth of standing water day by day. This is a measure taken up to improve the quality of drinking water and also to impound 50 mcft of irrigation water from the Sullurpet town to Vatambedu groyne along the Kalinga river. More accurate results can be had after monsoon season (September to February) since the residual sea water already entered will get diluted with flow of Kalangi river and subsequent standing water on u/s of groyne to a depth of 1.85m. But due to cyclonic weather prevailed during 3rd week of July there was run off in the river on 22/7/05 and water sample taken after 5days and analysed which has revealed that salinity in wells has reduced considerably as shown earlier.

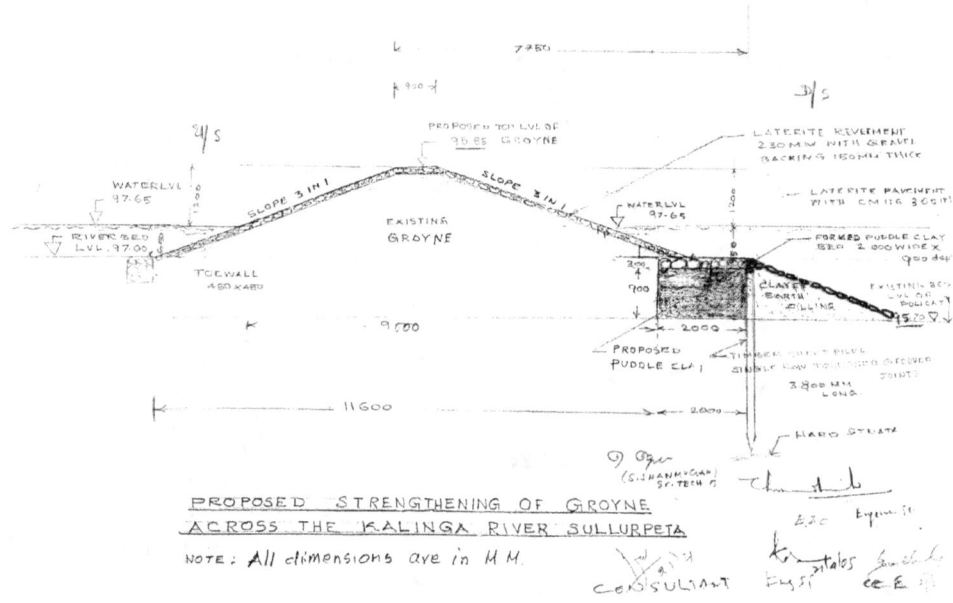

Figure 12.13 Strengthening of Groyne.

EFFORTS MADE TO IMPROVE THE QUANITY OF DRINKING WATER TO "SHAR" HOUSING COLONIES AND TO SULLURPET TOWN:

Rain water harvesting and artificial recharge will help to improve the ground water regime of the area. Massive water conservation methods have been taken up by Punjab and Rajasthan Governments and salt water content was considerably tackled (year 2001) and has now returned to the normal level.

The available drinking water supply for SHAR housing colonies and Sullurpet Town is 109 lpcd and 5.7 lpcd respectively during summer and during monsoon 181.8 lpcd for a population of 5500 and 3500 respectively i.e. 14.40 lakh/lit/day (monsoon) 8 lakh/lit/day (summer). With a view to improve the quantity of drinking water and also quality it is proposed to harvest the rain water run off going waste to sea. The period of shortage being from March to August when there is no surface flow in the river. But it is noticed that there is base flow at 0.5m below RBL during April and gradually decreasing to 6.0m below B.L by month of June. All the infiltration wells are tapping water from 1m to 3.5m below R.B.L. In order to circumvent this situation, it was decided to harvest the base flow and store water for subsequent use. A sub surface dam across river with impervious strata at bottom is proposed to be constructed at Sullupet limits ie 11.2km from confluence of Kalangi with Pulicat lake. The depth of sandy alluvium at the selected site being 7m. A rise of 0.66m is also at site to act as check dam to impound surface flow.

ONLY 100 MM RAINFALL FALLING ON A 1 HECTARE PLOT CAN YIELD UPTO 1 MILLION LITRES OF WATER

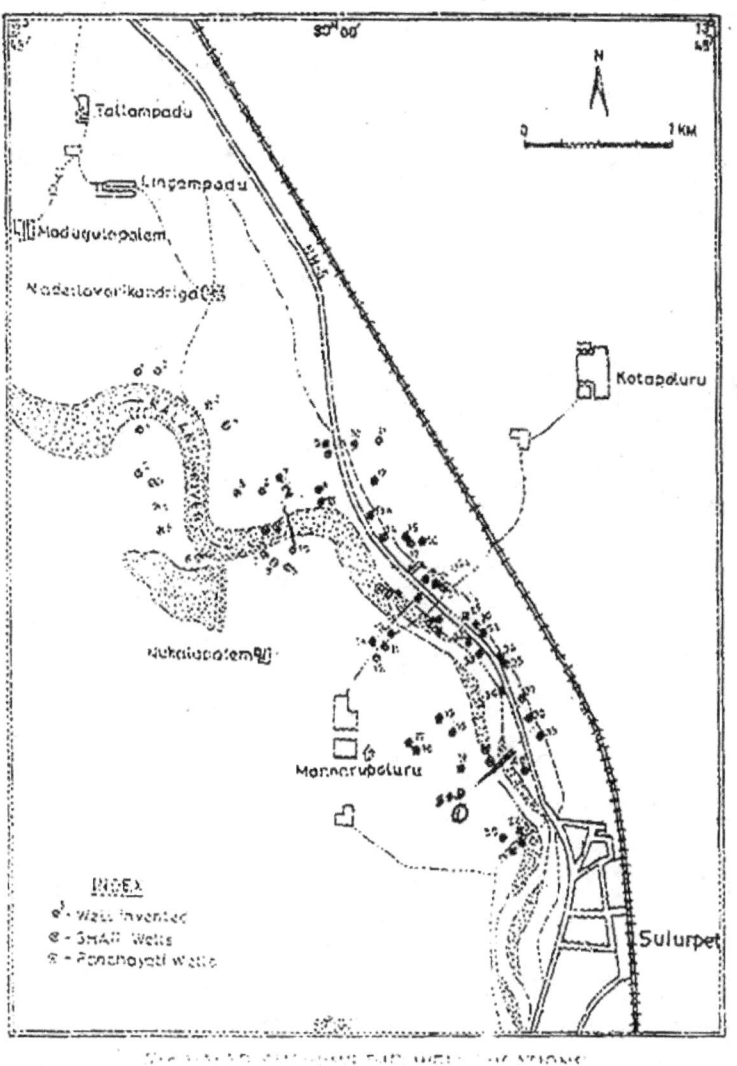

Figure 12.14 Location of S.S. dams and wells Invented along Kalangi River near Sullurpet Town, A.P.

Location Map

The total shortage of 27.5 lakh/lit/day is required to cater the above demand. The water supply now available during summer is 8 lakhs/lit/day. Hence a shortage of 19.5 lakh/lit/day is to be met with by constructing a water harvesting structure.

A site near Holy Cross School near Sullurpet town with a length of 127m is selected and geophysical survey has been carried out to know the depth to imperious strata and position of fractures and fissures if any underlying the sandy alluvium whose depth is 7m. To corroborate the results arrived by geophysical survey drilling of bore holes was taken and depths have been found to be correct. One jack well to a depth of 6m and 1.5m dia with two collector wells on right side of river was erected during formation of subsurface dam for pumping water to the already available DOS well No.4 which is located at 46m u/s of the jack well. The sub surface dam is constructed with clay quarried from a irrigation tank located at 5 km away. The width of clay wall of S.S. Dam is 0.9m. The wall is covered with LDPE film on all the sides except 1/3 width at bottom.

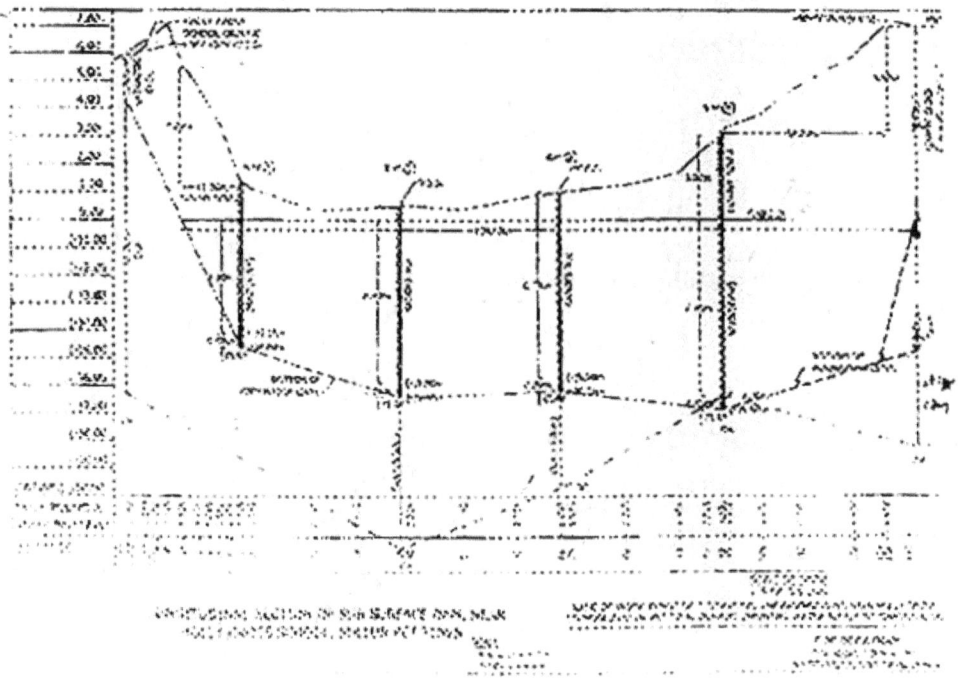

Figure 12.15   Cross Section of S.S. dams near Holy Cross School, Sullurpet Town, A.P.

The covering of LDPE film will help to retain the optimum moisture content imbibed in the puddle wall. The longevity of LDPE film under ground is tested as 50 years. Bed pitching on top of dam is made to avoid scouring as the dam is 0.66m above R.B.L. Revetment is also provided on both sides of river banks safe guard the bank from scouring. The specific yield is taken as 10%. The bed fall is 1:1000 and influencing zone is 6.4 km. The quantity of water stored is as follows:

$$= \left[\frac{(127 + 135/2)}{2} \times 64 + 0 \right] \times 6400 \times 60.1$$

$$= \frac{838.4 + 0}{2} \times 6400 \times 0.1$$

$$= 419.2 \times 6400 \times 0.1$$

$$= 268288 \text{ cum}$$

$$= \frac{268288 \times 35.33}{1000000} \text{ or } 2689 \text{ lkh/ltrs or } 7.47 \text{ lakh/ltrs/day}$$

Figure 12.16 Inspection of Director, ISRO, S.S. dam Construction

Director ISRO inspecting the S.S. Dam at Sullurpet during construction

Replanishment of water to ground water reservoir for 90 days as river flow is for 90 days (due to dynamic storage caused).

7.47 x 90= 672 lakh/ltrs or 1.87 lakh/ltrs/day or 2.37 mcft.

Mcft	Lakh/ltrs	Lakh/ltrs/day
9.479	2689	
2.370	672	
Total 11.85	3361	9.34

The S.S. Dam is raised above river Bed level by 0.66m to have more impounding capacity during summer.

The quantity being or 1.424 mcft

$$\left(\frac{95+90}{2}\right) \times 0.66 \times 660 = 40293\,cum \text{ or } 403.95 \text{ lakh/ltrs or } 1.122 \text{ lakh/ltrs/day}$$

Figure 12.17 Completed view of S.S. dam with Water

Completed view of S.S. Dam as on 22.7.05 on receipt of water in the river

Due to this construction of S.S. Dam extra storage achieved during monsoon period = 9.34+1.122=10.56 lakh/ltrs/day.

During summer static storage 7.47 lakh/lit/day

Dynamic storage 1.122 lakh/lit/day Total being 8.592 or 8.60 lakh/ltrs/day.

The total water made available after construction during monsoon − 14.4+10.56 = 24.96 lakh/lit/day and during summer 8+8.60 = 16.60/lakh/lit/day increase in water supply due to construction of sub surface dam per capita as follows

	Summer		Monsoon	
	Original	Revised	Original	Revised
Sullurpet Town	5.7 lpcd	24	12.6	40
Shar	109 lpcd	150 lpcd	181.8	200

The storage is calculated within the width of river. The groundwater basin also extends on both sides of river. Further another S.S. dam to obtain balance quantity of water is be located nearer to 8 bore wells of SHAR to effect recharge to the wells.

To overcome sea water and related hazards ingress it is necessary that the sea water intrusion has to be arrested by adopting the following precautionary measures:

1. Over exploitation of ground water causes salinity ingress in coastal aquifers connecting with the ocean floor or along with the river estuary or bay which contains sea water and as such ground water level should always be maintained at least one foot above mean sea level. Then salt water occurs underground not at the sea level but rather at a depth below sea level at 40 feet.

2. Prevention or reduction of sea water intrusion can be achieved economically by re arrangement of pumping draft or by shifting the pumping stations towards the interior away from the coast line. This enables the ground water level to rise above sea level and maintain a sea ward gradient. Such a procedure would no doubt results in not allowing full development and utilization of the available ground water storage capacity. However, this limitation is justifiable in protecting the entire aquifer from contamination.

3. Desalination of sea water within 5 to 10 km distance from coast line, by any of the cost effective processes can solve drinking water needs. But also by this approach the heavy extraction of ground water in the coastal alluvial formation could be minimized and salt water ingress could be arrested from further intrusion to a considerable extent. As the length of coast in Andhra Pradesh is 700 km and agricultural operations and population density is high within 10 to 20 km distance from the coast it is essential to install desalination plants along the coast. Desalination is no more only an experimental idea.

Several thousands desalination plants already exist primarily in Saudi Arabia, Israel and countries of middle east where there is scarcity for fresh water. Depending upon the salt content of the sea water in the coastal tracks of this state and quantity of water required for various sectoral demands within 5 to 10 km distance from the coast by any of cost effective processes such as osmosis, micro filtration or reverse osmosis, distillation etc could be used for desalination.

For example, Lakshadweep had no easy access to safe drinking water for its 60,000 residents till recently. All the had was salt water. During 4/2005 the ordeal of the local residents came to an end with the commissioning of a new desalination plant built by the Chennai based National Institute of ocean technology with a capacity to handle 1,00,000 liters of sea water every day, the plant now provides about 10 liters potable water for every individual by distillation method. The sea water is heated up to boiling point in a vacuum flash chamber. The resultant vapours, are condensed using cold water to produce safe drinking water.

REFERENCES

Central Ground water board 1994 - manual on artificial recharge of ground water.

Raju K.C.B. 1985 – Artificial recharge through percolation tanks and Sub-surface dykes – Proceedings of the international Seminar on artificial recharge of ground water – Ahmedabad India. Jan. PP 12c-1 to 21

UNDP – Ground water studies in Ghaggar river basin in Punjab Haryana and Rajasthan – Technical report of UNDP – 1978 – New York.

UNDP - Artificial recharge studies in Mehsana area and coastal Saurashtra Gujarat – Proceedings and recommendations – New York – 1986.

Don L. Warner 1985 – Recharge for sea water intrusion control – Proceedings of international seminar on artificial recharge of ground water – Ahmedabad India – PP 14-1 to 14-13.

Gupta S.K. and Sharma P 1996 – Rejuvenating our rivers – The Akshaydhara Concept – Current Science Vol. 70 No. 8, PP 694-696.

GWSDA- Proceedings of the all India Seminar on "Modern Technique – Water conservation and artificial recharge for drinking water – Afforestation – Horticulture and agriculture" Nov. 19-21, 1990 Pune.

THE SOLUTION LIES IN HARVESTING RAINWATER THROUGH CAPTURING, STORING AND RECHARGING IT AND LATER USING IT DURING PROLONGED PARCHED PERIODS

VOICES

CONSERVATION OF BASE FLOW BY CONSTRUCTION OF SUBSURFACE DAMS ACROSS SWARNAMUKHI RIVER, CHITTOOR Dt. A.P. BY RASHTRIYA SEVA SAMITHI, TIRUPATI. FUNDED BY – CAPART – NEW DELHI

The original plan that the quantum of base flow passing through sandy alluvium blocked by sub-surface dams would emerge as ground water into the shallow in filtration wells/filter points in and around the river bed and help the farmers in getting water for their crops during the month of April to May not affecting Rabi crop fag end of Crop period.

Results achieved through the execution of sub surface dams:

1. **Construction of sub surface dams across Swarnamukhi river near Munagalapelm (V), Yerpedu mandal, Chittoor Dist.**

   The work was completed during the year 2001, the position of water table below river bed before construction was 1.2^m during the month of January. The wells around sub surface dam are of two types namely shallow Tube Wells and Dug Wells which are pierced through the sand bed of present and ancient river couse. The maximum thickness of sand bed is 5^m.

   The additional ground water conserved by sub surface dam has helped to rise the ground water levels in sandy alluvium of river bed to 0.5 m below river bed [as measured on 2.2.2005] and thereby a sustainable basis through out the year.

   Case 1: The interaction made with the farmer by name V. Venkatachalapathy whose land is on left side of Sub Surface Dam at a distance of 250 m.

   (a) As per him 10 acres of his land is being cultivated. 6 acres Ground nut and 4 acres paddy crop.

   (b) He owns one dug well with 20 feet depth 10 feet diameter and 2 filter points with 20 ft. depth and a bore well 150 ft. depth. The static water lever in dug well is 2 m.

Figure 12.18 Well of Venkata Chalapathi on left side of S.S. dam

   (c) He states that a draw down of 3 ft. was observed in the dug well for a period of 7 hours pumping [because electricity not available for further period]. He also states that recuperation is within one hour.

   (d) The centrifugal motor fitted is 3 H.P. and discharge as observed is 2½.

   (e) There are 15 nos. of filter points around it within 100 feet. All are functioning.

   (f) Khariff crop is also being cultivated with 2 acres Paddy and 4 acres Ground nut crop.

There are about 120 filter points on left side within 1.5 k.m. from sub surface dam.

There is also an instance as on 2.2.05 in a dug well a bore is also drilled contains water in dug portion itself and being used for irrigating his land. Hither to the farmer use to go into bore well for pumping. Further there were instances that the delivery of filter points were being restricted by putting an orifice to effect continuous flow at fag end of crop season. But there are no such cases at present as per the voices of the farmers.

Figure 12.19   Farmer K. Venkataratnam.

Case – 2: Farmer Name: K. Venkataratnam, 50 years, on right back of sub surface dam.

Water was being exhausted by the month of March. Now, water is available for March, April, May and June. His Dug Well 32 ft. deep dried during March. Now water retains up to June/July and S.W.L. is 5 m. and draw down is 2 m. and being recuperated within ½ hour. The date of interaction on 2.02 his well contains 10 feet water column.

Further he states that 5 years back the value of one acre of land was Rs.72,000/- and due to availability of water the value now is Rs.1.5 lakh. He owns 9 acres and he has cultivated 5 acres Ground nut & 4 acres paddy, He states that to an extent of 4 acres Sugarcane crop is to be raised after harvesting of paddy sown.

Further the Dug Well of P. Krishnamma & Mr. Velavanam Naidu which lie 160m. distance has rejuvenated after construction of sub surface dam.

It was reported by farmers that there was surface flow in the river 25.01.2005.

Case 3: Farmer Name: Sudhakar Babu.

Owns a dug well being 30 ft. depth S.W.L. is 20 ft. He says continuous pumping is being restored and sufficient yield is now available.

There was surface flow in the river as per his version up to Dec. 15. He says that he could get water for Paddy & Groundnut during March & April now and getting crop produce to home. 2nd crop is also being cultivated for which he gets one month water ie. June / July month and the rains occur, river flows and with that again water is raised in wells and crop is cultivated. He finally states that the above benefits are due to S.S. Dam construction. Previously he use to cultivate only 50% of his land with great difficulty and with an orifice in delivery pipe.

Figure 12.20   Lady farmer Papamma, Bangarupalle (V), A.P.

Case 4: Farmer Name : Papamma, Bandarupalle.

Owns 10 acres land. The land lies on left side bank of river on upstream of S.S.Dam. She says due to S.S.Dam, it is possible to irrigate Groundnut & Sugarcane in her land. The S.W.L. is 7.6 m. in the dug well as the bank height is very high on left side of the river.

She says that land value has increased from Rs.70,000/- to Rs.1,00,000/- per acre. Now the "Billa" i.e., Orifice is not being adopted now for delivery pipe as yields have increased in the well inspite of less rainfall as S.S.Dam has obstructed the base flow

Figure 12.21   Measuring Water level in the Well of S.C. Corporation, Middle of Swarnamukhi River

Case 5: Farmer Name : Munikrishnaiah Chetty.

In his dug well whose depth is 45 ft. water column is 15 ft. He also says as above. It was also reported that a bore well which was on right side was not yielding much water earlier and now its fractures having hydraulic connection with sandy alluvium have been recharged and yielding continuous water flow. The benefits obtained by construction of sub surface dams are being appreciated by all farmers and there are attempts to drill new bore well also in isolated rocky aquifers since there is increase in yield of existing bore wells.

2. **Construction of Sub Surface Dam across Swarnamukhi River near Siddaiahgunta (V), Yerpedu Mandal, Chittoor District, A.P.**

This work was executed during the year 2001. The water level in the river bed was 3.13 m during January, 2001 before construction of S.S. Dam. The water table was measured in the existing S.C. Corporation dug wells constructed in the river bed. The present water table as on 2.2.05 is 2m. The additional ground water conserved has helped to rise water table in sandy alluvium of river bed and thereby make the wells to give more discharge on a sustainable basis through out the year. The interaction made with some of the farmers on 2.2.05 is presented below.

CASE 1: Farmer Name: Pujari Munikrishna Reddy, Siddiahgunta (V)

Figure 12.22   Farmer P. Muni Krishna Reddy of Siddiahgunta (V)

He owns 10 acres of land and the cultivation is 8 acres Paddy & 2 Acres Ground nut (G.L. 24 variety). He owns a dug well 20 ft. deep and 10.5 dia. Present S.W. is 12 ft. He also owns a filter point.

Case 4: Farmer Name : G. Sudhakar, Siddaiahgunta

Owns 10 acres and have 2 filter points at a distance of 700 m. upstream of S.S. Dam in the river bed and irrigation is being done Groundnut 5 acres, Paddy-5 acres.

He states that water help is there for the months of March & April due to construction of S.S.Dam. Previously at fag end for lack of water crop was foregoing. Now second crop is also being done.

Case 5: Farmer Name : Sankaraiah, Siddaiahgunta

He says his full extend of land is being sown now. Further he states that last two months wettings are being assured due to construction of S.S. Dam 7 hours pumping is resorted, as electricity is available for that period only. He says the F.P. can give continuous supply if electricity is available. He wants some more S.S.Dams to be constructed at some more places as they are much useful for them in less rainfall years. Particularly and at large in good seasons also to enhance their extend of cultivation.

Figure 12.23   Observing the Present Yield 2½ inches from the Well at Mungalapalem Village

Figure 12.24   Interaction With the Beneficiaries by the Consultant

Figure 12.25   Measuring Water Table in the Well at Mungalapalem (V), AP

Measuring the water table as on Feb 02, 2005 at Munagalapalem Village.

Personal visit of sites and interaction was made by P. Muniratnam, B.E., C.E. M.I.E, Former Technical Consultant – World Bank Aided Project, APERP for preparation of Technical Papers to be presented at National & International Seminars.

7 hours pumping is resorted, as electricity is available for that period only. He says the Filter Point can give continuous supply if electricity is available. He wants some more S.S. dams to be constructed at some more places, as they are much useful for them in less rainfall years. Particularly and at large in good seasons also to enhance their extend of cultivation.

SMALL IS ECOLOGICALLY EFFICIENT AS10 DAMS WITH 1 HECTARE CATCHMENT WILL STORE MORE WATER THAN 1 DAM OF 10 HECTARE.

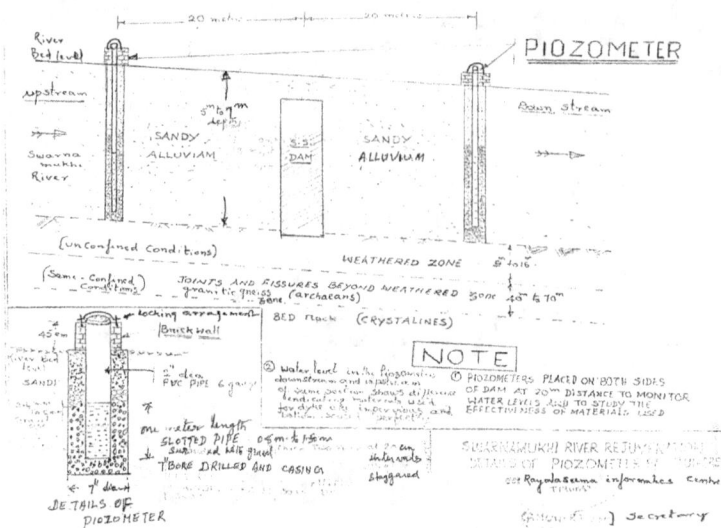

Figure 12.26   Construction Details of Piozometer at Swarnamukhi River Sites.

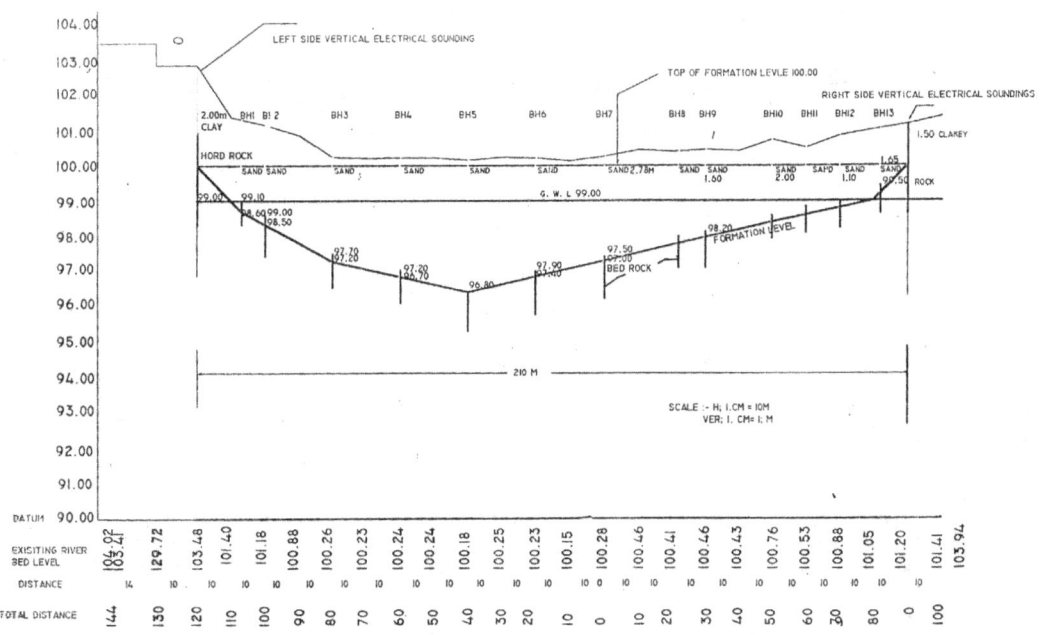

LONGITUDINAL SECTION OF SUB-SURFACE DAM ON SWARNAMUKHI RIVER NEAR MUNAGALAPALEM [V] YERPUDU [M] CHITTOOR DIST

Figure 12.27 Longitudinal Section of SS dam at Mungalapalem

SUBSURFACE DAMS TO HARVEST RAINWATER- A CASE STUDY OF THE SWARNAMUKHI RIVER BASIN, SOUTHERN INDIA

[1]N. Janardhana Raju, T.V.K. Reddy & [2]P. Munirathnam
Department of Geology, Sri Venkateswara University, 517502 India e-mail: rajunj7@yahoo.com
Department of Geology, Banaras Hindu University, Varanasi, 221005 India
Hydrogeology
Journal (2006) 14: 526-531

ABSTRACT

Declining water level trends and yields of wells, deterioration of groundwater quality and drying up of shallow wells are common in many parts of India. This is mainly attributed to the recurrence of drought years, over exploitation of groundwater, increase in the number of groundwater structures and explosion of population. In this subcontinent, the saving of water has to be done on the days it rains. India receives much of its rainfall in just 100 h in a year mostly during the monsoon period. If this water is not captured or stored, the rest of the year experiences a precarious situation manifest in water scarcity. The main objective behind the construction of subsurface dams in the Swarnamukhi River basin was to harvest the base flow infiltrating into sandy alluvium as waste to the sea and thereby to increase groundwater potential for meeting future water demands. An analysis of hydro graphs of piezometers of four sub-surface dams, monitored during October 2001-December 2002, reveals that there is an average rise of 1.44 m in post-monsoon and 1.80 m in the pre-monsoon period after the subsurface dams were constructed. Further, during the pre-monsoon month of June, much before construction of subsurface dams in October 2001, the water level was found fluctuating in the range of 3.1-10 m, in contrast to the fluctuation ranging from 0.4 to 3.1 m during the period following the construction of dams. Hence, the planning of rainwater harvesting structures entails thorough scientific investigations for identifying the most suitable locations for subsurface dams.

Received: 1 March 2004 / Accepted: 29 December 2004
Published online: 9 April 2005
© Springer-Verlag 2005
Hydrology Journal (2006), 14-526-5513
DOI 10 1007/S 1040-005-04285

SOCIAL MOBILISATION IS A PREREQUISITE FOR EFFECTIVE WATER HARVESTING ACTIVITY.

RESUME

La diminution des niveaux d'eau ainsi que des capacites hydrauliques des puits, la deterioration de la qualite de l' eau souterraine ainsi que l' assechement de puits de faibles profondeurs sont chose commune dans plusieurs regions de 1 'Inde. Cela est principalement du a la recurrence d'annees seches, a la surexploitation de la 'ressource en eau souterraine, a I' augmentation du nombre de captages d'eau souterraine, et a l'explosion demographique. Dans cette partie du globe, l'economie de l'eau do it se faire a partir des jours de pluie. L'inde recoit la majorite de sa precipitation annuelle en moins de 100 heures associees ala periode de la mousson. Si cette eau n'est pas captee et entreposee, le reste de l' annee fait face it une penurie et a une rarete de l'eau disponible. L'objectif principal de la construction de barrages souterrains dans le bassin de la riviere Swamamukhi est de favoriser l'infiltration des eaux de ruissellement en direction des sables alluvion-naires qui, autrement, seraient perdues en mer, ce qui per-met d' augmenter le potentiel en eau souterraine afin de repondre ala demande future en eau. L'analyse des hydro-graphes produits par des piezometres installes dans quatre (4) barrages souterrains, suivis d'octobre 2001 a december 2002, a revele une remontee moyenne des niveaux de 1,44 met de 1,80 m respectivement avant et apres la peri ode de la mousson, et ce apres l installation des barrages souterrains. De plus, durant le mois de juin, precedent la mousson, bien avant la construction des barrages souterrains en octobre 2001, Ies niveaux d'eau ont fluctue entre 3,1 met 10 m, en comparaison avec des fluctuations oscillant entre 0,4 m et 3,1 m lors de la periode suivant la construction des barrages souterrains. Par consequent, la planification d'une procedure de collecte des eaux de pluies par des structures adaptees impose des etudes scientifiques completes, permettant d'identifier les emplacements les plus favorables it la construction des barrages souterrains.

RESUMEN

La tendencia a la disminucion de niveles de agua y del rendimiento de pozos, el deterioro de la calidad del agua subterranea y la sequia de pozos de agua poco profundos son comunes en muchas partes de la India. Esto se atribuye principalmente ala recurrencia de años de sequia, la sobreexplotaci6n de ~guas subterraneas, el incremento en el mimero de estructuras que explotan aguas subterraneas y la explosi6n de la poblaci6n. En este subcontinente los ahorros de agua deben llevarse a cabo cuando llueve. La India recibe la mayor parte de lluvia en solo 100 horas cada año durante el perfodo del monzon. Si esta agua no se capta 0 almacena, el resto del año queda en una situacion precaria de escacez de agua. El principal objetivo al con-struir embalses a nivel de subsuperficie en la cuenca del rio Swarnamukhi es la cosecha del flujo de base que infiltra los aluvios de arena y que deberia generar un flujo de agua que se desecha en el mar. La consecuencia es el incremento del potencial de agua subterranea para satisfacer la demanda futura. El analisis de los hidrografos de piezometros de 4 embalses de subsuperficie que fueron monitoreados desde octubre 2001 hasta diciembre 2002 revela que hubo un in- cremento promedio de 1.44 metros despues del monzon y de 1.8 metros antes del monzon una vez que se construyeron los embalses. Adicionalmente durante el mes de junio, mu-cho antes de la construccion de embalses de subsuperficie en octubre del 2001, se observaron variaciones en el nivel de agua en un rango de 3.1 a 10 metros, en contraste a la variac ion de 0.4 a 3.1 metros durante el periodo pos-terior a la construccion de los embalses. En consecuencia el planeamiento de estructuras para la captacion de lluvia involucra investigacion cientffica exhaustiva para identi- ficar los lugares idoneos para la ubicacion de embalses de subsuperficie 0 subterraneos, Keywords Rainwater harvesting Water resources Sub-surface dam Swarnamukhi River Groundwater level fluctuations

INTRODUCTION

In India there are a few years of very good rainfall and many years of low rainfall. Even during low rainfall years there occurs high runoff causing flash floods of short duration. If these flash floods are harnessed for recharging the depleted aquifers, there will be sufficient improvement of water supply. For all the fairly large amount of rainfall in India, much of it is lost through runoff due to the absence of sufficient sites for storage and impounding. There is thus an imbalance between recharge and groundwater development in many parts of the country (Raju 1998). It is a paradox that the town of Chirapunji in Meghalaya State, northeast India, which has the distinction of an annual rainfall in excess of 11,000 mm, experiences water scarcity for nine months of the year. Conservation of a significant proportion of local rainfall during non-monsoon months by storage in under-ground cisterns is necessary in order to reduce dependence on imported water supplies in Bangalore City (Curtis 1998). Rainwater harvesting and groundwater recharge should be encouraged with the participation of every member of the village community.

Concerted efforts for promoting and accelerating groundwater recharge through the adoption of rainwater harvesting methods are to be preferred to the artificial recharge of groundwater. Some artificial recharge methods by the uninitiated invariably pose the great danger of polluting one of the few sources of unpolluted water in the country (Radhakrishna 2003).

Surface storage of water has serious disadvantages such as evaporation losses, pollution problems, submersion of valuable land, and silting up of dams. In recent years, groundwater dams (i.e. subsurface dams) have been constructed to solve many of the aforesaid problems. Such devices are not new. Groundwater dams were constructed on the island of Sardinia in Roman times and similar structures to dam groundwater were developed by old civilizations in Tunisia of North Africa. Sustainable management of groundwater is of vital importance to economically weaker communities in rural India. In watersheds or sub-basins in which around 50% of the available groundwater resources are being utilized, the emphasis is now being shifted from groundwater development to sustainable groundwater management. This shift ensures long-term availability of adequate quantity and acceptable quality of groundwater (Limaye 2003). In the over-exploited watersheds, it is necessary to impose pumpage control and to apply provisions of groundwater legislation. India's ancient tradition of community based water harvesting through the involvement of individual households is fading away (Agarwal and Narain 1997). This predicament has necessitated the increased role of governments (Central and State) in water management.

Water Harvesting

Rainwater harvesting by runoff conservation structures (gully plug, rockfill dams. check dam and bench trenching) is basically intended to slow or stop the running water (contour trenching and subsurface dams) and to infiltrate (percolation tank) into the underground. A survey of developments in the past decade shows that numerous projects have been undertaken to promote local water harvesting both in urban and rural areas in India. In drought prone areas rainwater harvesting is undertaken to address the serious problems of drought and water scarcity. The people of the State of Rajasthan, who have a long-standing tradition of rainwater harvesting (Agarwal and Narain, 1997) used drought relief aid for developing 100,000 water-harvesting structures. Water conservation measures in Wagarwadi watershed, Maharastra, are responsible for raising the water table by about 2 m in the wells located within 200 m on the down stream side of nala (i.e. stream) bunds and in other wells located adjacent to the water conservation structures. Water conservation structures of Wagarwadi watershed are responsible for groundwater recharge estimated at 6 ha m per year and a rise of 0.5 to 2.5 m of the water table at different locations (Gore et al. 1995).

The Government and non-governmental agencies in India are increasingly involved in water harvesting in rural areas. Areas of low to moderate rainfall occurring mostly in a single monsoon period do require all efforts to conserve rainwater. Various widely practiced methods of artificial recharge of groundwater such as gully plugging, contour trenching and bunding, rock fill dams, check dams, percolation dams and subsurface dams are being employed for increasing the contact area and resident time of surface water with the soil to maximize the quantity of water percolation and augment groundwater storage.

A subsurface dam, intended for arresting the ground-water flow in a natural aquifer, is constructed across a valley by digging a trench to bedrock. The trench has at its base an impervious wall which is covered with excavated material until the trench is completely concealed. The reservoir thus built can never be totally drained and can be used throughout the dry season, depending upon the storage volume, to meet the water demand. Sub-surface dams are often built in riverbeds that generally constitute highly permeable aquifers with good storage potential. The most common type of sub-surface dams are clay, concrete, stone masonry and brick walls built in excavated trenches of 3-10 m deep. The advantages of sub-surface dams include low-cost construction and very low evaporation, besides, the utilization of more land area for cultivation.

Case study of Swarnamukhi River subsurface dams   The Swarnamukhi River basin is essentially a semi-arid tract with an average annual rainfall of 1,000 mm. The river basin is occupied by granite and granitic gneisses. With alluvium thickness in the range of 3-7.5 m, the underlying weathered granite occurs at a depth ranging from 9 to 16 m, the depth up to which dug wells are pierced. The Swarnamukhi River has occasionally dried up, impinging on the irrigation and drinking needs. Impounding more base flows in the Swarnamukhi River basin necessitates the development of a watershed program constituting sub-surface dams and other groundwater structures scientifically designed. The purpose of construction of subsurface dams in the Swarnamukhi River basin is to harvest the base flow going in sandy alluvium and thereby to increase groundwater resource potential.

Need for the construction of sub-surface dams across swarnamukhi river

1. Sand quarrying along the river course has decreased the thickness of sandy alluvium in which maximum storage of water occurs.

2. Over-exploitation of groundwater in excess of the annual replenishment has caused continuous decline of water levels and a decrease in yields from shallow wells along the river courses.

3. Vagaries of monsoons are invariably marked by insufficient rainfall.

4. Construction of Kalyani Dam (a tributary to Swarnamukhi River) to meet drinking water needs to the Tirupati - Tirumala region has reduced stream flow at Swarnamukhi River.

5. An imperative for restoring the perennial nature of the river course through such basin development activities in order to augment the water table in the adjacent wells.

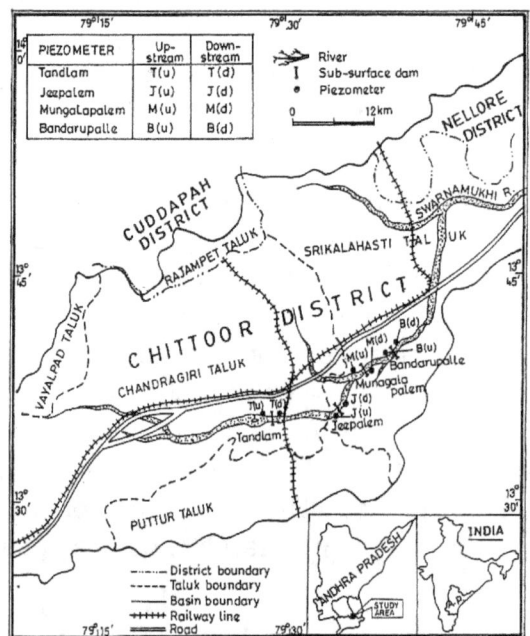

Figure 12.28 Location Map of Subsurface Dams Across Swarnamukhi River Basin

Results and Discussion

The construction of subsurface dams i one of the components envisaged for rejuvenation of the Swarnamukhi River basin. Like surface runoff, groundwater also leaves the basin as base flows if it is not harnessed. Hence, the movement of groundwater in the river course has been checked to utilize it for stabilization of Rabi season (October to December) crops like groundnut and paddy as well as early Kharif (June to September) paddy crop by construction of subsurface dams across Swarnamukhi River basin at appropriate places. Suitable sites for sub-surface dams (Fig. 1) have been selected by studying the topographic and geological maps. The data on the depth to rock was collected from resistivity soundings carried out along the course of proposed subsurface dam alignment. A few auger holes have been drilled to corroborate the results of the resistivity survey. The results of these investigations led to the construction of four subsurface dams. The dam wall occupies very little space without any buttresses. The Subsurface dam (Fig. 2) was constructed with puddled clay 0.9 m thick which has been keyed into a hardend connection on either side of the river bank. Details of length and depth of subsurface dams in the Swarnamukhi River basin are shown in Table 1. The top of subsurface dams has been limited to 1 m below riverbed level to allow excess flow downstream, to flush the salts accumulated and to leave unaffected the riparian rights of the farmers downstream. The dam has been keyed into weathered bedrock 0.55 m below the bedrock surface. The clay wall was compacted with wooden rammers in order to obtain proctors' density. The clay balls were so prepared as to imbibe optimum moisture content. In order to retain this moisture content for a longer period, 200 u low-density polyethylene film was emplaced on the upstream, downstream and top of the dam in order to protect the clay wall from yielding to cracks.

Table 12.1 Details of Length and Depth of Subsurface Dams in the Swarnamukhi River Basin

S.No.	Subsurface	Length (m)	Depth (m)	Slope in the Vicinity of Dam	Maximum Depth to Weathered Rock (m)	Cost of Construction in Rupees (US Dollars)
1.	Tandlam	300	5.00	1 in 500	6.00	1,800,000 (36,000)
2.	Jeepalem	340	6.30	1 in 360	7.30	3,700,000 (74,000)
3.	Munagalapalem	210	3.00	1 in 450	4.00	900,000 (18,000)
4.	Bandarupalle	193	3.00	1 in 900	4.00	1,500,000 (30,000)

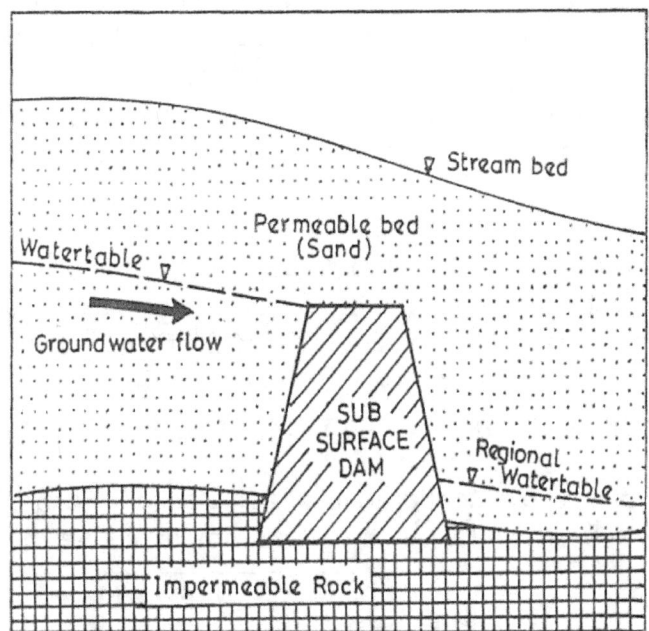

Figure 12.29 Schematic Diagram of Subsurface Dam

The imperviousness due to the dam has been proved by the study of upstream and downstream water levels monitored in the piezometers erected 200 m away from the dam alignment. In order to establish the trends in the water level fluctuations monitored at fortnightly intervals, well hydrographs are drawn for Tandlam, Jeepalem, Munagalapalem and Bandarapalli subsurface dams and are shown in Figs. 3 and 4.

The hydrograph of Tandlam subsurface dam (Fig. 3) shows that the water level fluctuations are around 1.5 m after the construction of the subsurface dam. This points to the improvement and stability of water levels in the up-stream side of the subsurface dam. The well density in the upstream side of the Tandlam subsurface dam is about 285. The very high density in the left bank with a thickness of sand ranging from 9 to 10m is in contrast to the low density of wells on the right bank marked by the shallowness of the sand bed. Following the construction of the dam the water level fluctuation in the right bank was around 1.5 m (very minimum) because of the low well density coupled with the limited overdraft in existing wells and absence of any cultivable land for development of agricultural activities. Hence there is an indirect benefit to the areas further away from the dam due to a rise in the water table. On the other hand, the water level fluctuations, as revealed by the hydrograph of a piezometer at the Jeepalem subsurface dam (Fig. 3), are around 7 m. The large water level fluctuations are due to the presence of 382 tube wells on the upstream side of the dam causing overdraft of groundwater for agricultural activity and to the small thickness of the sand bed in the upstream side of the subsurface dam.

Following the construction of the subsurface dam, there are minimum water level fluctuations (around 1.25 m), as indicated by- the hydrograph of Munagalapalem sub-surface dam (Fig. 4). The high recharge of water from Nakklavanka and Seethakalva streams (the tributaries of Swarnamukhi joining on the upstream side of the Munagapalem subsurface dam) and the thickness of the sandy alluvium (around 10 m) in the Seethakalva stream contribute to the stable water level. Around 640 wells are drilled near the Munagalapaem subsurface dam. Prior to dam construction some of them dried up in the summer months (May to July). It is significant that after constructions of the subsurface dam the storage capacity of water impounded is about 8.67 Mcft, which stabilized 520 acres of paddy crop during the summer season, besides offering improvement of irrigation land. The hydrograph of Bandarupalli subsurface dam (Fig. 4) reveals that the water level fluctuations are around 2.5 m after the construction

PROFICIENT WATER HARVESTING ALONG WITH POLITICAL WISDOM AND COMMUNITY ACCOUNTABILITY CAN HELP SOLVE RURAL POVERTY.

Table12.2   Water Levels Monitored in Piezometers Before and after Construction of Subsurface Dams in the Swarnamukhi River Basin

S.No.	Subsurface Dam	Observation Well Water Level Before Construction of Subsurface Dam		Water level in Upstream Piezometer after Construction		Rise in Water level in Corresponding Month after Construction	
		January (Post-monsoon) (m)	June (pre-monsoon) (m)	January (m)	June (m)	January (m)	June (m)
1.	Tandlam	4.2	6.3	2.8	3.2	1.4	3.1
2.	Jeepalem	6.0	10	3.8	9.6	2.2	0.4
3.	Munagalapalem	1.1	3.3	0.3	0.9	0.8	2.4
4.	Bandarupalle	1.3	3.2	0.0	2.0	1.3	1.2

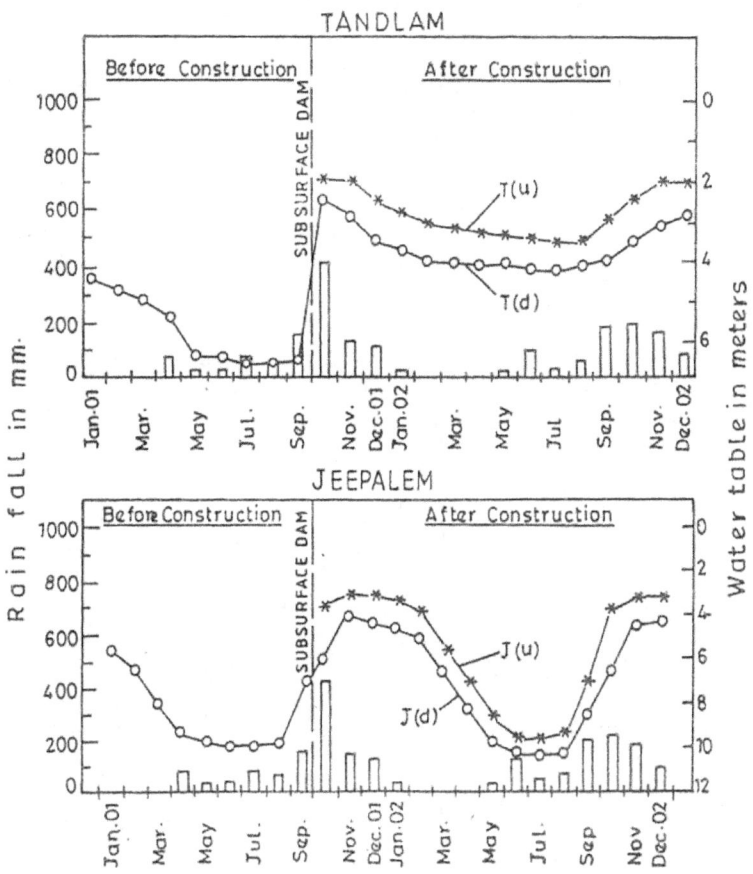

Figure 12.30 Well Hydrographs for Piezometers Upstream and Downstream of Tandlam and Jeepalam Subsurface Dams across Swarnamukhi River Basin

Fig.4 Well hydrographs for piezometers upstream and downstream of Munagalapalem and Bandarupalle subsurface dams across Swarnamukhi River basin of the subsurface dam. The very low well density around this subsurface dam is due to the thin sandy layer (the average being 3 m). In the case of the existing 105 wells in the upstream side of this dam area, the discernible development in the groundwater structures (i.e. wells) is made possible through the improvement of water levels by means of constructing the subsurface dam.

Water levels were recorded for 9 months (January 2001 to September 2001) in observation wells upstream and down-stream of the dam before and after construction (October 2001 to December 2002) (Munirathnam 2003). Analysis of water tables indicates that there was an average rise of 1.44 m in post-monsoon and 1.80 m in the pre-monsoon periods in the four subsurface dams. Further, the data recorded for the pre-monsoon month of June 2001 is significant in that the pre-construction fluctuation in water level was 3.1 to 10 m as opposed to the fluctuation of 0.4 to 3.1 m after construction (Table 2). It is also important to mention that farmers were restricting the diameter of the delivery pipe of pumps and irrigating as part of the already sown summer groundnut crop during the period preceding the construction of surface dams. Interestingly, after the construction of the subsurface dams, the irrigation activity has become a continuing feature in the summer season, thanks to the artificial recharge of the groundwater.

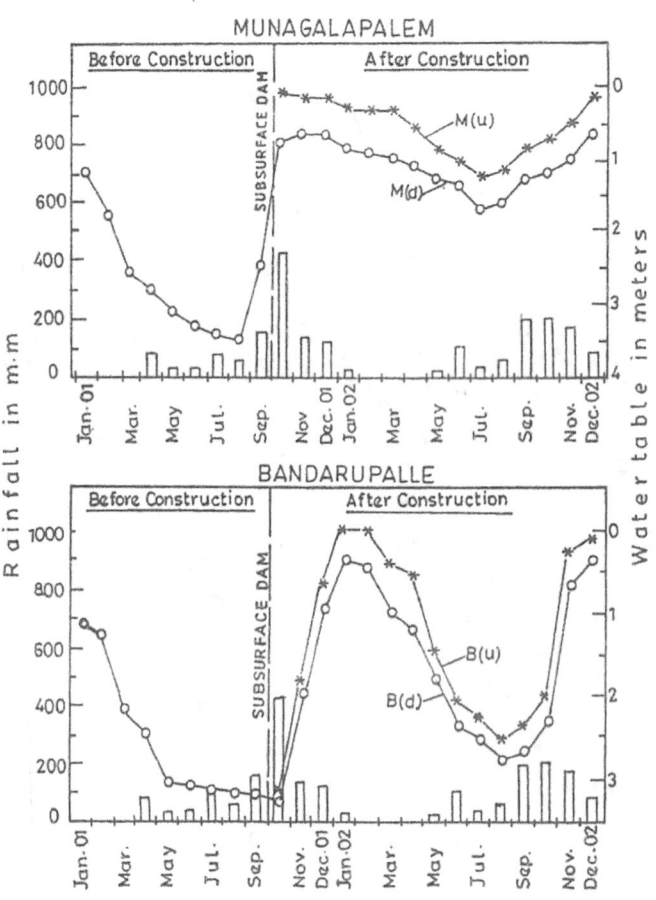

Figure 12.31   Well Hydrographs for Piezometers Upstream and Downstream of Munagalapalem and Bandarupalle Subsurface Dams Across Swarnamukhi River Basin

CONCLUSIONS

Rainwater harvesting and artificial recharge have become compulsory in the areas where the pursuits of groundwater development have surpassed the annual replenishment. Involvement of the community is necessary in water management and development by laying greater emphasis on extending and improving local water harvesting systems. It is imperative for the rural areas to participate in watershed development programmes of soil moisture conservation and groundwater recharge such as catchment area treatment through gully plug, rock fill dams, contour trenching and bunding, check dams, percolation tanks and subsurface dams.

This case study indicates that consideration of subsurface dams in the Swarnamukhi River basin is feasible and economical by using cheap material like clay and LDPE Film. From the well hydrographs, the water levels in the upstream piezometers after construction of the subsurface dams average 1.44 m in the post-monsoon and 1.80 m in the pre-monsoon periods. Construction of subsurface dams in the Swarnamukhi River basin increased groundwater storage and thus provided for an increase in land productivity. Watershed management of the basin has improved base flow and improved the storage capacity at subsurface dams and ultimately facilitated the rejuvenation of the river to the extent of becoming a perennial one. The assistance of financial institutions is now very much sought after by the farmers to construct such dams on a co-operative basis. The needs of the basin are to conserve every drop of water and recharge the depleted aquifers.

ACKNOWLEDGEMENTS

The first author N. Janardhana Raju is highly grateful to the Alexander von Humbolt Foundation, Germany for the Humboldt Fellowship and equipment donation for carrying out research studies of this kind. The authors are thankful to Prof. Y.Yagama Reddy, Center for Indo-china, for his critical and valuable suggestions during the preparation of the manuscript. Authors are also thankful to the two anonymous reviewers and to Prof. Zaisheng Han and Dr. Perry Olcott, managing editor whose thoughtful suggestions and specific comments significantly improved the manuscript.

REFERENCES

Agarwal A, Narain S (eds) (1997) Dying wisdom: Rise, fall and potential of India's traditional water harvesting systems. Centre for Science and Environment, New Delhi

Curtis LC (1998) Rainwater Harvesting-A Possible Seasonal Addition to Bangalore Water Supply. J Geol Soc f India 51:455-460.

Gore KP, Pendke MS, Mal BC, Gurunadha R3.o VVS, Gupta CP (1995) Assessment of impact of water conservation structures on groundwater regime in wagarwadi watershed, Parbhani, Maharashtra. Gondwana Geol Mag 10:63-77

Limaye SD (2003) Some aspects of sustainable development of groundwater in India. IGC'S Eighth Professor Jhingran Mem. Lect., 90th Indian Science Congress, Bangalore 22 pp

Munirathnam P (2003) Rejuvenation of Swarnamukhi River Basin in Chittoor District, Andhra Pradesh, India. Technical Report, pp 1-8.

Radhakrishna BP (2003) Groundwater Recharge. J Geol Soc India 62:135-138.

Raju KCB (1998) Importance of Recharging Depleted Aquifers: State of the Art of Artificial Recharge in India. J Geol Soc India 51:429-454.

MANAGED AQUIFER RECHARGE (MAR) BY THE CONSTRUCTION OF SUB-SURFACE DAMS IN THE SEMI-ARID REGIONS : A CASE STUDY OF THE KALANGI RIVER BASIN, ANDHRA PRADESH

N. JANARDHANA RAJU[1], TVK REDDY[2], P MUNIRATNAM[3], WOLFGANG GOSSEL[4] and PETER WYCISK[3]

[1]Jawaharlal Nehru University New Delhi, New Delhi - 110067
[2]Department of Geology, S.V. University, Tirupati - 517502
[3]Technical Consultant. Air by-pass road, Tirupati - 517502
[4]Institute for Geosciences. Martin Luther University, Halle (Saale), Germany
Email: njrajuI963@yahoo.com

Abstract: Overuse of groundwater in coastal areas, due to high population and agricultural activity results in seawater intrusion into the coastal aquifer. This paper presents the control measures taken to manage aquifer recharge (MAR) and also to overcome the problem of seawater intrusion into the coastal aquifer along the Kalangi river, Nellore district of Andhra Pradesh, India having connectivity with Pulicat (saltwater) lake estuary. Due to overexploitation of groundwater and less rainfall in past years, adjacent seawater has started intruding in the Kalangi river subsurface and deteriorating groundwater quality up to 11.6 km from the confluence of the river with Pulicat lake. To prevent this situation, sub-surface dams were constructed in traditional manner using local earth material in three different places across the Kalangi river near Sullurpet town. The water storage capacities calculated after the sub-surface dams' construction are 1.28 mcft at GK Engineering College. 6.23 mcft at Challamagudi and 3.143 mcft at Holy Cross School sites.

The Holy Cross School sub-surface dam IS the first full scale dam-cum-check dam constructed to prevent salt water intrusion in the Kalangi river at Sullurpet, Nellore district, Andhra Pradesh. At the Kalangi river estuary portion (at the mouth of sea) a groyne was reconstructed over old groyne site with the introduction of clay bed and wooden sheet piles at down stream. Apart from prevention of sea water entry into Kalangi river sub-surface (during seasons) the groyne top level was raised to prevent mixing of high sea water tides with fresh water and ensuring additional storage of fresh water at upstream side. The reconstructed groyne was serving the purpose of obstructing the surface seawater entry in the Kalangi river and water quality has improved in the river as well as in the wells. After construction of sub-surface dam, as per the Simpson ratio classification. there is substantial improvement of water quality in the SHAR infiltration well situated near the Holy Cross School sub-surface dam.

Keywords: Saltwater intrusion, Sub-surface dams, Groundwater, Rainwater harvesting, Kalangi River, Andhra Pradesh.

INTRODUCTION

Fresh water availability for sustainable development is a major challenge facing the global community. In many arid and semi-arid regions in South Asia and the Middle East, large amounts of water annually flood the sea or desert during extreme rainfall events. Dillon (2005) defined managed aquifer recharge (MAR) as the 'intentional banking and treatment of waters in aquifers'. The term MAR was introduced as an alternative to 'artificial recharge', which has the connotation that the use of the water was in some way unnatural. Managed aquifer recharge of reclaimed water can be an important water management tool in arid lands in three main ways, (I) by providing storage, (2) improving the quality of stored water through natural contaminant attenuation processes, and 3) protecting freshwater resources (e.g salinity barrier systems). In coastal margins of the fresh groundwater basin, the lowering of groundwater level beyond certain limit can result in seawater intrusion. On one side is prudence and thrift regarding water; on the other lies extensive use of technology to harness the maximum amount of safe drinking water for consumption. In order to sustain agriculture and general water supply, efforts to harvest rainwater are currently being employed in many arid and semi-arid regions throughout the world (Raju et al. 2006; Murad et al. 2007; Stiefel et al, 2009; Ibrahim 2009).Increase of population and subsidized electricity for agricultural pumps in India has encouraged the rural affluent class to lift more and more water for irrigation from deeper aquifers. Areas of low to moderate rainfall occurring mostly in single monsoon period do require all efforts to conserve rainwater to recharge the depleted shallow aquifers in arid and semi-arid regions. Rainwater harvesting structures in semi-arid areas to tap stream water was practiced by ancient Jordanians about 5,000 years ago to provide drinking water to the old city of Jawa (Abdelkhaleq and Ahmed 2007), in the Mediterranean region around 4,000 years ago (Joshi 2002) and archeological evidence reveals that RWH activities have been central to indigenous civilizations in India for the past 2,000 years (Gunnell et al. 2007). In the light of the decreasing recharge space and over-exploitation of groundwater resources by modem extraction techniques, there has been a resurgence of traditional rainwater harvesting activities in many parts ofIndia in recent decades (Sharma 2002). Several rainwater harvesting structures such as check dam, across the drainage system at Mandvi in Gujarat has been taken up to arrest the declining trend of water levels, deterioration of groundwater quantity and prevent seawater intrusion into coastal aquifers (CGWB 1994). The Komesu underground dam is the first full scale underground dam constructed to prevent saltwater intrusion in Japan (Nawa and Miyazaki 2009). Seawater intrusion is often associated with over pumping in coastal regions, resulting in overdraft conditions and creating an inland gradient of seawater (EIMoujabber et al. 2006). The intrusion of saltwater into coastal aquifers is a wide phenomenon, especially in Mediterranean regions where semi-arid conditions lead to excessive pumping, high extraction rate and low recharge (Petal as et al. 2009). Check-dams were constructed across seasonal streams in many states in India to make water available for agriculture and domestic use (Balooni et al. 2008). Small alluvial aquifers will have lower potential storage than larger ones which are seen as good source of irrigation water in the summer season to save the agricultural crops (Owen and Dahlin 2005; Moyce et al. 2006; Raju et al. 2006).

Groundwater flow rate varies depending on the geologic characteristics of the stratum holding the groundwater, the depth of stratum, and the permeability and shape of the basement. Generally, shallower groundwater runs at a relatively higher flow rate, the water can be stored by damming the upstream flow with an impermeable wall as is done with rivers (OGB 1986). Sub-surface dam is a facility that stores groundwater flow in the pores of permeable strata to enable sustainable use in agricultural as well as domestic purposes (Raju et al. 2006). The advantages of the underground dams, unlike a normal surface dam, is cheaper to construct and the thickness of the wall is very thin compared to normal dam, evaporation is very low, the reservoir area can be used for cultivation (no land is wasted) and the riparian rights of the downstream formers are not affected as the dam does not project out (lies 1m below the land surface). The sub-surface dam allows the development of water resources in the alluvial aquifer regions and prevention of seawater intrusion into the fresh water alluvial aquifers. The classification of storage dams and salt-water intrusion prevention dams, is important (Hanson and Nilsson 1986). Storage dams are used to raise the groundwater level where the groundwater table is originally low and salt water intrusion prevention dams arc used to prevent saline water infiltrating the aquifer from the sea coast (Fig.1).

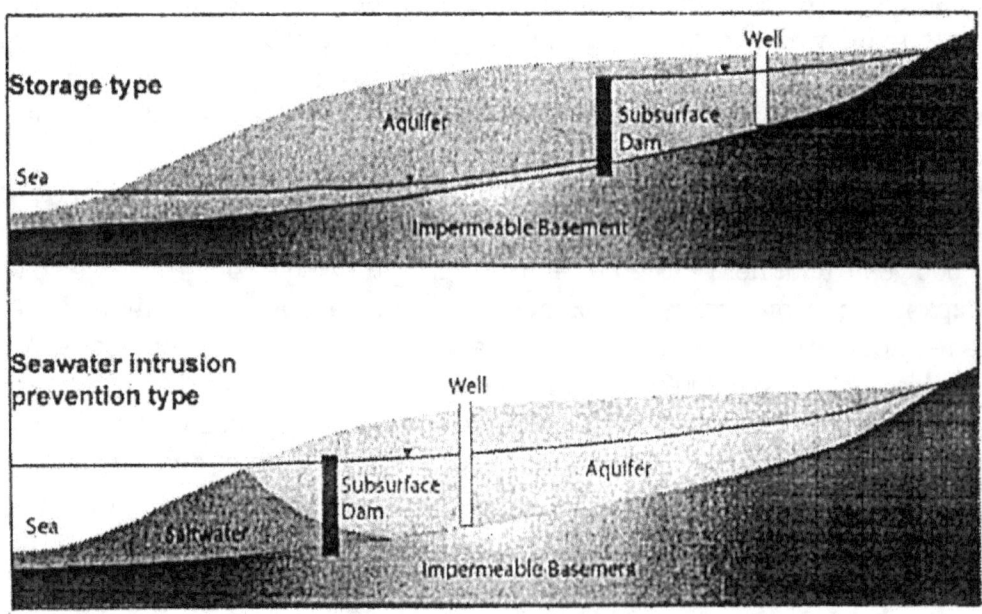

Figure 12.32   Schematic Diagram of Different Types of Subsurface Dams.

STUDY AREA

The alluvial tract of Andhra Pradesh is 700 km long and 5-25 km in extent from the coast line. The- Kalangi river rises near Narayanavanarn in Chittoor district, Andhra Pradesh and drains 355.75 km² in Chittoor district and 119.25 km² in Nellore district before joining Bay of Bengal (Fig. 2). The Kalangi river is an ephemeral river and the river course near Sullurpet is highly meandered. After draining a length of 86 km it enters Pulicat (saltwater) lake at Gradhagunta which is 11.6 krn from Sullurpet Town. In Kalangi river, continuous flow of surface water exists only for three months (90 days) during rainy season. The average thickness of sandy alluvium in the river is around 5 m. The Kalangi river is the main source of water supply for irrigation and drinking in and around Sullurpet area. The bed level of Kalangi river from Sullurpet to confluence with Pulicat lake runs below high tide level. Due to the presence of back waters of Pulicat lake, surface water in Kalangi river becomes saline and not useful for any purpose. Almost 500 acres of land is available on either side of Kalangi river in the stretch of around 10 km length from Sullurpet to confluence of the Pulicat lake. The irrigation was being perceived through dug wells and filter (tube wells) point wells. The average annual rainfall recorded at Sullurpet rain guage station is 1090 mm. The hydrograph of Sullurpet observation well is showing generally a declining trend in water level (Fig. 3).

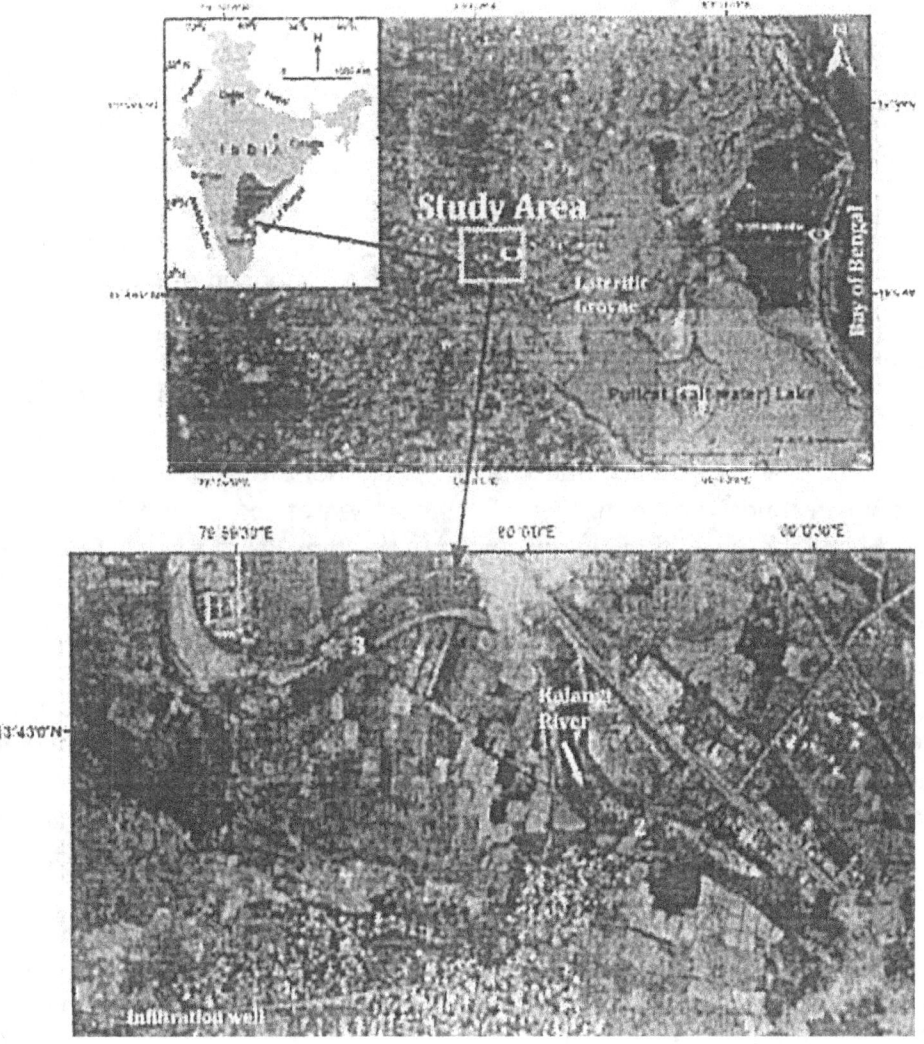

Figure 12.33   Location Map of the Kalangi River basin and Pulicat Lake Estuary

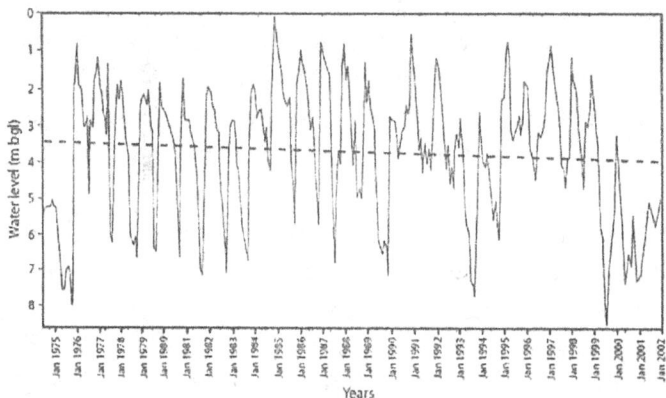

Figure 12.34   Water Level Fluctuation of the Observation Well of Sullurpet Area

GEOLOGY AND HYDROGEOLOGY

Geologically, the area is covered by unclassified crystalline rocks of Archean age which are overlain by recent alluvial deposits consisting of sand, clay, pebbles and boulders of different sizes. At few places laterite outcrops are observed along the course of Kalangi river. Most of the soils are derived by the disintegration of laterite and hence are classified by lateritic soils. Though the basement is of crystalline rocks in the area, no trace of exposure arc found along the river course. From the geological investigations, the sub-surface dam must rest either on crystalline rock or hard clay stratum to collect maximum storage of sub-surface water in the river. Detailed well inventory has been done for a length of5 km and a width of250 m on both the river banks around Sullurpet area and the cross sections of vertical thickness of wells are shown in fence diagrams (Fig. 4). Groundwater extraction is mainly from the dug and shallow tube (filter points) wells with some exceptions of deep bore wells in the area. Both dug wells and tube wells are pierced through the sand bed of the present and ancient river course (buried channels). From the well inventory data, it is found that there are two sandy types of aquifers (unconfined and confined) in which the movement of water is occurring (Fig. 4). The quality of water in confined aquifer is highly saline and is not potable. Hence, it is intended to store groundwater in unconfined freshwater aquifer by constructing sub-surface dams. The sand bed occurs lip to a maximum depth of 7 m in the unconfined aquifer. Groundwater from bore wells and open dug wells is extensively used for drinking and agriculture. The depth of groundwater level is too shallow and varies from 1-10m bgl. After thorough examination of geology and well inventory data, a few locations along the Kalangi river have been chosen to probe geophysically in order to know the sub-surface lithology, depth to basement and also the position of end connections. Hand augur drilling was done across the river bed in order to corroborate results of the geophysical studies. The-longitudinal cross sections of three sub-surface dams (i.e G.K Engineering College, Challammagudi and Holy Corss School are shown in the Fig. 5.

IN INDIAN CONTEXT, RAIN WATER HARVESTING IS INDISPENSABLE FOR FORTIFYING FOOD SECURITY AS RAINFED LAND CONSTITUTES THE BULK OF THE CULTIVATED LAND.

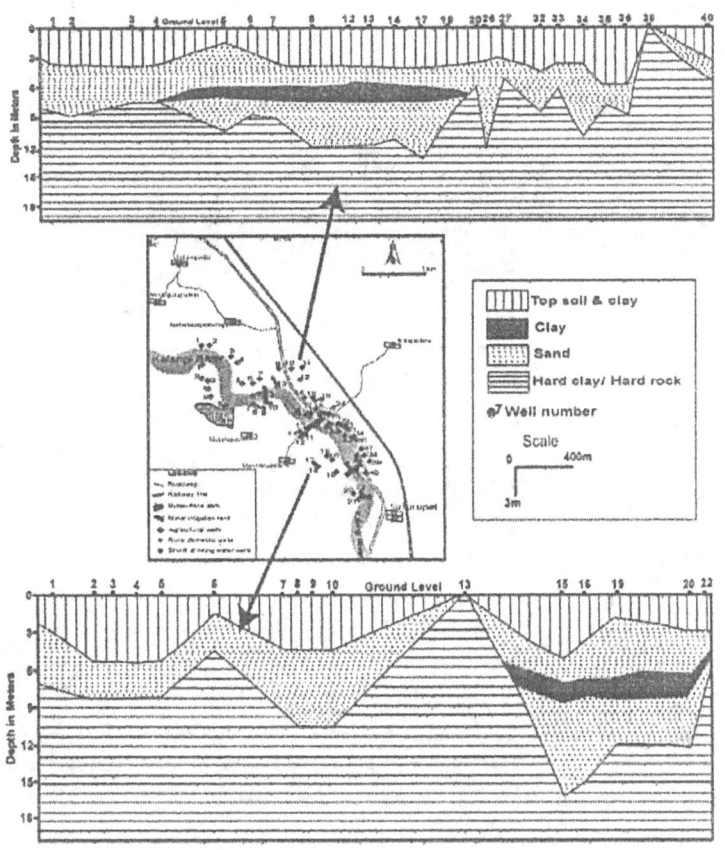

Figure 12.35   Fence diagram of well inventory along the Kalangi River course near Sullurpet environs

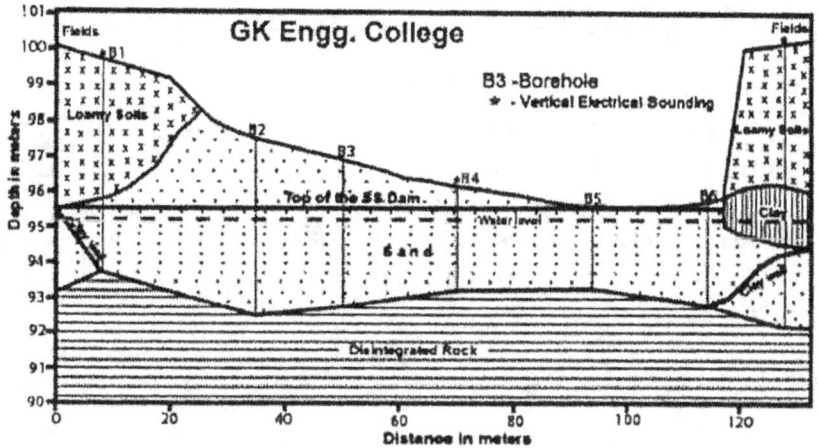

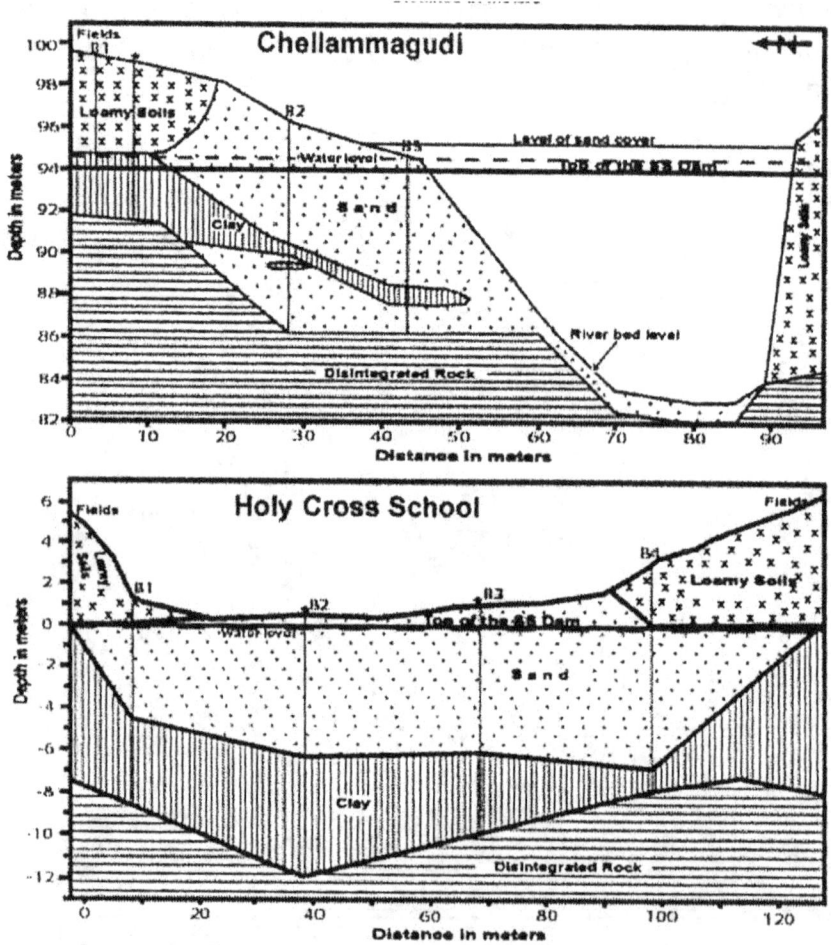

Figure 12.36 Lithological Cross Section of the Subsurface Dams
Constructed Across the Kalangi River

WATER RESOURCE SCENARIO IN THE STUDY AREA

To understand the overall salinity problems along the coastal area of Nellore district, approximately 60 groundwater samples were collected (2 km away from the coast line) and a rapid determination of total dissolved solids (TDS) made by measuring electrical conductivity of groundwater samples were made. Based on the TDS values of groundwater samples, a tentative saline zone is marked which is approximately 13 km away from Bay of Bengal sea coast near the Sullurpet area (Fig. 6). But Sullurpet town is still 9.5 krn away from this tentative saline zone. Then, the source of salinity in the Kalangi river up to Sullurpet town may be due to the contamination of Pulicat lake back waters. The majority of the wells reported in and around the Sullurpet area are contaminated with saline water intrusion. In order to understand the sources of salinity in the area, the river course for a length of -20 km has been examined to locate suitable sites for the construction of sub-surface dams.

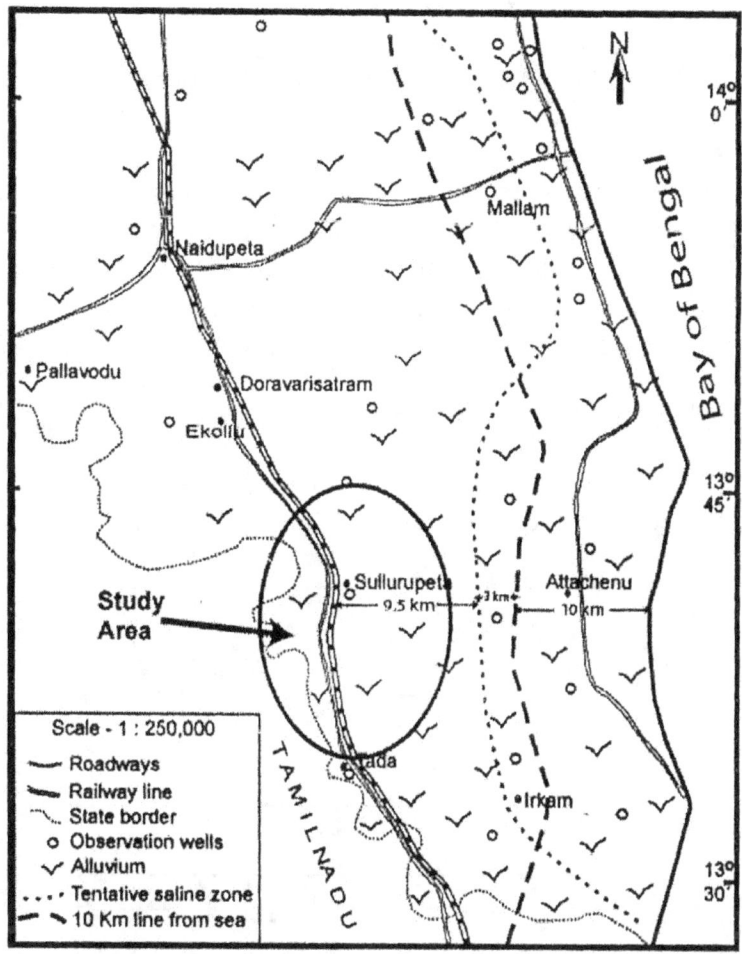

Figure 12.37   Demarcation of the tentative saline zone from coast of the Bay of Bengal

The Panchyat of Sullurpet has constructed infiltration wells in the Kalangi river bed for providing drinking water supply, but the quantity of water is grossly inadequate to supply for 30% of its present population at 40 li/capita/day in normal monsoon and 14% of population supplied 40 li/capita/day in summer months. Remaining population is depending on open wells for domestic consumption. Sriharikota Space Center (SHAR) in Sullurpet town has made its own arrangements for the supply of domestic water at the rate of 150 l/capita/day to their housing colonies. But in summer these colonies are also facing shortage of domestic water supply (<150 l/capita/day) from the infiltration wells constructed in the Kalangi river. In order to bridge the short supply for domestic consumption it is contemplated to tap base flows in the Kalangi river bed by arresting the flow by constructing sub-surface dams across the river to impound groundwater on the upstream of dams. Five kilometer stretch of Kalangi river upstream of Sullurpet town has been investigated for location of suitable sites for construction of sub-surface dams. Based on the detailed hydrogeological, geophysical, hydrological engineering surveys, three suitable sites have been located for construction of sub-surface dams across Kalangi river (Fig.2). The average specific yield of sandy alluvium in the river course is taken as 10%. The length of three sub-surface dams varies from 90 m to 130 m depending on the availability of end connections and average thickness of sandy alluvium at the sub-surface dam site varies from 2.8 m to 6.6 m. The calculated storage capacity of the Holi Cross School dam, Chellammagudi dam and GK Engineering Collge are 3.143 mcft, 6.23 mcft and 1.28 mcft, respectively (total storage capacity in three dams is 10.656 mcft) during non-flowing season of Kalangi River (Fig. 7). This will facilitate to recharge the existing infiltration wells in the river bed apart from stabilizing the bores drilled on the left bank of river for drinking water supply to the Sullurpet town.

Necessity for Rainwater Harvesting Structures across the Kalangi River

1. The water table (~3 m bgl) in the infiltration wells along Kalangi river (near Holy Cross School) in Sullurpet town has short supply of drinking water.

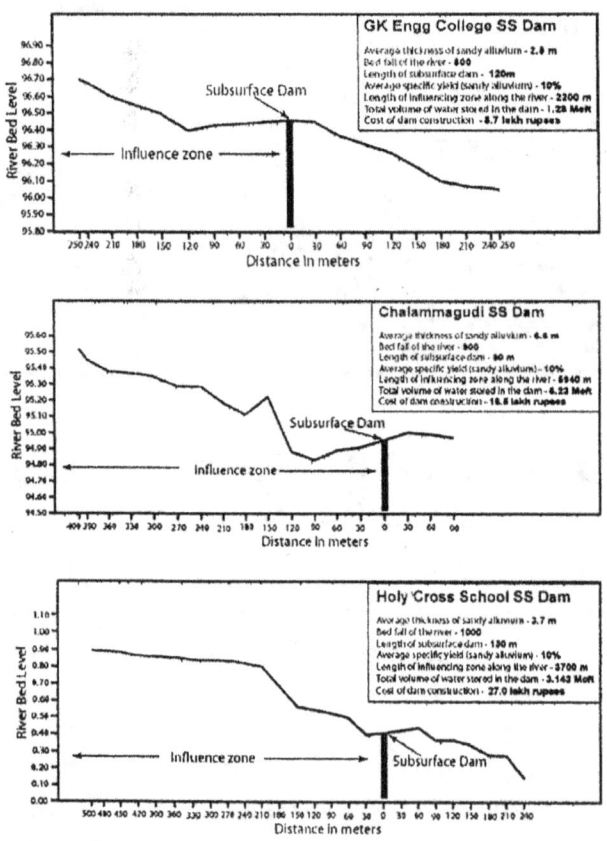

Figure 12.38 Influencing Zones of River Water Storage Upstream of the Subsurface Dams

2. On the Kalangi river bed, seawater contamination has been noticed up to a length of 10.2 km along the river course and it made the farmers stop pumping of surface water to their fields of 500 acres on both sides of the river.

3. Over-exploitation of groundwater in excess of the annual replenishment has caused continuous decline of water levels from shallow wells along the Kalangi river course.

4. Salinity has also been noticed in the groundwater from tube wells (filter points) and dug wells located in the agricultural fields.

5. Badly damaged rough stone laterite groyne near Vattembadu village allow surface seawater to enter into river Kalangi which contaminates the groundwater.

6. River bed level from Sullurpet town (10.2 km) to damaged laterite groyne is below high tide (depth of 0.8 m) at confluence, causing inundation of entire length of 10.2 km by seawater

7. When there is no surface river flow sand mining has taken place and a reverse gradient developed enhancing the velocity and depth of water to flow upstream towards Sullurpet town.

RESULTS AND DISCUSSION

A laterite rough groyne 1.2 m high above river bed level (RBL) was constructed in 1984 (at 1.4 km upstream of the Gradagunta confluence of Pulicat lake with river Kalangi near Vatembadu village), to overcome the problem of surface seawater entry into ephemeral Kalangi river from Pulicat lake which contains saline water. The top of the groyne was kept 0.1 m above high tide level (HTL) of Pulicat lake. In due course of time the groyne was damaged due to water force as well as human activities at many places resulting in the entry of surface saltwater of Pulicat lake into the Kalangi river when it was dry which percolated into the sand bed of the river. Another indication of saltwater ingress in the groundwater is due to the contamination of SHAR infiltration well in the year 2003, situated in the Kalangi river bed near Holy Cross School of Sullurpet town. The distance between groyne and the Sullurpet town is around 10.2 km and the analysis of surface water samples along the river course indicated salinity of surface water up to the Sullurpet town. The reason for the saltwater in the Kalangi river bed is due to breaching and badly damaged laterite groyne which allowed high tide Pulicat lake saltwater into the river course. The ingress of seawater into coastal aquifer along Kalangi river through estuary at Pulicat lake has affected the agricultural economy of Sullurpet area and also drinking water availability (Muniratnam, 2004).

Revelle (1941) pointed out that the increase in total dissolved solids (electrical conductivity) is not sufficient proof of occurrence of seawater intrusion. Seawater intrusion involves mixing of saline and freshwater. Owing to its considerable salt content, a small fraction of seawater would dominate the chemical composition of groundwater mixture. The most obvious indication of seawater intrusion is an increase in Cl concentration. Another important indicator related to the predominance of Cl in seawater is the ratio of chloride to bicarbonate and carbonate ions which is known as Simpson ratio (ElMoujabber et al, 2006) and carbonate ions are present only in very small amount in seawater. The Simpson classification ratio of $Cl^-/(HCO_3^-+CO_3^{2-})$

included five classes: 0.5 for good water quality, 1.3 for slightly contaminated water, 2.8 moderately contaminated, 6.6 injuriously contaminated and 15.5 highly contaminated water (Todd, 1980). In order to diagnose the seawater intrusion as evidenced by temporary increases of total dissolved solids, Simpson criteria used to understand the seawater intrusion. Groundwater samples from the infiltration well of SHAR near Holy Cross School of Sullurpet town which supplies the drinking water to the housing colonies was analyzed before construction of rainwater harvesting systems and found the ratio of chlorides and sum of carbonate plus bicarbonate is of the order of 10.47 and after construction the ratio is around 1.16 (Table 1).

Reconstruction of Groyne 10 Prevent Back Waters of Pulicat Lake

The reconstruction of groyne and construction of sub-surface dams were completed in the month of July 2005. The problem of seawater intrusion at estuary was not contemplated during original groyne construction even though there is 3.5 m depth of sand bed over a clay bed. This has resulted in the fresh groundwater becoming saline in the infiltration wells constructed for drinking water. The river bottom level (RBL) is high and width of the river is 140 m which is a feasible site for the reconstruction of the groyne to arrest the surface saline water from the Pulicat lake and longitudinal section is shown in Fig.8. The river bed level (RBL) on upstream side of groyne is 1.3 m above Pulicate bed level. So drilling has been taken up across the river and at a depth of 2.6 m clay bed is encountered in all the bores drilled on down stream of groyne (i.e at the bed level of Pulicat Lake). Therefore 3.8 m long wooden pile is selected to drive to the alluvium. The laterite rough stone groyne is reconstructed whose height is kept 1.2 m above

Table 12.3 Water Analysis in the Sriharikota Space Center (SHAR) Infiltration Drinking Water well in the Kalangi River near Sullurpet Town

Water sample analyzed	Chloride (me/l)	Bicarbonate (me/l)	Simpson ratio $Cl^-/ (HCO_3^- + CO_3^{2-})$	Remark
Before construction (April 2005)	33.08	3.16	10.47	Injuriously contaminated
During construction (21st June 2005)	30.45	3.77	8.07	Injuriously contaminated
After construction (27th July 2005)	11.5	6.03	1.90	Slightly contaminated
After construction (4th August 2005)	8.46	7.29	1.16	Good quality

Water level of Pulicat lake and timber piles of size 0.3 m x 0.15 m x 3.8 m with shoe at bottom and tongue and groove arrangement on both side of pile is made (Fig. 8). Bentonite clay slurry is poured into hole during augur drilling to make the sides of hole to retain. When the clay bed is encountered the pile is placed in position duly inserting the same in the tongue & groove arrangement made on side of previous pile and hammered with a over head mechanism till it intrudes into clay bed and the pile is kept 1.2 m above Pulicat lake bed level. To avoid cattle/vehicle movements (which will damage the structure) on the reconstructed groyne, small guard stones were placed at the top of this groyne. This procedure of drilling holes with augur and using bentonite clay slurry to retain the sides and also to seal the gap between piles has resulted in stopping the entry of seawater into groundwater. A change is seen on upstream of groyne to a stretch of 10.2 km by gradual decrease of standing saltwater in the river bed. Due to this measure the quality of drinking water has improved and also 50 mcft of irrigation water impounded from the Sullurpet town to Vatambedu groyne along the Kalangi River. There is a possibility of 1.85 m column of impounding river water on the upstream side of groyne in the subsequent monsoon. But due to cyclonic storms there was high runoff in the river in July 2005 which considerably reduced the salinity in the wells (Table 1). The contamination of fresh surface water by the tidal saltwater inflow and the seawater intrusion into groundwater is stopped by the reconstruction of laterite rough stone groyne by providing a wall of timber wooden piles.

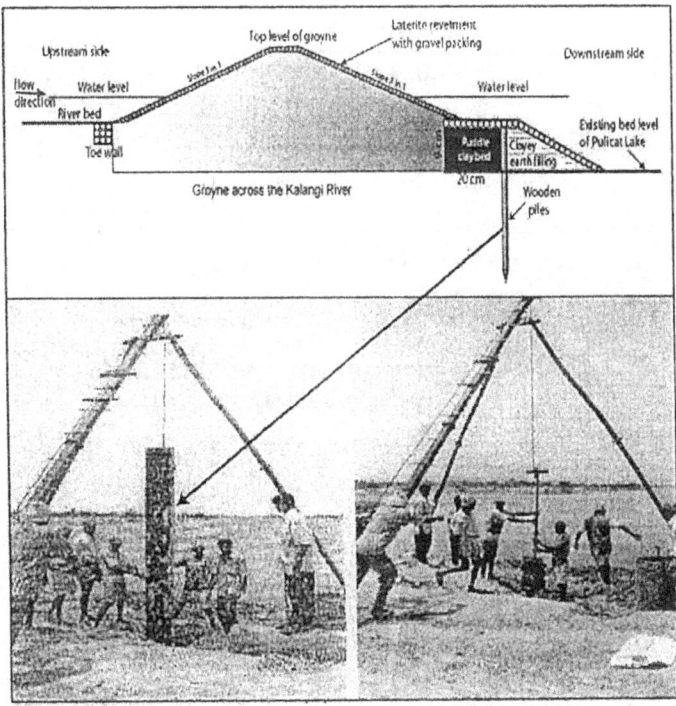

Figure 12.39 Strengthening of the Old groyne by Placing Wooden Piles in the Down Stream Side of the Kalangi River to Protect Saltwater Intrusion

Design and Construction of Sub-surface Dams

The location requirements for the construction of underground dams are (i) the distribution of geological layers (reservoir layers) must allow for effective porosity and hydraulic conductivity so groundwater can be stored and collected, (ii) there must be sufficient groundwater recharge to correspond to the amount of water to be stored. Sub-surface dams are often built in riverbeds that generally constitute highly permeable aquifers with good storage potential (Raju et al. 2006). A subsurface dam, intended for arresting the groundwater flow in a natural aquifer, is constructed a cross a valley by digging a trench to bedrock. The trench has at its base an impervious wall which is covered with excavated material until the trench is completely concealed (Raju et al. 2010).

The main water shortage period in the Kalangi river basin is from March to August when there is no surface water flow in the river. The well inventory studies reveals that there is a base flow at 0.5 m below river bed level (RBL) during April and gradually decreasing to 6.0 m below bed level by the month of June (Muniratnam 2004). All the infiltration wells for drinking water supply to Sullurpet and SHAR housing colonies are tapping water from 1 to 3.5 m below river bed level (RBL). In order to circumvent this situation and to improve the quantity and also quality of drinking water in the Kalangi river basin, three sub-surface dams (GK Engg College, Challarnmagudi and Holy Cross School) constructed to harvest the rainwater runoff going as a waste to sea in order to recharge the infiltration wells and tube wells in the river alluvium (Fig. 2).

The first suitable site for the subsurface dam at Holy Cross School in Sullurpet town with a length of 130 m is selected across Kalangi river. The distance of the first sub-surface dam site is around 11.6 km from the confluence of Kalangi river with Pulicat lake. Geophysical surveys were carried out to know the depth of impervious strata and structural weak planes underlying the sandy alluvium whose depth is 7 m. To substantiate the geophysical survey results few augur hand boreholes were drilled along the proposed site (Fig. 5).

Table 12.4   Engineering: properties of the soil sample used for construction of Clay wall

S.No	Property of soil	Value
1.	Grdvcl (%)	0
2.	Sand (%)	20
3.	Silt + Clay (%)	80
4.	Liquid limit (%)	41
5.	Plastic limit (%)	20

6.	Plasticity Index (%)	21
7.	I.S. Classification	CI*
8.	Maximum Dry Density MDD (gm/cc) (from Standard Proctor s Test)	1.632
9.	Optimum Moisture Content, OMC % (from Standard Proctor s Test)	16
10.	Coefficient of Permeability, K (cm/see) (al Proctor's Density)	1.75x10*
11.	Cohesion, C (Kg/cm^2) (at Proctors Density)	0.58
12.	Angle of Internal Friction, O (dcg.) (at Proctor s Density)	8

*CI-Inorganic clay of medium plasticity

The subsurface dam is constructed with clay quarried from an irrigation tank located at 5 km away and engineering properties of soil samples are presented in Table 2. The width of clay wall of subsurface dam is 0.9 m (Fig. 9). The dam has been keyed into weathered bedrock 0.5 m below the bedrock/hard clay surface, the clay wall was compacted with wooden rammers in order to obtain proctors density. The clay balls were so prepared as to imbibe optimum moisture content. In order to retain this moisture content for a longer period, 200 microns low-density polyethylene films (LDPE) was emplaced on the upstream, downstream and top of the dam in order to protect the clay wall from yielding cracks (Fig. 9). The longevity of LDPE film underground is tested as 50 years. Bed pitching on top of sub-surface dam is made to avoid scouring as the dam is 0.66 m above river bed level (RBL). Revetment is also provided on both sides of river banks to safe guard the bank from scouring. The storing of fresh water will dilute the seawater intruded before reconstruction of groyne. The river Kalangi has received water from its upper catchment and water has reached up to the Holy Cross School subsurface darn site in July 2005 and water has been impounded, since 0.66 m raise of wall (act as check dam) is made with bed pitching and revetment duly grouting the joints with cement mortar (Fig. 10). One jack well to a depth of 6 m and 1.5 m diameter with two collector wells on right side of river was erected during the construction of the Holy Cross School sub-surface dam (Figs. 9&10) for pumping drinking water from the already available infiltration well which is located at 46 m upstream of the jack well. The groundwater sample analysis of SHAR infiltration well in the month of July 2005 revealed that the intensity of salinity has been considerably reduced and is within the limits of drinking water standards (Table I). The available drinking water supply for Sullurpet town plus SHAR housing colonies before construction of sub-surface dam is around 14.40 lakh/lit/day (monsoon) and 8 lakh/lit/day (summer) for the population of around 40,000. After the construction of SS dam near Holy Cross School, it is 24.96 lakh/litre/day during monsoon and 16.60 lakh/lit/day day during summer season. The storage is calculated within the width of river. The groundwater basin also extends on both sides of the river.

Figure 12.40   Construction of Subsurface Dam with Clay Across the Kalangi River near Holy Cross School, Sullurpet Town

CONCLUSION

Rainwater harvesting and artificial recharge will help to improve the quality and quantity of groundwater of the overexploited/contaminated area. Developing countries are under pressure due to rapid increase in population, fast urbanization and deficient water services reflecting on improper management of water resources. The ingress of seawater into the coastal aquifer has affected the drinking water availability and agricultural economy in the Kalangi river basin. The SHAR infiltration drinking well depth is extended up to 4.3 m in the river bed from the surface near the

Holy Cross School in Sullurpet town due to rapid decline of water table in the river. Mean time sand mining has taken place in between damaged groyne and Sullurpet town when water was not available in the entire stretch (i.e upstream of groyne). This has resulted in the intrusion of surface seawater which has contaminated the infiltration well of SHAR situated near Holy Cross School in the Kalangi river bed of Sullurpet town.

Figure 12.41   Construction of Jack well upstream side of SS dam (A); Check Dam above the SS dam (B); Impounding of River Water (C)

A sub-surface dam is a facility that stores groundwater in the pores of strata and uses groundwater in a sustainable manner. The length of the influencing zone after the construction of three sub-surface dams along the river course is 2200 at GK Engineering College, 5940 m at Challammagudi and 3700 m at Holy Cross School. The quantity of groundwater availability has been substantially increased in the Sullurpet area by the construction of sub-surface dams. Due to the reconstruction of groyne seawater intrusion into the river course has been stopped and improved the quality of drinking water, Hence, a general increase not only in agricultural land productivity but also improvement of groundwater quality by these types of constructions in the Kalangi river basin, Sullurpet area.

The construction cost of sub-surface dams are around 8.7 lakh rupees, 16.5 lakh rupees and 27 lakh rupees at G.K Engineering College, Challammagudi and Holy Cross School, respectively. This study indicates that consideration for the construction of groyne and sub-surface dams in the Kalangi river basin is feasible and economical by using cheap materials like clay and low density polyethylene film. Watershed management of this type has enhanced base flow and the storage capacity at sub-surface dams and ultimately facilitated the improvement of domestic water supply and agricultural activities in the ephemeral river basin. Involvement of the community is necessary in water management by laying greater emphasis on extending and improving local water harvesting systems for the sustainable development of any region.

ACKNOWLEDGEMENTS

The first author is thankful to the Alexander von Humboldt Foundation, Germany for providing financial assistance under renewed research stay at Martin Luther University, Halle (Saale).

AN AVERAGE INDIAN VILLAGE REQUIRES JUST 1.14 HECTORS OF LAND TO MEET ITS DRINKING WATER NEEDS.

REFERENCES

ABDELKHALEQ, R.A. and AHMED.I.A. (2007) Rainwater harvesting in ancient civilization in Jordan. Water Sci. Technology. v.7, pp.85-93.

ALOONI, K., KALRO, A.H. and KAMALAMMA, A.G. (2008) Community initiatives in building and managing temporary check-dams across seasonal streams for water harvesting in south India. Agric. Water Mgmt., v.95, pp.1314-1322.

CGWS (1994) Manual on artificial recharge of groundwater, Central Groundwater Board.

DILLION, P. (2005) Future management of aquifer recharge. Hydrogeol, Joul. v.13. pp.313-316.

EL MOUJABBER, M., BOU SAMRA, B., DARWISH, T and ATALLAH, T (2006) Comparison of different indicators for groundwater contamination by seawater intrusion on the Lebanese coast. Water Resour, Managmt., v.20, pp.161-180.

GUNNELL, Y., ANUPAMA, K. and SULTAN, B. (2007) Response of the south Indian runoff-harvesting civilization to northeast monsoon rainfall variability during the last 2000 years: instrumental records and indirect evidence. Holocene, v.17(2), pp.207-215.

HANSON, G. and NILSSON, . A. (1986) Groundwater dams for rural water supplies in developing countries. Groundwater, v.24(4), pp.497-506.

IBRAHIM, M.B. (2009) Rainwater harvesting for urban areas: a success story from Gadari f city in central Sudan. Water Resource, Managmt. v.23, pp.2727-2736.

JOSHI, N.K. (2002) Impact assessment of small water harvesting structures in the Rupard River basin. Institute of Development Studies, Jaipur, India 32 p.

MOYCE, W., MANGEYA, P., OWEN, R. and LOVE, D. (2006) Alluvial aquifers in the Mzingwane Catchment: their distribution, properties, current usage and potential expansion. Physics and Chemistry of the Earth, v.31, pp.988-994.

MUNIRATHNAM, P. (2004) Construction of subsurface dams at three locations across the Kalangi River near Sullurpet town to augment drinking water supply, Technical Report 65p.

MURAD, A.A., AL NUAIMI, H. and AL HAMMADI, M. (2007) Compre (Received: II August 2012; Revised for hensive assessment of water resources in the United Arab Emirates (UEA). Water Resour, Managmt., v.21, pp.144-146.

NAWA, N. and MIYAZAKI, K. (2009) The analysis of saltwater intrusion through Komesu underground dam and water quality management for salinity. Paddy Water Environment, v.7(7), pp.71-82.

OGB (Okinawa General Bureau) (1986) Subsurface dam: A new technology for water resource development, pp.1-29.

OWEN, R. and DAHLIN, T. (2005) Alluvial aquifers at geological boundaries: geophysical investigations and groundwater resources. In: E. Bocanegra, M. Hernandez and E. Usunoff (Eds.), Groundwater and Human Development, AA Balkeura Publishers, Rotterdam, pp.233-246.

PETALAS, C., PISINARAS, Y., GEMTIZI, A., TSIHRINTZIS, V.A and OUZOUNIS, K. (2009) Current conditions of saltwater intrusion in the coastal Rhodope aqufier system, north-eastern Greece. Desalination. v.237, pp.22-41.

RAJU, N.J., REDDY, TY.K. and MUNIRATNAM, P. (2006) Subsurface dam to harvest rainwater - a case study of the Swarnamukhi River basin, South India. Hydrogcol. Jour., v.14. pp.526-53I.

RAW, N.J., KRISHNA REDDY TV. and MUNIRATN!\M. P. (2010). Rainwater catchment structures to rejuvenate seasonal river course for augmentation of groundwater in alluvial aquifer in semi-arid region of Chittoor district, Andhra Pradesh, India. Proc. Workshop on "Climate change and its impacts on water resources-adaptation issues, n.o.l, pp.39-52.

REVELLE. R. (1941) Criteria for recognition of seawater in ground-water. Trans. Amer, Geophys, Union, v.22, pp.593-597.

SHARMA, A. (2002) Does water harvesting help in water scarce regions: a case study of two villages in Alwar, Rajasthan. IWMI-Tata Water Policy Research Program, Gujarat, 24p.

STIEFEL, J.M., MELESSE, A.M., MCCLAIN. M.E., PRICE, R.M. ANDERSON, E.P. and CHAUHAN, N.K. (2009) Effects of rainwater harvesting induced artificial recharge on the groundwater of wells in Rajasthan, India. Hydrogeol, Jour., v.l7, pp.2061-2073.

TODD, O.K. (1980) Groundwater Hydrology. New York, United States, 2nd edn., John Wiley and Sons, Inc.

Briefing paper for members of parliament and state legislatures
An occasional paper from the centre for science and environment

DROUGHT

Try Capturing the Rain

By Anil Agarwal
It is possible to banish drought completely – and in ten years maximum if the government puts its mind to it

DROUGHT?
TRY CAPTURING THE RAIN
By **Anil Agarwal**
It is possible to banish drought completely — and in ten years maximum if the government puts its mind to it.

Dear Members of Parliament and State Legislatures,

Several parts of the country are witnessing a serious drought. The result is acute drinking water shortages in Gujarat, Rajasthan, western Madhya Pradesh, Orissa and Andhra Pradesh. According to newspaper reports, several towns in Saurashtra are getting an extremely irregular supply of water (see table 1: Drinking water availability in Gujarat towns in April 2000). Most of the dams and reservoirs in the region have dried up. The government is trying to deal with the problem by providing water through tankers and by deepening existing borewells. The Central government has promised to run water trains. But is all this inevitable? That we have to do this every time the rainfall is less than normal?

We don't think so and we wish to brief you on what can be done to solve this problem almost permanently unless, of course, there is a series of consecutive droughts for several years. The change we are advocating will come, we are afraid to say, only if our political leaders are prepared to promote a new approach to water management in this country.

During one of the meetings of the World Water Commission, which submitted its report in the Hague in April 2000 to a bevy of water ministers, a member had strongly emphasized the need for educating politicians about the importance of water. I, however, found that argument incorrect because I have rarely met a politician, especially in India, who does not emphasize the importance of water. But what I have indeed found is that hardly any of them know how to solve the water problem.

Not surprisingly, few government initiatives have been able to deliver the goods as the current unprecedented drought has shown in such starkness. The portents of this so-called once-in-a-century drought had become clear as far back as September 1999 when national elections were being held and Shri L.K. Advani had to face slogans like 'Pehle Paani Phir Advani' in Saurashtra. As far back as December, many villagers in north Gujarat had begun to leave home because of water shortage. Once the monsoon season was over, the government could not have done anything to solve the water problem except to provide some succour in terms of drought relief works, emergency water supply through tankers, and digging deeper borewells for some residual water in the bowels of the earth.

But this is not enough. Many will term what is happening in Gujarat and Rajasthan a 'natural disaster'. This is really far from the truth. It is truly a 'human-made' or rather a 'government- made' disaster. Over the last one hundred years or so, the world and India, too, have seen two major shifts in water management. One is that individuals and communities have steadily given over their role almost completely to the state even though more than 150 years ago no government anywhere in the world provided water. The second is that the simple technology of using rainwater has declined and in its place exploitation of rivers and groundwater through dams and tubewells has become the key source of water. As water in rivers and aquifers is only a small portion of the total rainwater, there is an inevitable and growing and, in many cases, unbearable stress on water from rivers and groundwater.

Table 12.5 Drinking Water Availability in Gujarat towns in April 2000

Place	Availability of Drinking Water
Rajkot (1)	30 minutes every alternate day
Jamnagar, Jasdan and Amreli (1)	20 minutes once in three days
Jodiya town, Jamnagar district (2)	20 minutes in 12 days
Dhrol town, Jamnagar district (2)	Half the population gets water once in eight days

Sources:

1. Janyala Sreenivas 2000, Forget the Sensex for a second, look what else is going down, Indian Express, New Delhi, April 19, p.1

2. Janyala Sreenivas 2000, once a fortnight, they get a few drops and that too for 20 minutes, Indian Express, New Delhi, April 21, p.1

Dependence on the state has also meant that costs of water supply are high; with cost recovery being poor the financial sustainability of water schemes has run aground; and, repairs and maintenance is abysmal. With people having no interest in using water carefully, the sustainability of water resources has itself become a question mark-problems we see across the board today. As a result, there are serious problems with government drinking water supply schemes. Despite all the government efforts, the number of 'problem villages' does not seem to go down. As N.C.Saxena, former rural development secretary puts it, "In our mathematics, 200,000 problem villages minus 200,000 problem villages is still 200,000 problem villages" **(see graph 1: UNENDING EXERCISE: Record of government rural drinking water supply schemes).**

Given the fact that India is one of the most well-endowed nations in the world in terms of aver- age annual rainfall, there is no reason why it should suffer from drought. This year or any other year. The most important lesson that our decision - makers should learn from the current crisis is how to drought-proof the nation in the years to come — a task that can easily be accomplished in less than a decade if the country puts its mind to it.

The government has indeed invested heavily on water resources development. But these programmes have focussed mainly on:

(a) Large-scale irrigation development for increasing Green Revolution-style agricultural production; and,

(b) Drinking water supply programmes.

Yet a large part of the country remains drought- prone. This is because no specific effort has been made to drought-proof rainfed areas which suffer from high rainfall variability from year to year and season to season. Moreover, the government has encouraged a massive intervention into the country's hydrological cycle but it has done precious little to sustain the integrity of the hydrological system. A fine example of the disintegration of the country's hydrological system is what we are doing to our groundwater. The country has been constantly encouraging exploitation of groundwater but has done little to recharge it. As a result, groundwater tables are falling all over the country. Considering the fact that over 90 per cent of rural Indians depend on groundwater to get their drinking water, the decline poses a serious problem which becomes an emergency in a year when the rains are low as in this year. The poor, of course, who depend on dug wells, which dry off first, as compared to tube wells or bore wells, are the first to suffer.

Graph 1 : UNENDING EXERCISE: Record of government rural drinking water supply schemes

The graph shows that even though a large number of villages are covered between two surveys the number of problem villages keeps growing. For instance, in 1980, there should not have been more than 56,000 problem villages, but there were 231,000. Obviously, the money pumped in and the methods used were unsustainable. As N C Saxena, former rural development secretary, puts it, "In our mathematics, 200,000 problem villages minus 200,000 problem villages is still 200,000 problem villages."

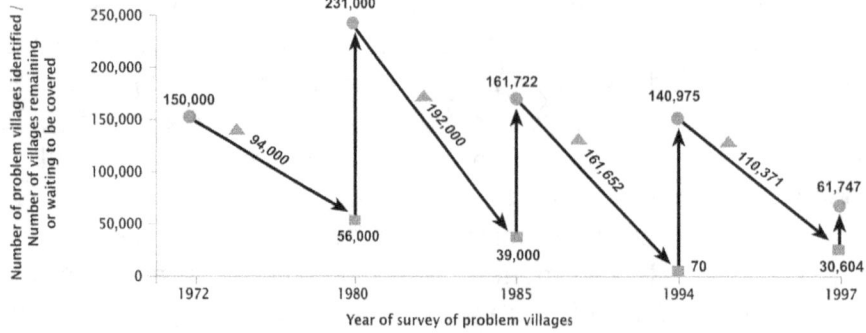

Figure 12.42 Number of Problem Villages

Number of Problem villages identified

Number of villages waiting to be covered

Number of villages covered during the specified survey period

SEVERAL VILLAGE COMMUNITIES HAVE SHOWN THEIR COMMITMENT POTENTIAL AND PERSEVERANCE IN DROUGHT PROOFING THEIR OWN VILLAGES EVEN DURING THE PRESENT CRISIS.

Year of Survey	Number of Problem Villages Identified	Number of Villages Covered till the Next Survey	Number of Villages not Covered before the Next Survey
1972	150,000	94,000	56,000
1980	231,000	192,000	39,000
1985	161,722	161,652	70
1994	140,975	110,371	30,604
1997	61,747		

Source: N C Saxena, Ministry of Rural Development, Government of India

1. **Communities and rain**

   Community-based rainwater harvesting — the paradigm of the past — has in it as much strength today as it ever did before. A survey conducted by the Centre for Science and Environment (CSE) of several villages facing drought in Gujarat and western Madhya Pradesh last December — Ghelhar Choti in Jhabua district, Thunthi-Kankasiya in Dahod district, Raj-Samadhiyala and Mandlikpur in Rajkot district, and Gandhigram in Kachchh district — found that all those villages which had undertaken rainwater harvesting and/or watershed development in earlier years had no drinking water problem whatsoever and even had some water to irrigate their crops. On the other hand, neighbouring villages were desperate for water and planning to migrate when the real summer hit them. This survey revealed that rainwater harvesting can meet even the acid test of a bad drought.

   In late March, we got further confirmation of our conviction. Going with president K R Narayanan in a helicopter to the Arvari watershed who was scheduled to give the Down to Earth-Joseph C John Award to village Bhaonta-Kolyala in late March, we could see nothing but barren fields all the way from Delhi to Alwar. This area is suffering from a second consecutive drought- year. But suddenly we came across green and golden fields and realised that we had reached the oasis of the Arvari watershed where several villages have over the last 5-10 years built hundreds of rainwater harvesting structures. Nobody needed to emphasize the importance of rainwater harvesting any more. The president saw a more or less dead Arvari river, unable to with- stand the burden of two years' drought, but wells were still full of water and, therefore, fields were rich and productive and villagers reasonably happy.

2. **Getting priorities right: Potential of rainwater harvesting**

What makes rainwater harvesting such a powerful technology? Just the simple richness of rainwater availability that few of us realise because of the speed with which water, the world's most fluid substance, disappears. Imagine you had a hectare of land in Barmer, one of India's driest places, and you received 100 mm of water in the year, common even for this area. That means that you received as much as one million litres of water — enough to meet drinking and cooking water needs of 182 people at a liberal 15 litres per day. Even if you are not able to capture all that water — this would depend on the nature of rainfall events and type of runoff surface, among other factors — you could still, even with rudimentary technology, capture at least half a million litres a year.

It is, in fact, only with this rudimentary technology that people came to inhabit the Thar desert and have made it the most densely populated desert in the world. And assuming you could capture the 2000 mm annual rainfall that is common in eastern India, you would need only 500 square metre of land (a 21 metre by 21 metre plot) to capture one million litres. It is also interesting to note that rural population density follows intensity of annual rainfall. Barmer, for instance, has less rainfall but few people and a lot of land available per person whereas 24-Parganas in West Bengal has much more rain but more people and less land available per person.

Even in the villages suffering from drought this year, it is not as if there was no rain. In many areas like Andhra Pradesh, Orissa and western Madhya Pradesh the rainfall levels, though lower than normal, were still more than 500 mm, which is an enormous amount of rainfall. The average annual rainfall in Saurashtra and Kachchh, the worst affected, is 578 mm. This year, according to newspaper reports, it was very depressed but still around a couple of hundred millimetres (mm). But the people of Saurashtra and Kachchh let the water go. It does not matter how much rain you get, if **you don't capture it you can still be short of water. It is unbelievable but it is true that Cherrapunji which gets 11,000 mm annual rainfall, still suffers from serious drinking water shortages.**

In fact, we have consistently argued that there is no village in India that cannot meet its basic drinking and cooking needs through rainwater harvesting. Figures speak for themselves. India's average annual rainfall is 1170 mm. It varies from 100 mm in the deserts of Western India to 15,000 mm in the high rainfall hills of the Northeast. Nearly 12 per cent of the country receives an average rainfall of less than 610 mm per annum while 8 per cent receives more than 2500 mm. But more than 50 per cent of this rain falls in about 15 days and less than 100 hours out of a total of 8760 hours in a year. The total number of rainy days can range from a low of five days in a year in the desert regions of Gujarat and Rajasthan — though on some of these days there can be high-intensity rainstorms — to 150 days in the Northeast. Therefore, it is very important to capture this rainwater which just comes and goes in a few hours[1].

Recognising this fact that almost all the rain comes down in a few years, our ancestors had learnt to harvest water in a variety of ways:

(a)   They harvested the rain drop directly. From rooftops, they collected water and stored it in tankas built in their courtyards. From open community lands, they collected the rain and stored it in artificial wells called kundis.

(b)   They harvested monsoon runoff by capturing water from swollen streams during the monsoon season and stored it in zings in Ladakh, ahars in Bihar, johads in Rajasthan and eris in Tamil Nadu, to name a few.

(c)   They harvested water from flooded rivers in places like north Bihar and West Bengal.

In 1991, India had 587,226 inhabited villages with a total population of 629 million giving us an average population of 1071 persons per village, up from 942 persons in 1981. Let us, therefore, assume that the average population of an Indian village today is about 1200. India's average annual rainfall is about 1170 mm. If even only half this water can be captured, though with technology inputs this can be greatly increased, an average Indian village needs 1.12 hectares of land to capture 6.57 million litres of water it will use in a year for cooking and drinking. If there is a drought and rainfall levels dip to half the normal, the land required would rise to a mere 2.24 hectares. The amount of land needed to meet the drinking water needs of an average village will vary from 0.10 hectares in Arunachal Pradesh (average population 236) where villages are small and rainfall high to 8.46 hectares in Delhi where villages are big (average population 4769) and rainfall is low. In Rajasthan, the land required will vary from 1.68-3.64 hectares in different meteorological regions and, in Gujarat, it will vary from 1.72-3.30 hectares (see table 2: EVERY VILLAGE IN INDIA CAN MEET ITS OWN WATER NEEDS: Land area needed per village in different states of India to capture enough rainwater to meet drinking and cooking water needs). And, of course, any more water the villagers catch can go for irrigation.

Does this sound like an impossible task? Is there any village that does not have this land availability? India's total land area is over 300 million hectares. Let us assume that India's 587,000 villages can harvest the runoff from 200 million hectares of land, excluding inaccessible forest areas, high mountains and other uninhabited terrains, that still gives every village on average access to 340 hectares or a rainfall endowment of 3.75 billion litres of water. These calculations show the potential of rainwater harvesting is enormous and undeniable. There is just no reason whatsoever for thirst in India.

Therefore, it is possible to drought proof the entire country. Not just drinking water, most of India's agricultural fields should also be able to get some irrigation water to grow less water-intensive crops every year through rainwater harvesting. The strategy for drought proofing would be to ensure that every village captures all the runoff resulting from the rain falling over its entire land and the associated government revenue and forest lands, especially during years when the rain was nor- mal, and store it in tanks or ponds or use it to recharge the depleting groundwater. It would then have enough water in its tanks or in its wells to cultivate substantial lands with water-saving crops like millets and maize.

Even with a national water grid or all the proposed dams like those on the Narmada being built, there just isn't going to be enough river water to provide irrigation to every single village of India. A substantial part of India's agricultural lands will remain rainfed. This is confirmed by official statistics. The National Commission for Integrated Water Resources Development Plan has estimated that the ultimate irrigation potential is as much as 140 million hectares. Some 75.9 million hectares will be irrigated by surface water schemes — of which 58.47 million hectares will be irrigated by major and medium projects and 17.38 million hectares will have to be irrigated by minor irrigation schemes like tanks[2] — while 64.1 million hectares will be irrigated by groundwater[3]. But all this is still a dream. As of 1992-93, 119.3 million hectares were rainfed, of which 78.43 million hectares were under foodgrains[4].

But there is enough rain which with a good combination of rainwater harvesting and ground- water recharge, can increase and stabilise the productivity of this rainfed land. These lands today support some of the poorest people in the country. And if we don't do this, people living in rainfed areas will not prosper.

Table 2 EVERY VILLAGE IN INDIA CAN MEET ITS OWN WATER NEEDS: Land area needed per village in different states of India to capture enough rainwater to meet drinking and cooking water needs (in hectares)

State	Meteorological Divisions	Average Annual Rainfall (millimetres)	Estimated no. of Villagers per Village, 2001	Land Area needed per Village to Meet Drinking and Cooking Water needs Assuming Half of Normal Rainfall is Captured (hectares)	Land Area needed per Village to Meet Drinking and Cooking Water needs Assuming Severe Drought Conditions, that is, 50 per cent decline in Normal Rainfall (hectares)
INDIA	—	1,170	1220	1.14	2.28
Andaman and Nicobar Islands	Andaman and Nicobar Islands	2,967	408	0.16	0.32
Arunachal Pradesh	Arunachal Pradesh	2782	236	0.10	0.20
Assam	Assam and Meghalaya	2818	807*	0.32	0.64
Meghalaya	Assam and Meghalaya	2818	311	0.12	0.24
Nagaland	Nagaland, Manipur, Mizoram and Tripura	1881	1153	0.68	1.36
Manipur	Nagaland, Manipur, Mizoram and Tripura	1881	726	0.42	0.84
Mizoram	Nagaland, Manipur, Mizoram and Tripura	1881	549	0.32	0.64

Tripura	Nagaland, Manipur, Mizoram and Tripura	1881	3496	2.04	4.08
West Bengal	1. Sub-Himalayan West Bengal and Sikkim	2739	1602	1.10-1.22*	2.20-2.44*
	2. Gangetic West Bengal	1439			
Sikkim	Sub-Himalayan West Bengal and Sikkim	2739	1132	0.46	0.92
Orissa	Orissa	1489	683	1.10	2.20
Bihar	1. Bihar Plateau	1326	1367	1.12-1.26*	2.24-2.52*
	2. Bihar Plains	1186			
Uttar Pradesh	1. Uttar Pradesh	1025	1026	0.68-1.26*	1.36-2.52*
	2. Plain of West Uttar Pradesh	896			
	3. Hills of West Uttar Pradesh	1667			
Haryana	Haryana, Chandigarh and Delhi	617	2.258	4.00	8.00
Delhi	Haryana, Chandigarh and Delhi	617	4769*	8.46	16.92

Chandigarh	Haryana, Chandigarh and Delhi	617	2647*	4.70	9.40
Punjab	Punjab	649	1345	2.26	4.52
Himachal Pradesh	Himachal Pradesh	1251	328	0.28	0.56
Jammu Kashmir	Jammu and Kashmir	1011	1140	1.24	2.48
Rajasthan	1. West Rajasthan	313	1039	1.68-3.64*	3.36-7.28*
	2. East Rajasthan	675			
Madhya Pradesh	1. Madhya Pradesh	1017	867	0.70-0.94*	1.40-1.88*
	2. East Madhya Pradesh	1338			
Gujarat	1. Gujarat Region	1107	1741	1.72-3.30*	3.44-6.60*
	2. Saurashtra and Kachchh	578			
Goa	Konkan and Goa	3005	1816*	0.66	1.32
Maharashtra	1. Konkan and Goa	3005	1389	0.50-1.72*	1.00-3.44*
	2. Madhya Maharashtra	901			

State	Sub-division				
Andhra Pradesh	3. Marathwada	882			
	4. Vidarba	1034			
	1. Coastal Andhra Pradesh	1094	2231	2.34-3.60*	4.68-7.20*
	2. Telengana	961			
	3. Rayalaseema	680			
Tamilnadu	Tamil Nadu and Pondicherry	998	2627	2.88	5.76
	TamilNadu	998	1106*	1.32	2.64
Pondicherry					
Karnataka	1. Coastal Karnataka	3456	1343	0.42-2.02*	0.84-4.04*
	2. North Interior Karnataka	731			
	3. South Interior Karnataka	1126			
Kerala	Kerala	3055	14083	5.04	10.08
Lakshadweep	Lakshadweep	1515	3.228	2.34	4.68

Notes: *Calculation based on the assumption that average village population in different meteorological sub-divisions is the same as that of the state.

Source: India Meteorological Department for normal rainfall data and projections of average population in 2000 based on Census of India data for 1981 and 1991.

3. **Drought-proofing Vs large-scale irrigation**

But for this to happen, our planners and politicians will have to stop confusing irrigation for drought-proofing with large-scale irrigation for Green Revolution-style agricultural development. Otherwise, the country will get its priorities wrong and tens of millions of poor people will continue to suffer the horrors of drought. Depending on the availability of money and resolution of problems like rehabilitation, both can be attempted but the priority must go to drought-proofing measures which require little money in comparison and will bring results very quickly, within 5-10 years.

Drought-proofing and large scale irrigation development are not a substitute of each other. What one can do, the other cannot. Firstly, because even after all the proposed dams are built to promote large-scale irrigation development and interlinking of rivers takes place not every piece of the country's cultivated land will see the benefit of canal irrigation. These lands will have to depend either on groundwater or local water harvesting. These two will also have to go together because heavy use of groundwater can only be sustained if there are local efforts to keep recharging the groundwater. Therefore, large-scale irrigation development is no substitute for drought-proofing based on local water harvesting systems and sustain- able use of groundwater.

The second argument against large-scale irrigation development follows from the first. Big dams can only help to create pockets of Green Revolutions-style agricultural production (with water-intensive crops) but they cannot drought proof the whole country. As a result they can at best create 'national' food security as they have done up till now — which means that few districts of country generate a huge agricultural surplus which is then used to feed the ones which are doing agriculturally poorly, especially during drought years.

But they cannot create 'local' food security — which means that all areas of the country have water management strategies to ensure that local food production is as productive as possible and stable even during water-short years. Local food security is as important as national food security. Which Bhil or Oraon adivasi, for instance, wants to depend only on grains from Punjab? All of them would like to grow enough grain at least to feed themselves. Water harvesting and groundwater together can definitely drought-proof the country and create local food security which big dams cannot. Then India's poor people and poor lands will not have to suffer the ignominy of the kind they had to suffer this year. But the government has to realise that Indians cannot survive on a single-track water management policy.

In addition, it is important to realise that India's future food security even from its so-called Green Revolution areas will depend heavily on a nation-wide groundwater recharge programme, which can only be taken by individual communities through rainwater harvesting. If this is not done, agriculture will suffer even in current irrigated areas because of the increasing overexploitation of groundwater and lowering of groundwater tables across the country. With more than 17 million tubewells and borewells energised by diesel and electricity, groundwater is now used to irrigate more than half of the country's irrigated area. And as areas irrigated by groundwater show higher productivity than those irrigated by canals, the contribution of groundwater to India's total agricultural output from irrigated areas is much more than that of canals. During drought years, when rivers dry up, groundwater becomes the main source of water both for drinking and irrigation[5]. Indian agriculture and rural life is today heavily dependent on ground- water. Further lowering of groundwater tables can seriously threaten India's hard-earned food security at a time when India will need to produce more food to feed its growing population.

4.   **Small means even more water**

Let us look at the relevance of village-based rainwater harvesting from yet another point of view. The key component of water management is 'storage' especially in a country like India where the monsoon gives us on average about one hundred hours of rain and then nothing for the remaining 8,660 hours in a year.

This water can be captured in:

(a)   large reservoirs with large catchments by building large dams,

(b)   in small tanks and ponds with small catchments, or

(c)   by storing it in a way that it percolates down into the ground and gets stored as groundwater.

In fact, there is strong scientific evidence to show that village-scale rainwater harvesting will yield much more water than big or medium dams, making the latter an extremely cost-ineffective and unscientific way of providing key water needs especially in dry areas. Some very instructive lessons can be learnt from the work of Israeli scientist Michael Evenari who has produced the best corpus of knowledge on this subject from the bone-dry Negev desert where the average annual rainfall is a mere 105 mm.

Evenari was intrigued by the fact that the ancient Israeli civilisation had built towns right in the middle of the Negev desert with their own agriculture and water supply systems — much like the towns of Jodhpur and Jaisalmer that the enterprising Marwaris developed in the Thar desert. Both the Israelis and the Marwaris used the rain they received with great ingenuity to meet their food and water needs. In his effort to reconstruct the ancient farms of the Negev, Evenari came up with a very surprising finding: Water harvested from small watersheds per hectare of watershed area was much more in quantity than that collected over large watersheds. On hindsight this makes eminent sense because water collected over larger watersheds will have to run over a larger area before it is collected and a large part will get lost in small puddles and depressions, as soil moisture and through evaporation.

This loss of water can be stunningly high. While a one ha watershed in the Negev yielded as much as 95 cubic metres of water per hectare per year, a 345 ha water- shed yielded only 24 cum/ha/year. In other words, as much as

75 per cent of the water that could be collected was lost. The loss was even higher during a drought year. After years of research, Evenari summed his findings as follows: "…during drought years with less than 50 mm of rainfall (normal rainfall in the Negev desert is about 105 mm) watersheds larger than 50 ha will not produce any appreciable water yield while small natural watersheds will yield 20-40 cubic metres per hectare and microcatchments (smaller than 0.1 hectare) as much as 80-100 cubic metres per hectare" (see table 3).

Table 3   THE SURPRISING EFFECT OF SIZE: The effect of the size of catchments on the quantity of water harvested as found in the Negev desert (in case of catchments with a 10 per cent slope and a 105 mm rainfall year)

A number of factors determine how much rain falling over a watershed will turn into runoff which can be collected by villagers for their drinking and irrigation needs. This table shows that all other factors like slope remaining the same, the larger the size of the catchment, the less runoff (water) can be collected from it. This is because in large catchments water has to run over larger distances before it gets collected and during that period, a lot of the water is lost in puddles and small depressions, in evaporation or through infiltration into the soil. Therefore, small catchments give the maximum water. And the difference can be quite high. As the table shows, 3000 microcatchments of 0.1 hectare each will give five times more water together than one catchment of 300 hectares even though the total land area from which the rain is harvested remains the same. In simpler words, in drought- prone areas, 10 dams with a one hectare catchment each will give substantially more water than one large dam with a 10 hectare catchment.

S.No.	Size of Catchment (hectares)	Quality of Water Harvested (cubic metres / hectare)	Percentage of Annual Rainfall Collected
1.	Microcatchment (a)	160 cubic metres / hectare	15.21%
2.	20 hectares	100 cubic metres / hectare	9.52%
3.	300 hectares	50 cubic metres / hectare	3.33%

Notes: (a) A microcatchment is a very small catchment of size upto 1000 square metres or 0.1 hectare.

Source: Michael Evenari *et al* 1971, The Negev: The Challenge of a Desert, Oxford University Press, UK.

Table 4 EVEN GREATER EFFECT DURING DROUGHT: The effect of the size of catchments on the quantity of water harvested as found in the Negev desert during drought years with less than 50 millimetres rainfall

The table below shows clearly that in a desert very little water can be collected from large catchments in a drought year.

S.No.	Size of Catchment (hectares)	Quality of Water Harvested (cubic metres / hectare)	Percentage of Annual Rainfall Collected
1.	Microcatchment (a)	80-100 cubic metres / hectare	16-20%
2.	Small natural watersheds	20-40 cubic metres / hectare	4-8%
3.	Larger than 50 hectares	No appreciable water yield	0%

Notes: (a) A microcatchment is a very small catchment of size upto 1000 square metres or 0.1 hectare.

Source: Michael Evenari et al 1971, The Negev: The Challenge of a Desert, Oxford University Press, UK.

INDIA RECEIVES MOST OF ITS RAINFALL IN JUST 100 HOURS OUT OF 8760 HOURS IN A YEAR. IF THIS WATER IS NOT CAPTURED OR STORED, THEIR WILL BE NO WATER FOR THE REST OF THE YEAR.

Effect of Size

The effect of the size of catchments on the quality of water harvested as found in the Negev desert in case of catchments with a 10 percent slope and a 105 mm rainfall year and table 4.

Even greater effect during drought   The effect of the size of catchments on the quantity of water harvested found in the Negev desert during drought years with less than 50 mm of rainfall.

These were critical findings because the amount of rainwater one can collect depends on the amount of land from which the runoff can be harvested. But Evenari was finding that even if you have the same amount of land you will collect more water if you break up the land into many small catchments than if you collect water from it as one catchment. Several studies conducted in India by the Central Soil and Water Conservation Research and Training Institute in Dehradun also show a clear relationship between size of catchment and amount of runoff that can be captured. One study shows that just increasing the size of the catchment from 1 ha to about 2 ha reduces the water collected per hectare by as much as 20 per cent (see graph 2: EFFECTS OF SIZE ON QUANTITY OF WATER HARVESTED: Studies from India). Several other studies conducted by the Central Soil and Water Conservation Research Institute in Agra, Bellary and Kota and another study conducted in the high rainfall region of Shillong, have all found that smaller watersheds give higher amounts of water per hectare of catchment area[6]. **In simple words, all this means that in a drought-prone area where water is scarce, 10 tiny dams with a catchment of 1 ha each will collect much more water than one larger dam with a catchment of 10 ha (see diagram 1: SMALLER CATCHMENTS GIVE MUCH MORE WATER: One dam with a catchment of 10 hectares will collect much less water than 10 dams with one hectare catchement each).**

It should not be surprising that the large number of medium-size dams that have been constructed in Saurashtra stored very little water in this drought year and started going dry by December 1999. But then the answer to drought-proofing of the area lies not in mega-water harvesting projects with medium and large dams. It lies in small water harvesting structures which are constructed at the farm and village-level.

To demonstrate his findings, Evenari even developed an orchard in the middle of the bone-dry Negev desert by creating a separate microcatchment — a plot of land ranging from 15.6 sq m to 1000 sq m — for each tree to maximise the quantity of harvested water. Therefore, look at it any way, community-based, small-scale rainwater harvesting is not just capable of providing more than drinking water needs even in the worst of drought situations but is also the most efficient way to collect water.

5. **Ministers are now talking of rainwater harvesting**

In the last few weeks, since the media storm on drought hit our politicians, several of them, including the prime minister and the ministers for rural

AN AVERAGE INDIAN VILLAGE REQUIRES JUST 1.14 HECTARE OF LAND TO MEET ITS DRINKING WATER NEEDS

urban development and water resources, have made statements regarding the importance of a community-based rainwater harvesting strategy to drought-proof the country.

The first statement came from the rural development minister, Sunderlal Patwa, that the government hopes to gradually replace a government-oriented programme by a people-oriented, decentralised and demand-driven rural water supply programme, the budget for the programme is being increased by 8 per cent, from Rs.1800 crore last year to Rs. 1960 crore in 2000-01, and a fifth of this budget will be given to state governments which undertake community-based programmes so that villagers have a decisive role. A part of the total capital cost and operation and maintenance cost will be borne by the users. Some 58 districts have been identified to try out this scheme[7]. Meanwhile, a steering committee has also been set up to study the recharge of drinking water sources.

The second statement came from the prime minister who told the parliament that only six per cent of the country's rain is being conserved and that greater emphasis will henceforth be placed on water harvesting[9].

The third statement has come from Jagmohan, minister for urban development, who has suggested that water harvesting should be made compulsory under the urban building bye-laws of the country[10].

The fourth statement has come from the minister of water resources, C P Thakur, who in an interview in late April had merely talked about water harvesting in a few buildings in Delhi like the Shram Shakti Bhawan and the Indira Gandhi National Open University and some nine buildings in Jaipur as his key initiatives in this area, but by early April he was announcing a whopping Rs. 550 crore plan to recharge groundwater by constructing about 10,000 water harvesting structures in water-scarcity regions with these funds going directly to village beneficiary groups and water users associations[11,12].

There have been developments in the states, too. Andhra Pradesh's chief minister, N Chandrababu Naidu, has announced and already held the first meeting of a Water Conservation Mission to revamp its existing watershed development programme[13,14]. The chief minister told the Indian Express in an interview, "Till now, we have dumped crores of rupees into the watershed programme. But the results are very disappointing. Unless farmers are involved, the situation cannot be reversed[15]. The Andhra Pradesh government plans to spend Rs.400 crores through the new mission and bring 10 million hectares of land under good watershed management. In the areas chosen, villagers will be given Rs.20 lakh to develop a 500-hectare watershed[16].

SMALL IS ECOLOGICALLY EFFICIENT AS 10 DAMS WITH 1 HECTARE CATCHMENT WILL STORE MORE WATER THAN 1 DAM OF 10 HECTARE

Diagram 1: SMALLER CATCHMENTS GIVE MUCH MORE WATER: One dam with a catchment of 10 hectares will collect much less water than 10 dams with one hectare catchment each

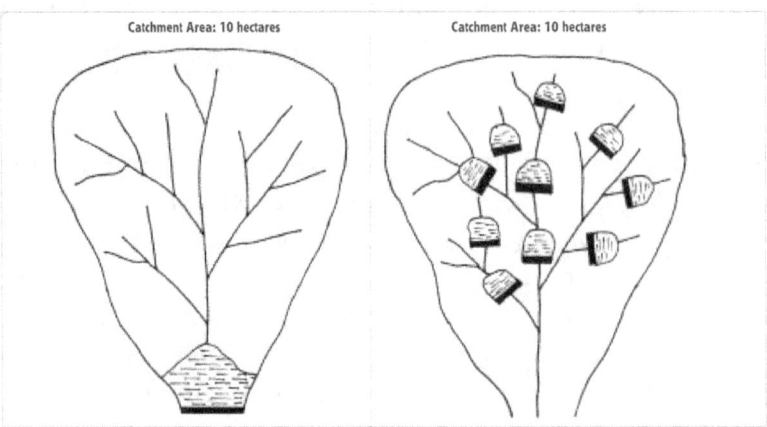

Figure 12.43   Smaller and Bigger Catchments and Annual run of Collected

Graph 2: EFFECT OF SIZE ON QUANTITY OF WATER HARVESTED: Studies from India

Several studies carried out is India show the same results that were found in the Negev desert: Larger catchments resulted in a smaller percentage of the annual rainfall falling on the catchment which resulted in runoff being captured.

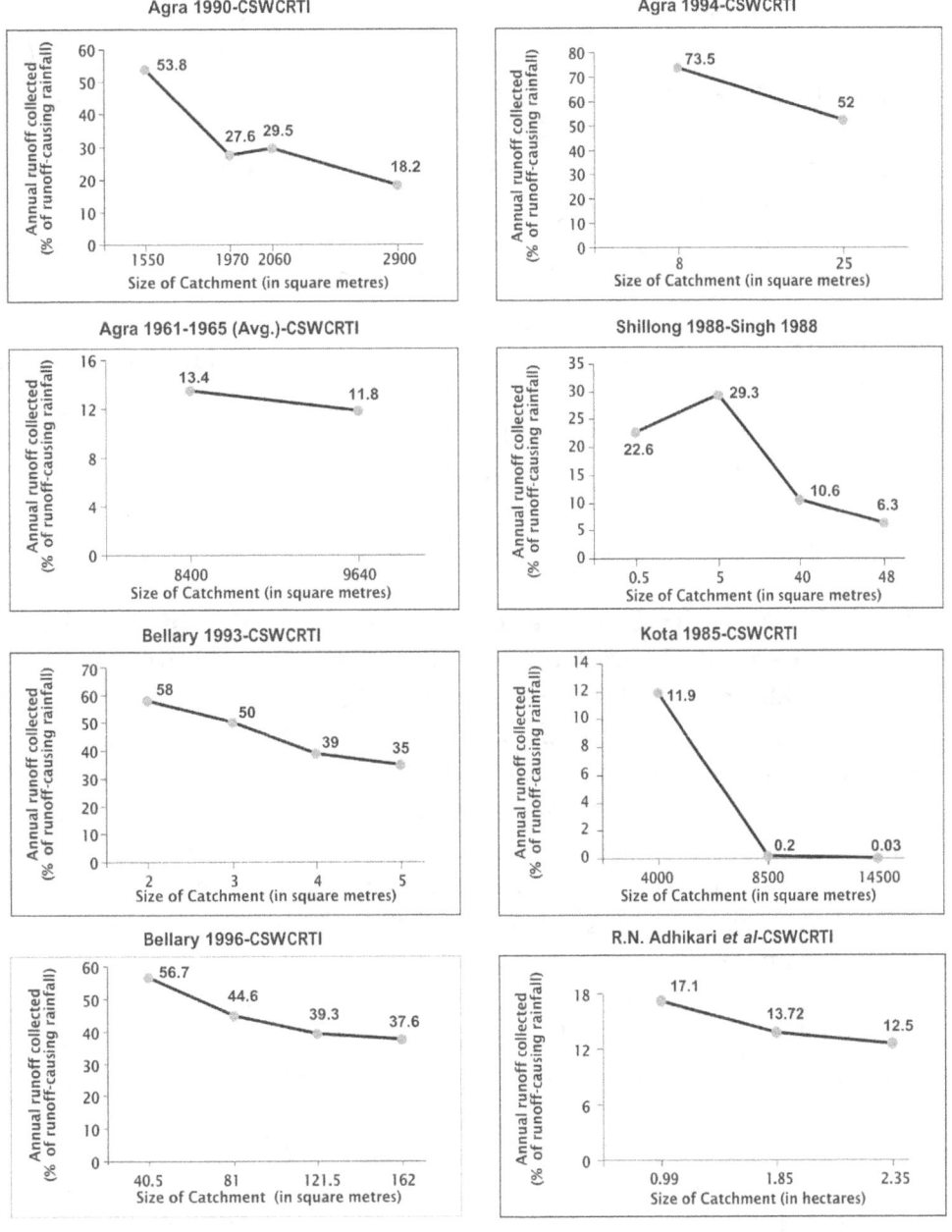

Figure 12.44   Size of Catchment and Annual run off Collected

Source: J S Samra, Director, Central Soil and Water Conservation Research Institute (CSWCRTI), Dehradun, personal communication

In the urban context, the Delhi government has expressed interest in water harvesting. Given the fact that groundwater levels have declined by 4-10 metres in several parts of the Capital over the last decade, the Delhi government is contemplating a law to make it mandatory for all new group housing societies to harvest that rain falling within their complexes[17,18]. Meanwhile, the New Delhi Municipal Corporation has sanctioned a special water harvesting project which consists of water ponds being constructed in four major parks: Talkatora Garden, Nehru Park, Lodi Garden and Kushak Nullah[19].

In Gujarat, irrigation minister Nitin Patel said in late April that the government is planning to construct 10,000 check dams across the state under its Sardar Patel Sahbhagi Jal Sanchay Yojana. Villagers will have to provide 40 per cent of the cost of the check dam. The government has received over 16,000 applications of which 8,563 have been approved and, to begin the exercise, the government has sanctioned Rs.100 crore for the construction of 2,000 check dams[20]. This is a remarkable change as compared to the past. Between 1991 and 1999-2000, various departments of the Gujarat government only built a total of 1,341 check dams[21].

In Madhya Pradesh, a workshop held in Neemuch district in late April emphasized the importance of water conservation through water harvesting and the district collector promised to promote rooftop water harvesting in the town and draw up a plan for restoring and developing ponds in each gram panchayat[22].

If all the commitments made by the Central Minister and the State government of Gujarat and Andhra Pradesh, come true, a sum of Rs.1500-2000 crore has been committed for rainwater harvesting. This is indeed heartening. Equally heartening are reports of people promoting rainwater harvesting. Under the inspiration of the local Ananda Baba Ashram, a Lakhota Jal Sanchay Abhiyan Samiti has been set up to collect money from the residents of Jamnagar town to desilt the huge Lakhota Lake that had been constructed by the former royal family. The people of the town have given their full support to this exercise[23]. Under the leadership of the Swaminarayan gurukul in Rajkot, foundation stones were laid for check dams in more than 15 villages in just five days around the middle of April[24]. In Ahmedabad city itself, the Khadia Itihas Samiti headed by the state's health minister, has completed a survey of the underground water tanks (tanka) that were traditionally built in each house to harvest roof-top rainwater and is planning a campaign to revive them[25] Care Today, an organisation set up the magazine India Today, has announced that it has identified two projects, one each in Rajasthan and Gujarat, to help villagers desilt their traditional water harvesting structures[26]. The Supreme Court, hearing a case on water shortage in Delhi, has ordered the Delhi Development Authority (DDA) to harvest rain water starting off with the water-thirsty DDA colonies of Vasant Kunj in New Delhi[27].

These developments show that community-based rainwater harvesting may well become a widely adopted paradigm in the years to come both in the urban and rural areas. But the question is: Will this all lead to effective results, especially in the rural context?

6. **Structures with a Social Process**

Building water harvesting structures is a very easy task — any contractor endowed with a bit of money can do so. But building an effective structure which starts off a process of self-management in village communities is a much more difficult task. This is possible only if each structure is the result of a cooperative social process-the ability of a community to work in cooperation. Water is a strange natural resource: It can unite a community as easily as it can divide it. Therefore, it is essential that a strong social process precede each structure to build what economists call the 'social capital'. **This is an area where the track- record of government agencies is literally non-existent and inflexible government rules militate against the very principle of social mobilisation.**

In other words, the above statements by state and Central officials are as much a cause of worry as of applause. Fixed annual targets will prove to be a total disaster: money literally going down the drain. At least the first two years of any water harvesting programme will have to be spent on social mobilisation. This will mean, firstly, creating awareness and confidence in the people that water harvesting works. Once this is achieved, it means sitting with the people to create village institutions which will decide where, when and how the water harvesting structures will be built, who will build the structures, how much the villagers will provide to share the cost of the structures, and once the structure is built, how will its benefit, that is, water, will be shared amongst the villagers, especially in the early years when water is scarce, and how will its use be regulated. Every part of the community will have to be involved by making each section — from the landed to the landless and women's groups — appreciate the benefits it will derive from the exercise. And by making efforts to ensure that benefits do indeed flow to each section of the community.

It is for this reason that water harvesting works best when combined with watershed development. It is in the nature of structures to benefit mainly those who have land leaving the landless without any benefits and therefore alienated from the exercise. But development of watersheds to conserve both water and soil not only increases soil and water conservation and leaf and grass production on what are usually common lands, which can greatly benefit landless households, but also increases the life and effectiveness of the structures that benefit the landed by reducing siltation. Contractors must be totally kept out and all wage benefits should go to the landless.

Nothing works better than when villagers see all this in actual practice. For this purpose, it is important to have funds to take interested villagers to see villages where such principles are being observed and how water harvesting has changed their lives. Under the Madhya Pradesh Rajiv Gandhi Watershed Development Mission (RGWDM), busloads of tribals from Jhabua were taken to see Ralegan Siddhi. Tarun Bharat Sangh (TBS) also regularly organises Paani Yatras in neighbouring areas so that villagers who have done it can talk face to face with those who have not.

All this means that the progress in the first few years will be slow. This was the case both with RGWDM and TBS. But the experience of both also shows that replication comes very rapidly once 'the idea' catches on. As R Gopalakrishnan, who oversees RGWDM in Madhya Pradesh, puts it, "In the first two years of the programme, the results were so slow that we kept wondering whether we are getting it right." In other words, governments must be prepared to accept that their first year's effort will bring nothing, second and third year maybe something, and fourth and fifth years hopefully a lot. It is a gradual exercise. The number of villages participating in the water harvesting programme of TBS also grew slowly at first

SOCIAL MOBILISATION IS A PREREQUISITE FOR EFFECTIVE WATER HARVESTING ACTIVITY

and then very rapidly in the later years as the villagers became more and more confident of the value of what they were doing (see graph 3: Promoting environmental self-reliance).

Social mobilisation is essential for the success of water harvesting for several other reasons. Firstly, the community must be closely involved in the construction of the water harvesting structures to ensure that they are built with technical competence; that is, the site is chosen properly, the technical parameters are correct, etc. Badly built structures will not deliver water and can get easily washed away. Secondly, even in properly built structures which deliver water, once the water starts getting available either as increasing levels of groundwater or as surface water in a tank, the community will have to start managing the available water which in the earlier years may not be enough to irrigate lands of all the farmers. In those years, the farmers will have to share the water and use it on crops that don't use too much water. This will happen if there is already a community process associated with the water harvesting structure. Otherwise a few people will grab the water leading the rest of the community alienated.

Graph 3 : Promoting environmental self-reliance

In the first three years, few villages got involved in the water harvesting programme of Tarun Bharat Sangh. With passage of time, the numbers, however, grew rapidly as villagers gained confidence in the programme and were able to see a positive outcome of their efforts.

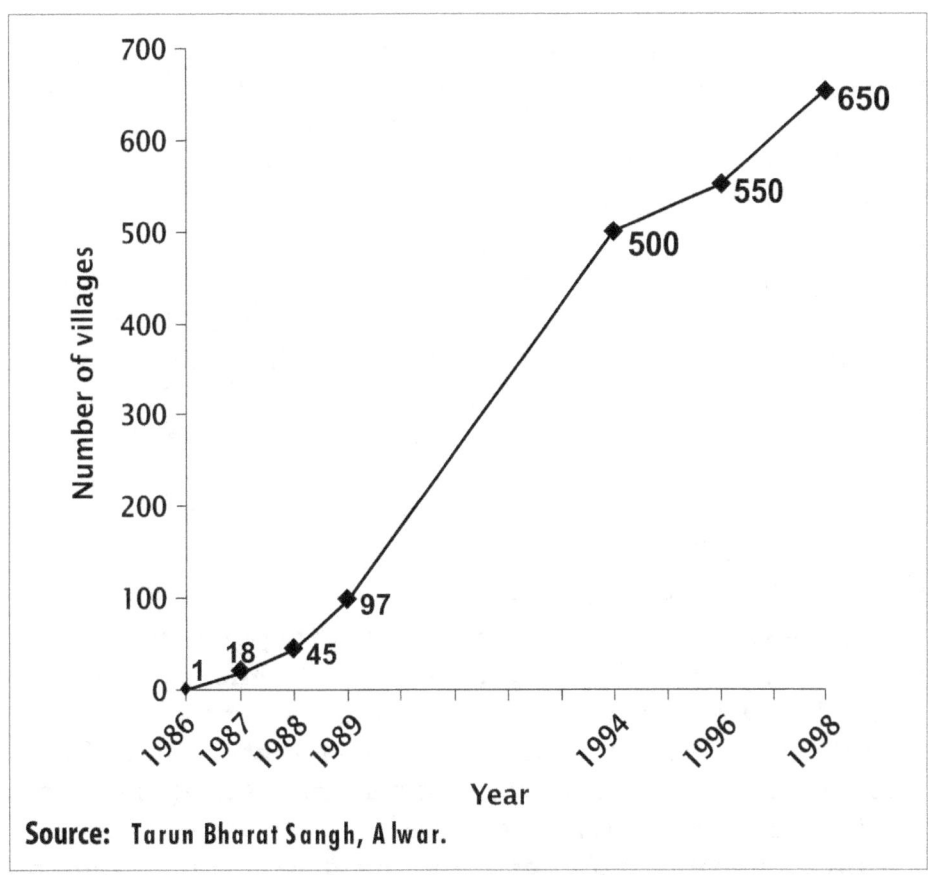

Figure 12.45 Progress of Water Harvesting Achievement Increment of Villages year wise

The government will also have to review and revise old British-time laws like the Indian Easement Act which prevent public participation in water management. Management experts point out, 'God is in the details'. **Patwa, Thakur and Chandrababu Naidu and others like them will have to make sure the culture of rigid and inflexible government rules is changed to fit the task of social mobilisation.** And as government officials are not social workers, all this will happen only if they hand over work to NGOs with a track record, howsoever few they might be, and wait patiently for results to come. But if they want a government programme, it can still be done, but then they themselves will have to oversee the implementation of the programme to keep their warring and errant officials in place. Inter-departmental coordination — between the departments of revenue, soil conservation, irrigation, forests and agriculture — is essential for water harvesting-watershed development programmes to succeed. The direct supervision of the Madhya Pradesh chief minister of the RGWDM was a key element of the programme's success. It is not surprising that despite the drought this year, newspaper reports from western Madhya Pradesh were few. Otherwise the result will be a lot of wasted mud, bricks and mortar and alienated people, with water harvesting getting a bad name. Already, newspaper reports are showing that there has been considerable corruption in the check dams sponsored by the government of Gujarat[28]. **We must make sure that water harvesting does not become money harvesting.**

7. **Reviving systems which have gone into disuse**

    It is not enough to build new water harvesting structures. Efforts must be made also to revive the vast treasure that already exists but has gone into disuse. According to official estimates, there were 15.13 lakh tanks in India in 1986-87 — 95 per cent of which were in eight states, Andhra Pradesh, Karnataka, Kerala, Madhya Pradesh, Maharashtra, Tamil Nadu, Uttar Pradesh and West Bengal. But because of silting and poor maintenance, the gross area irrigated by them had come down from 4.78 million hectares in 1962-63 to 3.07 million hectares in 1985-86 though many new tanks were constructed in this period. This represents a capital loss of Rs.5,000 crore[29]. It is obvious that efforts must be made to restore these structures to their full potential.

8. **Rainwater harvesting can eradicate rural poverty**

    If water harvesting-watershed development programmes are handled well, the experience of villages like Sukhomajri, Ralegan Siddhi and several villages in Alwar district has clearly shown that rainwater harvesting is not just the starting point for meeting drinking water needs but the starting point of an effort to eradicate rural poverty itself, generate massive rural employment and reduce distress migration from rural areas to urban areas. Increased and assured water availability means increased and stable agricultural production and improved animal husbandry — both of which together form the fulcrum of the rural economy. Rainwater harvesting has helped Ralegan Siddhi to transform itself from one of the most destitute villages of the country in the 1970s to one of the richest villages today. In all villages which have regenerated their local economy with the help of good management of their natural resources, distress rural-urban migration has been greatly reduced or has been totally eliminated.

    Another interesting dimension of community-based rainwater harvesting is that it helps the generate a community spirit within the village — something that is getting lost across the country — and build up what economists call the 'social capital'. In fact, we can argue that if we want Panchayati Raj to work, then the first thing that panchayats should be asked to take up is water harvesting — har gaon ka apna talaab (a tank in every village).

    PROFICIENT WATER HARVESTING ALONG POLITICAL WISDOM AND COMMUNITY ACCOUNTABILITY CAN HELP SOLVE RURAL POVERTY.

    If it all makes so much sense then why is this paradigm not accepted by the government and spread across the country? President K R Narayanan even called for a national movement for water harvesting in his last Republic Day address to the nation.

The problem is really mind-sets. Rainwater harvesting demands a new approach to governance itself — a participatory form of governance rather than a top-down bureaucratic one. Unfortunately, our political leaders have created a culture of dependence on the government and love to make promises, howsoever hollow they may be, that they will provide everything to the people through government largesse. Given this political mind-set, the water bureaucracy, too, has developed a culture of providing services, howsoever poor and abysmal they may be, rather than one of empowering people to develop their own water supplies. And it is still locked into the big dam, pumps, pipes and borewell paradigm. Water resources minister, C P Thakur, today does not hesitate in accepting the importance of rainwater harvesting but when push comes to a shove, as it did, for example, at the recent World Water Forum in the Hague, he and the entire Indian delegation could do nothing but talk about big dams. We are not here going into the merits of the small versus large dam debate — in certain situations large dams could be needed — but there can be no doubt that the small has an extremely important role to play and this is being totally neglected. A balance between the big and the small is essential.

But, nonetheless, there has been a remarkable change in the public discourse on the relevance of the country's rainwater harvesting traditions in the last few years, especially since Centre for Science and Environment (CSE) published Dying Wisdom: The rise, fall and potential of India's traditional water harvesting systems in 1997, a compelling account of how these traditions still work in many parts of the country even in the face of all odds. The recent statements by the Central and state ministers are encouraging but the government needs to go beyond them. It should heed the president's advice and prepare a concrete plan of action to develop a mass movement for water harvesting.

(The author was a member of the World Water Commission and is the director of the Centre for Science and Environment, New Delhi)

Note: If any member of parliament or state legislature is interested in understanding this issue further, please do not hesitate to get in touch with the author or with Indira Khurana, CSE's coordinator for its Campaign to Make Water Everybody's Business.

TRADITIONALLY, OUR ANCESTORS HAVE HARVESTED WATER THROUGH TANKAS, ZINGS, AHARS AND JOHADS

REFERENCES

Anon 1999, Water an Overview — Issues and Concerns, National Commission for Integrated Water Resources Development Plan, National Commission for IWRDP, New Delhi, pp 6-7.

Anon 1999, Development and Management Issues Irrigation, Flood Control, Hydropower and Navigation, National Commission for Integrated Water Resources Development Plan, National Commission for IWRDP, New Delhi, pp 9-10.

Anon 1999, Water Availability and Requirements, National Commission for Integrated Water Resources Development Plan, National Commission for IWRDP, New Delhi, p 36.

Anon 1999, Water Availability and Requirements, National Commission for Integrated Water Resources Development Plan, National Commission for IWRDP, New Delhi, p 38.

World Bank/Central Groundwater Board 1999, India: Water Resources Management — Groundwater Regulation and Management, World Bank/Allied Publishers, Washington DC/New Delhi, p 15.

J S Samra 2000, Central Soil and Water Conservation Research and Training Institute, Dehradun, personal correspondence.

Anon 2000, Flood Of Promises To Deal With Lack of Water, Times of India, New Delhi, April 20, p.12.

The Hindu, New Delhi, April 22, p.7.

Anon 2000, PM Calls for Moratorium on Bickerings, The Hindustan Times, April 26.

Anon 2000, Water Harvesting may Become the Law, The Hindustan Times, New Delhi, May 6.

Sonu Jain 2000, Water in India is Like Draupadi, There Are Five Ministries That Deal With It, The Indian Express, New Delhi, April 30.

H S Bartwal 2000, Centre Readies Rs. 550 Crore Plan To Recharge Water Level, The Hindustan Times, New Delhi, May 5, p.12.

Anon 2000, Andhra Forms Conservation Panel, The Indian Express, New Delhi, April 28.

Anon 2000, Water Conservation Mission Launched, The Hindu, New Delhi, May 1.

S Ramakrishna 2000, AP Set to Revamp Watershed Project, The Indian Express, April 30.

Anon 2000, AP Government to Spend Rs 4,000 Crore to Conserve Rainwater, The Pioneer, New Delhi, April 30.

Anon 2000, Depleting Water Table is Now a Capital Woe, The Hindustan Times, New Delhi, April 29.

Anon 2000, Drafting of Bill on Harvesting Rainwater Begins, The Indian Express, May 8.

Arati Bhargava 2000, Harvesting of Water: NDMC Gives Approval, The Hindustan Times, May 5.

Anon 2000, Work on 10,000 Check Dams Begins, Times of India, Ahmedabad, April 28.

Darshan Desai 2000, Drought and Deja vu, The Indian Express, New Delhi, May 2.

Anon 2000, Minister's Call to Support Water Conservation Scheme, Central Chronicle, Bhopal, April 30.

Rathin Das 2000, Jamnagar Cleans up Lakhota Lake, The Hindustan Times, New Delhi, April 29.

Anon 2000, More Check Dam Projects take off in Amreli District, Times of India, Ahmedabad, April 18.

Tanvir Siddiqui 2000, Ahmedabad Falls Back on Old Way of Conserving Rainwater, Indian Express, New Delhi, May 2.

Anon 2000, Fight the Drought: New Innings, India Today, New Delhi, May 15.

Anon 2000, Harvest Rainwater, SC orders DJB, The Hindustan Times, New Delhi, May 11.

Anon 2000, Damning evidence against Junagadh check dams sites, Times of India, New Delhi, May 17.

Anon 1999, Local Water Resources Development and Management, National Commission for Integrated Water Resources Development Plan, National Commission for IWRDP, New Delhi, p 2.

APPENDIX-1

SOME USEFUL CONVERSION FACTORS

Length

1 cm	=	0.394 inch	1 inch	=	2.54 cm
1 m	=	3.28 ft	1 ft	=	0.3048 m
1 m	=	1.094 yard	1 yard	=	0.9144 m = 3ft
1 km	=	0.6215 mile	1 mile	=	1.609 km

Area

$1\ cm^2$	=	$0.155\ inch^2$	$1\ inch^2$	=	$6.4516\ cm^2$
$1\ m^2$	=	$10.76\ ft^2$	$1\ ft^2$	=	$0.0929\ m^2$
$1\ m^2$	=	$1.1961\ yard^2$	$1\ yard^2$	=	$0.8361\ m^2$
				=	$9\ ft^2$
$1\ km^2$	=	$0.386\ mile^2 = 2.471\ acre$	$1\ mile^2$	=	$2.59\ km^2$
				=	640 acre
1 hectare	=	$10000\ m^2 = 2.471\ acre$	1 acre	=	0.4047 hect

	=	11961 yard² = 107650 ft²		=	4047 m²
				=	4860 yard²
				=	43560 ft²

Volume

1 cm³	=	0.061 inch³	1 inch³	=	16.387 cm³
1 m³	=	35.515 ft³ = 1000 litre=1KL	1 ft³	=	0.0283 m³
1 litre	=	0.0353 ft³ = 1000 cc	1 ft³	=	28.3 litres
1 litre	=	0.22 imp. Gal (UK)	1 imp.gal	=	4.546 litres

Mass

1 kg	=	2.205 lb	1 lb	=	0.4536 kg

Force

1 kg (f)	=	2.205 lb (f)	1 lb (f)	=	0.4536 kg (f)
1 kg (f)	=	9.807 N	1N	=	0.1020 kg (f)
				=	105 dynes

Pressure

1 kg (f)/cm²	=	14.2255 lb(f)/inch²	1 lb (f)/inch²	=	0.0703 kg(f)/cm²
1 kg (f)/cm²	=	0.205 lb (f) /ft²	1 lb (f)/ft²	=	4.882kg(f)/cm²
1 N/m²	=	0.102 kg (f) /m+	1 kg (f)/m²	=	9.807 N/m²
1 N/m²	=	0.0209 lb(f)/ft²	1 lb (f)/ft²	=	47.88 N/m²

APPENDIX-I

1 Pascal (Pa) = 1.02×10^3 kg/cm²		=	1 Mega Pa = 10.1972 kg/cm²
Pa=1 N/m²			(MPa)=1N/mm²

1 Bar = 1.02 kg/cm^2 = 1450 lb/inch^2

Atmospheric = 1.033 kg/cm^2 = 14.7 lb/inch^2 =   34 ft head of water

Pressure   =   10.32 m head of water

Pressure (English)

Atmospheric = 1.00 kg/cm^2 = 14.22 lb/inch^2 =   32.8 ft head of water

Pressure (Metric)   =   10.00 m head of water

Density

1 kg /m^3 = 0.06243 lb/ft^3 1 lb/ft^3 = 16.0181 kg/m^3

1 gm/cc = 62.43 lb/ft^3 1 gm/cc = 0.01602 gm/cc

Temperature

°C = 5/9 (°F-32°)   °F = 1.8 x °C + 32°

Work, Energy, Heat

1 kg (f) m = 7.233 ft-lb (f)   1 ft-lb(f) = 0.1382 kg (f) m

1 Joule (J) = 0.1020 kg (f) m   1 kg (f) m = 9.807 J

 = N.m

1 J = 0.7376 ft-lb (f)   1 ft-lb(f) = 1.356 J

1 KWh=3.60 x 10^6 J=0.367 x 106 kg (f) m

1 BTU=1055 J = 0.293 Wh = 778 ft-lb(f)

1 Kcal = 426.7 kg(f)m   1 kg (f) m = 2.344 cal

Power

1 Watt (W) = 1 J/s = 0.102 kgm/s = 0.727 lb-ft/s

1 HP (FPS)=746W=76.04 kg(f)m/s=550 ft-lb (f)/s

1 HP (Metric) = 735.5 W = 75 kg (f) m/s = 542.3 ft-lb(f)/s

Angle

1 rad = 57.296°   1° = 0.017453 rad

Concentration

1 ppm in water = 1 mg / litre

Discharge
(1 imperial UK gallon = 4.54609 litres)

Discharge	MGD	Cusec	gal/s	Gal/min	gal/h	gal/d	lit/s	lit/m	lit/h	lit/d	KLD	MLD	M³/s
1 MGD	1.0	1.8584	11.574	694.44	41666.67	10^6	52.6167	3157.01	189420.42	4546090	4546.09	4.54609	0.0526167
1 gal/sec	0.5381	1.0	6.23	373.6667	22420.52	538092.6	28.31	1698.76	101925.73	2446217.4	2446.2174	2.4462174	0.02831
1 cft/sec (cusec)	0.0864	0.16051	1.0	60.0	3600.0	86400.0	4.54609	272.765	16365.924	392782.18	392.78218	0.392782	0.00454609
1 gal/min	0.00144	0.002676	0.01666	1.0	60.0	1440.0	0.07577	4.54609	272.765	6546.37	6.54637	0.0665464	0.0000758
1 gal/hr	0.000024	0.000045	0.000278	0.016667	1.0	24.0	0.001263	0.07577	4.54609	109.106	0.109106	0.0001091	0.00000126
1 gal/day	0.000001	0.000002	0.000012	0.000694	0.04167	10	0.000053	0.00316	0.1894	4.54609	0.00455	0.00000046	0.000000005
1 lit/sec	0.019005	0.035323	0.219969	13.1978	791.89	1911.40	1.0	60.0	3600.0	86400.0	86.4	0.0864	0.001
1 lit/min	0.000317	0.000589	0.003666	0.219969	13.1978	316.76	0.016667	1.0	60.0	1440.0	1.44	0.00144	0.00001667
1 lit/hr	0.0000053	0.000009	0.000061	0.003667	0.219969	5.28	0.000278	0.01667	1.0	24.0	0.024	0.000024	0.00000028
1 lit/day	0.0000002	0.0000004	0.000003	0.000153	0.00917	0.219969	0.000012	0.000069	0.0417	1.0	0.001	0.000001	0.000000001
1 KLD	0.0002199	0.00040089	0.002546	0.152765	9.1659	219.97	0.011574	0.69444	41.6667	1000.0	1.0	0.001	0.00001157
1 MLD	0.219969	0.0408879	2.5459	152.765	9165.9	219973.6	11.574	694.44	41666.67	10^6	1000.0	1.0	0.011574
1 m³/sec	19.008	35.424	219.97	13197.83	791889.3	19011407	1000.0	60000	3600000	86400000	86400	86.4	1.0

LASER LEVELING

RESOURCE CONSERVATION THROUGH LASER LEVELING

INTRODUCTION

Shrinking water resources owing to over exploitation of ground water in Punjab threatens the maintenance of agricultural productivity. As a result, the water table is falling in 90% area of the state. Most of this area falls in the Central part of the state. With the inception of Green revolution in the Sixties, the water table started declining and the area having water table below 30 feet depth has increased from 3% in 1973 to 90% in 2004. During 1993-2003, the average fall of water table in the Central Punjab was 50 cm per annum. However, in some of the areas, the fall of water table is even more than 80-100 cm per annum. Out of 141 blocks of the state more that 100 blocks are over exploited. In the Central part, out of 70 blocks, the water table in 40 blocks has gone down below 50 feet depth and in these blocks, submersible pumps are being installed to replace centrifugal pumps. It is projected that by 2023 in Central Punjab, the water table depth will be below 70 feet in 66% area, below 100 feet in 34% area and below 130 feet in 7% area. Correspondingly in each district, the percent area below 70 feet depth will be 100% in Moga ft. Sangrur, 80% in Patiala, 70% in Ludhiana, 60% in Kapurthala & Jalandhar.

To arrest this dangerous trend of ground water exploitation, there is an urgent need to conserve irrigation water through various on-farm water conservation practices. **Land Leveling through Laser Leveler** is one such proven technology that is highly useful in conservation of irrigation water.

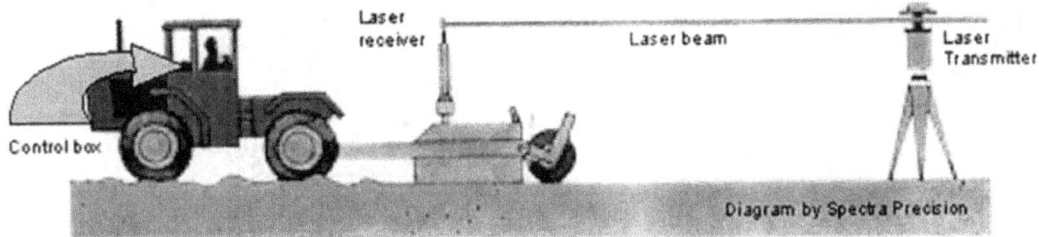

LASER GUIDED LAND LEVELING

As per studies, a significant (20-25%) amount of irrigation water is lost during its application at the farm due to poor farm designing and unevenness of the fields. This problem is more pronounced in the case of rice fields. Fields that are not level, have uneven crop stands, increased weed burden and uneven maturing of crops. All these factors lead to reduced yield & poor grain quality.

Laser land leveling is leveling the field within certain degree of desired slope using a guided laser beam throughout the field. Unevenness of the soil surface has a significant impact on the germination, stand and yield of crops. Farmers also recognize this and therefore devote considerable time resources in leveling their fields properly. However, traditional methods of leveling land are cumbersome, time consuming as well as expensive.

Why Laser Levelling?

Land looks leveled but even then wide topographic variation exists

- Wide variability in crop yields at field / village / block / district / regional level
- For Better distribution of water
- For Water savings
- For Improvement in nutrient use efficiencies
- Option for Precision Farming
- Higher crop productivity

Equipment Details

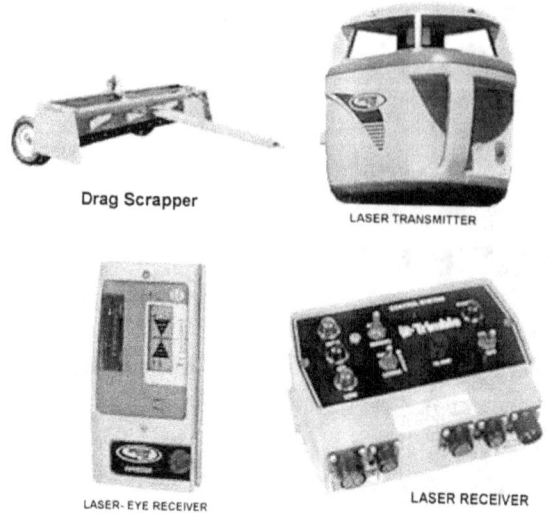

Drag Scrapper

LASER TRANSMITTER

LASER- EYE RECEIVER

LASER RECEIVER

LASER LEVELING

LEVELER SETTING

OBJECTIVES OF LAND LEVELING

Effective land leveling is meant to optimize water use efficiency, improve crop establishment, reduce the irrigation time & effort required to manage crop.

Laser leveled land:

- Saves 25-30% of water
- Improves crop establishment and improves Yield.
- Reduces weed problems
- Improves uniformity of crop maturity
- Decreases the time to complete tasks
- Reduces the amount of water required for land preparation.
1. **YIELD:** Research conducted by PAU has shown a large increase in rice yield due to proper field leveling. The following table is self-explanatory:

LAND LEVELING IN PROGRESS

Year	Rice Yield (t/ha)		
	Leveled Fields	Unleveled Fields	
1996	3.40	2.67	21.47%
1997	2.27	1.46	35.68%
1998	2.72	2.36	13.24%
1999	2.34	2.00	14.53%
Average	**2.72**	**2.19**	**84.92%**

Source: PAU, Ldh.

It clearly shows that for the same rice varieties and average increase in crop yield was 24% or 530 kg h/a.

Yield and irrigation water saving for Laser leveled and traditionally leveled plots for rice crop under replicated experiments at PAU, Ludhiana

S.No.	Leveled (t/ha)	Unleveled (t/ha)	Percentage increase in yield	Percentage Saving in irrigation time / water
Site 1	8.78 ± 0.33	7.73 ± 0.21	13.60	26.15
Site 2	8.30 ± 0.46	7.53 ± 0.39	10.30	-
Site 3	7.60 ± 0.21	7.00 ± 0.25	8.57	25.00
Mean			10.82	25.57

2. **WEED CONTROL:** Improved water coverage from better land leveling reduces weeds by up to 40%. This reduction in weeds results in less time for crop weeding.

3. **FARM OPERATION:** Laser leveling makes possible the use of larger fields. Larger fields increase farming area and improve operational efficiency. This increase in farming area gives the farmer the option to reduce operating time by 10% to 15%.

4. **SEEDING PRACTICES:** Laser leveled larger fields reduce the time taken for planting, for transplantation and for direct seeding.

5. **EFFICIENT WATER MANAGEMENT:** An unleveled field means extra water storage in fields to accomplish puddling in paddy field. Moreover, land

leveling effectively terraces fields allowing water in the higher fields to be used in the lower fields for land preparation, plant establishment & irrigation.

6. **ECONOMICS:** The initial cost of laser land leveling is high but if the appropriate ploughing techniques are used, re-levelling the whole field should not be necessary for at least eight to ten years. Measurements taken in fields in the second and third year after leveling have shown very little variation in surface topography. Other benefits are:

   ○ Being able to direct seed

   ○ Plough the field on time

   ○ Harvest evenly ripened crop

   ○ Reduced weeding cost.

To sum up the objectives of laser land leveling:

○ More level and smooth soil surface

○ Reduction in time and water required to irrigate the field

○ More uniform moisture environment for crops

○ Reduced consumption of seeds, fertilizers, chemicals and fuel

○ Improved field trafficability (for subsequent operations)

THE PROPOSAL

One laser leveler costs about Rs. 4 Lacs and a 50 HP Tractor costs about Rs.4 Lacs. Thus, the cost 1 Laser Leveler & Tractor Set is Rs. 8 Lacs. Individual farmers can not afford to incus this heavy cost. Therefore, it is proposed that a group of farmers, i.e., Farmers' Interest Group (FIG) will be form who will be assisted to our chase the equipment. An assistance of 75% of the cost is proposed on the equipment comprising of 1 Laser Leveler & a 50 HP Tractor. The FIG shall level their own fields at also can do custom hiring to level the fields of other farmers. 500 Laser levelers & Tractors are required to level more than 2 lakh hectares area in the next 5 years.

Project Cost

The total cost of the Project for 5 years shall be 40 Crores. The State Govt. shall provide 30 Crores as 75% assistance for 500 machines (1 Laser Leveler & a 50 HP Tractor) and the beneficiaries shall bear the balance cost of Rs. 10 Crores. These machines shall help in precision leveling of more than lakh hectares of cropped area in the next 5 years.

For the Year 2007-08, the requirement of funds is proposed as under:

No. of Sets (Laser Leveler + Tractor 50 HP)	100
Total Cost of 100 Sets (@ Rs.8 Lacs per set)	Rs. 8 Crores
Assistance proposed under ACA @ 75% of cost	Rs. 6 Crores.
Share of FIGs	Rs. 2 Crores.

Benefits of the Project

The assistance shall be provided to the FIGs who shall not only use the machines on their own farm but shall also operate on custom hiring on the fields of other farmers. Each machine shall be able level an area of more than 400 hectares benefiting a total area of 40000 hectares through 1 machines during the year 2007-08. The major benefits of the Project shall be as follows:

1. **Water Saving**

   The Laser Leveling of fields saves 25-30% of water. Taking into consideration the Paddy-wheat rotation, there is a saving of 2625 cum per hectare after leveling the field. Therefore, there shall a saving of 105 MCM (Million Cubic Metres) of water by leveling 40,000 hectares through machines in one year.

2. **Increase in Yield**

   The Laser Leveling of fields helps in increased yields by 10%. This increased yield shall help in additional income of Rs.6000/- per hectare. Therefore, there shall be a cumulative additional income of Rs.24 crores in 1 year by 100 machines.

3. **Energy saving**

   Saving of 105 MCM of water on 40000 hectares shall lead to less requirement of irrigation to the fields thereby saving of 125 lakh Units of Electricity valued at Rs.5 Crores in 1 year.

4. **Other benefits**

   Apart from above, other benefits of Laser Leveling include Weed Control, Labour saving, Time saving, Land saving etc. as described earlier.

LASER LEVELED LAND

PRECIPITATION - ANANTAPUR DISTRICT (1901-2002 - 102 Years)

Year	Jan	Feb	Mar	Apr	May	Jun	Jul	Aug	Sep	Oct	Nov	Dec
1901	0.054	12.422	0.003	13.646	73.475	43.486	43.741	34.063	130.792	93.757	66.966	9.941
1902	0.205	0	0	32.725	51.185	54.83	27.823	61.025	127.487	196.235	37.085	11.047
1903	1.423	0	0	6.271	76.354	61.566	80.052	106.341	203.903	163.795	151.939	6.299
1904	2.098	0	5.337	25.83	103.551	39.952	70.543	17.818	67.199	153.618	0.631	0.977
1905	2.097	2.703	7.379	18.921	50.693	54.772	25.485	149.098	24.322	115.605	7.65	0.011
1906	3.46	0	0.054	0.433	13.72	84.41	69.466	145.448	158.651	127.688	15.93	50.403
1907	1.396	0	4.121	91.809	9.545	50.209	96.527	42.264	122.505	13.962	45.06	6.31
1908	0.94	1.88	13.07	9.366	43.865	37.346	47.959	34.737	183.554	41.266	0.534	0.033
1909	2.739	0	0.608	45.657	96.961	32.999	45.549	164.143	146.404	59.808	8.791	2.188
1910	0	0.012	3.492	11.607	67.261	48.851	162.111	171.442	176.634	137.112	103.382	0
1911	0	0	1.051	56.085	92.334	54.526	36.192	35.157	97.238	100.402	25.008	14.357
1912	0.44	2.05	0	19.583	42.396	51.25	70.044	123.024	200.7	127.922	97.705	0
1913	0	0	0	4.843	83.974	35.463	81.738	21.585	209.33	138.844	0.599	9.2
1914	0.025	0.136	0.896	11.447	47.48	46.43	51.825	131.858	134.607	63.782	45.301	3.761
1915	0.869	1.168	32.883	6.119	85.233	65.949	114.259	29.181	231.953	56.281	110.374	0.984
1916	0	0	0.109	8.878	81.772	36.017	116.008	112.455	154.433	203.316	94.985	0.079
1917	0.303	5.555	8.059	13.735	104.974	79.889	46.647	177.066	191.426	162.836	88.23	0.507
1918	0.419	0.101	25.777	14.873	78.151	20.968	20.033	42.681	168.599	4.587	119.505	6.272
1919	1.382	0	6.209	54.36	80.895	91.708	88.125	15.89	328.15	73.54	121.237	5.589
1920	2.204	0	0	17.184	65.547	34.072	25.527	69.32	151.876	89.483	14.062	0
1921	2.669	0	1.906	84.345	12.902	60.811	90.182	67.811	58.872	314.973	79.786	0.073
1922	1.209	0	0	14.761	65.731	33.04	32.416	42.433	30.443	105.018	158.07	1.492
1923	0.816	5.189	16.035	17.439	39.88	49.737	46.595	20.301	129.578	35.394	3.117	0.395
1924	0.104	0	0.122	48.396	91.473	26.514	77.178	88.388	261.386	46.588	65.301	1.344

Year												
1925	0	0	1.675	41.126	151.148	19.842	46.449	81.816	113.184	104.44	60.629	49.547
1926	2.452	0.437	1.018	73.34	33.3	94.054	55.391	36.533	173.28	87.127	6.116	0.792
1927	0.065	1.371	2.996	3.21	48.349	68.462	110.598	42.516	207.353	51.199	140.255	0.007
1928	0.061	10.769	1.415	56.579	54.999	77.357	90.816	100.613	97.407	162.651	16.604	4.79
1929	0.175	1.245	0	33.022	49.841	34.99	29.478	26.125	210.466	118.081	83.253	2.528
1930	2.026	2.727	12.849	13.623	138.251	49.256	65.279	40.526	168.534	290.599	45.142	13.769
1931	0	0	0.796	16.157	65.871	67.753	36.506	23.464	162.86	51.519	65.733	6.42
1932	0	2.807	0	3.043	41.193	65.564	40.795	194.297	101.055	135.817	122.377	2.488
1933	0	2.884	14.071	44.635	166.688	18.537	110.433	148.276	86.883	163.168	32.989	34.393
1934	0.636	0	0	50.383	32.035	47.477	102.147	24.232	34.273	102.038	30.516	0.612
1935	1.177	5.136	0.511	47.543	24.562	112.605	48.781	249.442	69.557	137.161	0.682	0.479
1936	0.105	10.508	11.034	29.167	84.422	52.237	45.858	42.727	145.572	53.506	107.354	1.593
1937	0	7.944	11.741	75.101	50.346	64.824	60.065	26.565	149.571	198.272	20.1	4.2
1938	0	0.383	5.775	12.993	52.963	33.646	46.659	204.767	195.871	9.012	2.384	0.11
1939	0.226	0	22.427	81.533	8.912	62.856	35.074	123.202	176.637	145.055	62.28	0.295
1940	0.419	0.131	1.079	47.426	143.598	58.217	41.966	54.919	116.851	269.339	98.994	4.692
1941	0.781	7.877	0	18.226	49.268	45.479	43.824	89.184	191.985	92.907	15.386	36.065
1942	0	4.785	0	16.054	65.864	94.227	26.561	67.732	78.634	37.808	17.875	4.964
1943	0.62	0.007	3.45	30.875	156.691	31.888	28.861	66.864	127.462	197.916	64.51	0.018
1944	0	7.84	23.417	9.651	46.078	98.387	86.017	18.95	125.606	194.408	38.136	0.177
1945	0.034	0	0	49.96	76.027	16.004	128.212	70.403	107.851	44.232	35.697	0.79
1946	0.054	2.879	10.132	49.416	74.492	26.302	45.577	115.32	114.836	146.34	164.453	18.775
1947	1.423	9.309	0.025	22.557	29.606	47.684	57.166	186.977	189.308	79.823	14.418	10.747
1948	0.41	0	3.996	37.382	115.394	14.742	56.745	97.065	94.196	57.417	141.929	0.567
1949	0	0	0.016	6.699	107.344	59.686	54.383	116.239	189.028	115.23	19.815	0
1950	0	4.995	8.237	0.353	86.502	44.545	56.358	101.851	128.805	136.347	41.435	0

1951	0	0	14.94	47.633	155.615	56.762	73.092	20.414	133.601	91.795	8.294	0
1952	0	2.096	1.482	5.901	131.744	24.232	58.325	56.143	53.676	142.359	0.462	32.811
1953	0.047	0	0	60.898	16.705	63.142	146.955	34.908	186.616	337.638	1.416	0
1954	0.724	0	4.612	18.988	76.478	34.127	117.703	98.968	47.19	146.336	0.462	11.568
1955	0.058	0.001	11.111	39.893	177.616	43.025	47.532	117.465	170.028	151.035	28.92	2.548
1956	0.112	0.487	0.001	117.808	103.515	74.608	112.235	32.113	167.871	267.765	111.325	11.282
1957	0	0.002	2.195	26.183	79.991	143.196	61.172	73.46	75.86	134.544	25.678	0
1958	0.048	2.845	4.849	54.911	60.638	60.48	38.495	134.597	130.415	145.605	38.34	4.8
1959	0	3.911	0	45.558	86.242	118.519	76.474	80.587	168.428	59.637	19.037	10.378
1960	0	0	6.116	15.991	44.74	37.668	86.415	13.115	307.697	76.527	66.973	0.022
1961	0.306	5.311	0.225	19.021	128.872	89.82	69.67	74.412	39.297	147.76	25.799	0.007
1962	0	4.6	3.942	58.197	105.094	40.832	54.011	90.004	141.797	176.578	30.725	58.427
1963	0.389	0	4.243	97.871	56.79	34.725	70.177	166.569	85.508	156.21	0.566	5.412
1964	0	0	0	17.475	12.343	76.643	189.478	61.041	292.445	69.572	58.091	2.157
1965	0.871	0.372	0.697	35.917	25.655	55.507	36.944	99.948	109.246	3.574	27.911	31.816
1966	1.842	0	0.181	10.245	87.151	109.998	106.512	150.874	150.331	129.793	140.131	23.661
1967	4.476	0	8.507	19.833	60.855	63.167	159.355	35.139	101.855	123.543	5.154	13.14
1968	0	10.628	25.83	66.248	74.172	51.147	53.773	11.234	255.184	120.813	45.811	11.459
1969	0	0	0.204	10.751	100.564	54.694	58.498	155.552	29.944	182.821	38.6	14.138
1970	0	0	0.474	28.078	154.75	41.376	68.132	94.596	177.213	161.961	1.802	0.001
1971	0	1.095	6.125	41.591	68.565	32.234	31.956	107.583	66.037	183.223	10.885	0.935
1972	0.912	0	0	17.416	131.378	55.271	25.833	12.227	199.067	135.514	32.64	17.074
1973	0	0	0	5.716	39.55	70.122	37.024	121.193	141.085	270.398	43.788	3.031
1974	0	0	0.097	20.776	153.421	44.787	58.634	36.409	290.557	148.508	15.37	0
1975	0.087	0.001	4.101	16.605	71.592	27.346	115.097	79.308	146.146	319.802	77.958	2.14
1976	0.04	0	0.026	56.263	23.477	27.888	33.681	104.679	37.284	62.427	53.031	0.007

Year												
1977	0	0.246	6.828	69.614	169.652	80.754	84.975	115.723	86.637	140.713	80.129	0.234
1978	0.163	5.71	0.28	68.208	52.925	38.108	110.121	44.517	216.012	82.653	66.813	28.998
1979	0	14.591	1.295	9.189	91.475	64.118	66.484	52.426	195.454	93.465	122.022	0.262
1980	0	0	1.6	26.152	67.235	34.932	36.759	72.389	114.484	75.243	42.703	3.304
1981	0.576	0	17.747	44.258	45.242	18.217	69.692	76.893	244.789	133.89	61.912	6.474
1982	0	0	1.154	10.306	37.549	72.725	56.575	21.003	107.553	79.099	141.997	0.295
1983	0	0	0.699	16.89	86.193	96.896	67.928	150.977	220.753	85.994	8.008	17.6
1984	0.417	5.191	17.331	40.384	20.631	31.459	127.939	17.887	79.89	67.252	13.05	10.134
1985	0.39	0	8.609	9.54	39.025	55.643	58.758	40.234	74.008	91.469	45.385	5.306
1986	2.7	4.758	0	32.644	34.875	82.071	45.182	80.358	212.298	78.514	73.511	18.186
1987	0.294	0.012	11.752	31.163	60.776	50.456	48.449	91.054	74.556	142.84	33.532	20.477
1988	0	5.184	0	63.944	79.531	30.646	145.136	133.283	241.516	25.119	16.907	19.389
1989	0.007	0	7.822	11.56	39.555	70.253	149.355	30.803	153.504	163.337	8.119	3.716
1990	0.171	0.087	4.621	18.278	188.382	47.417	35.823	83.135	71.909	141.507	39.751	1.636
1991	0.057	0.007	0	37.639	96.796	116.598	57.386	85.942	110.071	182.089	29.541	0
1992	0.037	0	0	10.272	74.621	93.266	89.227	92.005	112.125	114.898	45.597	0.269
1993	0	0	8.609	27.848	60.953	73.998	72.192	85.835	132.107	182.427	14.523	26.912
1994	0.759	1.331	2.523	29.156	41.106	41.339	73.752	65.062	52.916	228.128	41.452	2.002
1995	2.938	0	9.6	18.458	116.136	61.954	80.11	122.375	154.098	246.497	30.537	0.03
1996	0	0.012	0	74.871	57.405	109.911	54.142	169.863	158.99	153.586	53.504	5.664
1997	0.963	0	8.409	64.054	43.915	44.666	43.287	99.633	125.113	218.99	109.439	38.582
1998	0	0	0	45.315	61.334	39.021	139.294	191.88	138.115	221.133	33.981	11.4
1999	0	2.955	0	38.832	147.02	44.108	75.649	60.609	97.503	146.897	28.982	1.165
2000	0	5.876	0	45.422	74.744	76.743	72.593	185.413	118.197	176.051	7.881	4.431
2001	0.748	0	8.177	56.759	28.46	50.801	55.809	73.125	127.757	155.67	19.56	2.472
2002	0.823	0.108	2.03	7.985	120.932	100.99	26.657	57.483	50.547	155.442	15.902	1.171

TEMPERATURE OF ANANTAPUR DISTRICT (1901-2002)

Year	Jan	Feb	Mar	Apr	May	Jun	Jul	Aug	Sep	Oct	Nov	Dec
1901	23.56	24.679	26.802	29.559	28.083	26.304	25.034	24.823	25.98	25.072	23.22	21.344
1902	22.33	23.959	27.815	29.301	29.593	27.234	25.955	26.013	24.869	24.791	22.548	22.246
1903	22.719	24.589	27.829	29.661	28.557	27.242	25.015	24.657	24.821	24.164	21.865	20.815
1904	21.669	23.465	26.955	29.549	28.171	25.499	24.82	25.21	25.175	24.829	22.151	21.08
1905	22.13	24.462	27.277	28.577	29.03	27.226	25.86	25.208	25.218	24.894	23.351	21.328
1906	23.297	25.194	26.869	30.646	30.334	26.436	25.33	24.945	24.366	24.605	23.216	21.906
1907	22.063	24.692	27.086	28.442	28.489	26.185	25.316	24.047	25.329	25.275	23.541	21.658
1908	22.229	24.201	26.712	30.249	29.014	27.211	24.638	24.583	25.006	24.882	22.378	20.857
1909	22.801	24.739	27.839	28.547	28.573	26.118	24.595	25.168	24.601	24.917	23.606	22.835
1910	22.578	24.322	27.69	30.071	29.526	26.558	25.694	24.738	24.139	24.364	21.767	20.56
1911	22.52	23.099	27.197	29.651	29.242	26.409	25.262	24.583	25.898	25.147	23.68	22.135
1912	22.215	25.48	28.007	29.797	30.006	27.502	25.174	24.864	25.382	24.72	22.54	20.942
1913	21.297	24.952	27.62	30.169	29.181	26.328	24.831	25.192	25.939	24.623	22.95	22.689
1914	21.706	25.152	27.715	29.144	29.333	27.064	25.013	24.653	25.157	24.654	23.226	22.571
1915	22.649	24.899	26.903	29.475	29.201	27.391	25.768	25.53	25.272	25.126	23.658	20.911
1916	21.959	24.795	27.72	29.845	28.923	25.387	25.497	24.808	24.937	24.582	22.742	20.658
1917	21.323	23.359	26.142	28.62	28.2	25.232	25.534	25.061	24.677	23.965	22.76	21.271
1918	21.381	23.274	26.236	28.993	27.24	26.154	26.407	25.253	25.497	25.646	24.162	22.225
1919	23.577	25.558	27.634	30.054	28.773	25.864	25.501	25.176	25.456	24.777	23.213	21.702
1920	22.025	24.61	27.867	29.125	29.12	26.909	25.498	25.568	25.55	25.357	23.357	21.528
1921	23.095	24.278	28.163	29.227	30.248	26.56	25.336	25.086	24.695	24.341	21.773	21.148
1922	22.45	24.012	27.847	29.859	28.824	26.472	25.087	25.262	25.003	24.361	22.842	20.627
1923	21.335	23.717	26.287	29.146	28.895	27.864	24.679	25.103	24.679	24.394	23.124	22.021
1924	22.533	25.174	27.432	30.441	29.797	26.898	25.454	25.432	24.893	24.337	22.136	20.953

Year												
1925	22.129	23.954	27.04	29.995	28.061	26.529	25.384	24.925	25.527	24.247	22.681	21.034
1926	22.745	24.632	27.783	28.719	29.234	28.196	26.327	25.397	25.269	24.878	22.818	21.887
1927	23.493	25.296	28.505	30.386	29.413	26.595	24.82	25.279	25.514	24.603	22.039	21.979
1928	22.93	24.68	27.21	29.649	30.039	26.934	25.453	25.114	25.496	24.645	23.045	21.618
1929	23.186	24.458	27.218	28.603	29.338	26.509	25.668	25.78	25.669	24.725	23.076	22.663
1930	23.035	23.955	27.43	29.274	28.709	27.116	25.439	25.28	25.412	24.653	22.557	22.372
1931	22.279	24.972	27.766	30.256	29.937	27.049	25.337	25.342	24.984	24.821	22.731	21.478
1932	21.269	23.892	26.255	29.446	28.384	26.494	24.854	25.514	24.91	25.094	23.339	21.51
1933	22.19	24.96	26.793	29.071	27.371	25.461	24.529	24.517	24.755	23.907	23.146	21.263
1934	21.957	23.838	26.769	29.463	29.554	26.808	25.183	24.699	25.548	25.07	21.875	20.81
1935	22.585	24.774	27.293	28.79	30.012	26.998	24.845	24.878	24.536	24.831	22.06	22.062
1936	23.473	24.765	26.739	29.484	29.096	25.285	25.134	24.682	25.533	24.831	23.04	22.154
1937	22.032	25.249	27.466	28.934	29.24	26.859	24.811	25.472	25.533	24.422	22.704	21.66
1938	23.12	25.171	28.07	29.727	29.206	25.108	24.643	24.55	24.553	23.745	22.202	21.463
1939	21.88	23.973	26.679	29.025	29.45	26.747	24.619	24.88	24.9	24.871	22.027	20.905
1940	20.934	24.134	27.215	28.824	28.628	26.435	24.547	24.546	25.278	24.733	23.016	21.898
1941	22.9	24.759	28.452	30.345	29.702	26.658	26.1	25.477	25.232	25.172	23.496	22.586
1942	22.688	24.854	27.825	29.691	29.771	26.766	25.025	24.654	24.843	24.628	22.68	22.438
1943	23.12	23.54	27.492	28.878	27.782	25.434	24.551	25.023	24.068	23.794	23.006	20.838
1944	22.795	24.292	26.038	28.599	29.564	26.718	24.759	25.426	25.163	24.719	23.005	21.642
1945	22.498	24.892	27.507	28.837	29.536	26.591	25.128	24.626	24.725	24.63	22.147	21.29
1946	21.236	23.895	27.713	29.138	28.908	26.068	25.036	24.734	25.073	25.248	23.263	22.124
1947	23.107	24.183	27.646	28.901	29.84	26.962	24.835	25.078	24.526	23.883	22.432	21.728
1948	22.91	25.067	27.067	29.331	29.169	26.796	25.278	24.777	24.721	24.934	24.01	20.755
1949	21.849	23.9	27.561	29.326	28.511	26.066	24.933	24.997	24.402	24.948	22.037	21.159
1950	22.048	24.37	26.924	29.567	29.36	26.218	24.755	25.281	24.372	23.989	22.26	21.222

Year												
1951	22.413	24.247	27.594	29.284	28.932	25.963	25.024	25.322	25.81	25.309	23.896	21.305
1952	22.669	25.081	27.498	30.161	29.437	26.908	25.771	25.034	25.216	24.434	22.167	22.171
1953	22.624	24.641	28.771	29.209	30.718	27.721	25.203	24.891	24.848	24.867	22.431	21.196
1954	22.587	25.291	27.301	30.175	29.717	26.711	24.603	24.743	24.388	23.901	21.812	22.033
1955	22.415	25.178	27.57	29.293	28.127	26.18	25.016	23.97	25.252	24.73	21.913	20.799
1956	22.166	23.98	28.107	30.174	28.485	25.831	23.88	24.51	24.548	23.907	22.951	21.356
1957	22.404	24.681	26.528	29.125	29.316	26.676	25.03	24.524	25.183	25.375	23.437	22.533
1958	22.815	25.435	27.967	29.689	29.084	28.021	24.869	25.108	24.807	24.736	23.755	21.927
1959	23.175	25.675	27.854	29.877	29.91	26.323	24.551	24.87	25.224	25.111	23.432	22.258
1960	23.33	24.68	27.818	29.699	29.094	26.738	25.751	25.852	25.057	24.82	22.804	22.158
1961	22.848	25.213	28.941	30.316	28.567	26.165	24.581	25.031	24.921	24.516	22.661	21.902
1962	21.967	24.605	27.434	28.797	28.519	27.171	25.466	24.881	24.887	24.658	22.709	21.654
1963	21.682	23.686	27.422	28.572	29.238	26.203	25.734	24.741	25.43	24.713	23.401	22.168
1964	22.72	25.329	28.387	30.48	30.142	26.956	25.643	24.552	24.308	24.978	22.213	21.186
1965	22.063	24.428	27.521	28.905	29.428	26.9	25.054	25.21	25.447	26.004	24.136	22.874
1966	23.566	25.793	28.699	30.194	29.316	27.323	25.874	25.958	25.108	24.685	23.613	21.987
1967	22.53	24.359	27.086	29.998	29.815	27.136	24.623	24.612	24.728	24.067	22.881	22.839
1968	22.259	25.31	27.263	29.125	29.688	26.965	25.203	25.393	25.149	24.368	23.301	22.635
1969	22.575	25.766	29.085	30.653	29.391	27.443	25.405	25.6	24.66	24.714	23.706	21.76
1970	23.046	25.336	28.486	30.238	29.158	26.061	25.523	24.576	25.01	24.67	22.62	20.767
1971	22.242	25.247	27.051	29.409	28.113	25.481	25.494	25.11	25.262	23.887	22.734	21.069
1972	21.641	24.809	27.314	29.411	28.516	27.729	25.827	26.204	25.757	25.138	23.83	22.879
1973	23.523	25.597	28.283	30.857	30.311	26.396	26.041	24.61	25.472	24.795	22.498	21.362
1974	22.39	24.582	27.957	30.111	28.911	26.107	25.336	25.158	24.916	24.314	23.344	21.082
1975	21.945	25.443	28.017	30.616	29.373	26.503	24.7	25.245	24.619	24.428	22.385	20.851
1976	21.526	23.947	27.856	29.002	28.386	27.122	25.316	24.986	25.708	25.868	24.111	23.146

1977	22.479	25.5	28.013	29.581	27.503	25.712	25.103	24.506	25.911	24.662	24.031	21.861
1978	23.105	25.165	27.694	29.549	28.754	25.676	24.786	24.375	24.936	25.171	24.138	22.397
1979	23.253	25.626	28.118	30.347	28.826	27.153	25.257	25.155	24.741	25.163	23.377	22.756
1980	22.963	25.895	28.107	29.664	30.207	26.136	25.287	24.372	26.026	26.153	23.998	23.003
1981	23.221	25.635	27.838	30.095	29.093	26.393	25.924	24.302	24.513	24.814	23.462	21.649
1982	22.97	26.206	28.664	29.961	29.288	26.936	25.707	25.742	25.82	26.116	23.799	22.2
1983	22.582	26.179	29.024	30.733	30.109	27.367	26.379	25.248	24.961	24.987	23.143	21.992
1984	23.748	24.585	27.425	29.413	30.814	26.819	24.96	25.489	25.732	25.045	23.572	23.235
1985	23.703	26.069	28.599	30.565	30.186	26.273	25.527	25.363	26.453	24.748	23.349	23.257
1986	22.157	25.159	28.77	31.106	30.306	26.92	26.118	25.463	26.141	25.713	24.234	23.391
1987	23.527	24.642	27.786	30.545	29.877	27.421	26.748	25.68	26.879	25.233	24.035	22.573
1988	22.99	26.547	29.27	29.435	30.178	27.784	25.085	24.838	24.9	25.684	23.633	22.275
1989	22.742	24.818	27.086	29.927	30.215	26.791	24.98	25.264	25.284	25.985	23.969	22.637
1990	23.234	25.497	28.417	30.719	27.533	25.942	25.284	24.966	26.293	24.805	23.766	22.387
1991	23.74	25.637	29.261	30.011	30.368	26.528	25.038	25.224	26.272	25.594	22.993	22.09
1992	21.933	25.592	28.462	30.502	29.601	27.294	26.126	25.023	26.135	25.563	24.071	21.549
1993	22.961	24.413	28.302	30.661	30.257	27.636	26.17	25.431	25.351	24.894	23.593	21.337
1994	23.007	25.564	28.772	29.615	30.432	27.195	25.128	25.493	26.684	24.714	22.648	21.311
1995	22.599	25.891	28.502	30.88	28.862	28.576	25.817	25.563	25.683	25.209	24.803	22.712
1996	23.936	25.419	28.904	29.58	30.634	27.044	25.944	24.971	25.161	24.472	24.338	21.827
1997	22.779	25.215	28.089	28.606	29.764	28.177	26.167	25.551	26.056	25.695	24.989	23.334
1998	24.275	26.483	29.312	31.066	30.525	28.377	26.215	25.738	25.62	25.308	24.207	22.208
1999	22.84	25.7	28.984	30.313	28.257	26.327	25.575	25.391	25.776	25.386	23.905	22.283
2000	23.756	25.812	27.966	30.417	28.867	25.853	25.239	25.096	25.758	25.471	24.206	22.107
2001	23.682	26.829	28.694	30.124	30.006	26.693	26.026	25.099	26.305	25.024	24.458	22.5
2002	23.563	25.433	28.968	30.791	29.894	26.906	26.349	25.342	26.434	25.778	23.831	22.929

SPI INPUT DATA FOR ANANTAPUR DISTRICT, A.P.

Year	Month	1st Month	3rd Month	6th Month	9th Month	12th Month
1901	1	10	23	110	231	523
1902	2	12	12	96	239	600
1903	3	7	8	91	339	858
1904	4	1	4	138	267	488
1905	5	1	5	82	312	459
1906	6	51	54	69	368	670
1907	7	7	8	114	303	484
1908	8	1	3	70	190	415
1909	9	3	5	149	391	606
1910	10	0	1	83	465	882
1911	11	15	15	164	290	513
1912	12	0	3	65	309	736
1913	13	10	10	99	237	586
1914	14	4	4	64	294	538
1915	15	1	4	128	337	736
1916	16	1	1	91	356	809
1917	17	1	7	134	437	880
1918	18	7	7	126	210	502
1919	19	6	7	149	345	868
1920	20	0	3	85	214	470
1921	21	1	3	102	321	775
1922	22	2	3	84	192	485
1923	23	1	7	80	197	365
1924	24	2	2	142	334	707

Year	Month	1st Month	3rd Month	6th Month	9th Month	12th Month
1925	25	50	50	244	392	670
1926	26	1	4	112	298	564
1927	27	1	2	56	278	677
1928	28	5	16	129	398	675
1929	29	3	4	87	178	590
1930	30	14	19	184	339	843
1931	31	7	7	90	217	498
1932	32	3	6	50	351	710
1933	33	35	38	263	540	823
1934	34	1	2	84	258	425
1935	35	1	7	80	491	698
1936	36	2	13	137	278	585
1937	37	5	13	150	301	669
1938	38	1	1	73	358	565
1939	39	1	1	114	335	719
1940	40	5	6	198	353	838
1941	41	37	45	113	291	591
1942	42	5	10	92	281	415
1943	43	1	1	192	320	710
1944	44	1	9	88	291	649
1945	45	1	1	127	342	530
1946	46	19	22	156	343	769
1947	47	11	22	74	366	650
1948	48	1	1	158	327	620

Year	Month	1st Month	3rd Month	6th Month	9th Month	12th Month
1949	49	0	0	115	345	669
1950	50	0	5	101	303	610
1951	51	0	0	219	369	603
1952	52	33	35	175	313	510
1953	53	0	1	78	323	849
1954	54	12	13	113	364	558
1955	55	3	3	232	440	790
1956	56	12	12	234	453	1000
1957	57	0	1	109	387	623
1958	58	5	8	129	362	677
1959	59	11	15	147	422	669
1960	60	1	1	67	205	656
1961	61	1	6	154	388	601
1962	62	59	64	231	416	765
1963	63	6	6	165	437	679
1964	64	3	3	32	360	780
1965	65	32	34	96	288	429
1966	66	24	26	124	491	911
1967	67	14	18	107	365	596
1968	68	12	23	189	305	727
1969	69	15	15	126	395	646
1970	70	1	1	184	388	729
1971	71	1	3	119	291	551
1972	72	18	18	167	261	628

Year	Month	1st Month	3rd Month	6th Month	9th Month	12th Month
1973	73	4	4	49	277	732
1974	74	0	0	175	315	769
1975	75	3	3	95	317	861
1976	76	1	1	80	247	399
1977	77	1	1	247	529	836
1978	78	29	35	157	350	715
1979	79	1	15	117	300	711
1980	80	4	4	99	243	475
1981	81	7	8	115	280	720
1982	82	1	1	50	200	529
1983	83	18	18	122	438	752
1984	84	11	16	95	272	432
1985	85	6	6	63	218	429
1986	86	19	26	94	301	666
1987	87	21	21	125	315	566
1988	88	20	25	169	478	761
1989	89	4	4	63	314	639
1990	90	2	2	214	380	633
1991	91	0	1	135	395	717
1992	92	1	1	86	360	633
1993	93	27	27	125	357	686
1994	94	3	5	77	258	580
1995	95	1	3	148	412	843
1996	96	6	6	138	472	838
1997	97	39	40	156	344	798
1998	98	12	12	119	489	882
1999	99	2	5	190	371	644
2000	100	5	11	131	466	768
2001	101	3	4	97	277	580
2002	102	2	3	134	319	541

SPI OUTPUT DATA FOR ANANTAPUR DISTRICT

Year	Month	1st Month	3rd Month	6th Month	9th Month	12th Month
1901	1	0.12	0	0	0	0
1901	2	2.26	0	0	0	0
1901	3	−0.43	1	0	0	0
1901	4	−0.72	−0.31	0	0	0
1901	5	0.15	−0.44	0	0	0
1901	6	−0.41	−0.66	−0.5	0	0
1901	7	−0.73	−0.68	−0.88	0	0
1901	8	−0.96	−1.43	−1.6	0	0
1901	9	0	−0.97	−1.24	−1.16	0
1901	10	−0.31	−0.96	−1.11	−1.21	0
1901	11	0.65	−0.2	−0.94	−1.08	0
1901	12	0.52	−0.06	−0.83	−1.01	−0.96
1902	1	0.12	0.62	−0.66	−0.87	−0.97
1902	2	−0.1	0.25	−0.22	−0.95	−1.08
1902	3	−0.69	−1.25	−0.17	−0.92	−1.08
1902	4	0.24	−0.08	0.37	−0.72	−0.92
1902	5	−0.42	−0.54	−0.56	−0.5	−1.13
1902	6	0.07	−0.47	−0.64	−0.47	−1.09
1902	7	−1.49	−1.28	−1.34	−0.74	−1.25
1902	8	−0.24	−0.99	−1.3	−1.31	−1.11
1902	9	−0.05	−0.86	−1.05	−1.13	−1.11
1902	10	0.91	0.37	−0.23	−0.29	−0.15
1902	11	0.08	0.48	−0.13	−0.35	−0.37
1902	12	0.67	0.75	−0.08	−0.28	−0.34

Year	Month	1st Month	3rd Month	6th Month	9th Month	12th Month
1903	1	1.09	0.11	0.27	−0.27	−0.34
1903	2	−0.1	0.5	0.49	−0.12	−0.33
1903	3	−0.69	−0.89	0.7	−0.15	−0.33
1903	4	−1.35	−1.82	−0.73	−0.03	−0.53
1903	5	0.22	−0.57	−0.51	0.18	−0.35
1903	6	0.34	−0.33	−0.48	0.39	−0.32
1903	7	0.48	0.4	−0.19	−0.25	0.09
1903	8	0.58	0.64	0.18	0.19	0.46
1903	9	0.95	1.06	0.78	0.72	1.04
1903	10	0.58	1.13	1.06	0.81	0.73
1903	11	1.69	1.74	1.71	1.44	1.45
1903	12	0.24	1.41	1.7	1.44	1.41
1904	1	1.94	1.74	1.74	1.63	1.41
1904	2	−0.1	0.15	1.72	1.69	1.41
1904	3	0.47	0.38	1.46	1.7	1.44
1904	4	−0.05	−0.13	1.55	1.66	1.55
1904	5	0.77	0.57	0.52	1.78	1.73
1904	6	−0.6	0.18	0.19	1.34	1.66
1904	7	0.2	0.31	0.15	1.37	1.62
1904	8	−1.67	−1.3	−0.76	−0.79	1.11
1904	9	−1.18	−1.79	−1.42	−1.41	0.04
1904	10	0.47	−1.21	−0.81	−0.87	−0.03
1904	11	−2.09	−0.99	−1.56	−1.22	−1.25
1904	12	−0.75	−0.25	−1.52	−1.27	−1.28

Year	Month	1st Month	3rd Month	6th Month	9th Month	12th Month
1905	1	1.94	−1.9	−1.7	−1.22	−1.28
1905	2	0.56	−0.18	−1.05	−1.61	−1.25
1905	3	0.72	0.91	−0.2	−1.48	−1.23
1905	4	−0.41	−0.21	−1.48	−1.77	−1.29
1905	5	−0.45	−0.73	−0.85	−1.36	−1.8
1905	6	0.07	−0.83	−0.69	−0.66	−1.75
1905	7	−1.61	−1.35	−1.47	−2.13	−2.24
1905	8	1.17	0.38	−0.14	−0.22	−1.05
1905	9	−2.61	−1.1	−1.43	−1.36	−1.45
1905	10	0.01	−0.6	−1.1	−1.18	−1.66
1905	11	−1.06	−2.06	−1.27	−1.49	−1.55
1905	12	−0.75	−0.68	−1.38	−1.56	−1.52
1906	1	2.67	−1.22	−1.01	−1.44	−1.52
1906	2	−0.1	−0.46	−2.15	−1.34	−1.54
1906	3	−0.43	−0.2	−0.76	−1.44	−1.6
1906	4	−2.66	−2.82	−2.51	−1.41	−1.76
1906	5	−2.09	−3.44	−3.52	−3.3	−2.16
1906	6	1.11	−1.5	−1.58	−1.49	−1.96
1906	7	0.17	−0.54	−1.27	−1.8	−1.54
1906	8	1.12	1.26	−0.04	−0.14	−1.74
1906	9	0.4	0.9	0.19	0.16	−0.42
1906	10	0.16	0.78	0.39	0.09	−0.28
1906	11	−0.61	−0.09	0.6	−0.14	−0.2
1906	12	2.43	0.22	0.73	0.2	0.18

Year	Month	1st Month	3rd Month	6th Month	9th Month	12th Month
1907	1	1.09	0.46	0.77	0.44	0.15
1907	2	−0.1	2.43	0.3	0.89	0.15
1907	3	0.32	0.12	0.19	0.71	0.18
1907	4	1.82	1.73	1.2	1.17	0.8
1907	5	−2.44	−0.02	0.78	0.2	0.78
1907	6	−0.09	−0.18	−0.21	0	0.57
1907	7	0.89	−0.77	0.28	0.36	0.77
1907	8	−0.72	−0.2	−0.29	0.23	−0.01
1907	9	−0.12	−0.28	−0.4	−0.42	−0.31
1907	10	−2.46	−2.05	−1.97	−1.35	−1.23
1907	11	0.26	−1.53	−1.33	−1.28	−0.93
1907	12	0.24	−1.74	−1.34	−1.27	−1.29
1908	1	0.12	0.16	−1.84	−1.87	−1.31
1908	2	0.32	0.15	−1.55	−1.35	−1.29
1908	3	1.31	1.17	−1.59	−1.27	−1.21
1908	4	−1.04	−0.4	−0.23	−1.95	−1.96
1908	5	−0.67	−1.01	−1.05	−1.93	−1.66
1908	6	−0.7	−1.74	−1.46	−2.31	−1.88
1908	7	−0.57	−1.4	−1.6	−1.33	−2.42
1908	8	−0.96	−1.47	−2.02	−2.03	−2.76
1908	9	0.72	−0.24	−1.03	−0.94	−2
1908	10	−1.35	−0.95	−1.42	−1.52	−1.54
1908	11	−2.09	−0.94	−1.6	−1.89	−1.92
1908	12	−0.75	−2.37	−1.54	−2.03	−1.95

Year	Month	1st Month	3rd Month	6th Month	9th Month	12th Month
1909	1	1.94	−1.9	−1.44	−1.83	−1.94
1909	2	−0.1	−0.62	−1.04	−1.68	−1.95
1909	3	−0.43	−0.39	−2.51	−1.61	−2.07
1909	4	0.69	0.44	−0.92	−1.33	−1.71
1909	5	0.64	0.71	0.56	−0.63	−1.26
1909	6	−0.98	0.3	0.21	−1.57	−1.38
1909	7	−0.65	−0.4	−0.26	−1.03	−1.41
1909	8	1.35	0.56	0.83	0.72	−0.31
1909	9	0.23	0.74	0.74	0.7	−0.63
1909	10	−0.92	0.23	−0.02	0.02	−0.41
1909	11	−0.99	−1.08	−0.53	−0.26	−0.34
1909	12	−0.28	−1.63	−0.38	−0.27	−0.31
1910	1	−0.32	−1.28	−0.22	−0.38	−0.34
1910	2	0.08	−0.62	−1.18	−0.61	−0.33
1910	3	0.17	−0.2	−1.74	−0.43	−0.3
1910	4	−0.87	−0.92	−1.73	−0.44	−0.56
1910	5	0.01	−0.57	−0.77	−1.4	−0.81
1910	6	−0.18	−0.73	−0.81	−1.83	−0.72
1910	7	2.23	1.34	0.93	0.16	0.21
1910	8	1.44	2.15	1.69	1.59	0.27
1910	9	0.63	2.08	1.64	1.62	0.52
1910	10	0.29	1.22	1.53	1.36	1.01
1910	11	1.17	0.97	1.91	1.63	1.59
1910	12	−1.29	0.7	1.94	1.58	1.57

Year	Month	1st Month	3rd Month	6th Month	9th Month	12th Month
1911	1	−0.32	1.02	1.42	1.71	1.56
1911	2	−0.1	−1.89	0.86	1.83	1.55
1911	3	−0.2	−0.89	0.65	1.9	1.53
1911	4	1.01	0.81	1.16	1.53	1.79
1911	5	0.56	0.86	0.63	1.08	1.95
1911	6	0.07	0.84	0.72	0.89	2.07
1911	7	−1.03	−0.23	0.1	0.66	1.3
1911	8	−0.93	−1.32	−0.54	−0.72	0.36
1911	9	−0.55	−1.56	−0.84	−0.91	−0.29
1911	10	−0.2	−1.27	−1.13	−0.94	−0.55
1911	11	−0.24	−0.97	−1.55	−1.07	−1.18
1911	12	0.88	−0.43	−1.52	−0.97	−1.03
1912	1	0.12	−0.13	−1.31	−1.21	−1.03
1912	2	0.56	0.86	−0.88	−1.47	−1
1912	3	−0.69	−0.39	−0.52	−1.59	−1.02
1912	4	−0.35	−0.56	−0.57	−1.47	−1.33
1912	5	−0.7	−1.17	−0.93	−1.37	−1.82
1912	6	−0.05	−1.09	−1.19	−1.08	−1.95
1912	7	0.2	−0.6	−0.9	−0.98	−1.63
1912	8	0.83	0.6	−0.13	−0.05	−0.91
1912	9	0.92	1.1	0.53	0.49	0.02
1912	10	0.16	0.95	0.52	0.38	0.23
1912	11	1.09	1.03	1.09	0.69	0.73
1912	12	−1.29	0.55	1.09	0.66	0.63

Year	Month	1st Month	3rd Month	6th Month	9th Month	12th Month
1913	1	–0.32	0.94	1.14	0.75	0.62
1913	2	–0.1	–1.89	0.92	1	0.6
1913	3	–0.69	–1.76	0.46	1.03	0.6
1913	4	–1.61	–2.11	0.23	0.86	0.49
1913	5	0.37	–0.44	–0.74	0.65	0.77
1913	6	–0.81	–0.83	–1.03	–0.05	0.69
1913	7	0.51	0.09	–0.56	0.11	0.77
1913	8	–1.47	–1.08	–1.32	–1.51	0.01
1913	9	1.02	0.29	–0.16	–0.25	0.09
1913	10	0.3	0.22	0.17	–0.11	0.17
1913	11	–2.09	0.37	–0.28	–0.46	–0.56
1913	12	0.52	–0.32	–0.09	–0.4	–0.47
1914	1	0.12	–1.28	–0.23	–0.19	–0.47
1914	2	0.08	0.34	0.35	–0.28	–0.46
1914	3	–0.43	–0.61	–0.42	–0.15	–0.45
1914	4	–0.87	–1.13	–1.87	–0.47	–0.39
1914	5	–0.54	–1.23	–1.2	–0.16	–0.69
1914	6	–0.27	–1.3	–1.43	–1.09	–0.64
1914	7	–0.42	–1	–1.49	–2.01	–0.91
1914	8	0.94	0.38	–0.37	–0.38	–0.01
1914	9	0.06	0.34	–0.3	–0.35	–0.65
1914	10	–0.84	–0.17	–0.6	–0.83	–1.23
1914	11	0.26	–0.73	–0.4	–0.8	–0.81
1914	12	–0.13	–0.85	–0.36	–0.79	–0.84

Year	Month	1st Month	3rd Month	6th Month	9th Month	12th Month
1915	1	0.12	0.09	−0.23	−0.63	−0.85
1915	2	0.32	−0.18	−0.79	−0.44	−0.84
1915	3	2.58	2.38	−0.47	−0.14	−0.57
1915	4	−1.35	0.27	0.04	−0.22	−0.61
1915	5	0.42	0.38	0.26	−0.6	−0.33
1915	6	0.49	−0.05	0.5	−0.5	−0.2
1915	7	1.3	1.14	1.1	0.81	0.28
1915	8	−1.14	0.1	0.22	0.14	−0.58
1915	9	1.26	0.92	0.76	1	0.26
1915	10	−0.99	−0.3	0.27	0.27	0.19
1915	11	1.25	0.81	0.66	0.69	0.65
1915	12	−0.75	−0.08	0.55	0.44	0.63
1916	1	−0.32	1.14	0.2	0.62	0.62
1916	2	−0.1	−1.39	0.7	0.57	0.6
1916	3	−0.43	−1.25	−0.2	0.48	0.38
1916	4	−1.13	−1.48	0.55	−0.07	0.4
1916	5	0.33	−0.37	−0.64	0.46	0.37
1916	6	−0.76	−0.75	−0.93	−0.63	0.16
1916	7	1.35	0.66	0.13	0.78	0.18
1916	8	0.67	0.85	0.49	0.34	0.85
1916	9	0.35	1	0.55	0.48	0.24
1916	10	0.98	1.1	1.14	0.92	1.23
1916	11	1.05	1.25	1.4	1.18	1.11
1916	12	−0.75	1.22	1.51	1.15	1.11

Year	Month	1st Month	3rd Month	6th Month	9th Month	12th Month
1917	1	0.12	0.92	1.27	1.31	1.11
1917	2	1.18	−0.06	1.21	1.37	1.14
1917	3	0.83	1.08	1.32	1.56	1.19
1917	4	−0.72	−0.26	0.62	1.17	1.22
1917	5	0.79	0.41	0.32	1.23	1.37
1917	6	0.96	0.73	0.87	1.45	1.7
1917	7	−0.61	0.6	0.37	0.78	1.27
1917	8	1.5	1.31	1.33	1.27	1.83
1917	9	0.81	1.29	1.44	1.51	2.06
1917	10	0.57	1.57	1.52	1.42	1.62
1917	11	0.97	1.17	1.58	1.57	1.55
1917	12	−0.75	0.81	1.41	1.49	1.55
1918	1	0.12	0.83	1.65	1.62	1.54
1918	2	0.08	−0.82	1.09	1.51	1.51
1918	3	2.18	1.93	1.01	1.54	1.6
1918	4	−0.65	0.27	0.73	1.64	1.6
1918	5	0.27	0.25	0.06	1.05	1.46
1918	6	−1.78	−1.09	−0.59	0.4	1.14
1918	7	−1.93	−1.62	−1.45	−0.62	0.97
1918	8	−0.72	−2.33	−1.73	−1.87	−0.03
1918	9	0.53	−0.66	−1.16	−0.94	−0.21
1918	10	−3.29	−1.51	−1.97	−1.9	−1.54
1918	11	1.36	−0.18	−1.28	−1.12	−1.21
1918	12	0.24	−0.57	−0.98	−1.3	−1.14

Year	Month	1st Month	3rd Month	6th Month	9th Month	12th Month
1919	1	1.09	1.36	−0.58	−1.16	−1.13
1919	2	−0.1	0.05	−0.22	−1.31	−1.14
1919	3	0.59	0.38	−0.6	−1	−1.3
1919	4	0.96	0.89	1.52	−0.37	−0.95
1919	5	0.31	0.7	0.63	0.05	−0.96
1919	6	1.32	1.24	1.25	0.14	−0.41
1919	7	0.69	1.07	1.31	1.84	0.13
1919	8	−1.78	−0.11	0.26	0.21	−0.1
1919	9	2.17	1.44	1.81	1.82	1.16
1919	10	−0.64	0.66	0.96	1.09	1.5
1919	11	1.38	1.78	1.41	1.5	1.48
1919	12	0.13	0.3	1.19	1.46	1.48
1920	1	1.94	1.38	1.13	1.36	1.47
1920	2	−0.1	0.05	1.75	1.38	1.47
1920	3	−0.69	−0.61	0.23	1.15	1.42
1920	4	−0.47	−0.85	0.96	0.95	1.19
1920	5	−0.04	−0.57	−0.64	1.44	1.11
1920	6	−0.86	−0.98	−1.11	−0.33	0.76
1920	7	−1.61	−1.47	−1.86	−0.29	0.31
1920	8	−0.07	−1.26	−1.55	−1.59	0.76
1920	9	0.3	−0.47	−0.94	−1.01	−0.72
1920	10	−0.37	−0.37	−0.95	−1.14	−0.51
1920	11	−0.66	−0.59	−1.2	−1.36	−1.4
1920	12	−1.29	−0.99	−1.1	−1.37	−1.43

Year	Month	1st Month	3rd Month	6th Month	9th Month	12th Month
1921	1	1.94	–0.95	–0.74	–1.25	–1.44
1921	2	–0.1	–0.82	–0.69	–1.28	–1.43
1921	3	–0.2	–0.2	–1.09	–1.16	–1.41
1921	4	1.68	1.51	0.26	–0.29	–0.83
1921	5	–2.17	–0.18	–0.39	–0.78	–1.3
1921	6	0.3	–0.05	–0.12	–0.98	–1.14
1921	7	0.74	–0.62	0.23	–0.38	–0.58
1921	8	–0.11	0.23	0	–0.12	–0.67
1921	9	–1.41	–0.86	–0.83	–0.87	–1.54
1921	10	1.87	0.85	0.43	0.75	0.43
1921	11	0.85	1.25	1.09	0.93	0.87
1921	12	–0.75	1.96	1.04	0.9	0.88
1922	1	1.09	0.7	0.95	0.57	0.86
1922	2	–0.1	–0.82	1.17	1.02	0.86
1922	3	–0.69	–0.89	1.97	0.99	0.85
1922	4	–0.65	–1.06	0.14	0.74	0.38
1922	5	–0.04	–0.65	–0.88	0.83	0.74
1922	6	–0.92	–1.09	–1.24	1.43	0.57
1922	7	–1.23	–1.33	–1.79	–0.98	0.15
1922	8	–0.72	–1.71	–1.98	–2.12	–0.06
1922	9	–2.33	–2.76	–2.91	–3	–0.29
1922	10	–0.13	–2.05	–2.26	–2.48	–2.13
1922	11	1.76	–0.16	–1.02	–1.22	–1.32
1922	12	–0.48	0.94	–0.9	–1.22	–1.28

Year	Month	1st Month	3rd Month	6th Month	9th Month	12th Month
1923	1	0.12	1.74	−0.62	−1.09	−1.3
1923	2	1.18	0.05	−0.2	−1.05	−1.24
1923	3	1.56	1.68	1.1	−0.77	−1.09
1923	4	−0.47	0.23	1.66	−0.61	−1.06
1923	5	−0.8	−0.81	−0.88	−0.58	−1.31
1923	6	−0.14	−1.27	−0.86	0.43	−1.24
1923	7	−0.61	−1.23	−1.13	0.56	−1.12
1923	8	−1.52	−1.53	−1.94	−1.97	−1.46
1923	9	−0.01	−1.14	−1.69	−1.49	−0.49
1923	10	−1.53	−1.94	−2.13	−2.06	−1.03
1923	11	−1.45	−1.72	−2.27	−2.44	−2.48
1923	12	−0.75	−2.46	−2.4	−2.62	−2.45
1924	1	0.12	−1.78	−2.4	−2.53	−2.46
1924	2	−0.1	−1.06	−1.86	−2.38	−2.52
1924	3	−0.43	−0.89	−2.68	−2.5	−2.69
1924	4	0.78	0.54	−0.81	−2.17	−2.34
1924	5	0.54	0.68	0.48	−1.29	−1.85
1924	6	−1.34	0.14	0	−1.76	−2.19
1924	7	0.4	0	0.15	−0.63	−1.89
1924	8	0.29	−0.15	0.21	0.07	−1.36
1924	9	1.56	1.39	1.29	1.24	−0.12
1924	10	−1.22	0.47	0.34	0.4	−0.02
1924	11	0.63	0.59	0.35	0.51	0.44
1924	12	−0.48	−0.83	0.54	0.49	0.45

Year	Month	1st Month	3rd Month	6th Month	9th Month	12th Month
1925	1	−0.32	0.44	0.49	0.38	0.43
1925	2	−0.1	−1.06	0.49	0.26	0.43
1925	3	−0.2	−0.89	−0.97	0.48	0.44
1925	4	0.56	0.34	0.39	0.5	0.39
1925	5	1.55	1.56	1.41	1.1	0.8
1925	6	−1.86	1	0.88	−0.26	0.78
1925	7	−0.61	0.38	0.41	0.45	0.56
1925	8	0.16	−0.91	0.32	0.18	0.55
1925	9	−0.27	−0.53	0.02	−0.05	−0.7
1925	10	−0.14	−0.49	−0.25	−0.22	−0.16
1925	11	0.55	−0.33	−0.81	−0.1	−0.19
1925	12	2.4	0.44	−0.13	0.22	0.18
1926	1	1.94	1.16	0.06	0.17	0.19
1926	2	0.08	2.47	0.09	−0.44	0.2
1926	3	−0.2	−0.03	0.42	−0.17	0.2
1926	4	1.43	1.28	1.52	0.36	0.42
1926	5	−1.03	0.05	0.85	0.03	−0.45
1926	6	1.41	0.81	0.76	0.68	0.11
1926	7	−0.28	−0.23	0.39	1	0.18
1926	8	−0.9	−0.25	−0.29	0.24	−0.2
1926	9	0.59	−0.24	0.14	0.12	0.31
1926	10	−0.4	−0.51	−0.55	−0.24	0.16
1926	11	−1.14	−0.45	−0.56	−0.55	−0.24
1926	12	−0.75	−1.15	−1	−0.58	−0.6

Year	Month	1st Month	3rd Month	6th Month	9th Month	12th Month
1927	1	0.12	−1.5	−0.97	−0.93	−0.63
1927	2	0.32	−0.62	−0.55	−0.63	−0.62
1927	3	0	−0.03	−1.24	−1.05	−0.61
1927	4	−1.78	−1.58	−2.31	−1.28	−1.18
1927	5	−0.51	−1.4	−1.62	−1.12	−1.07
1927	6	0.59	−0.9	−0.96	−1.57	−1.38
1927	7	1.22	0.55	0	−0.73	−0.89
1927	8	−0.72	0.27	−0.56	−0.67	−0.93
1927	9	1	0.78	0.28	0.26	−0.59
1927	10	−1.1	−0.47	−0.15	−0.41	−0.83
1927	11	1.58	0.82	0.75	0.3	0.23
1927	12	−0.75	0.21	0.62	0.26	0.24
1928	1	0.12	1.53	0.33	0.46	0.23
1928	2	1.99	0.42	0.81	0.75	0.3
1928	3	−0.2	0.91	0.26	0.66	0.29
1928	4	1.01	1.11	1.77	0.55	0.65
1928	5	−0.33	0.13	0.14	0.75	0.69
1928	6	0.89	0.57	0.67	0.42	0.78
1928	7	0.74	0.47	0.88	1.67	0.64
1928	8	0.49	0.89	0.79	0.78	1.13
1928	9	−0.55	0.03	0.24	0.3	0.28
1928	10	0.57	0.13	0.27	0.48	1.03
1928	11	−0.57	−0.35	0.19	0.17	0.17
1928	12	0.01	0.11	0	0.16	0.2

Year	Month	1st Month	3rd Month	6th Month	9th Month	12th Month
1929	1	0.12	−0.73	−0.21	−0.01	0.2
1929	2	0.32	−0.06	−0.4	0.15	0.13
1929	3	−0.69	−0.61	0.03	−0.06	0.12
1929	4	0.28	0.04	−0.73	−0.25	−0.05
1929	5	−0.48	−0.57	−0.66	−0.68	−0.09
1929	6	−0.86	−0.98	−1.11	−0.52	−0.45
1929	7	−1.38	−1.78	−1.72	−1.99	−0.98
1929	8	−1.25	−2.15	−2.26	−2.31	−1.84
1929	9	1.03	−0.22	−0.72	−0.78	−0.69
1929	10	0.05	0.08	−0.66	−0.69	−0.99
1929	11	0.91	0.93	−0.15	−0.37	−0.41
1929	12	−0.28	0.34	−0.01	−0.37	−0.41
1930	1	1.94	0.82	0.33	−0.37	−0.41
1930	2	0.56	0.05	0.89	−0.18	−0.4
1930	3	1.23	1.32	0.45	0.06	−0.29
1930	4	−0.72	−0.21	0.52	0.23	−0.45
1930	5	1.36	1.09	1.04	1.22	0.21
1930	6	−0.14	0.81	0.99	0.7	0.32
1930	7	0.05	0.96	0.75	0.98	0.59
1930	8	−0.77	−0.77	0.04	−0.01	0.74
1930	9	0.53	−0.13	0.23	0.33	0.41
1930	10	1.69	1.33	1.46	1.37	1.53
1930	11	0.26	1.64	1.02	1.28	1.26
1930	12	0.82	1.63	1.13	1.26	1.33

Year	Month	1st Month	3rd Month	6th Month	9th Month	12th Month
1931	1	–0.32	0.28	1.21	1.38	1.3
1931	2	–0.1	0.5	1.65	1.02	1.28
1931	3	–0.43	–1.25	1.62	1.07	1.2
1931	4	–0.53	–0.85	–0.27	1.03	1.22
1931	5	–0.04	–0.57	–0.51	1.33	0.76
1931	6	0.56	–0.22	–0.39	1.35	0.91
1931	7	–1.03	–0.5	–0.9	–0.69	0.72
1931	8	–1.38	–1.3	–1.59	–1.54	0.64
1931	9	0.45	–0.78	–0.85	–0.94	0.59
1931	10	–1.1	–1.22	–1.23	–1.42	–1.38
1931	11	0.63	–0.32	–0.98	–1.16	–1.16
1931	12	0.24	–0.68	–1.14	–1.13	–1.2
1932	1	–0.32	0.53	–0.93	–1.02	–1.2
1932	2	0.56	0.15	–0.35	–1	–1.17
1932	3	–0.69	–0.61	–0.79	–1.22	–1.18
1932	4	–1.78	–1.82	–0.23	–1.26	–1.29
1932	5	–0.74	–1.76	–1.76	–1.03	–1.53
1932	6	0.49	–1.17	–1.3	–1.35	–1.63
1932	7	–0.86	–0.96	–1.6	–1.01	–1.61
1932	8	1.69	1.27	0.36	0.32	–0.14
1932	9	–0.48	0.54	–0.05	–0.11	–0.66
1932	10	0.26	0.78	0.24	–0.02	0.07
1932	11	1.39	0.47	1.02	0.5	0.48
1932	12	–0.28	0.89	0.94	0.49	0.46

Year	Month	1st Month	3rd Month	6th Month	9th Month	12th Month
1933	1	−0.32	1.32	1.2	0.69	0.45
1933	2	0.56	−0.31	0.4	0.97	0.45
1933	3	1.4	1.25	1	1	0.55
1933	4	0.66	0.92	1.49	1.34	0.82
1933	5	1.77	2	1.92	1.26	1.61
1933	6	−1.94	1.29	1.44	1.42	1.38
1933	7	1.22	1.58	1.79	2.18	1.84
1933	8	1.16	1	2.07	1.99	1.68
1933	9	−0.76	0.63	1.13	1.21	1.55
1933	10	0.58	0.49	1.07	1.2	1.59
1933	11	−0.04	−0.29	0.3	1.04	0.99
1933	12	1.87	0.61	0.79	1.13	1.2
1934	1	0.12	0.46	0.51	1.07	1.2
1934	2	−0.1	1.77	−0.05	0.47	1.18
1934	3	−0.69	−1.25	0.53	0.73	1.08
1934	4	0.84	0.57	0.52	0.58	1.11
1934	5	−1.07	−0.57	−0.01	−0.34	0.22
1934	6	−0.22	−0.66	−0.83	0.1	0.44
1934	7	1.03	−0.24	−0.06	0.09	0.38
1934	8	−1.34	−0.44	−0.83	−0.49	−0.67
1934	9	−2.17	−1.69	−1.83	−1.93	−1.13
1934	10	−0.17	−2.32	−1.87	−1.72	−1.57
1934	11	−0.1	−1.74	−1.63	−1.76	−1.55
1934	12	−0.75	−0.53	−1.68	−1.76	−1.84

Year	Month	1st Month	3rd Month	6th Month	9th Month	12th Month
1935	1	1.09	−0.35	−2.35	−1.98	−1.84
1935	2	1.18	0.05	−1.77	−1.66	−1.78
1935	3	−0.43	0.38	−0.55	−1.7	−1.76
1935	4	0.75	0.69	−0.04	−2.05	−1.78
1935	5	−1.43	−0.84	−0.91	−2.05	−1.89
1935	6	1.89	0.49	0.5	−0.27	−1.36
1935	7	−0.53	−0.19	0.06	−0.29	−1.9
1935	8	2.22	2.42	1.88	1.82	0.07
1935	9	−1.13	0.85	0.93	0.94	0.37
1935	10	0.29	0.99	0.71	0.79	0.59
1935	11	−2.09	−1.17	0.73	0.4	0.37
1935	12	−0.75	−0.46	0.27	0.37	0.38
1936	1	0.12	−2.21	0.45	0.27	0.36
1936	2	1.99	0.42	−1.16	0.72	0.4
1936	3	1.13	1.68	−0.27	0.38	0.47
1936	4	0.12	0.63	−0.81	0.53	0.35
1936	5	0.4	0.39	0.4	−0.9	0.76
1936	6	−0.01	0.14	0.44	−0.25	0.36
1936	7	−0.65	−0.24	−0.02	−0.83	0.34
1936	8	−0.72	−1.04	−0.7	−0.69	−1.53
1936	9	0.22	−0.63	−0.54	−0.38	−0.76
1936	10	−1.05	−1.17	−1.07	−0.93	−1.42
1936	11	1.22	−0.04	−0.62	−0.47	−0.48
1936	12	−0.48	−0.14	−0.66	−0.57	−0.45

Year	Month	1st Month	3rd Month	6th Month	9th Month	12th Month
1937	1	–0.32	1.11	–0.51	–0.57	–0.47
1937	2	1.53	0.15	–0.07	–0.64	–0.49
1937	3	1.13	1.4	–0.02	–0.57	–0.49
1937	4	1.48	1.71	1.69	0	–0.14
1937	5	–0.45	0.62	0.58	0.15	–0.4
1937	6	0.45	0.6	0.81	0.21	–0.34
1937	7	–0.11	–0.38	0.56	1.08	–0.22
1937	8	–1.25	–0.84	–0.37	–0.41	–0.39
1937	9	0.28	–0.6	–0.26	–0.14	–0.34
1937	10	0.93	0.27	0.02	0.43	0.76
1937	11	–0.41	0.55	–0.01	0.17	0.14
1937	12	0.01	0.54	–0.1	0.09	0.18
1938	1	–0.32	–0.62	–0.05	–0.24	0.17
1938	2	0.08	–0.31	0.48	–0.06	0.12
1938	3	0.47	0.12	0.52	–0.13	0.07
1938	4	–0.79	–0.73	–1.1	–0.24	–0.39
1938	5	–0.39	–0.89	–1.05	0.07	–0.39
1938	6	–0.92	–1.5	–1.52	–0.25	–0.67
1938	7	–0.61	–1.3	–1.65	–1.88	–0.8
1938	8	1.79	1.09	0.48	0.39	0.64
1938	9	0.86	1.56	0.87	0.86	0.99
1938	10	–2.75	0.58	–0.05	–0.21	–0.49
1938	11	–1.59	–1.17	–0.24	–0.55	–0.61
1938	12	–0.75	–3.78	–0.11	–0.61	–0.63

Year	Month	1st Month	3rd Month	6th Month	9th Month	12th Month
1939	1	0.12	−1.9	0.07	−0.46	−0.63
1939	2	−0.1	−1.06	−1.3	−0.33	−0.63
1939	3	1.99	1.68	−2.86	0	−0.5
1939	4	1.61	1.89	0.35	0.59	0.03
1939	5	−2.54	0.13	−0.09	−1.16	−0.31
1939	6	0.38	−0.16	0.17	−2.05	−0.11
1939	7	−1.08	−1.96	−0.44	−1.19	−0.19
1939	8	0.83	0.27	0.22	0.08	−0.97
1939	9	0.63	0.53	0.35	0.49	−1.12
1939	10	0.38	0.89	0.03	0.51	0.09
1939	11	0.58	0.69	0.64	0.59	0.52
1939	12	−0.75	0.38	0.56	0.42	0.52
1940	1	0.12	0.38	0.85	0.07	0.52
1940	2	0.08	−0.82	0.6	0.57	0.52
1940	3	−0.2	−0.39	0.33	0.51	0.38
1940	4	0.75	0.57	0.45	0.91	0.14
1940	5	1.44	1.53	1.39	1.18	1.05
1940	6	0.23	1.61	1.54	1.03	1.07
1940	7	−0.81	0.81	0.91	0.82	1.12
1940	8	−0.4	−0.79	0.38	0.26	0.68
1940	9	−0.22	−0.91	0.1	0.05	0.2
1940	10	1.54	0.85	1	1.05	1.06
1940	11	1.11	1.49	0.87	1.33	1.27
1940	12	0.01	1.8	0.86	1.31	1.29

Year	Month	1st Month	3rd Month	6th Month	9th Month	12th Month
1941	1	0.12	1.04	1.11	1.23	1.28
1941	2	1.53	0.5	1.5	0.88	1.32
1941	3	−0.69	0.38	1.86	0.85	1.31
1941	4	−0.41	−0.35	0.71	0.99	1.12
1941	5	−0.48	−0.98	−0.91	1.07	0.52
1941	6	−0.31	−1.09	−1.06	1.31	0.44
1941	7	−0.73	−1.16	−1.36	−0.35	0.46
1941	8	0.31	−0.38	−1	−0.97	0.76
1941	9	0.81	0.42	−0.14	−0.14	1.31
1941	10	−0.32	0.26	−0.29	−0.4	−0.05
1941	11	−0.61	−0.11	−0.35	−0.69	−0.69
1941	12	1.95	−0.38	−0.03	−0.41	−0.41
1942	1	−0.32	0.13	0.17	−0.32	−0.43
1942	2	0.99	2.02	0.18	−0.11	−0.45
1942	3	−0.69	−0.2	−0.45	−0.07	−0.45
1942	4	−0.53	−0.61	−0.33	0	−0.46
1942	5	−0.04	−0.59	0.1	−0.14	−0.35
1942	6	1.41	0.34	0.27	−0.28	0.01
1942	7	−1.55	−0.17	−0.5	−0.48	−0.12
1942	8	−0.11	−0.22	−0.64	−0.23	−0.34
1942	9	−0.93	−1.51	−1.12	−1.16	−1.38
1942	10	−1.47	−1.97	−1.6	−1.75	−1.76
1942	11	−0.53	−2.3	−1.85	−1.97	−1.69
1942	12	0.01	−1.89	−2.45	−1.93	−1.97

Year	Month	1st Month	3rd Month	6th Month	9th Month	12th Month
1943	1	0.12	−0.69	−2.17	−1.82	−1.97
1943	2	0.08	−0.18	−2.35	−1.89	−2
1943	3	0.17	−0.03	−1.98	−2.5	−1.96
1943	4	0.16	0.04	−0.7	−2.15	−1.81
1943	5	1.63	1.5	1.42	−1.06	−1.01
1943	6	−1.03	1.1	1.06	−0.76	−1.66
1943	7	−1.44	0.37	0.26	−0.27	−1.66
1943	8	−0.13	−1.32	0	−0.07	−1.84
1943	9	−0.05	−0.78	−0.11	−0.14	−1.3
1943	10	0.92	0.42	0.46	0.42	0.13
1943	11	0.61	0.73	−0.04	0.51	0.47
1943	12	−0.75	0.91	0.1	0.46	0.45
1944	1	−0.32	0.4	0.43	0.48	0.43
1944	2	1.53	0.05	0.69	−0.07	0.48
1944	3	2.06	2.17	1.14	0.28	0.62
1944	4	−1.04	0.27	0.32	0.43	0.47
1944	5	−0.57	−0.65	−0.72	0.35	−0.33
1944	6	1.52	−0.11	0.36	0.92	0.17
1944	7	0.64	0.62	0.59	0.57	0.6
1944	8	−1.62	0.01	−0.46	−0.5	0.26
1944	9	−0.08	−0.67	−0.7	−0.45	0.24
1944	10	0.89	−0.08	0.17	0.18	0.21
1944	11	0.11	0.46	0.32	0.05	0.02
1944	12	−0.75	0.64	−0.07	−0.14	0.03

Year	Month	1st Month	3rd Month	6th Month	9th Month	12th Month
1945	1	0.12	–0.16	–0.24	0.03	0.03
1945	2	–0.1	–1.06	0.35	0.23	–0.03
1945	3	–0.69	–1.25	0.57	–0.15	–0.2
1945	4	0.81	0.54	0	–0.16	0.09
1945	5	0.22	0.39	0.18	0.42	0.3
1945	6	–2.12	–0.38	–0.55	0.25	–0.33
1945	7	1.59	0.45	0.56	0.23	0
1945	8	–0.05	0.19	0.31	0.17	0.42
1945	9	–0.37	0.23	–0.03	–0.11	0.28
1945	10	–1.27	–1.42	–0.88	–0.78	–0.93
1945	11	0.03	–1.45	–1.02	–0.83	–0.93
1945	12	–0.75	–1.42	–0.74	–0.84	–0.91
1946	1	0.12	–0.24	–1.49	–1	–0.91
1946	2	0.56	–0.46	–1.53	–1.09	–0.89
1946	3	1.04	1	–1.33	–0.7	–0.79
1946	4	0.81	0.95	0.21	–1.19	–0.79
1946	5	0.18	0.57	0.42	–1.11	–0.82
1946	6	–1.34	–0.2	–0.06	–1.23	–0.78
1946	7	–0.65	–0.98	–0.46	–0.68	–1.55
1946	8	0.72	–0.23	0.07	–0.03	–1.25
1946	9	–0.25	–0.12	–0.27	–0.2	–1.15
1946	10	0.39	0.28	–0.2	–0.02	–0.19
1946	11	1.82	1.03	0.69	0.79	0.74
1946	12	1.13	1.48	1	0.81	0.86

Year	Month	1st Month	3rd Month	6th Month	9th Month	12th Month
1947	1	1.09	1.99	1.23	0.72	0.86
1947	2	1.85	1.53	1.18	0.82	0.9
1947	3	−0.43	0.81	1.55	1.02	0.83
1947	4	−0.2	−0.04	1.85	1.17	0.65
1947	5	−1.2	−1.47	−0.85	0.62	0.35
1947	6	−0.22	−1.47	−1.32	0.86	0.51
1947	7	−0.21	−1.25	−1.29	0.76	0.6
1947	8	1.6	1.17	0.34	0.56	1.19
1947	9	0.79	1.45	0.76	0.8	1.71
1947	10	−0.54	0.98	0.33	0.27	1.12
1947	11	−0.66	−0.28	0.41	−0.05	0.08
1947	12	0.6	−0.98	0.51	0	0.03
1948	1	0.12	−0.58	0.63	0.07	0.02
1948	2	−0.1	0.34	−0.29	0.4	−0.06
1948	3	0.17	−0.2	−1.07	0.48	−0.03
1948	4	0.42	0.27	−0.47	0.63	0.08
1948	5	0.99	0.96	0.95	0.13	0.67
1948	6	−2.31	0.16	0.08	−0.84	0.46
1948	7	−0.24	−0.17	−0.14	−0.57	0.46
1948	8	0.44	−0.54	0.11	0.09	−0.26
1948	9	−0.61	−0.44	−0.37	−0.41	−1.11
1948	10	−0.96	−1.07	−0.95	−0.92	−1.19
1948	11	1.59	−0.17	−0.49	−0.12	−0.13
1948	12	−0.75	0.28	−0.2	−0.16	−0.19

Year	Month	1st Month	3rd Month	6th Month	9th Month	12th Month
1949	1	−0.32	1.53	−0.13	−0.21	−0.21
1949	2	−0.1	−1.39	−0.3	−0.59	−0.21
1949	3	−0.43	−1.25	0.19	−0.28	−0.23
1949	4	−1.35	−1.7	0.99	−0.43	−0.46
1949	5	0.84	0.17	−0.07	−0.27	−0.54
1949	6	0.27	0.28	0.13	0.22	−0.21
1949	7	−0.31	0.45	−0.12	0.97	−0.23
1949	8	0.73	0.4	0.36	0.21	−0.1
1949	9	0.79	0.78	0.77	0.7	0.68
1949	10	0.01	0.69	0.71	0.47	1.02
1949	11	−0.45	0.14	0.26	0.25	0.17
1949	12	−1.29	−0.52	0.18	0.22	0.17
1950	1	−0.32	−0.86	0.3	0.4	0.16
1950	2	0.99	−0.46	0.06	0.2	0.2
1950	3	0.83	0.91	−0.47	0.21	0.25
1950	4	−2.66	−1.06	−1.48	0.07	0.21
1950	5	0.44	−0.25	−0.42	−0.12	0.06
1950	6	−0.36	−0.63	−0.5	−0.8	−0.07
1950	7	−0.24	−0.15	−0.61	−1.09	−0.05
1950	8	0.5	0	−0.25	−0.35	−0.2
1950	9	−0.03	0.01	−0.35	−0.29	−0.73
1950	10	0.27	0.19	0.05	−0.16	−0.48
1950	11	0.17	−0.04	−0.1	−0.22	−0.29
1950	12	−1.29	0.04	−0.07	−0.31	−0.27

Year	Month	1st Month	3rd Month	6th Month	9th Month	12th Month
1951	1	−0.32	−0.13	0.02	−0.08	−0.28
1951	2	−0.1	−1.89	−0.17	−0.2	−0.32
1951	3	1.4	1	0.1	−0.03	−0.27
1951	4	0.75	0.92	0.26	0.2	0.08
1951	5	1.61	1.89	1.72	0.74	0.56
1951	6	0.15	1.77	1.87	0.94	0.66
1951	7	0.29	1.44	1.66	1.18	0.79
1951	8	−1.52	−0.86	0.63	0.48	0.2
1951	9	0.05	−0.71	0.34	0.4	0.24
1951	10	−0.34	−1.12	−0.13	0.04	−0.11
1951	11	−0.99	−0.85	−1.2	−0.25	−0.35
1951	12	−1.29	−1.06	−1.34	−0.38	−0.34
1952	1	−0.32	−1.5	−1.56	−0.52	−0.34
1952	2	0.56	−0.82	−0.96	−1.29	−0.32
1952	3	−0.2	−0.2	−1.16	−1.39	−0.42
1952	4	−1.47	−1.39	−2.19	−1.87	−0.75
1952	5	1.25	0.64	0.46	−0.6	−0.96
1952	6	−1.48	0.03	−0.04	−0.98	−1.31
1952	7	−0.18	0.33	−0.19	−0.9	−1.47
1952	8	−0.35	−1.06	−0.53	−0.66	−1.25
1952	9	−1.55	−1.57	−1.35	−1.39	−2.05
1952	10	0.35	−1.03	−0.67	−0.92	−1.36
1952	11	−2.09	−1.32	−1.7	−1.3	−1.39
1952	12	1.79	0.01	−1.16	−1.04	−1.08

Year	Month	1st Month	3rd Month	6th Month	9th Month	12th Month
1953	1	0.12	–0.32	–1.18	–0.83	–1.07
1953	2	–0.1	1.68	–1.03	–1.46	–1.1
1953	3	–0.69	–1.25	–0.09	–1.26	–1.11
1953	4	1.12	0.86	0.1	–0.93	–0.65
1953	5	–1.88	–0.73	–0.18	–1.33	–1.67
1953	6	0.42	–0.42	–0.59	–0.38	–1.38
1953	7	1.94	0.53	0.81	0.36	–0.61
1953	8	–0.96	0.58	0.05	0.32	–0.89
1953	9	0.76	0.85	0.53	0.46	0.29
1953	10	2.03	1.77	1.66	1.77	1.58
1953	11	–1.79	1.79	1.73	1.42	1.55
1953	12	–1.29	1.55	1.66	1.38	1.35
1954	1	0.12	–2.21	1.23	1.23	1.34
1954	2	–0.1	–1.39	1.7	1.65	1.34
1954	3	0.32	–0.03	1.57	1.65	1.36
1954	4	–0.41	–0.5	–1.82	1.09	1.1
1954	5	0.22	–0.15	–0.42	1.52	1.48
1954	6	–0.86	–0.68	–0.73	1.13	1.36
1954	7	1.37	0.57	0.26	–0.57	1.18
1954	8	0.46	0.66	0.42	0.27	1.73
1954	9	–1.72	–0.26	–0.62	–0.64	0.7
1954	10	0.39	–0.57	–0.22	–0.35	–0.82
1954	11	–2.09	–1.35	–0.64	–0.7	–0.8
1954	12	0.67	–0.19	–0.43	–0.66	–0.68

Year	Month	1st Month	3rd Month	6th Month	9th Month	12th Month
1955	1	0.12	−1.16	−0.97	−0.56	−0.69
1955	2	0.08	0.5	−1.31	−0.62	−0.68
1955	3	1.13	0.91	−0.14	−0.39	−0.62
1955	4	0.49	0.63	−0.52	−0.82	−0.45
1955	5	1.91	2.04	2.08	−0.01	0.29
1955	6	−0.41	1.78	1.87	0.78	0.36
1955	7	−0.57	1.19	1.3	0.52	−0.18
1955	8	0.75	0.09	1.4	1.4	−0.04
1955	9	0.56	0.53	1.3	1.35	0.94
1955	10	0.45	0.84	1.16	1.23	0.89
1955	11	−0.15	0.38	0.29	1.05	1.06
1955	12	−0.28	0.09	0.36	0.96	1
1956	1	0.12	−0.38	0.55	0.93	0.99
1956	2	0.08	−0.46	0.31	0.23	0.99
1956	3	−0.43	−0.61	0.02	0.31	0.92
1956	4	2.31	2.18	1.02	1.14	1.4
1956	5	0.77	1.94	1.85	1.15	0.95
1956	6	0.8	2.29	2.22	1.19	1.19
1956	7	1.26	1.51	2.44	1.76	1.63
1956	8	−1.03	0.24	1.45	1.35	1.16
1956	9	0.52	0.29	1.4	1.36	1.12
1956	10	1.52	1.08	1.49	1.96	1.76
1956	11	1.27	1.94	1.71	2.24	2.2
1956	12	0.67	1.93	1.65	2.26	2.24

Year	Month	1st Month	3rd Month	6th Month	9th Month	12th Month
1957	1	−0.32	1.29	1.44	1.79	2.23
1957	2	0.08	0.42	1.94	1.7	2.22
1957	3	0	−0.39	1.96	1.62	2.23
1957	4	−0.01	−0.17	1.05	1.35	1.71
1957	5	0.29	0.05	0.06	1.81	1.59
1957	6	2.61	1.61	1.54	2.44	2.07
1957	7	−0.08	1.42	1.21	1.72	1.78
1957	8	0.01	1.01	0.86	0.85	2.21
1957	9	−1	−0.94	0.08	0.03	1.56
1957	10	0.25	−0.67	0.16	0.08	0.53
1957	11	−0.24	−0.82	−0.06	−0.09	−0.09
1957	12	−1.29	−0.18	−0.91	−0.13	−0.17
1958	1	0.12	−0.58	−0.93	−0.09	−0.17
1958	2	0.56	−0.62	−0.92	−0.13	−0.15
1958	3	0.32	0.38	−0.2	−0.92	−0.14
1958	4	0.96	0.92	−0.02	−0.68	0.07
1958	5	−0.17	0.28	0.1	−0.76	−0.08
1958	6	0.3	0.32	0.32	−0.09	−0.77
1958	7	−0.94	−0.7	−0.25	−0.66	−0.98
1958	8	0.98	0.44	0.45	0.34	−0.51
1958	9	0	0.19	0.26	0.27	−0.02
1958	10	0.38	0.59	0.17	0.33	0.07
1958	11	0.11	0.04	0.21	0.24	0.17
1958	12	0.01	0.16	0.16	0.21	0.22

Year	Month	1st Month	3rd Month	6th Month	9th Month	12th Month
1959	1	−0.32	−0.08	0.41	0.05	0.2
1959	2	0.79	0.05	0	0.18	0.21
1959	3	−0.69	−0.39	0.1	0.11	0.17
1959	4	0.69	0.54	0.06	0.47	0.11
1959	5	0.44	0.51	0.44	0.15	0.29
1959	6	2.04	1.63	1.55	0.87	0.71
1959	7	0.37	1.38	1.43	1.01	0.99
1959	8	0.14	0.98	1.08	1.03	0.65
1959	9	0.53	0.43	1.13	1.1	0.94
1959	10	−0.92	−0.39	0.33	0.39	0.25
1959	11	−0.45	−0.68	0.01	0.15	0.11
1959	12	0.6	−1.24	−0.49	0.2	0.17
1960	1	−0.32	−0.45	−0.63	0.1	0.16
1960	2	−0.1	0.25	−0.69	0	0.13
1960	3	0.59	0.12	−1.31	−0.52	0.18
1960	4	−0.59	−0.56	−0.86	−0.79	−0.03
1960	5	−0.64	−1.01	−1.02	−1.13	−0.36
1960	6	−0.7	−1.53	−1.55	−1.97	−1.07
1960	7	0.64	−0.52	−0.82	−1.09	−1
1960	8	−1.91	−1.1	−1.69	−1.69	−1.79
1960	9	1.99	1.22	0.51	0.49	−0.41
1960	10	−0.59	0.48	0.14	0	−0.23
1960	11	0.65	1.23	0.51	0.15	0.13
1960	12	−0.75	−0.39	0.61	0.08	0.07

Year	Month	1st Month	3rd Month	6th Month	9th Month	12th Month
1961	1	0.12	0.46	0.51	0.2	0.07
1961	2	1.18	−0.06	1.19	0.47	0.11
1961	3	−0.43	0.25	−0.42	0.6	0.07
1961	4	−0.35	−0.35	0.1	0.38	0.1
1961	5	1.2	0.82	0.74	1.38	0.68
1961	6	1.26	1.42	1.41	0.37	1.06
1961	7	0.17	1.46	1.2	1.1	0.95
1961	8	0.03	0.44	0.8	0.73	1.48
1961	9	−1.99	−1.34	−0.32	−0.32	−0.67
1961	10	0.4	−0.92	0.01	−0.1	−0.04
1961	11	−0.24	−1.11	−0.63	−0.3	−0.34
1961	12	−0.75	−0.01	−1.03	−0.32	−0.33
1962	1	−0.32	−0.58	−1.17	−0.24	−0.34
1962	2	0.99	−0.31	−1.18	−0.69	−0.35
1962	3	0.17	0.38	−0.02	−1.05	−0.33
1962	4	1.06	1.05	0.08	−0.85	−0.03
1962	5	0.81	1.14	1.04	−0.49	−0.21
1962	6	−0.55	0.86	0.87	0.35	−0.63
1962	7	−0.31	0.09	0.52	0	−0.77
1962	8	0.32	−0.26	0.45	0.37	−0.7
1962	9	0.16	0.01	0.37	0.38	0.2
1962	10	0.72	0.58	0.47	0.65	0.4
1962	11	−0.1	0.37	0.1	0.48	0.43
1962	12	2.68	0.94	0.62	0.82	0.83

Year	Month	1st Month	3rd Month	6th Month	9th Month	12th Month
1963	1	0.12	0.83	0.77	0.65	0.82
1963	2	−0.1	2.67	0.78	0.46	0.79
1963	3	0.32	−0.03	0.93	0.59	0.8
1963	4	1.94	1.85	1.55	1.21	1.04
1963	5	−0.28	0.99	1.74	1.08	0.73
1963	6	−0.86	0.57	0.52	1.01	0.71
1963	7	0.2	−0.66	0.45	0.77	0.84
1963	8	1.38	0.93	1.33	1.8	1.46
1963	9	−0.79	0.4	0.56	0.54	1.04
1963	10	0.5	0.58	0.17	0.62	0.82
1963	11	−2.09	−0.74	−0.06	0.27	0.6
1963	12	0.13	−0.14	0.1	0.26	0.24
1964	1	−0.32	−1.67	0.08	−0.23	0.22
1964	2	−0.1	−0.31	−0.81	−0.11	0.22
1964	3	−0.69	−1.76	−0.27	0.02	0.19
1964	4	−0.47	−0.85	−1.92	−0.13	−0.4
1964	5	−2.17	−2.44	−2.56	−1.68	−0.76
1964	6	0.86	−1.27	−1.49	−0.94	−0.46
1964	7	2.67	1.34	0.94	0.11	0.46
1964	8	−0.24	1.58	0.5	0.42	−0.39
1964	9	1.85	2.33	1.74	1.68	1.26
1964	10	−0.72	0.71	1.12	0.95	0.56
1964	11	0.51	0.99	1.59	0.98	0.94
1964	12	−0.28	−0.57	1.46	0.97	0.92

Year	Month	1st Month	3rd Month	6th Month	9th Month	12th Month
1965	1	0.12	0.34	0.67	1.08	0.92
1965	2	0.08	−0.46	0.92	1.54	0.93
1965	3	−0.43	−0.61	−0.69	1.42	0.93
1965	4	0.35	0.12	0.19	0.63	1.04
1965	5	−1.38	−1.17	−1.35	0.43	1.14
1965	6	0.11	−1.01	−1.14	−1.18	1.04
1965	7	−1.03	−1.67	−1.58	−1.16	−0.07
1965	8	0.47	−0.17	−0.89	−1	0.23
1965	9	−0.34	−0.48	−0.96	−1.03	−1.41
1965	10	−3.46	−1.55	−2.03	−2	−1.9
1965	11	−0.18	−2.18	−1.74	−2.07	−2.15
1965	12	1.75	−1.82	−1.54	−1.77	−1.83
1966	1	1.09	0.32	−1.32	−1.84	−1.83
1966	2	−0.1	1.68	−1.8	−1.5	−1.83
1966	3	−0.43	−0.61	−1.98	−1.62	−1.83
1966	4	−0.95	−1.3	−0.36	−1.61	−2.06
1966	5	0.46	−0.18	0.28	−1.74	−1.5
1966	6	1.81	0.91	0.81	−0.89	−1.08
1966	7	1.13	1.69	1.21	1.03	−0.49
1966	8	1.19	1.99	1.69	1.89	−0.1
1966	9	0.29	1.22	1.46	1.42	0.25
1966	10	0.19	0.77	1.34	1.13	1.11
1966	11	1.58	0.99	1.84	1.65	1.78
1966	12	1.39	1.18	1.63	1.75	1.73

Year	Month	1st Month	3rd Month	6th Month	9th Month	12th Month
1967	1	3.31	1.83	1.5	1.92	1.74
1967	2	−0.1	1.44	1.12	1.93	1.74
1967	3	0.83	0.91	1.26	1.67	1.78
1967	4	−0.35	−0.26	1.61	1.4	1.82
1967	5	−0.17	−0.41	−0.03	0.85	1.69
1967	6	0.42	−0.36	−0.23	0.94	1.47
1967	7	2.18	1.41	1.17	2.16	1.82
1967	8	−0.93	0.77	0.38	0.57	1.16
1967	9	−0.48	0.12	−0.12	−0.07	0.78
1967	10	0.11	−0.94	−0.03	−0.12	0.67
1967	11	−1.23	−0.88	−0.26	−0.43	−0.31
1967	12	0.82	−0.4	−0.28	−0.41	−0.37
1968	1	−0.32	−0.86	−1.26	−0.33	−0.42
1968	2	1.99	1.22	−0.73	−0.16	−0.33
1968	3	2.18	2.43	−0.06	−0.05	−0.2
1968	4	1.27	1.87	0.59	−0.58	0.14
1968	5	0.18	1.13	1.33	−0.14	0.24
1968	6	−0.05	0.64	1.14	0.2	0.15
1968	7	−0.35	−0.3	0.75	0.11	−0.71
1968	8	−2.05	−1.53	−0.46	−0.28	−1.01
1968	9	1.5	0.38	0.58	0.84	0.35
1968	10	0.07	0.39	0.15	0.61	0.3
1968	11	0.26	1	0.13	0.5	0.58
1968	12	0.67	0.04	0.21	0.37	0.57

Year	Month	1st Month	3rd Month	6th Month	9th Month	12th Month
1969	1	−0.32	0.24	0.33	0.13	0.56
1969	2	−0.1	0.34	0.99	0.12	0.49
1969	3	−0.43	−1.25	−0.07	0.13	0.32
1969	4	−0.95	−1.3	−0.45	0.08	−0.08
1969	5	0.71	0.11	0.1	0.91	0.1
1969	6	0.07	0.12	−0.04	−0.09	0.12
1969	7	−0.18	0.31	−0.19	−0.17	0.16
1969	8	1.25	0.89	0.77	0.76	1.28
1969	9	−2.38	−0.51	−0.44	−0.52	−0.54
1969	10	0.78	0.2	0.26	0.03	0.01
1969	11	0.11	−0.64	−0.02	−0.03	−0.04
1969	12	0.88	0.66	0.05	0.05	−0.01
1970	1	−0.32	0.16	0.13	0.2	−0.01
1970	2	−0.1	0.58	−0.61	0	−0.01
1970	3	−0.43	−1.25	0.59	−0.02	−0.01
1970	4	0.08	−0.21	−0.14	0.03	0.12
1970	5	1.6	1.4	1.44	0.1	0.5
1970	6	−0.5	1.21	1.08	1.02	0.41
1970	7	0.14	1.13	0.91	0.68	0.49
1970	8	0.39	0.03	0.87	0.88	0.02
1970	9	0.64	0.58	1.04	0.98	1.18
1970	10	0.56	0.79	1.1	1.01	0.94
1970	11	−1.79	0.29	0.19	0.66	0.68
1970	12	−0.75	−0.13	0.25	0.63	0.59

Year	Month	1st Month	3rd Month	6th Month	9th Month	12th Month
1971	1	−0.32	−2.21	0.26	0.66	0.58
1971	2	0.32	−0.82	0.2	0.11	0.59
1971	3	0.59	0.38	−0.14	0.24	0.63
1971	4	0.56	0.57	−0.86	0.33	0.71
1971	5	0.03	0.21	0.02	0.2	0.12
1971	6	−0.98	−0.38	−0.36	−0.4	0.05
1971	7	−1.28	−1.3	−0.98	−1.72	−0.24
1971	8	0.6	−0.49	−0.39	−0.52	−0.17
1971	9	−1.21	−1.01	−1.13	−1.12	−1.16
1971	10	0.79	0.1	−0.47	−0.38	−0.85
1971	11	−0.86	−0.53	−0.75	−0.66	−0.75
1971	12	−0.75	0.23	−0.62	−0.73	−0.73
1972	1	0.12	−1.22	−0.33	−0.82	−0.73
1972	2	−0.1	−1.06	−0.65	−0.85	−0.75
1972	3	−0.69	−1.25	0.14	−0.71	−0.8
1972	4	−0.47	−0.85	−1.64	−0.54	−0.99
1972	5	1.25	0.82	0.63	−0.25	−0.5
1972	6	0.11	0.86	0.72	0.47	−0.34
1972	7	−1.61	0.3	−0.11	−0.76	−0.4
1972	8	−1.98	−2.07	−1.06	−1.21	−1.35
1972	9	0.91	−0.58	−0.1	−0.18	−0.1
1972	10	0.26	0.01	0.1	−0.09	−0.46
1972	11	−0.04	0.54	−0.5	−0.19	−0.27
1972	12	1.07	0.13	−0.42	−0.07	−0.13

Year	Month	1st Month	3rd Month	6th Month	9th Month	12th Month
1973	1	−0.32	0.09	−0.07	0.03	−0.15
1973	2	−0.1	0.8	0.58	−0.45	−0.15
1973	3	−0.69	−1.76	0.02	−0.51	−0.14
1973	4	−1.47	−1.96	−0.78	−0.39	−0.23
1973	5	−0.8	−1.76	−1.48	−0.07	−0.98
1973	6	0.66	−1.03	−1.24	−0.56	−0.91
1973	7	−0.99	−0.96	−1.62	−1.4	−0.81
1973	8	0.8	0.38	−0.58	−0.5	0.07
1973	9	0.16	0.16	−0.37	−0.46	−0.41
1973	10	1.54	1.59	0.98	0.72	0.64
1973	11	0.22	1.27	1.19	0.67	0.7
1973	12	−0.13	1.38	1.09	0.67	0.61
1974	1	−0.32	0.02	1.37	0.85	0.6
1974	2	−0.1	−0.62	1.2	1.12	0.6
1974	3	−0.43	−1.25	1.35	1.03	0.61
1974	4	−0.3	−0.61	−0.45	1.22	0.71
1974	5	1.58	1.26	1.12	1.57	1.44
1974	6	−0.36	1.1	0.97	1.65	1.33
1974	7	−0.18	1.01	0.67	0.41	1.48
1974	8	−0.9	−1.06	−0.04	−0.15	0.98
1974	9	1.83	1.02	1.37	1.31	2.02
1974	10	0.41	1.14	1.32	1.18	1.07
1974	11	−0.61	1.26	0.55	0.93	0.87
1974	12	−1.29	−0.13	0.6	0.89	0.85

Year	Month	1st Month	3rd Month	6th Month	9th Month	12th Month
1975	1	0.12	−1	0.71	0.98	0.84
1975	2	0.08	−1.06	1.18	0.47	0.85
1975	3	0.32	0.12	−0.17	0.57	0.87
1975	4	−0.53	−0.56	−1.3	0.56	0.84
1975	5	0.11	−0.32	−0.56	0.94	0.29
1975	6	−1.28	−1.03	−1.06	−0.73	0.17
1975	7	1.32	0.33	0.01	−0.59	0.59
1975	8	0.13	0.29	−0.02	−0.16	0.97
1975	9	0.23	0.59	0.05	0.03	−0.19
1975	10	1.9	1.68	1.5	1.37	1.06
1975	11	0.82	1.92	1.71	1.49	1.43
1975	12	−0.28	1.99	1.88	1.44	1.44
1976	1	0.12	0.69	1.68	1.56	1.44
1976	2	−0.1	−0.62	1.85	1.65	1.43
1976	3	−0.43	−0.89	2.01	1.84	1.4
1976	4	1.01	0.78	0.83	1.77	1.63
1976	5	−1.48	−0.62	−0.82	1.53	1.36
1976	6	−1.28	−1.25	−1.41	1.43	1.41
1976	7	−1.18	−2.62	−1.88	−1.06	0.88
1976	8	0.55	−0.59	−1	−1.12	1.13
1976	9	−2.06	−1.46	−1.95	−2.04	0.29
1976	10	−0.86	−1.66	−2.51	−2.23	−1.92
1976	11	0.42	−1.96	−1.88	−2.01	−2.1
1976	12	−0.75	−0.78	−1.72	−2.02	−2.09

Year	Month	1st Month	3rd Month	6th Month	9th Month	12th Month
1977	1	–0.32	0.18	–1.5	–2.34	–2.11
1977	2	0.08	–1.06	–2.1	–1.98	–2.09
1977	3	0.59	0.25	–0.82	–1.74	–2.02
1977	4	1.34	1.3	0.73	–1.05	–1.89
1977	5	1.81	2.26	2.14	–0.34	–0.62
1977	6	0.99	2.62	2.62	0.87	–0.23
1977	7	0.59	2.1	2.46	2.01	0.16
1977	8	0.72	1.04	2.29	2.17	0.26
1977	9	–0.76	0.01	1.38	1.39	0.65
1977	10	0.32	–0.04	0.95	1.19	1.12
1977	11	0.86	–0.03	0.51	1.34	1.27
1977	12	–0.75	0.52	0.28	1.27	1.27
1978	1	0.12	0.7	0.16	1.05	1.27
1978	2	1.18	–0.06	–0.08	0.47	1.3
1978	3	–0.43	0.25	0.51	0.27	1.26
1978	4	1.32	1.26	1.1	0.44	1.25
1978	5	–0.39	0.32	0.22	–0.01	0.5
1978	6	–0.65	–0.01	–0.02	0.37	0.2
1978	7	1.22	0.1	0.66	0.83	0.4
1978	8	–0.66	–0.14	–0.02	–0.09	–0.17
1978	9	1.1	0.89	0.74	0.74	0.88
1978	10	–0.49	–0.03	–0.02	0.24	0.4
1978	11	0.65	0.52	0.29	0.33	0.29
1978	12	1.62	0.04	0.6	0.5	0.5

Year	Month	1st Month	3rd Month	6th Month	9th Month	12th Month
1979	1	−0.32	0.91	0.28	0.24	0.48
1979	2	2.51	2.1	0.79	0.52	0.54
1979	3	−0.2	1.17	0.12	0.65	0.55
1979	4	−1.04	−0.35	0.57	0.16	0.14
1979	5	0.54	−0.09	0.56	0.64	0.41
1979	6	0.45	0.12	0.29	0.08	0.61
1979	7	0.08	0.47	0.21	0.65	0.29
1979	8	−0.45	−0.3	−0.41	0	0.37
1979	9	0.86	0.31	0.27	0.35	0.2
1979	10	−0.31	−0.05	0.13	0.02	0.27
1979	11	1.39	0.92	0.56	0.45	0.66
1979	12	−0.75	0.46	0.46	0.42	0.48
1980	1	−0.32	1.29	0.5	0.57	0.47
1980	2	−0.1	−1.39	0.81	0.47	0.37
1980	3	−0.2	−0.89	0.4	0.4	0.36
1980	4	−0.01	−0.26	1.02	0.39	0.48
1980	5	0.01	−0.25	−0.51	0.61	0.31
1980	6	−0.86	−0.71	−0.86	−0.06	0.09
1980	7	−1.03	−1.16	−1.32	−0.06	−0.14
1980	8	−0.01	−0.99	−1.11	−1.28	0.01
1980	9	−0.25	−0.77	−1.08	−1.16	−0.69
1980	10	−0.61	−0.91	−1.29	−1.37	−0.77
1980	11	0.2	−0.86	−1.28	−1.32	−1.44
1980	12	−0.13	−0.71	−1.15	−1.32	−1.38

Year	Month	1st Month	3rd Month	6th Month	9th Month	12th Month
1981	1	0.12	0.02	−0.93	−1.3	−1.38
1981	2	−0.1	−0.46	−0.94	−1.35	−1.38
1981	3	1.64	1.32	−0.58	−1.07	−1.23
1981	4	0.66	0.92	0.37	−0.68	−1.07
1981	5	−0.6	0.02	−0.14	−0.91	−1.28
1981	6	−1.94	−1.22	−0.93	−1.19	−1.51
1981	7	0.17	−1.28	−0.73	−0.74	−1.21
1981	8	0.07	−0.61	−0.61	−0.72	−1.3
1981	9	1.39	1.06	0.45	0.55	−0.09
1981	10	0.24	0.97	0.31	0.47	0.36
1981	11	0.56	1.15	0.63	0.55	0.49
1981	12	0.24	0.31	0.91	0.44	0.52
1982	1	−0.32	0.46	0.95	0.36	0.5
1982	2	−0.1	−0.18	1.1	0.58	0.5
1982	3	−0.2	−0.89	0.23	0.85	0.39
1982	4	−0.95	−1.21	−0.18	0.72	0.15
1982	5	−0.88	−1.57	−1.69	0.51	0.09
1982	6	0.73	−0.9	−1.06	−0.3	0.49
1982	7	−0.24	−0.56	−1.06	−0.68	0.4
1982	8	−1.47	−0.86	−1.76	−1.83	−0.04
1982	9	−0.37	−1.31	−1.65	−1.73	−1.27
1982	10	−0.54	−1.61	−1.55	−1.79	−1.64
1982	11	1.59	0.18	−0.34	−0.83	−0.88
1982	12	−0.75	0.52	−0.56	−0.86	−0.91

Year	Month	1st Month	3rd Month	6th Month	9th Month	12th Month
1983	1	−0.32	1.53	−0.51	−0.7	−0.92
1983	2	−0.1	−1.39	0.06	−0.44	−0.92
1983	3	−0.43	−1.25	0.44	−0.65	−0.92
1983	4	−0.53	−0.85	1.13	−0.72	−0.87
1983	5	0.44	−0.06	−0.32	−0.04	−0.49
1983	6	1.46	0.77	0.63	0.68	−0.33
1983	7	0.11	0.92	0.52	1.41	−0.24
1983	8	1.19	1.43	1.2	1.06	0.78
1983	9	1.14	1.49	1.64	1.59	1.6
1983	10	−0.44	0.99	1.16	0.99	1.5
1983	11	−0.99	0.04	0.8	0.69	0.61
1983	12	1.07	−0.86	0.6	0.77	0.73
1984	1	0.12	−0.55	0.65	0.89	0.73
1984	2	1.18	1.22	0.16	0.89	0.76
1984	3	1.64	1.75	−0.64	0.71	0.87
1984	4	0.53	0.97	0.04	0.82	1.02
1984	5	−1.64	−0.67	−0.34	−0.18	0.6
1984	6	−1.03	−1.68	−1.19	−1.46	0.14
1984	7	1.57	−0.3	0.13	−0.32	0.59
1984	8	−1.67	−0.41	−0.86	−0.67	−0.49
1984	9	−0.91	−0.75	−1.51	−1.3	−1.84
1984	10	−0.76	−2.27	−1.87	−1.57	−1.84
1984	11	−0.7	−1.85	−1.68	−1.85	−1.74
1984	12	0.6	−1.21	−1.45	−1.93	−1.77

Year	Month	1st Month	3rd Month	6th Month	9th Month	12th Month
1985	1	0.12	−0.62	−2.42	−2.06	−1.78
1985	2	−0.1	0.34	−1.83	−1.68	−1.83
1985	3	0.83	0.5	−1.22	−1.46	−1.92
1985	4	−1.04	−0.79	−1.13	−2.61	−2.22
1985	5	−0.8	−1.29	−1.26	−2.31	−2.06
1985	6	0.11	−1.33	−1.27	−1.79	−1.92
1985	7	−0.18	−0.83	−1.21	−1.51	−2.69
1985	8	−0.77	−0.79	−1.55	−1.54	−2.67
1985	9	−1.02	−1.49	−2.02	−2	−2.68
1985	10	−0.34	−1.64	−1.7	−1.88	−2.16
1985	11	0.26	−1.12	−1.38	−1.77	−1.78
1985	12	0.13	−0.4	−1.45	−1.81	−1.8
1986	1	1.94	0.18	−1.48	−1.61	−1.79
1986	2	0.99	0.5	−1.09	−1.36	−1.74
1986	3	−0.69	0.25	−0.44	−1.48	−1.82
1986	4	0.24	0.12	0.04	−1.46	−1.59
1986	5	−0.99	−1.01	−0.93	−1.51	−1.67
1986	6	1.05	−0.22	−0.23	−0.57	−1.5
1986	7	−0.65	−0.64	−0.65	−0.57	−1.64
1986	8	0.14	0.09	−0.57	−0.55	−1.39
1986	9	1.06	0.56	0.35	0.34	−0.09
1986	10	−0.55	0.24	−0.11	−0.13	−0.18
1986	11	0.76	0.51	0.4	0.04	0.04
1986	12	1.13	−0.05	0.29	0.15	0.15

Year	Month	1st Month	3rd Month	6th Month	9th Month	12th Month
1987	1	0.12	0.88	0.5	0.15	0.12
1987	2	0.08	0.99	0.58	0.46	0.1
1987	3	1.13	0.91	0	0.32	0.18
1987	4	0.2	0.37	0.82	0.52	0.18
1987	5	−0.17	−0.06	0.12	0.45	0.36
1987	6	−0.09	−0.38	−0.25	−0.27	0.12
1987	7	−0.53	−0.7	−0.57	0.04	0.15
1987	8	0.34	−0.19	−0.3	−0.19	0.24
1987	9	−1.02	−0.89	−1.02	−0.96	−0.96
1987	10	0.35	−0.39	−0.67	−0.62	−0.33
1987	11	−0.02	−0.64	−0.67	−0.7	−0.64
1987	12	1.23	0.25	−0.52	−0.64	−0.6
1988	1	−0.32	0.18	−0.39	−0.66	−0.62
1988	2	1.18	1.33	−0.48	−0.55	−0.58
1988	3	−0.69	−0.03	0.22	−0.56	−0.67
1988	4	1.19	1.11	0.6	−0.13	−0.41
1988	5	0.29	0.71	0.97	−0.16	−0.28
1988	6	−1.09	0.28	0.23	0.24	−0.47
1988	7	1.92	0.99	1.35	1.07	0.29
1988	8	0.97	1.38	1.56	1.69	0.64
1988	9	1.36	2.15	2.08	2.07	1.85
1988	10	−1.87	0.5	0.81	0.99	0.93
1988	11	−0.57	−0.28	0.55	0.71	0.8
1988	12	1.18	−1.84	0.84	0.81	0.8

Year	Month	1st Month	3rd Month	6th Month	9th Month	12th Month
1989	1	0.12	–0.24	0.28	0.61	0.8
1989	2	–0.1	0.99	–0.19	0.6	0.76
1989	3	0.72	0.38	–1.86	0.83	0.81
1989	4	–0.87	–0.73	–0.75	0.09	0.46
1989	5	–0.8	–1.26	–0.96	–0.72	0.18
1989	6	0.66	–0.88	–0.86	–2	0.47
1989	7	2	1.05	0.69	0.29	0.51
1989	8	–1.1	0.66	–0.1	0	–0.32
1989	9	0.33	0.51	0.03	0.03	–1.11
1989	10	0.58	0.01	0.45	0.29	0.11
1989	11	–0.99	0.15	0.42	0	0.06
1989	12	–0.13	0.01	0.29	–0.05	–0.05
1990	1	0.12	–1.16	–0.41	0.1	–0.06
1990	2	0.08	–0.31	0.08	0.37	–0.05
1990	3	0.32	0.12	–0.02	0.27	–0.07
1990	4	–0.41	–0.45	–1.34	–0.56	–0.02
1990	5	2.06	1.81	1.72	0.89	1
1990	6	–0.22	1.69	1.67	0.81	0.88
1990	7	–1.08	1.23	0.94	0.2	0.06
1990	8	0.2	–0.58	0.75	0.66	0.48
1990	9	–1.09	–1.23	–0.07	–0.09	–0.19
1990	10	0.33	–0.52	0.16	0.03	–0.33
1990	11	0.13	–0.62	–0.87	–0.03	–0.08
1990	12	–0.48	0.09	–0.88	–0.07	–0.09

Year	Month	1st Month	3rd Month	6th Month	9th Month	12th Month
1991	1	0.12	−0.1	−0.63	0.03	−0.09
1991	2	0.08	−0.62	−0.72	−0.95	−0.09
1991	3	−0.69	−0.89	0	−0.96	−0.13
1991	4	0.42	0.15	−0.18	−0.64	0.01
1991	5	0.64	0.55	0.38	−0.45	−0.71
1991	6	1.99	1.63	1.52	0.79	−0.18
1991	7	−0.21	1.22	1.13	0.73	−0.01
1991	8	0.24	0.78	0.92	0.81	−0.01
1991	9	−0.32	−0.38	0.5	0.44	0.32
1991	10	0.78	0.3	0.76	0.73	0.59
1991	11	−0.12	0.12	0.47	0.57	0.51
1991	12	−1.29	0.41	−0.06	0.55	0.5
1992	1	0.12	−0.45	0.03	0.52	0.5
1992	2	−0.1	−1.39	0	0.37	0.49
1992	3	−0.69	−1.25	0.33	−0.14	0.49
1992	4	−0.95	−1.39	−1.23	−0.24	0.3
1992	5	0.18	−0.51	−0.8	−0.28	0.14
1992	6	1.38	0.37	0.23	0.38	−0.04
1992	7	0.72	1.02	0.52	0.05	0.21
1992	8	0.36	0.98	0.54	0.39	0.27
1992	9	−0.29	0.1	0.2	0.13	0.29
1992	10	−0.01	−0.28	0.23	−0.01	−0.24
1992	11	0.26	−0.4	0.21	−0.01	−0.11
1992	12	−0.75	−0.17	−0.14	−0.03	−0.09

Year	Month	1st Month	3rd Month	6th Month	9th Month	12th Month
1993	1	−0.32	0	−0.37	0.12	−0.1
1993	2	−0.1	−1.39	−0.52	0.12	−0.1
1993	3	0.83	0.38	−0.18	−0.15	−0.03
1993	4	0.04	0.08	−0.14	−0.4	0.09
1993	5	−0.17	−0.22	−0.49	−0.65	−0.02
1993	6	0.76	0.03	0.04	−0.22	−0.18
1993	7	0.26	0.19	0.11	−0.07	−0.32
1993	8	0.24	0.41	0.14	−0.01	−0.43
1993	9	0.03	0.05	0	0	−0.24
1993	10	0.78	0.5	0.45	0.42	0.3
1993	11	−0.66	0.19	0.31	0.17	0.08
1993	12	1.53	0.54	0.33	0.27	0.27
1994	1	0.12	−0.1	0.32	0.31	0.27
1994	2	0.32	1.49	0.35	0.43	0.29
1994	3	0	−0.03	0.51	0.3	0.25
1994	4	0.12	0	−0.27	0.26	0.26
1994	5	−0.74	−0.81	−0.34	−0.04	0.12
1994	6	−0.5	−1.11	−1.16	−0.11	−0.13
1994	7	0.29	−0.77	−0.82	−0.9	−0.12
1994	8	−0.15	−0.34	−0.88	−0.62	−0.32
1994	9	−1.57	−1.22	−1.67	−1.71	−1.04
1994	10	1.2	0	−0.36	−0.41	−0.54
1994	11	0.17	0.12	−0.15	−0.44	−0.31
1994	12	−0.28	1	−0.07	−0.46	−0.48

Year	Month	1st Month	3rd Month	6th Month	9th Month	12th Month
1995	1	1.94	0.02	−0.11	−0.43	−0.48
1995	2	−0.1	−0.31	0.05	−0.2	−0.49
1995	3	0.94	0.81	1.06	−0.05	−0.43
1995	4	−0.41	−0.26	−0.29	−0.22	−0.52
1995	5	1	0.75	0.63	0.31	0.04
1995	6	0.34	0.71	0.79	1.2	0.18
1995	7	0.48	1.04	0.81	0.52	0.24
1995	8	0.82	0.84	1.1	1.02	0.71
1995	9	0.35	0.75	0.95	0.99	1.45
1995	10	1.35	1.51	1.64	1.55	1.44
1995	11	−0.1	1.08	1.26	1.38	1.34
1995	12	−0.75	1.04	1.2	1.29	1.33
1996	1	−0.32	−0.41	1.18	1.39	1.3
1996	2	0.08	−1.06	0.99	1.18	1.31
1996	3	−0.69	−1.25	1	1.15	1.24
1996	4	1.46	1.26	0.32	1.41	1.57
1996	5	−0.25	0.51	0.3	1.06	1.23
1996	6	1.81	1.48	1.36	1.51	1.59
1996	7	−0.31	0.45	0.96	0.45	1.44
1996	8	1.41	1.64	1.69	1.57	1.89
1996	9	0.4	0.99	1.53	1.48	1.89
1996	10	0.47	1.19	1.13	1.34	1.13
1996	11	0.42	0.52	1.28	1.32	1.25
1996	12	0.13	0.42	0.93	1.32	1.29

Year	Month	1st Month	3rd Month	6th Month	9th Month	12th Month
1997	1	0.12	0.3	1.09	1.07	1.28
1997	2	−0.1	−0.18	0.46	1.24	1.28
1997	3	0.83	0.5	0.44	0.93	1.33
1997	4	1.22	1.21	0.76	1.31	1.26
1997	5	−0.67	0.21	0.1	0.45	1.19
1997	6	−0.36	−0.16	−0.12	0.23	0.79
1997	7	−0.73	−1.33	−0.56	−0.41	0.72
1997	8	0.47	−0.23	−0.17	−0.24	0.22
1997	9	−0.08	−0.2	−0.32	−0.3	−0.05
1997	10	1.11	0.88	0.21	0.45	0.44
1997	11	1.24	1.26	0.89	0.85	0.82
1997	12	2.02	1.76	1.2	1.02	1.03
1998	1	−0.32	1.6	1.45	0.82	1.02
1998	2	−0.1	1.9	1.45	1.06	1.02
1998	3	−0.69	−1.76	1.75	1.14	0.96
1998	4	0.69	0.4	1.56	1.46	0.83
1998	5	−0.14	0	0.54	1.32	0.96
1998	6	−0.6	−0.29	−0.48	1.46	0.96
1998	7	1.81	0.76	0.79	1.67	1.6
1998	8	1.65	2.03	1.82	2.06	2.36
1998	9	0.12	1.75	1.47	1.4	2.41
1998	10	1.14	1.72	1.71	1.72	2.21
1998	11	−0.02	0.77	1.69	1.56	1.72
1998	12	0.67	0.95	1.85	1.6	1.56

Year	Month	1st Month	3rd Month	6th Month	9th Month	12th Month
1999	1	−0.32	−0.03	1.48	1.53	1.55
1999	2	0.56	0.58	0.78	1.7	1.57
1999	3	−0.69	−0.61	0.91	1.82	1.56
1999	4	0.46	0.27	−0.06	1.47	1.51
1999	5	1.5	1.43	1.47	1.28	2.03
1999	6	−0.36	1.31	1.21	1.35	2.15
1999	7	0.35	1.17	1.13	0.77	1.78
1999	8	−0.26	−0.34	0.61	0.63	0.99
1999	9	−0.55	−0.63	0.12	0.06	0.67
1999	10	0.39	−0.44	0.19	0.19	0.07
1999	11	−0.15	−0.4	−0.57	0.03	0.04
1999	12	−0.48	0.02	−0.53	0.02	−0.02
2000	1	−0.32	−0.45	−0.67	−0.03	−0.03
2000	2	1.18	−0.06	−0.45	−0.61	−0.01
2000	3	−0.69	−0.03	−0.02	−0.57	0
2000	4	0.69	0.6	−0.16	−0.55	0.05
2000	5	0.18	0.28	0.18	−0.35	−0.51
2000	6	0.86	0.71	0.67	0.27	−0.29
2000	7	0.26	0.48	0.63	0.15	−0.32
2000	8	1.59	1.65	1.58	1.52	0.68
2000	9	−0.19	0.93	1.11	1.09	0.84
2000	10	0.72	1.18	1.14	1.19	0.97
2000	11	−1.06	−0.08	0.86	0.85	0.82
2000	12	0.01	0.16	0.71	0.85	0.84

Year	Month	1st Month	3rd Month	6th Month	9th Month	12th Month
2001	1	0.12	−1.16	0.73	0.78	0.84
2001	2	−0.1	−0.31	−0.15	0.81	0.8
2001	3	0.83	0.5	0.17	0.71	0.85
2001	4	1.01	1	−0.23	0.9	0.92
2001	5	−1.24	−0.29	−0.44	−0.33	0.63
2001	6	−0.09	−0.54	−0.5	−0.19	0.46
2001	7	−0.28	−1.25	−0.65	−1.23	0.34
2001	8	0.01	−0.36	−0.6	−0.69	−0.61
2001	9	−0.05	−0.35	−0.63	−0.61	−0.51
2001	10	0.49	0.09	−0.47	−0.25	−0.63
2001	11	−0.45	−0.08	−0.32	−0.45	−0.51
2001	12	−0.28	0.04	−0.32	−0.52	−0.51
2002	1	0.12	−0.69	−0.23	−0.72	−0.52
2002	2	0.08	−0.46	−0.16	−0.38	−0.51
2002	3	0	−0.2	−0.02	−0.37	−0.55
2002	4	−1.23	−1.3	−1.45	−0.5	−0.94
2002	5	1.07	0.49	0.34	0	−0.23
2002	6	1.57	1.27	1.21	0.57	0.14
2002	7	−1.55	0.87	0.37	−0.17	−0.08
2002	8	−0.33	−0.28	−0.02	−0.12	−0.24
2002	9	−1.63	−2.17	−0.95	−0.99	−0.92
2002	10	0.49	−0.9	−0.3	−0.53	−0.83
2002	11	−0.61	−0.99	−0.99	−0.76	−0.83
2002	12	−0.48	−0.02	−1.55	−0.79	−0.82

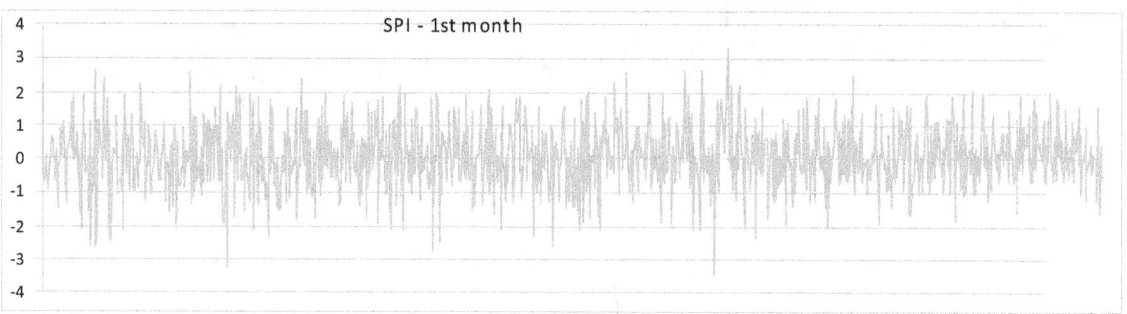

1900–2002

Description	Category	# of times in 102 years	Severity of Event
Extremely wet	2+	22	1 in 5 years
Very wet	1.5–1.99	68	1 in 2 years
Moderate wet	1.00–1.49	105	1 in 1 year
Near normal	(–)0.99 to 0.99	880	0
Moderate dry	(–)1 to (–)1.49	76	1 in 1 year
Severely dry	(–)1.5 to (–) 1.99	32	1 in 3 years
Extremely dry	less than –2	26	1 in 4 years

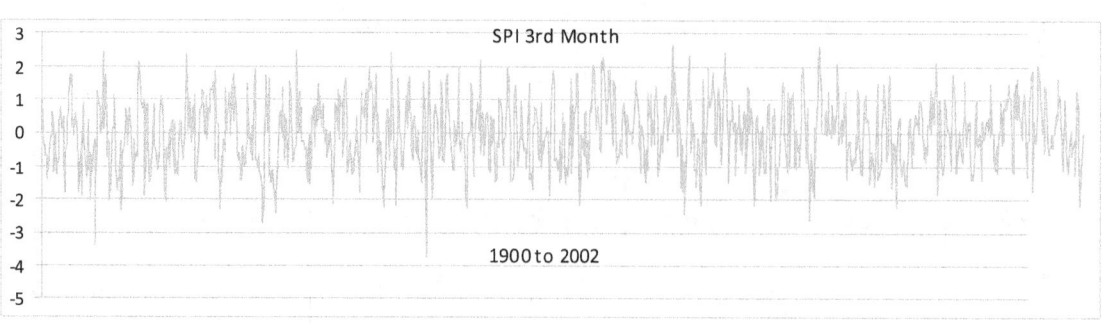

Description	Category	# of times in 102 years	Severity of Event
Extremely wet	2+	20	1 in 5 years
Very wet	1.5–1.99	62	1 in 2 years
Moderate wet	1.00–1.49	115	1 in 1 year
Near normal	(–)0.99 to 0.99	802	0
Moderate dry	(–)1 to (–)1.49	76	1 in 1 year
Serverly dry	(–)1.5 to (–) 1.99	56	1 in 2 years
Extremely dry	less than –2	23	1 in 4 years

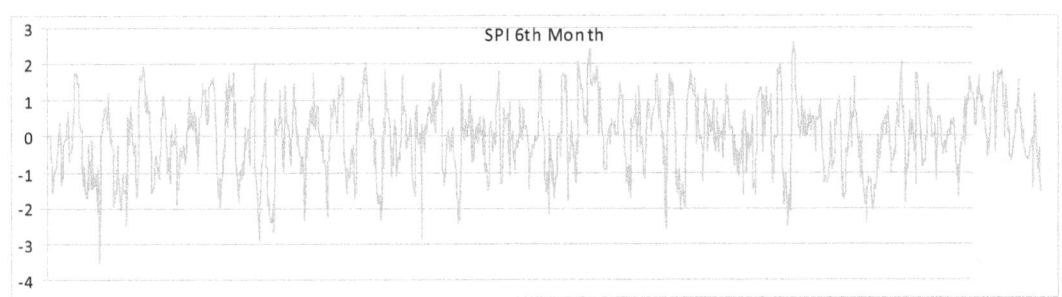

1900–2002

Description	Category	# of Times in 102 Years	Severity of Event
Extremely wet	2+	10	1 in 10 years
Very wet	1.5 – 1.99	77	1 in 1 years
Moderate wet	1.00 – 1.49	122	1 in 1 year
Near normal	(–)0.99 to 0.99	796	0
Moderate dry	(–)1 to (–)1.49	111	1 in 1 year
Serverly dry	(–)1.5 to (–) 1.99	73	1 in 1 years
Extremely dry	less than –2	26	1 in 4 years

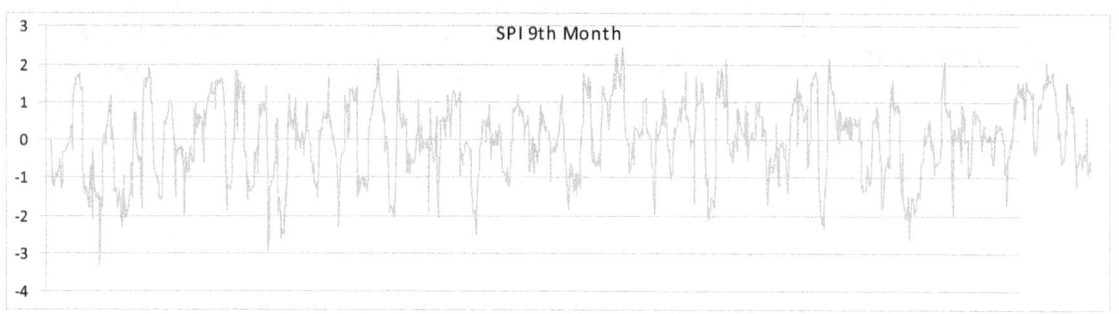

1900–2002

Description	Category	# of times in 102 years	Severity of Event
Extremely wet	2+	9	1 in 11 years
Very wet	1.5–1.99	65	1 in 2 years
Moderate wet	1.00–1.49	137	1 in 1 year
Near normal	(–)0.99 to 0.99	791	0
Moderate dry	(–)1 to (–)1.49	111	1 in 1 year
Serverly dry	(–)1.5 to (–) 1.99	65	1 in 2 years
Extremely dry	less than –2	34	1 in 3 years

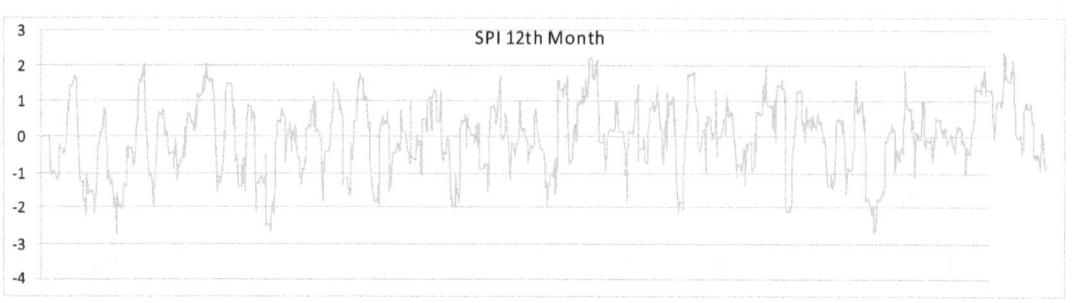

1900–2002

Description	Category	# of times in 102 years	Severity of Event
Extremely wet	2+	15	1 in 7 years
Very wet	1.5–1.99	61	1 in 2 years
Moderate wet	1.00–1.49	126	1 in 1 year
Near normal	(–)0.99 to 0.99	801	0
Moderate dry	(–)1 to (–)1.49	105	1 in 1 year
Serverly dry	(–)1.5 to (–) 1.99	81	1 in 1 years
Extremely dry	less than –2	29	1 in 4 years

	Jan	feb	march	april	may	june	Jul	Aug	Sep	Oct	nov	dec
2000	-0.32	1.18	-0.69	0.69	0.18	0.86	0.26	1.59	-0.19	0.72	-1.06	0.01
2001	0.12	-0.1	0.83	1.01	-1.24	-0.09	-0.28	0.01	-0.05	0.49	-0.45	-0.28
2002	0.12	0.08	0	-1.23	1.07	1.57	-1.55	-0.33	-1.63	0.49	-0.61	-0.48

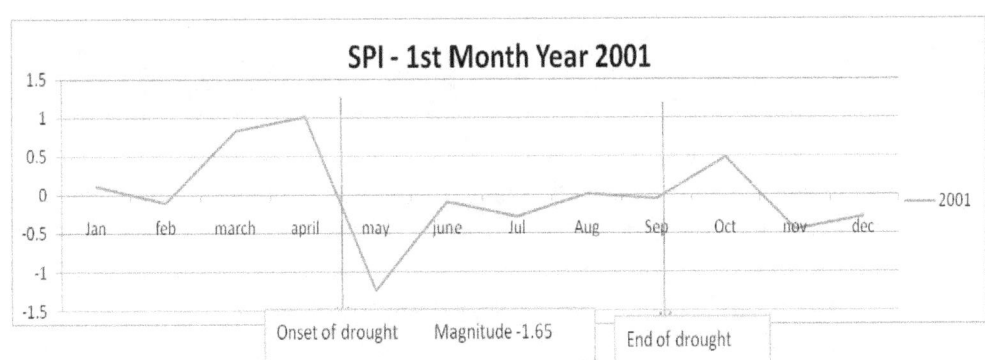

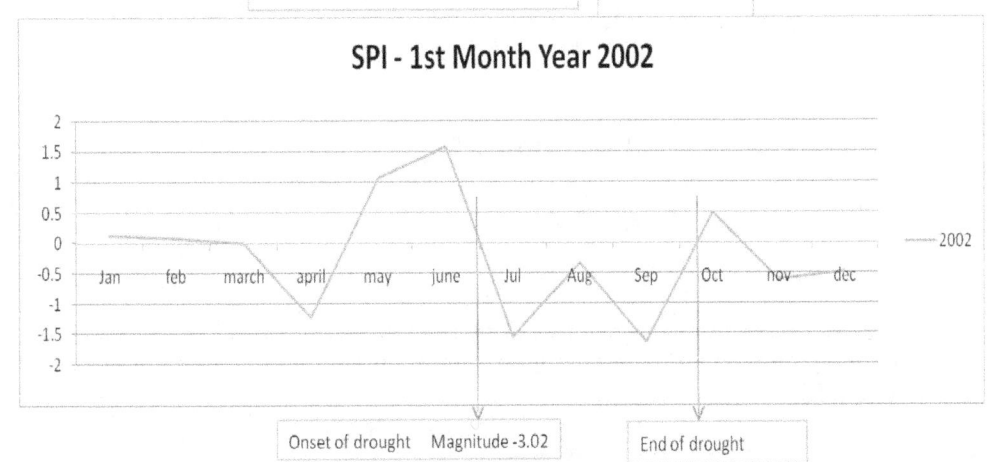

	Jan	feb	march	april	may	june	Jul	Aug	Sep	Oct	nov	dec
2000	-0.45	-0.06	-0.03	0.6	0.28	0.71	0.48	1.65	0.93	1.18	-0.08	0.16
2001	-1.16	-0.31	0.5	1	-0.29	-0.54	-1.25	-0.36	-0.35	0.09	-0.08	0.04
2002	-0.69	-0.46	-0.2	-1.3	0.49	1.27	0.87	-0.28	-2.17	-0.9	-0.99	-0.02

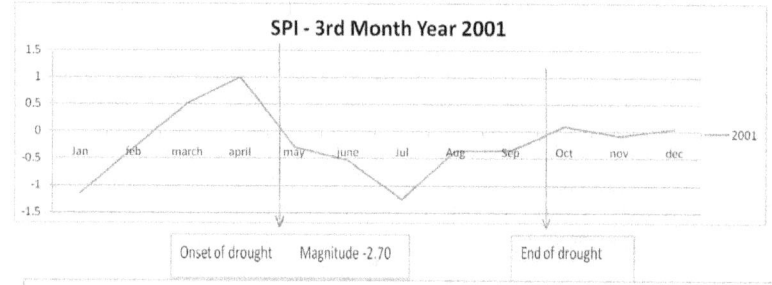

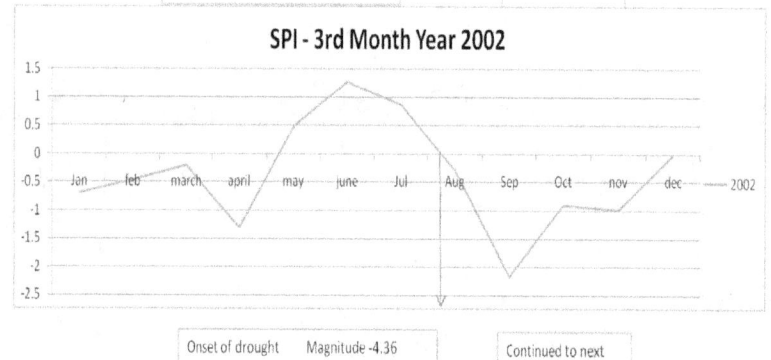

	Jan	feb	march	april	may	june	Jul	Aug	Sep	Oct	nov	dec
2001	0.73	-0.15	0.17	-0.23	-0.44	-0.5	-0.65	-0.6	-0.63	-0.47	-0.32	-0.32
2002	-0.23	-0.16	-0.02	-1.45	0.34	1.21	0.37	-0.02	-0.95	-0.3	-0.99	-1.55

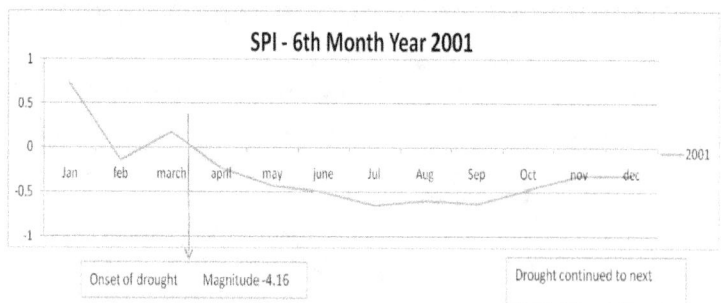

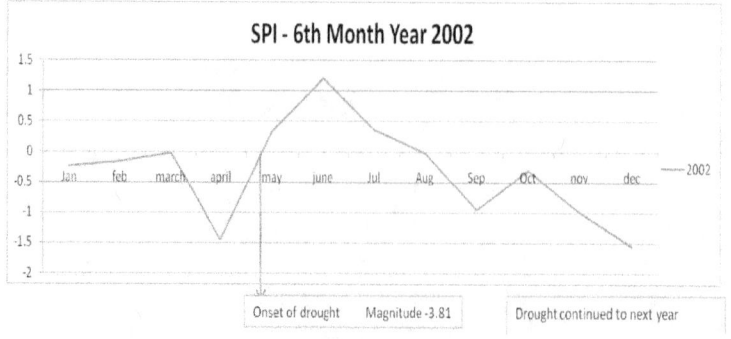

	Jan	feb	march	april	may	june	Jul	Aug	Sep	Oct	nov	dec
2001	0.78	0.81	0.71	0.90	-0.33	-0.19	-1.23	-0.69	-0.61	-0.25	-0.45	-0.52
2002	-0.72	-0.38	-0.37	-0.50	0.00	0.57	-0.17	-0.12	-0.99	-0.53	-0.76	-0.79

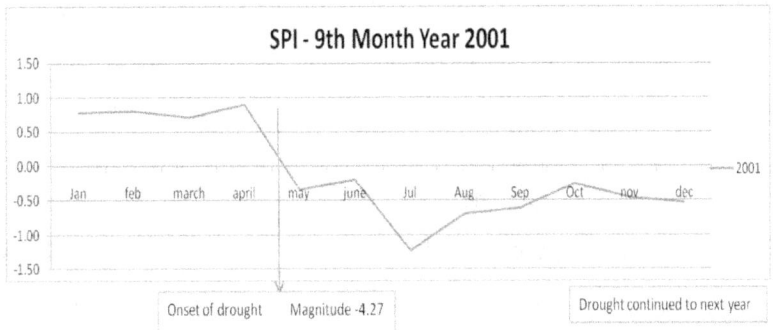

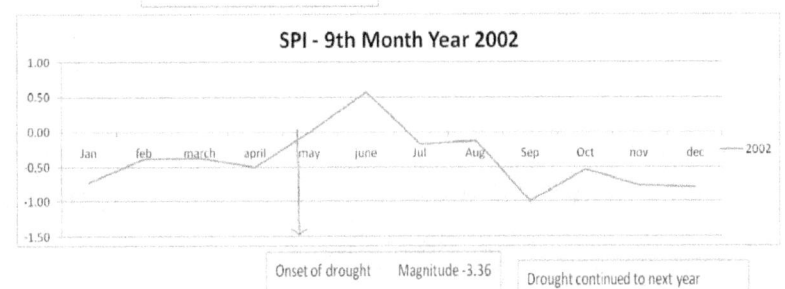

	Jan	feb	march	april	may	june	Jul	Aug	Sep	Oct	nov	dec
2001	0.84	0.8	0.85	0.92	0.63	0.46	0.34	-0.61	-0.51	-0.63	-0.51	-0.51
2002	-0.52	-0.51	-0.55	-0.94	-0.23	0.14	-0.08	-0.24	-0.92	-0.83	-0.83	-0.82

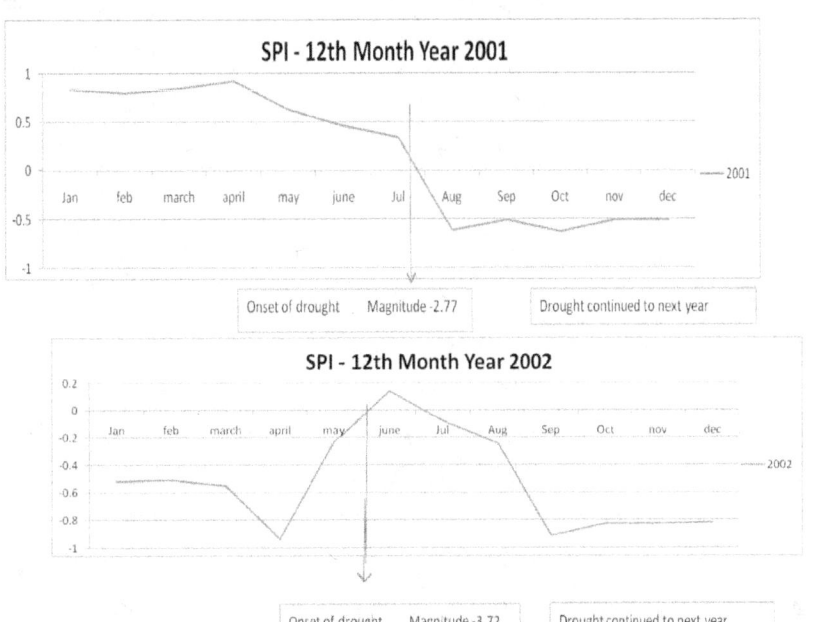

SPEI INPUT DATA OF ANANTAPUR DISTRICT, AP

Ananthpur

77.5814	2.026;23.035	0;21.967
1900;01	0;22.279	0.389;21.682
12	0;21.269	0;22.72
0.054;23.56	0;22.19	0.871;22.063
0.205;22.33	0.636;21.957	1.842;23.566
1.423;22.719	1.177;22.585	4.476;22.53
2.098;21.669	0.105;23.473	0;22.259
2.097;22.13	0;22.032	0;22.575
3.46;23.297	0;23.12	0;23.046
1.396;22.063	0.226;21.88	0;22.242
0.94;22.229	0.419;20.934	0.912;21.641
2.739;22.801	0.781;22.9	0;23.523
0;22.578	0;22.688	0;22.39
0;22.52	0.62;23.12	0.087;21.945
0.44;22.215	0;22.795	0.04;21.526
0;21.297	0.034;22.498	0;22.479
0.025;21.706	0.054;21.236	0.163;23.105
0.869;22.649	1.423;23.107	0;23.253
0;21.959	0.41;22.91	0;22.963
0.303;21.323	0;21.849	0.576;23.221
0.419;21.381	0;22.048	0;22.97
1.382;23.577	0;22.413	0;22.582
2.204;22.025	0;22.669	0.417;23.748
2.669;23.095	0.047;22.624	0.39;23.703
1.209;22.45	0.724;22.587	2.7;22.157
0.816;21.335	0.058;22.415	0.294;23.527
0.104;22.533	0.112;22.166	0;22.99
0;22.129	0;22.404	0.007;22.742
2.452;22.745	0.048;22.815	0.171;23.234
0.065;23.493	0;23.175	0.057;23.74
0.061;22.93	0;23.33	0.037;21.933
0.175;23.186	0.306;22.848	0;22.961

0.759;23.007

2.938;22.599

0;23.936

0.963;22.779

0;24.275

0;22.84

0;23.756

0.748;23.682

0.823;23.563

12.422;24.679

0;23.959

0;24.589

0;23.465

2.703;24.462

0;25.194

0;24.692

1.88;24.201

0;24.739

0.012;24.322

0;23.099

2.05;25.48

0;24.952

0.136;25.152

1.168;24.899

0;24.795

5.555;23.359

0.101;23.274

0;25.558

0;24.61

0;24.278

0;24.012

5.189;23.717

0;25.174

0;23.954

0.437;24.632

1.371;25.296

10.769;24.68

1.245;24.458

2.727;23.955

0;24.972

2.807;23.892

2.884;24.96

0;23.838

5.136;24.774

10.508;24.765

7.944;25.249

0.383;25.171

0;23.973

0.131;24.134

7.877;24.759

4.785;24.854

0.007;23.54

7.84;24.292

0;24.892

2.879;23.895

9.309;24.183

0;25.067

0;23.9

4.995;24.37

0;24.247

2.096;25.081

0;24.641

0;25.291

0.001;25.178

0.487;23.98

0.002;24.681

2.845;25.435

3.911;25.675

0;24.68

5.311;25.213

4.6;24.605

0;23.686

0;25.329

0.372;24.428

0;25.793

0;24.359

10.628;25.31

0;25.766

0;25.336

1.095;25.247

0;24.809

0;25.597

0;24.582

0.001;25.443

0;23.947

0.246;25.5

5.71;25.165

14.591;25.626

0;25.895

0;25.635

0;26.206

0;26.179

5.191;24.585

0;26.069

4.758;25.159

0.012;24.642

5.184;26.547

0;24.818

0.087;25.497

0.007;25.637

0;25.592

0;24.413

1.331;25.564

0;25.891

0.012;25.419

0;25.215

0;26.483

2.955;25.7

5.876;25.812	0;26.769	0.474;28.486
0;26.829	0.511;27.293	6.125;27.051
0.108;25.433	11.034;26.739	0;27.314
0.003;26.802	11.741;27.466	0;28.283
0;27.815	5.775;28.07	0.097;27.957
0;27.829	22.427;26.679	4.101;28.017
5.337;26.955	1.079;27.215	0.026;27.856
7.379;27.277	0;28.452	6.828;28.013
0.054;26.869	0;27.825	0.28;27.694
4.121;27.086	3.45;27.492	1.295;28.118
13.07;26.712	23.417;26.038	1.6;28.107
0.608;27.839	0;27.507	17.747;27.838
3.492;27.69	10.132;27.713	1.154;28.664
1.051;27.197	0.025;27.646	0.699;29.024
0;28.007	3.996;27.067	17.331;27.425
0;27.62	0.016;27.561	8.609;28.599
0.896;27.715	8.237;26.924	0;28.77
32.883;26.903	14.94;27.594	11.752;27.786
0.109;27.72	1.482;27.498	0;29.27
8.059;26.142	0;28.771	7.822;27.086
25.777;26.236	4.612;27.301	4.621;28.417
6.209;27.634	11.111;27.57	0;29.261
0;27.867	0.001;28.107	0;28.462
1.906;28.163	2.195;26.528	8.609;28.302
0;27.847	4.849;27.967	2.523;28.772
16.035;26.287	0;27.854	9.6;28.502
0.122;27.432	6.116;27.818	0;28.904
1.675;27.04	0.225;28.941	8.409;28.089
1.018;27.783	3.942;27.434	0;29.312
2.996;28.505	4.243;27.422	0;28.984
1.415;27.21	0;28.387	0;27.966
0;27.218	0.697;27.521	8.177;28.694
12.849;27.43	0.181;28.699	2.03;28.968
0.796;27.766	8.507;27.086	13.646;29.559
0;26.255	25.83;27.263	32.725;29.301
14.071;26.793	0.204;29.085	6.271;29.661

25.83;29.549

18.921;28.577

0.433;30.646

91.809;28.442

9.366;30.249

45.657;28.547

11.607;30.071

56.085;29.651

19.583;29.797

4.843;30.169

11.447;29.144

6.119;29.475

8.878;29.845

13.735;28.62

14.873;28.993

54.36;30.054

17.184;29.125

84.345;29.227

14.761;29.859

17.439;29.146

48.396;30.441

41.126;29.995

73.34;28.719

3.21;30.386

56.579;29.649

33.022;28.603

13.623;29.274

16.157;30.256

3.043;29.446

44.635;29.071

50.383;29.463

47.543;28.79

29.167;29.484

75.101;28.934

12.993;29.727

81.533;29.025

47.426;28.824

18.226;30.345

16.054;29.691

30.875;28.878

9.651;28.599

49.96;28.837

49.416;29.138

22.557;28.901

37.382;29.331

6.699;29.326

0.353;29.567

47.633;29.284

5.901;30.161

60.898;29.209

18.988;30.175

39.893;29.293

117.808;30.174

26.183;29.125

54.911;29.689

45.558;29.877

15.991;29.699

19.021;30.316

58.197;28.797

97.871;28.572

17.475;30.48

35.917;28.905

10.245;30.194

19.833;29.998

66.248;29.125

10.751;30.653

28.078;30.238

41.591;29.409

17.416;29.411

5.716;30.857

20.776;30.111

16.605;30.616

56.263;29.002

69.614;29.581

68.208;29.549

9.189;30.347

26.152;29.664

44.258;30.095

10.306;29.961

16.89;30.733

40.384;29.413

9.54;30.565

32.644;31.106

31.163;30.545

63.944;29.435

11.56;29.927

18.278;30.719

37.639;30.011

10.272;30.502

27.848;30.661

29.156;29.615

18.458;30.88

74.871;29.58

64.054;28.606

45.315;31.066

38.832;30.313

45.422;30.417

56.759;30.124

7.985;30.791

73.475;28.083

51.185;29.593

76.354;28.557

103.551;28.171

50.693;29.03

13.72;30.334

9.545;28.489

43.865;29.014

96.961;28.573

67.261;29.526	74.492;28.908	37.549;29.288
92.334;29.242	29.606;29.84	86.193;30.109
42.396;30.006	115.394;29.169	20.631;30.814
83.974;29.181	107.344;28.511	39.025;30.186
47.48;29.333	86.502;29.36	34.875;30.306
85.233;29.201	155.615;28.932	60.776;29.877
81.772;28.923	131.744;29.437	79.531;30.178
104.974;28.2	16.705;30.718	39.555;30.215
78.151;27.24	76.478;29.717	188.382;27.533
80.895;28.773	177.616;28.127	96.796;30.368
65.547;29.12	103.515;28.485	74.621;29.601
12.902;30.248	79.991;29.316	60.953;30.257
65.731;28.824	60.638;29.084	41.106;30.432
39.88;28.895	86.242;29.91	116.136;28.862
91.473;29.797	44.74;29.094	57.405;30.634
151.148;28.061	128.872;28.567	43.915;29.764
33.3;29.234	105.094;28.519	61.334;30.525
48.349;29.413	56.79;29.238	147.02;28.257
54.999;30.039	12.343;30.142	74.744;28.867
49.841;29.338	25.655;29.428	28.46;30.006
138.251;28.709	87.151;29.316	120.932;29.894
65.871;29.937	60.855;29.815	43.486;26.304
41.193;28.384	74.172;29.688	54.83;27.234
166.688;27.371	100.564;29.391	61.566;27.242
32.035;29.554	154.75;29.158	39.952;25.499
24.562;30.012	68.565;28.113	54.772;27.226
84.422;29.096	131.378;28.516	84.41;26.436
50.346;29.24	39.55;30.311	50.209;26.185
52.963;29.206	153.421;28.911	37.346;27.211
8.912;29.45	71.592;29.373	32.999;26.118
143.598;28.628	23.477;28.386	48.851;26.558
49.268;29.702	169.652;27.503	54.526;26.409
65.864;29.771	52.925;28.754	51.25;27.502
156.691;27.782	91.475;28.826	35.463;26.328
46.078;29.564	67.235;30.207	46.43;27.064
76.027;29.536	45.242;29.093	65.949;27.391

36.017;25.387	24.232;26.908	30.646;27.784
79.889;25.232	63.142;27.721	70.253;26.791
20.968;26.154	34.127;26.711	47.417;25.942
91.708;25.864	43.025;26.18	116.598;26.528
34.072;26.909	74.608;25.831	93.266;27.294
60.811;26.56	143.196;26.676	73.998;27.636
33.04;26.472	60.48;28.021	41.339;27.195
49.737;27.864	118.519;26.323	61.954;28.576
26.514;26.898	37.668;26.738	109.911;27.044
19.842;26.529	89.82;26.165	44.666;28.177
94.054;28.196	40.832;27.171	39.021;28.377
68.462;26.595	34.725;26.203	44.108;26.327
77.357;26.934	76.643;26.956	76.743;25.853
34.99;26.509	55.507;26.9	50.801;26.693
49.256;27.116	109.998;27.323	100.99;26.906
67.753;27.049	63.167;27.136	43.741;25.034
65.564;26.494	51.147;26.965	27.823;25.955
18.537;25.461	54.694;27.443	80.052;25.015
47.477;26.808	41.376;26.061	70.543;24.82
112.605;26.998	32.234;25.481	25.485;25.86
52.237;25.285	55.271;27.729	69.466;25.33
64.824;26.859	70.122;26.396	96.527;25.316
33.646;25.108	44.787;26.107	47.959;24.638
62.856;26.747	27.346;26.503	45.549;24.595
58.217;26.435	27.888;27.122	162.111;25.694
45.479;26.658	80.754;25.712	36.192;25.262
94.227;26.766	38.108;25.676	70.044;25.174
31.888;25.434	64.118;27.153	81.738;24.831
98.387;26.718	34.932;26.136	51.825;25.013
16.004;26.591	18.217;26.393	114.259;25.768
26.302;26.068	72.725;26.936	116.008;25.497
47.684;26.962	96.896;27.367	46.647;25.534
14.742;26.796	31.459;26.819	20.033;26.407
59.686;26.066	55.643;26.273	88.125;25.501
44.545;26.218	82.071;26.92	25.527;25.498
56.762;25.963	50.456;27.421	90.182;25.336

32.416;25.087

46.595;24.679

77.178;25.454

46.449;25.384

55.391;26.327

110.598;24.82

90.816;25.453

29.478;25.668

65.279;25.439

36.506;25.337

40.795;24.854

110.433;24.529

102.147;25.183

48.781;24.845

45.858;25.134

60.065;24.811

46.659;24.643

35.074;24.619

41.966;24.547

43.824;26.1

26.561;25.025

28.861;24.551

86.017;24.759

128.212;25.128

45.577;25.036

57.166;24.835

56.745;25.278

54.383;24.933

56.358;24.755

73.092;25.024

58.325;25.771

146.955;25.203

117.703;24.603

47.532;25.016

112.235;23.88

61.172;25.03

38.495;24.869

76.474;24.551

86.415;25.751

69.67;24.581

54.011;25.466

70.177;25.734

189.478;25.643

36.944;25.054

106.512;25.874

159.355;24.623

53.773;25.203

58.498;25.405

68.132;25.523

31.956;25.494

25.833;25.827

37.024;26.041

58.634;25.336

115.097;24.7

33.681;25.316

84.975;25.103

110.121;24.786

66.484;25.257

36.759;25.287

69.692;25.924

56.575;25.707

67.928;26.379

127.939;24.96

58.758;25.527

45.182;26.118

48.449;26.748

145.136;25.085

149.355;24.98

35.823;25.284

57.386;25.038

89.227;26.126

72.192;26.17

73.752;25.128

80.11;25.817

54.142;25.944

43.287;26.167

139.294;26.215

75.649;25.575

72.593;25.239

55.809;26.026

26.657;26.349

34.063;24.823

61.025;26.013

106.341;24.657

17.818;25.21

149.098;25.208

145.448;24.945

42.264;24.047

34.737;24.583

164.143;25.168

171.442;24.738

35.157;24.583

123.024;24.864

21.585;25.192

131.858;24.653

29.181;25.53

112.455;24.808

177.066;25.061

42.681;25.253

15.89;25.176

69.32;25.568

67.811;25.086

42.433;25.262

20.301;25.103

88.388;25.432

81.816;24.925

36.533;25.397

42.516;25.279

100.613;25.114	61.041;24.552	185.413;25.096
26.125;25.78	99.948;25.21	73.125;25.099
40.526;25.28	150.874;25.958	57.483;25.342
23.464;25.342	35.139;24.612	130.792;25.98
194.297;25.514	11.234;25.393	127.487;24.869
148.276;24.517	155.552;25.6	203.903;24.821
24.232;24.699	94.596;24.576	67.199;25.175
249.442;24.878	107.583;25.11	24.322;25.218
42.727;24.682	12.227;26.204	158.651;24.366
26.565;25.472	121.193;24.61	122.505;25.329
204.767;24.55	36.409;25.158	183.554;25.006
123.202;24.88	79.308;25.245	146.404;24.601
54.919;24.546	104.679;24.986	176.634;24.139
89.184;25.477	115.723;24.506	97.238;25.898
67.732;24.654	44.517;24.375	200.7;25.382
66.864;25.023	52.426;25.155	209.33;25.939
18.95;25.426	72.389;24.372	134.607;25.157
70.403;24.626	76.893;24.302	231.953;25.272
115.32;24.734	21.003;25.742	154.433;24.937
186.977;25.078	150.977;25.248	191.426;24.677
97.065;24.777	17.887;25.489	168.599;25.497
116.239;24.997	40.234;25.363	328.15;25.456
101.851;25.281	80.358;25.463	151.876;25.55
20.414;25.322	91.054;25.68	58.872;24.695
56.143;25.034	133.283;24.838	30.443;25.003
34.908;24.891	30.803;25.264	129.578;24.679
98.968;24.743	83.135;24.966	261.386;24.893
117.465;23.97	85.942;25.224	113.184;25.527
32.113;24.51	92.005;25.023	173.28;25.269
73.46;24.524	85.835;25.431	207.353;25.514
134.597;25.108	65.062;25.493	97.407;25.496
80.587;24.87	122.375;25.563	210.466;25.669
13.115;25.852	169.863;24.971	168.534;25.412
74.412;25.031	99.633;25.551	162.86;24.984
90.004;24.881	191.88;25.738	101.055;24.91
166.569;24.741	60.609;25.391	86.883;24.755

34.273;25.548	177.213;25.01	153.618;24.829
69.557;24.536	66.037;25.262	115.605;24.894
145.572;25.533	199.067;25.757	127.688;24.605
149.571;25.533	141.085;25.472	13.962;25.275
195.871;24.553	290.557;24.916	41.266;24.882
176.637;24.9	146.146;24.619	59.808;24.917
116.851;25.278	37.284;25.708	137.112;24.364
191.985;25.232	86.637;25.911	100.402;25.147
78.634;24.843	216.012;24.936	127.922;24.72
127.462;24.068	195.454;24.741	138.844;24.623
125.606;25.163	114.484;26.026	63.782;24.654
107.851;24.725	244.789;24.513	56.281;25.126
114.836;25.073	107.553;25.82	203.316;24.582
189.308;24.526	220.753;24.961	162.836;23.965
94.196;24.721	79.89;25.732	4.587;25.646
189.028;24.402	74.008;26.453	73.54;24.777
128.805;24.372	212.298;26.141	89.483;25.357
133.601;25.81	74.556;26.879	314.973;24.341
53.676;25.216	241.516;24.9	105.018;24.361
186.616;24.848	153.504;25.284	35.394;24.394
47.19;24.388	71.909;26.293	46.588;24.337
170.028;25.252	110.071;26.272	104.44;24.247
167.871;24.548	112.125;26.135	87.127;24.878
75.86;25.183	132.107;25.351	51.199;24.603
130.415;24.807	52.916;26.684	162.651;24.645
168.428;25.224	154.098;25.683	118.081;24.725
307.697;25.057	158.99;25.161	290.599;24.653
39.297;24.921	125.113;26.056	51.519;24.821
141.797;24.887	138.115;25.62	135.817;25.094
85.508;25.43	97.503;25.776	163.168;23.907
292.445;24.308	118.197;25.758	102.038;25.07
109.246;25.447	127.757;26.305	137.161;24.831
150.331;25.108	50.547;26.434	53.506;24.831
101.855;24.728	93.757;25.072	198.272;24.422
255.184;25.149	196.235;24.791	9.012;23.745
29.944;24.66	163.795;24.164	145.055;24.871

269.339;24.733

92.907;25.172

37.808;24.628

197.916;23.794

194.408;24.719

44.232;24.63

146.34;25.248

79.823;23.883

57.417;24.934

115.23;24.948

136.347;23.989

91.795;25.309

142.359;24.434

337.638;24.867

146.336;23.901

151.035;24.73

267.765;23.907

134.544;25.375

145.605;24.736

59.637;25.111

76.527;24.82

147.76;24.516

176.578;24.658

156.21;24.713

69.572;24.978

3.574;26.004

129.793;24.685

123.543;24.067

120.813;24.368

182.821;24.714

161.961;24.67

183.223;23.887

135.514;25.138

270.398;24.795

148.508;24.314

319.802;24.428

62.427;25.868

140.713;24.662

82.653;25.171

93.465;25.163

75.243;26.153

133.89;24.814

79.099;26.116

85.994;24.987

67.252;25.045

91.469;24.748

78.514;25.713

142.84;25.233

25.119;25.684

163.337;25.985

141.507;24.805

182.089;25.594

114.898;25.563

182.427;24.894

228.128;24.714

246.497;25.209

153.586;24.472

218.99;25.695

221.133;25.308

146.897;25.386

176.051;25.471

155.67;25.024

155.442;25.778

66.966;23.22

37.085;22.548

151.939;21.865

0.631;22.151

7.65;23.351

15.93;23.216

45.06;23.541

0.534;22.378

8.791;23.606

103.382;21.767

25.008;23.68

97.705;22.54

0.599;22.95

45.301;23.226

110.374;23.658

94.985;22.742

88.23;22.76

119.505;24.162

121.237;23.213

14.062;23.357

79.786;21.773

158.07;22.842

3.117;23.124

65.301;22.136

60.629;22.681

6.116;22.818

140.255;22.039

16.604;23.045

83.253;23.076

45.142;22.557

65.733;22.731

122.377;23.339

32.989;23.146

30.516;21.875

0.682;22.06

107.354;23.04

20.1;22.704

2.384;22.202

62.28;22.027

98.994;23.016

15.386;23.496

17.875;22.68

64.51;23.006

38.136;23.005

35.697;22.147

164.453;23.263	141.997;23.799	0.079;20.658
14.418;22.432	8.008;23.143	0.507;21.271
141.929;24.01	13.05;23.572	6.272;22.225
19.815;22.037	45.385;23.349	5.589;21.702
41.435;22.26	73.511;24.234	0;21.528
8.294;23.896	33.532;24.035	0.073;21.148
0.462;22.167	16.907;23.633	1.492;20.627
1.416;22.431	8.119;23.969	0.395;22.021
0.462;21.812	39.751;23.766	1.344;20.953
28.92;21.913	29.541;22.993	49.547;21.034
111.325;22.951	45.597;24.071	0.792;21.887
25.678;23.437	14.523;23.593	0.007;21.979
38.34;23.755	41.452;22.648	4.79;21.618
19.037;23.432	30.537;24.803	2.528;22.663
66.973;22.804	53.504;24.338	13.769;22.372
25.799;22.661	109.439;24.989	6.42;21.478
30.725;22.709	33.981;24.207	2.488;21.51
0.566;23.401	28.982;23.905	34.393;21.263
58.091;22.213	7.881;24.206	0.612;20.81
27.911;24.136	19.56;24.458	0.479;22.062
140.131;23.613	15.902;23.831	1.593;22.154
5.154;22.881	9.941;21.344	4.2;21.66
45.811;23.301	11.047;22.246	0.11;21.463
38.6;23.706	6.299;20.815	0.295;20.905
1.802;22.62	0.977;21.08	4.692;21.898
10.885;22.734	0.011;21.328	36.065;22.586
32.64;23.83	50.403;21.906	4.964;22.438
43.788;22.498	6.31;21.658	0.018;20.838
15.37;23.344	0.033;20.857	0.177;21.642
77.958;22.385	2.188;22.835	0.79;21.29
53.031;24.111	0;20.56	18.775;22.124
80.129;24.031	14.357;22.135	10.747;21.728
66.813;24.138	0;20.942	0.567;20.755
122.022;23.377	9.2;22.689	0;21.159
42.703;23.998	3.761;22.571	0;21.222
61.912;23.462	0.984;20.911	0;21.305

32.811;22.171	14.138;21.76	18.186;23.391
0;21.196	0.001;20.767	20.477;22.573
11.568;22.033	0.935;21.069	19.389;22.275
2.548;20.799	17.074;22.879	3.716;22.637
11.282;21.356	3.031;21.362	1.636;22.387
0;22.533	0;21.082	0;22.09
4.8;21.927	2.14;20.851	0.269;21.549
10.378;22.258	0.007;23.146	26.912;21.337
0.022;22.158	0.234;21.861	2.002;21.311
0.007;21.902	28.998;22.397	0.03;22.712
58.427;21.654	0.262;22.756	5.664;21.827
5.412;22.168	3.304;23.003	38.582;23.334
2.157;21.186	6.474;21.649	11.4;22.208
31.816;22.874	0.295;22.2	1.165;22.283
23.661;21.987	17.6;21.992	4.431;22.107
13.14;22.839	10.134;23.235	2.472;22.5
11.459;22.635	5.306;23.257	1.171;22.929

OUTPUT FOR SPEI ANALYSIS OF ANANTAPUR DISTRICT

		1st Month	3rd Month	6th Month	9th Month	12th Month
1901	1	−1.17962	−0.83953	−0.01907	0.099312	0.099312
1901	2	−1.11361	−0.17194	0.155816	0.105687	0.105687
1901	3	−0.20892	0.228871	0.291828	0.104406	0.104406
1901	4	0.395789	0.380621	0.328826	0.064845	0.064845
1901	5	0.479529	0.399086	0.223407	−0.041067	−0.041067
1901	6	0.249995	0.362451	0.046489	−0.221192	−0.221192
1901	7	0.400276	0.283969	−0.12489	−0.316105	−0.316105
1901	8	0.418207	0.030312	−0.43699	−0.372647	−0.372647
1901	9	0.000732	−0.51455	−0.92713	−0.31612	−0.31612
1901	10	−0.46418	−1.05317	−0.99661	−0.159997	−0.159997
1901	11	−1.14306	−1.28086	−0.66952	−0.083718	−0.083718

		1st Month	3rd Month	6th Month	9th Month	12th Month
1901	12	−1.1547	−1.27428	−0.1907	0.066969	0.066969
1902	1	−1.18175	−0.83997	0.130064	0.124891	0.124891
1902	2	−1.1212	−0.2158	0.187072	0.133962	0.133962
1902	3	−0.20575	0.273214	0.335128	0.134628	0.134628
1902	4	0.307682	0.617529	0.362958	0.095068	0.095068
1902	5	0.661995	0.483271	0.280372	0.005107	0.005107
1902	6	0.778277	0.402185	0.067649	−0.199426	−0.199426
1902	7	−0.01215	0.121353	−0.2254	−0.411735	−0.411735
1902	8	0.486627	0.050673	−0.42036	−0.423261	−0.423261
1902	9	−0.06481	−0.52625	−0.92361	−0.486571	−0.486571
1902	10	−0.41673	−1.02828	−1.02335	−0.395212	−0.395212
1902	11	−1.10944	−1.27565	−0.77873	−0.233094	−0.233094
1902	12	−1.16673	−1.25052	−0.42557	−0.041333	−0.041333
1903	1	−1.18175	−0.91806	−0.18203	0.035777	0.035777
1903	2	−1.02085	−0.37581	0.00257	0.050818	0.050818
1903	3	−0.37771	−0.06057	0.192703	0.047305	0.047305
1903	4	0.086627	0.164637	0.263154	0.023319	0.023319
1903	5	0.168668	0.273252	0.224343	−0.03867	−0.03867
1903	6	0.315429	0.445732	0.126753	−0.157107	−0.157107
1903	7	0.328982	0.372199	−0.0534	−0.244482	−0.244482
1903	8	0.651667	0.188844	−0.3194	−0.240656	−0.240656
1903	9	0.08402	−0.42833	−0.87794	−0.321381	−0.321381
1903	10	−0.36794	−1.00385	−0.93761	−0.244361	−0.244361
1903	11	−1.0947	−1.27172	−0.56062	−0.14403	−0.14403
1903	12	−1.1667	−1.27838	−0.23264	−0.026731	−0.026731

		1st Month	3rd Month	6th Month	9th Month	12th Month
1904	1	−1.18175	−0.76874	0.006381	0.04308	0.04308
1904	2	−1.12226	−0.06494	0.110549	0.06298	0.06298
1904	3	−0.08771	0.216043	0.216773	0.064275	0.064275
1904	4	0.520106	0.387358	0.237645	0.010042	0.010042
1904	5	0.256133	0.248437	0.121889	−0.119677	−0.119677
1904	6	0.397464	0.235106	−0.00822	−0.265814	−0.265814
1904	7	0.108784	0.103861	−0.21974	−0.385784	−0.385784
1904	8	0.252317	−0.00466	−0.457	−0.411857	−0.411857
1904	9	0.021683	−0.43469	−0.88021	−0.391504	−0.391504
1904	10	−0.29767	−0.96792	−0.94809	−0.295127	−0.295127
1904	11	−1.08471	−1.26574	−0.71169	−0.16055	−0.16055
1904	12	−1.15578	−1.27491	−0.3287	−0.014016	−0.014016
1905	1	−1.18175	−0.8242	−0.0698	0.057148	0.057148
1905	2	−1.12226	−0.28533	0.088877	0.062654	0.062654
1905	3	−0.17917	0.081342	0.232149	0.061451	0.061451
1905	4	0.147883	0.29441	0.269357	0.016474	0.016474
1905	5	0.324258	0.360998	0.20649	−0.054818	−0.054818
1905	6	0.437603	0.387134	0.065407	−0.206209	−0.206209
1905	7	0.293948	0.254516	−0.14925	−0.312906	−0.312906
1905	8	0.425548	0.040496	−0.43012	−0.378511	−0.378511
1905	9	0.040586	−0.50397	−0.92168	−0.361402	−0.361402
1905	10	−0.49578	−1.0709	−0.95757	−0.322497	−0.322497
1905	11	−1.14306	−1.28232	−0.68859	−0.165077	−0.165077
1905	12	−1.17047	−1.27609	−0.25835	−0.01963	−0.01963

		1st Month	3rd Month	6th Month	9th Month	12th Month
1906	1	−1.1697	−0.74528	−0.07584	0.048369	0.048369
1906	2	−1.12226	−0.24234	0.087166	0.051267	0.051267
1906	3	−0.05421	0.179218	0.225286	0.04999	0.04999
1906	4	0.135897	0.248387	0.238991	−0.007529	−0.007529
1906	5	0.483166	0.329031	0.17398	−0.079624	−0.079624
1906	6	0.141937	0.284254	−0.00912	−0.265435	−0.265435
1906	7	0.312172	0.241468	−0.15911	−0.327803	−0.327803
1906	8	0.399467	0.009081	−0.45059	−0.348065	−0.348065
1906	9	0.005044	−0.53413	−0.93642	−0.351589	−0.351589
1906	10	−0.53015	−1.07541	−0.97537	−0.284788	−0.284788
1906	11	−1.14306	−1.27573	−0.60762	−0.175224	−0.175224
1906	12	−1.13789	−1.2692	−0.23255	−0.060077	−0.060077
1907	1	−1.18175	−0.77641	−0.02571	−0.007635	−0.007635
1907	2	−1.12226	−0.12996	0.073207	−0.001779	−0.001779
1907	3	−0.10031	0.212378	0.172902	−0.003383	−0.003383
1907	4	0.399093	0.340729	0.173009	−0.058395	−0.058395
1907	5	0.363117	0.229362	0.056789	−0.174379	−0.174379
1907	6	0.269477	0.157768	−0.11082	−0.342935	−0.342935
1907	7	0.064939	0.026012	−0.30336	−0.471119	−0.471119
1907	8	0.204807	−0.12464	−0.54884	−0.506827	−0.506827
1907	9	−0.12683	−0.60046	−0.96231	−0.481809	−0.481809
1907	10	−0.51912	−1.07611	−1.04381	−0.397533	−0.397533
1907	11	−1.14306	−1.27688	−0.78421	−0.303229	−0.303229
1907	12	−1.15553	−1.23975	−0.38904	−0.128584	−0.128584

		1st Month	3rd Month	6th Month	9th Month	12th Month
1908	1	–1.1664	–0.91324	–0.17143	–0.068336	–0.068336
1908	2	–1.01084	–0.38294	–0.08211	–0.060311	–0.060311
1908	3	–0.38055	–0.01553	0.081461	–0.061842	–0.061842
1908	4	0.071335	0.178618	0.125613	–0.08957	–0.08957
1908	5	0.291407	0.133849	0.070592	–0.157112	–0.157112
1908	6	0.228464	0.193825	–0.07009	–0.315152	–0.315152
1908	7	–0.07299	0.091653	–0.24977	–0.417276	–0.417276
1908	8	0.485849	0.027132	–0.42583	–0.454774	–0.454774
1908	9	–0.07798	–0.54487	–0.94247	–0.548196	–0.548196
1908	10	–0.50499	–1.0438	–1.00201	–0.4858	–0.4858
1908	11	–1.02399	–1.24335	–0.81677	–0.412852	–0.412852
1908	12	–1.17047	–1.26862	–0.50509	–0.323731	–0.323731
1909	1	–1.14397	–0.85982	–0.29435	–0.300121	–0.300121
1909	2	–1.12226	–0.44974	–0.22112	–0.290993	–0.290993
1909	3	–0.25454	–0.15976	–0.15113	–0.289491	–0.289491
1909	4	–0.12173	–0.0292	–0.18101	–0.338811	–0.338811
1909	5	0.031218	–0.05582	–0.22408	–0.402956	–0.402956
1909	6	0.130122	–0.10904	–0.34164	–0.53842	–0.53842
1909	7	–0.22486	–0.29636	–0.53016	–0.686576	–0.686576
1909	8	–0.08831	–0.39252	–0.74153	–0.733744	–0.733744
1909	9	–0.47032	–0.78563	–1.07788	–0.662032	–0.662032
1909	10	–0.59326	–1.08209	–1.12972	–0.66238	–0.66238
1909	11	–1.03332	–1.25631	–0.97089	–0.604186	–0.604186
1909	12	–1.17047	–1.25819	–0.57003	–0.552571	–0.552571

		1st Month	3rd Month	6th Month	9th Month	12th Month
1910	1	−1.18175	−1.08344	−0.50913	−0.526981	−0.526981
1910	2	−1.04412	−0.66213	−0.44306	−0.52127	−0.52127
1910	3	−0.64792	−0.24964	−0.42262	−0.519113	−0.519113
1910	4	−0.27325	−0.24742	−0.41396	−0.538995	−0.538995
1910	5	0.191472	−0.29066	−0.46347	−0.602302	−0.602302
1910	6	−0.50841	−0.5218	−0.64709	−0.80425	−0.80425
1910	7	−0.43163	−0.53606	−0.72267	−0.829482	−0.829482
1910	8	−0.45666	−0.637	−0.91936	−0.8184	−0.8184
1910	9	−0.53345	−0.86894	−1.13489	−0.686362	−0.686362
1910	10	−0.82175	−1.13998	−1.13701	−0.659229	−0.659229
1910	11	−0.92237	−1.2244	−0.90886	−0.524449	−0.524449
1910	12	−1.16684	−1.27843	−0.57183	−0.459924	−0.459924
1911	1	−1.18175	−1.04137	−0.48915	−0.449395	−0.449395
1911	2	−1.12226	−0.59307	−0.35714	−0.42802	−0.42802
1911	3	−0.54336	−0.24397	−0.31114	−0.429495	−0.429495
1911	4	−0.18944	−0.23573	−0.32775	−0.45083	−0.45083
1911	5	0.068861	−0.19181	−0.36083	−0.516715	−0.516715
1911	6	−0.42541	−0.32866	−0.5116	−0.67489	−0.67489
1911	7	−0.13563	−0.38154	−0.57891	−0.722866	−0.722866
1911	8	−0.29071	−0.52901	−0.84136	−0.768628	−0.768628
1911	9	−0.63826	−0.86278	−1.10378	−0.751548	−0.751548
1911	10	−0.51195	−1.03581	−1.10294	−0.65341	−0.65341
1911	11	−1.09193	−1.24258	−0.91439	−0.605444	−0.605444
1911	12	−1.07416	−1.21686	−0.6665	−0.551723	−0.551723

		1st Month	3rd Month	6th Month	9th Month	12th Month
1912	1	−1.18175	−1.07412	−0.51063	−0.507985	−0.507985
1912	2	−1.00653	−0.58998	−0.4475	−0.495889	−0.495889
1912	3	−0.64863	−0.39076	−0.429	−0.491379	−0.491379
1912	4	−0.14325	−0.25326	−0.39274	−0.501945	−0.501945
1912	5	−0.25337	−0.33956	−0.4533	−0.580237	−0.580237
1912	6	−0.16022	−0.41982	−0.52442	−0.670354	−0.670354
1912	7	−0.44496	−0.49012	−0.65465	−0.775599	−0.775599
1912	8	−0.46018	−0.5476	−0.82126	−0.744302	−0.744302
1912	9	−0.3141	−0.69099	−0.96925	−0.634469	−0.634469
1912	10	−0.69902	−1.01909	−1.0794	−0.633826	−0.633826
1912	11	−0.83545	−1.14916	−0.85312	−0.507986	−0.507986
1912	12	−1.00359	−1.13748	−0.56847	−0.427864	−0.427864
1913	1	−1.18148	−1.0421	−0.49047	−0.399483	−0.399483
1913	2	−0.81239	−0.55896	−0.35463	−0.398736	−0.398736
1913	3	−0.68395	−0.29591	−0.29962	−0.415088	−0.415088
1913	4	−0.12138	−0.23732	−0.26619	−0.426848	−0.426848
1913	5	−0.02733	−0.2081	−0.33277	−0.500259	−0.500259
1913	6	−0.38852	−0.26404	−0.45952	−0.641362	−0.641362
1913	7	−0.11991	−0.25979	−0.53652	−0.693043	−0.693043
1913	8	−0.16105	−0.44252	−0.79041	−0.821591	−0.821591
1913	9	−0.38247	−0.83427	−1.11297	−0.905846	−0.905846
1913	10	−0.81783	−1.1991	−1.1888	−0.854238	−0.854238
1913	11	−1.14306	−1.28564	−1.0903	−0.761543	−0.761543
1913	12	−1.17047	−1.27402	−0.89623	−0.646549	−0.646549

		1st Month	3rd Month	6th Month	9th Month	12th Month
1914	1	−1.18171	−1.09009	−0.71873	−0.587384	−0.587384
1914	2	−1.10176	−0.81302	−0.61714	−0.584994	−0.584994
1914	3	−0.6344	−0.65471	−0.52733	−0.589056	−0.589056
1914	4	−0.55521	−0.535	−0.48539	−0.61369	−0.61369
1914	5	−0.54661	−0.48375	−0.49962	−0.643778	−0.643778
1914	6	−0.25542	−0.37539	−0.52248	−0.673079	−0.673079
1914	7	−0.45933	−0.39327	−0.64293	−0.787974	−0.787974
1914	8	−0.23878	−0.45629	−0.7995	−0.840183	−0.840183
1914	9	−0.24448	−0.78987	−1.03621	−0.879388	−0.879388
1914	10	−0.92691	−1.23653	−1.23082	−0.857408	−0.857408
1914	11	−1.12768	−1.2816	−1.11285	−0.791982	−0.791982
1914	12	−1.17047	−1.15978	−0.87612	−0.680093	−0.680093
1915	1	−1.18175	−1.1402	−0.71282	−0.665854	−0.665854
1915	2	−0.7115	−0.84699	−0.65235	−0.645723	−0.645723
1915	3	−0.90252	−0.67239	−0.58471	−0.665851	−0.665851
1915	4	−0.5611	−0.50883	−0.56949	−0.660509	−0.660509
1915	5	−0.44044	−0.52081	−0.57025	−0.668554	−0.668554
1915	6	−0.2997	−0.46787	−0.63491	−0.749645	−0.749645
1915	7	−0.62576	−0.58833	−0.74562	−0.856001	−0.856001
1915	8	−0.2797	−0.57448	−0.81095	−0.922235	−0.922235
1915	9	−0.7009	−0.96014	−1.08461	−1.027692	−1.027692
1915	10	−0.67092	−1.07857	−1.12567	−0.905938	−0.905938
1915	11	−1.13288	−1.06559	−1.1601	−0.883705	−0.883705
1915	12	−0.97244	−1.05303	−1.02016	−0.78034	−0.78034

		1st Month	3rd Month	6th Month	9th Month	12th Month
1916	1	−0.64982	−1.07854	−0.81475	−0.74846	−0.74846
1916	2	−1.12226	−1.04329	−0.79862	−0.767003	−0.767003
1916	3	−0.86836	−0.87963	−0.7135	−0.771071	−0.771071
1916	4	−0.81976	−0.66597	−0.67715	−0.783764	−0.783764
1916	5	−0.73844	−0.64597	−0.67046	−0.795886	−0.795886
1916	6	−0.16106	−0.52848	−0.67688	−0.830391	−0.830391
1916	7	−0.7676	−0.64569	−0.84454	−0.93098	−0.93098
1916	8	−0.41252	−0.63801	−0.94227	−0.920225	−0.920225
1916	9	−0.48398	−0.95903	−1.18667	−0.970698	−0.970698
1916	10	−1.04326	−1.28288	−1.21317	−0.951711	−0.951711
1916	11	−1.14306	−1.28462	−1.09765	−0.841885	−0.841885
1916	12	−1.16734	−1.27851	−0.93983	−0.826711	−0.826711
1917	1	−1.18148	−1.05905	−0.81793	−0.807396	−0.807396
1917	2	−1.12226	−0.82393	−0.70772	−0.791615	−0.791615
1917	3	−0.57336	−0.70544	−0.72732	−0.79569	−0.79569
1917	4	−0.61186	−0.67688	−0.74832	−0.835611	−0.835611
1917	5	−0.64989	−0.61949	−0.77011	−0.886287	−0.886287
1917	6	−0.50977	−0.70268	−0.82219	−0.95947	−0.95947
1917	7	−0.51071	−0.77684	−0.92932	−1.116111	−1.116111
1917	8	−0.87569	−0.92078	−1.17263	−1.265916	−1.265916
1917	9	−0.72641	−1.03762	−1.2316	−1.292148	−1.292148
1917	10	−0.85923	−1.21378	−1.44511	−1.32462	−1.32462
1917	11	−1.14306	−1.28444	−1.49012	−1.290499	−1.290499
1917	12	−1.16659	−1.27832	−1.36429	−1.192086	−1.192086

		1st Month	3rd Month	6th Month	9th Month	12th Month
1918	1	−1.18163	−1.62009	−1.28315	−1.213033	−1.213033
1918	2	−1.12226	−1.35728	−1.18971	−1.201592	−1.201592
1918	3	−1.50616	−1.18183	−1.11307	−1.20377	−1.20377
1918	4	−0.98782	−1.08662	−1.11721	−1.19311	−1.19311
1918	5	−1.0123	−1.08799	−1.14813	−1.242687	−1.242687
1918	6	−1.00141	−1.02231	−1.1943	−1.302829	−1.302829
1918	7	−1.0365	−1.10936	−1.25153	−1.269739	−1.269739
1918	8	−0.83348	−1.18285	−1.39602	−1.349155	−1.349155
1918	9	−1.40343	−1.59794	−1.54651	−1.285564	−1.285564
1918	10	−1.39243	−1.39636	−1.27555	−1.114879	−1.114879
1918	11	−1.09983	−1.27423	−1.37982	−1.11542	−1.11542
1918	12	−1.17047	−1.26938	−1.14943	−1.138147	−1.138147
1919	1	−1.18175	−1.07839	−0.98964	−1.171755	−1.171755
1919	2	−1.08469	−1.21701	−1.00812	−1.169057	−1.169057
1919	3	−0.62184	−0.95125	−1.05912	−1.152637	−1.152637
1919	4	−1.28038	−0.88715	−1.14663	−1.206248	−1.206248
1919	5	−0.65468	−0.88176	−1.14382	−1.210886	−1.210886
1919	6	−0.51868	−1.10246	−1.24566	−1.348386	−1.348386
1919	7	−1.1813	−1.34843	−1.43979	−1.551496	−1.551496
1919	8	−1.29531	−1.46906	−1.59704	−1.570389	−1.570389
1919	9	−1.48639	−1.5584	−1.51397	−1.478695	−1.478695
1919	10	−1.55405	−1.28361	−1.5128	−1.404117	−1.404117
1919	11	−0.55214	−1.09428	−1.49677	−1.370683	−1.370683

		1st Month	3rd Month	6th Month	9th Month	12th Month
1919	12	−1.16609	−1.24759	−1.43897	−1.277038	−1.277038
1920	1	−1.1163	−1.69459	−1.35929	−1.242774	−1.242774
1920	2	−1.07962	−1.47691	−1.30415	−1.226296	−1.226296
1920	3	−1.67584	−1.26864	−1.2028	−1.230432	−1.230432
1920	4	−1.10998	−1.15809	−1.14007	−1.193704	−1.193704
1920	5	−1.05257	−1.20301	−1.14793	−1.205948	−1.205948
1920	6	−1.07225	−1.11755	−1.18432	−1.236854	−1.236854
1920	7	−1.25835	−1.08233	−1.18848	−1.195035	−1.195035
1920	8	−0.80466	−0.98993	−1.1394	−1.158716	−1.158716
1920	9	−0.87641	−1.23933	−1.20033	−1.336397	−1.336397
1920	10	−1.25576	−1.23401	−1.15009	−1.421714	−1.421714
1920	11	−1.12195	−1.03676	−1.2109	−1.377692	−1.377692
1920	12	−0.79958	−0.96369	−1.35242	−1.213564	−1.213564
1921	1	−0.74722	−0.97736	−1.39086	−1.23872	−1.23872
1921	2	−0.89009	−1.13062	−1.32103	−1.232713	−1.232713
1921	3	−0.74537	−1.26736	−1.1778	−1.243574	−1.243574
1921	4	−1.11517	−1.3747	−1.23114	−1.287275	−1.287275
1921	5	−1.40853	−1.36402	−1.24537	−1.334146	−1.334146
1921	6	−1.3101	−1.07236	−1.20671	−1.291711	−1.291711
1921	7	−1.16046	−1.03287	−1.16052	−1.243148	−1.243148
1921	8	−0.51465	−0.9696	−1.19833	−1.262227	−1.262227
1921	9	−1.30949	−1.48728	−1.42427	−1.539785	−1.539785
1921	10	−1.18901	−1.29254	−1.34069	−1.494199	−1.494199
1921	11	−1.14202	−1.24188	−1.43299	−1.434201	−1.434201

		1st Month	3rd Month	6th Month	9th Month	12th Month
1921	12	−1.03044	−1.14174	−1.55075	−1.407084	−1.407084
1922	1	−1.18112	−1.31403	−1.46066	−1.376945	−1.376945
1922	2	−0.79774	−1.30675	−1.34299	−1.357748	−1.357748
1922	3	−1.15457	−1.4115	−1.34409	−1.370889	−1.370889
1922	4	−1.19782	−1.36645	−1.33054	−1.389987	−1.389987
1922	5	−1.49684	−1.32652	−1.35627	−1.442666	−1.442666
1922	6	−1.10271	−1.26325	−1.32597	−1.411798	−1.411798
1922	7	−1.126	−1.25864	−1.37335	−1.447666	−1.447666
1922	8	−1.3655	−1.35292	−1.53107	−1.561648	−1.561648
1922	9	−0.97981	−1.35425	−1.35927	−1.472971	−1.472971
1922	10	−1.42853	−1.35247	−1.43118	−1.605667	−1.605667
1922	11	−1.14306	−1.21701	−1.61586	−1.486184	−1.486184
1922	12	−0.95887	−1.16362	−1.50371	−1.364067	−1.364067
1923	1	−1.17289	−1.44548	−1.57373	−1.415294	−1.415294
1923	2	−0.96131	−1.55882	−1.39883	−1.405268	−1.405268
1923	3	−1.29848	−1.35878	−1.29953	−1.405509	−1.405509
1923	4	−1.48521	−1.45775	−1.35433	−1.424894	−1.424894
1923	5	−1.13673	−1.30497	−1.35418	−1.440868	−1.440868
1923	6	−1.57528	−1.22326	−1.41522	−1.520329	−1.520329
1923	7	−1.00186	−1.20689	−1.35074	−1.46032	−1.46032
1923	8	−0.90174	−1.38341	−1.56064	−1.574766	−1.574766
1923	9	−1.80184	−1.92055	−1.73309	−1.653895	−1.653895
1923	10	−1.74074	−1.45441	−1.55588	−1.564593	−1.564593
1923	11	−0.90075	−1.21934	−1.60708	−1.519919	−1.519919

		1st Month	3rd Month	6th Month	9th Month	12th Month
1923	12	−1.17047	−1.27848	−1.56217	−1.481455	−1.481455
1924	1	−1.18175	−1.60964	−1.50565	−1.454614	−1.454614
1924	2	−1.11816	−1.54628	−1.43177	−1.44312	−1.44312
1924	3	−1.49107	−1.38369	−1.39939	−1.445122	−1.445122
1924	4	−1.32286	−1.34623	−1.37598	−1.431892	−1.431892
1924	5	−1.23605	−1.35637	−1.40437	−1.461012	−1.461012
1924	6	−1.26656	−1.3837	−1.46461	−1.51958	−1.51958
1924	7	−1.35407	−1.37463	−1.47583	−1.522416	−1.522416
1924	8	−1.35418	−1.43497	−1.52967	−1.634698	−1.634698
1924	9	−1.15238	−1.54645	−1.3862	−1.493132	−1.493132
1924	10	−1.77685	−1.33315	−1.39188	−1.566187	−1.566187
1924	11	−1.11423	−1.00164	−1.66915	−1.607623	−1.607623
1924	12	−0.60842	−0.99179	−1.46371	−1.56569	−1.56569
1925	1	−0.85778	−1.38143	−1.53343	−1.576441	−1.576441
1925	2	−1.12226	−1.75042	−1.55136	−1.56682	−1.56682
1925	3	−1.27421	−1.37016	−1.51258	−1.558133	−1.558133
1925	4	−1.75233	−1.42918	−1.5258	−1.590119	−1.590119
1925	5	−0.95047	−1.44525	−1.50968	−1.584617	−1.584617
1925	6	−1.45787	−1.58707	−1.65323	−1.747834	−1.747834
1925	7	−1.67169	−1.59579	−1.71008	−1.798195	−1.798195
1925	8	−1.46775	−1.57599	−1.70185	−1.719689	−1.719689
1925	9	−1.43587	−1.64148	−1.53148	−1.68738	−1.68738
1925	10	−1.7649	−1.42302	−1.61342	−1.754412	−1.754412
1925	11	−0.77283	−1.0924	−1.67227	−1.761663	−1.761663

		1st Month	3rd Month	6th Month	9th Month	12th Month
1925	12	−1.17047	−1.18457	−1.7055	−1.703175	−1.703175
1926	1	−0.86497	−1.76869	−1.72438	−1.719685	−1.719685
1926	2	−1.12226	−1.70177	−1.68305	−1.683728	−1.683728
1926	3	−1.90137	−1.54764	−1.61825	−1.664902	−1.664902
1926	4	−1.35676	−1.54927	−1.62914	−1.664859	−1.664859
1926	5	−1.42289	−1.64156	−1.66281	−1.61516	−1.61516
1926	6	−1.64165	−1.64849	−1.72355	−1.64694	−1.64694
1926	7	−1.67189	−1.68766	−1.74314	−1.512887	−1.512887
1926	8	−1.49549	−1.6657	−1.42984	−1.542646	−1.542646
1926	9	−1.90876	−1.73551	−1.10158	−1.493905	−1.493905
1926	10	−1.24507	−1.14636	−0.73791	−1.546883	−1.546883
1926	11	−0.4595	−0.18414	−1.20122	−1.683809	−1.683809
1926	12	−1.15495	−0.24249	−1.39797	−1.673488	−1.673488
1927	1	1.053673	−0.26237	−1.55253	−1.688629	−1.688629
1927	2	−0.75661	−1.66846	−1.7547	−1.764695	−1.764695
1927	3	−0.90138	−1.54491	−1.69979	−1.761233	−1.761233
1927	4	−1.89384	−1.7132	−1.7604	−1.813922	−1.813922
1927	5	−1.39114	−1.74552	−1.76577	−1.772966	−1.772966
1927	6	−1.76854	−1.78646	−1.87852	−1.847791	−1.847791
1927	7	−1.8786	−1.79335	−1.88024	−1.647477	−1.647477
1927	8	−1.58741	−1.77521	−1.65608	−1.574867	−1.574867
1927	9	−1.84997	−1.93197	−1.28494	−1.52843	−1.52843
1927	10	−1.92907	−1.26254	−0.80875	−1.590849	−1.590849
1927	11	−0.62871	−0.44096	−1.12599	−1.608977	−1.608977

		1st Month	3rd Month	6th Month	9th Month	12th Month
1927	12	−0.6814	−0.38162	−1.37874	−1.499467	−1.499467
1928	1	0.420366	−0.29609	−1.57637	−1.583627	−1.583627
1928	2	−0.49094	−1.394	−1.64344	−1.574059	−1.574059
1928	3	−0.44161	−1.48063	−1.52759	−1.558323	−1.558323
1928	4	−1.80132	−1.72674	−1.65976	−1.629377	−1.629377
1928	5	−1.49147	−1.6911	−1.59744	−1.61316	−1.61316
1928	6	−1.778	−1.5386	−1.59057	−1.636033	−1.636033
1928	7	−1.63186	−1.56604	−1.47619	−1.479045	−1.479045
1928	8	−1.00205	−1.37276	−1.32815	−1.378545	−1.378545
1928	9	−2.1849	−1.53149	−1.3421	−1.417971	−1.417971
1928	10	−0.41399	−0.56944	−0.9129	−1.379745	−1.379745
1928	11	−0.05722	−0.63493	−1.22714	−1.375597	−1.375597
1928	12	−0.71931	−0.91802	−1.36227	−1.489547	−1.489547
1929	1	−0.59757	−1.16437	−1.50182	−1.432766	−1.432766
1929	2	−0.99705	−1.40311	−1.38925	−1.38807	−1.38807
1929	3	−1.10093	−1.29172	−1.4502	−1.372531	−1.372531
1929	4	−1.33354	−1.44624	−1.40612	−1.382542	−1.382542
1929	5	−1.17861	−1.35287	−1.37051	−1.36552	−1.36552
1929	6	−1.61027	−1.53381	−1.40167	−1.395148	−1.395148
1929	7	−1.07172	−1.33211	−1.27724	−1.239881	−1.239881
1929	8	−1.74136	−1.34726	−1.2947	−1.294436	−1.294436
1929	9	−0.58243	−0.65798	−0.73399	−1.155784	−1.155784
1929	10	−0.43607	−0.69774	−0.82519	−1.351596	−1.351596
1929	11	−0.48146	−0.61189	−1.09284	−1.504072	−1.504072

		1st Month	3rd Month	6th Month	9th Month	12th Month
1929	12	−0.64608	−0.67178	−1.31668	−1.574825	−1.574825
1930	1	−0.14938	−0.8423	−1.43837	−1.563865	−1.563865
1930	2	−0.74635	−1.23648	−1.51659	−1.592393	−1.592393
1930	3	−0.92167	−1.32524	−1.55902	−1.536616	−1.536616
1930	4	−1.22994	−1.45843	−1.5745	−1.567304	−1.567304
1930	5	−1.38762	−1.58162	−1.65183	−1.589972	−1.589972
1930	6	−1.5031	−1.6909	−1.64386	−1.412601	−1.412601
1930	7	−1.64995	−1.66474	−1.63647	−1.358352	−1.358352
1930	8	−1.7713	−1.74073	−1.47534	−1.246747	−1.246747
1930	9	−1.20577	−1.12416	−0.41545	−1.148385	−1.148385
1930	10	−2.17915	−0.79298	−0.45476	−1.33986	−1.33986
1930	11	0.54205	−0.10926	−0.5502	−1.45821	−1.45821
1930	12	−0.56037	0.357555	−1.12898	−1.47097	−1.47097
1931	1	0.088781	−0.04322	−1.40162	−1.387563	−1.387563
1931	2	1.257393	−0.81359	−1.56394	−1.431288	−1.431288
1931	3	−1.46153	−1.52916	−1.61252	−1.529707	−1.529707
1931	4	−1.36671	−1.63227	−1.51183	−1.534567	−1.534567
1931	5	−1.52727	−1.77109	−1.55372	−1.560573	−1.560573
1931	6	−1.76289	−1.65106	−1.51447	−1.432558	−1.432558
1931	7	−1.84228	−1.34929	−1.38593	−1.400635	−1.400635
1931	8	−1.15093	−1.05348	−1.02478	−1.321282	−1.321282
1931	9	−0.14563	−0.67871	−0.55528	−1.397741	−1.397741
1931	10	−1.86406	−1.05614	−1.15377	−1.675473	−1.675473
1931	11	0.015934	−0.60369	−1.46264	−1.812124	−1.812124

		1st Month	3rd Month	6th Month	9th Month	12th Month
1931	12	−0.82453	−0.29019	−1.64524	−1.83813	−1.83813
1932	1	−0.47845	−1.15954	−1.71432	−1.866297	−1.866297
1932	2	0.630354	−1.72958	−1.80197	−1.796628	−1.796628
1932	3	−2.35302	−1.74785	−1.83536	−1.794056	−1.794056
1932	4	−1.80359	−1.66392	−1.82897	−1.701901	−1.701901
1932	5	−1.41279	−1.78676	−1.79112	−1.712034	−1.712034
1932	6	−1.66841	−1.88417	−1.81518	−1.744828	−1.744828
1932	7	−2.04447	−1.97544	−1.69977	−1.68503	−1.68503
1932	8	−1.80238	−1.77245	−1.40264	−1.555778	−1.555778
1932	9	−2.09651	−1.07512	−0.73554	−1.603857	−1.603857
1932	10	−0.26363	0.250921	−0.47119	−1.535593	−1.535593
1932	11	0.6891	0.115867	−1.09207	−1.760252	−1.760252
1932	12	0.517666	−0.22514	−1.78679	−1.888101	−1.888101
1933	1	−0.83703	−1.21604	−1.88866	−1.953527	−1.953527
1933	2	−0.22286	−1.74446	−1.87426	−1.865634	−1.865634
1933	3	−1.48893	−1.88521	−1.88495	−1.880438	−1.880438
1933	4	−1.87137	−1.82094	−1.90497	−1.904117	−1.904117
1933	5	−1.9221	−1.87332	−1.8688	−1.912828	−1.912828
1933	6	−1.50772	−1.87863	−1.87326	−1.879029	−1.879029
1933	7	−1.95415	−1.9817	−1.95258	−1.946312	−1.946312
1933	8	−2.00855	−1.83896	−1.78537	−1.817588	−1.817588
1933	9	−1.88699	−1.42285	−1.1108	−1.784201	−1.784201
1933	10	−0.18592	−0.63912	−1.05262	−1.679064	−1.679064
1933	11	−0.70346	−0.58925	−1.53173	−1.635559	−1.635559

		1st Month	3rd Month	6th Month	9th Month	12th Month
1933	12	−0.58085	−0.57596	−1.91692	−1.704376	−1.704376
1934	1	0.031674	−1.39038	−1.81018	−1.723811	−1.723811
1934	2	−0.72413	−1.82581	−1.64345	−1.707829	−1.707829
1934	3	−2.02565	−1.89437	−1.68657	−1.659404	−1.659404
1934	4	−1.59292	−1.71172	−1.67035	−1.64218	−1.64218
1934	5	−1.95176	−1.54508	−1.66545	−1.578232	−1.578232
1934	6	−1.29966	−1.51118	−1.48862	−1.332615	−1.332615
1934	7	−1.0466	−1.60585	−1.52092	−1.191728	−1.191728
1934	8	−1.93196	−1.83981	−1.5109	−1.042736	−1.042736
1934	9	−1.70303	−1.14744	−0.63406	−0.769931	−0.769931
1934	10	−0.93406	−0.58154	−0.23426	−0.985322	−0.985322
1934	11	0.466361	0.045066	−0.08844	−1.094483	−1.094483
1934	12	−0.89726	0.022952	−0.58282	−1.213299	−1.213299
1935	1	0.777865	0.172825	−1.04604	−1.131745	−1.131745
1935	2	0.363151	−0.21328	−1.25203	−1.161076	−1.161076
1935	3	−0.35981	−0.82104	−1.32737	−1.109714	−1.109714
1935	4	−0.22735	−1.36147	−1.30451	−1.122131	−1.122131
1935	5	−1.23257	−1.59778	−1.37955	−1.124569	−1.124569
1935	6	−1.9569	−1.61311	−1.24897	−0.997881	−0.997881
1935	7	−1.40289	−1.20867	−0.84504	−0.605583	−0.605583
1935	8	−1.32954	−0.88992	−0.28546	−0.404579	−0.404579
1935	9	−0.16069	0.233114	0.268116	−0.164956	−0.164956
1935	10	−0.13193	0.270102	0.19171	−0.193976	−0.193976
1935	11	1.008011	0.627281	0.019499	−0.351837	−0.351837

		1st Month	3rd Month	6th Month	9th Month	12th Month
1935	12	0.028444	0.337421	−0.30133	−0.641696	−0.641696
1936	1	0.942645	0.198344	−0.35576	−0.844049	−0.844049
1936	2	0.28883	−0.55316	−0.65559	−0.899605	−0.899605
1936	3	−0.41741	−0.65499	−0.84176	−0.914645	−0.914645
1936	4	−0.78981	−0.60545	−1.02445	−0.833162	−0.833162
1936	5	−0.47931	−0.68051	−0.98475	−0.578527	−0.578527
1936	6	−0.32833	−0.94173	−1.03658	−0.562191	−0.562191
1936	7	−0.97031	−1.34933	−0.97959	−0.688408	−0.688408
1936	8	−1.20429	−1.35063	−0.4379	−0.799066	−0.799066
1936	9	−2.0186	−1.1863	−0.01129	−0.844549	−0.844549
1936	10	−0.01261	0.285168	0.244416	−0.547699	−0.547699
1936	11	0.111505	0.992248	−0.20266	−0.812328	−0.812328
1936	12	0.854071	0.951314	−0.67381	−0.921427	−0.921427
1937	1	1.668792	0.284549	−0.80597	−0.835103	−0.835103
1937	2	−0.03358	−1.32919	−1.25068	−1.044583	−1.044583
1937	3	−1.15565	−1.42442	−1.26652	−1.060504	−1.060504
1937	4	−1.48838	−1.16083	−1.03422	−0.940715	−0.940715
1937	5	−1.33573	−1.18825	−0.95267	−0.774093	−0.774093
1937	6	−0.30278	−1.10761	−0.78878	−0.520248	−0.520248
1937	7	−1.46232	−0.85358	−0.6993	−0.74284	−0.74284
1937	8	−1.14538	−0.46604	−0.18693	−0.351936	−0.351936
1937	9	1.226642	0.182953	0.293599	−0.46511	−0.46511
1937	10	−1.05727	−0.21611	−0.53039	−0.971309	−0.971309
1937	11	−0.28893	0.212075	−0.23266	−0.729112	−0.729112

		1st Month	3rd Month	6th Month	9th Month	12th Month
1937	12	0.760636	0.434044	−0.70647	−0.905732	−0.905732
1938	1	0.334167	−0.74442	−1.13576	−1.022387	−1.022387
1938	2	0.397515	−0.58177	−0.92741	−1.008315	−1.008315
1938	3	−1.90481	−1.16096	−1.12158	−1.044715	−1.044715
1938	4	0.103371	−1.16095	−1.0379	−0.812587	−0.812587
1938	5	−1.45143	−1.0542	−1.11014	−0.768098	−0.768098
1938	6	−1.43817	−1.05474	−0.94635	−0.40008	−0.40008
1938	7	0.153301	−0.86146	−0.45542	0.016749	0.016749
1938	8	−1.48713	−1.13632	−0.36201	−0.096585	−0.096585
1938	9	−0.86863	−0.4912	0.421619	−0.293795	−0.293795
1938	10	0.150383	0.434915	0.736817	−0.455643	−0.455643
1938	11	−0.14747	0.82921	0.638162	−0.357867	−0.357867
1938	12	1.127183	1.110196	−0.16879	−0.421764	−0.421764
1939	1	1.248621	1.016941	−0.75468	−0.61479	−0.61479
1939	2	0.91925	0.362971	−0.73665	−0.732546	−0.732546
1939	3	0.714643	−1.03137	−0.8589	−0.736993	−0.736993
1939	4	−0.38941	−1.45344	−1.01577	−0.837148	−0.837148
1939	5	−1.91919	−1.22294	−1.04658	−0.686282	−0.686282
1939	6	−1.33659	−0.67131	−0.52015	−0.078988	−0.078988
1939	7	0.228681	−0.40915	−0.17964	0.077787	0.077787
1939	8	−0.64588	−0.66688	0.038767	−0.394703	−0.394703
1939	9	−0.72014	−0.08328	0.491402	−0.675122	−0.675122
1939	10	−0.18095	0.226947	0.435432	−0.774776	−0.774776
1939	11	0.93037	0.871319	−0.14354	−0.988922	−0.988922

		1st Month	3rd Month	6th Month	9th Month	12th Month
1939	12	0.079019	0.965265	−0.88024	−1.175462	−1.175462
1940	1	1.474831	0.679002	−1.05394	−1.170954	−1.170954
1940	2	1.132692	−1.09147	−1.37256	−1.150358	−1.150358
1940	3	−0.93771	−1.61605	−1.48825	−1.202144	−1.202144
1940	4	−1.90929	−1.56085	−1.44256	−1.046602	−1.046602
1940	5	−1.52031	−1.44626	−1.15305	−0.837744	−0.837744
1940	6	−1.09559	−1.36626	−0.89023	−0.312731	−0.312731
1940	7	−1.45717	−1.28023	−0.45973	−0.197515	−0.197515
1940	8	−1.33304	−0.52036	0.069805	−0.059287	−0.059287
1940	9	−0.4197	0.571247	0.925049	0.369239	0.369239
1940	10	1.246561	1.15592	0.849931	0.180924	0.180924
1940	11	0.672229	0.754701	0.271078	−0.178042	−0.178042
1940	12	1.279481	1.17686	0.300765	−0.539025	−0.539025
1941	1	0.08019	0.503526	−0.3228	−0.794437	−0.794437
1941	2	1.54112	−0.24424	−0.51014	−0.836067	−0.836067
1941	3	−0.71781	−0.50898	−0.99541	−0.937964	−0.937964
1941	4	−1.24765	−0.72706	−1.04453	−0.941483	−0.941483
1941	5	0.506737	−0.62662	−0.99258	−0.838739	−0.838739
1941	6	−1.18289	−1.28387	−1.14788	−1.064435	−1.064435
1941	7	−0.90277	−1.29196	−1.08143	−1.021773	−1.021773
1941	8	−1.54957	−1.43457	−1.0591	−1.126531	−1.126531
1941	9	−1.17209	−0.52489	−0.47946	−1.170695	−1.170695
1941	10	−0.81957	−0.16403	−0.43896	−0.778209	−0.778209
1941	11	0.929725	0.106611	−0.68231	−0.925203	−0.925203

		1st Month	3rd Month	6th Month	9th Month	12th Month
1941	12	−0.51362	−0.30656	−1.38929	−1.118254	−1.118254
1942	1	0.066972	−0.61043	−0.92445	−1.171545	−1.171545
1942	2	0.005494	−1.17583	−1.09962	−1.199961	−1.199961
1942	3	−1.09951	−1.51543	−1.18299	−1.09537	−1.09537
1942	4	−1.29244	−0.96165	−1.2147	−1.024307	−1.024307
1942	5	−1.66099	−1.03681	−1.1933	−0.920706	−0.920706
1942	6	0.918352	−0.85722	−0.78001	−0.567568	−0.567568
1942	7	−1.3975	−1.41249	−1.04231	−0.644537	−0.644537
1942	8	−1.30039	−1.37711	−0.71893	−0.460415	−0.460415
1942	9	−1.37447	−0.46077	−0.13991	−0.608301	−0.608301
1942	10	−1.06574	0.291774	0.396815	−0.594411	−0.594411
1942	11	1.268654	0.61935	0.334688	−0.512329	−0.512329
1942	12	0.330124	0.256121	−0.62447	−0.701104	−0.701104
1943	1	0.186912	0.558676	−0.86494	−0.734247	−0.734247
1943	2	0.548855	0.009363	−0.82437	−0.733223	−0.733223
1943	3	0.762533	−0.98322	−0.88221	−0.747254	−0.747254
1943	4	−0.85064	−1.33578	−0.99826	−0.791354	−0.791354
1943	5	−1.66643	−1.17366	−0.94167	−0.707815	−0.707815
1943	6	−1.02623	−0.75794	−0.57003	−0.358086	−0.358086
1943	7	−0.44259	−0.54404	−0.22863	−0.221635	−0.221635
1943	8	−0.62928	−0.46325	−0.0725	−0.200343	−0.200343
1943	9	−0.2618	−0.03096	0.120765	−0.122493	−0.122493
1943	10	0.033664	0.356005	0.070923	−0.222392	−0.222392
1943	11	0.428568	0.432161	0.012732	−0.289078	−0.289078

		1st Month	3rd Month	6th Month	9th Month	12th Month
1943	12	0.760471	0.323945	−0.11927	−0.407321	−0.407321
1944	1	0.331149	−0.14161	−0.42829	−0.50028	−0.50028
1944	2	0.065045	−0.37439	−0.51929	−0.525106	−0.525106
1944	3	−0.40699	−0.41595	−0.59496	−0.459997	−0.459997
1944	4	−0.33568	−0.53466	−0.57842	−0.464895	−0.464895
1944	5	−0.29797	−0.57061	−0.56254	−0.334194	−0.334194
1944	6	−0.72924	−0.69362	−0.4603	−0.318547	−0.318547
1944	7	−0.52735	−0.58048	−0.38135	−0.202589	−0.202589
1944	8	−0.6478	−0.48289	−0.07484	−0.191568	−0.191568
1944	9	−0.241	0.141472	0.11302	−0.250984	−0.250984
1944	10	−0.03254	0.048241	0.140725	−0.355875	−0.355875
1944	11	0.843626	0.454199	0.041134	−0.448513	−0.448513
1944	12	−0.50414	0.135324	−0.38084	−0.583373	−0.583373
1945	1	1.052301	0.313222	−0.47706	−0.530245	−0.530245
1945	2	−0.01431	−0.34674	−0.70232	−0.57277	−0.57277
1945	3	−0.05424	−0.63945	−0.73387	−0.57517	−0.57517
1945	4	−0.50512	−0.81015	−0.72213	−0.573748	−0.573748
1945	5	−0.85442	−0.85035	−0.63323	−0.478127	−0.478127
1945	6	−0.75467	−0.76623	−0.50954	−0.281727	−0.281727
1945	7	−0.73519	−0.58534	−0.34673	−0.218476	−0.218476
1945	8	−0.62053	−0.216	−0.03655	−0.140723	−0.140723
1945	9	0.057846	0.164246	0.264381	0.018359	0.018359
1945	10	0.5206	0.147884	0.118791	0.043799	0.043799
1945	11	−0.00717	0.155245	−0.0734	−0.080606	−0.080606

		1st Month	3rd Month	6th Month	9th Month	12th Month
1945	12	0.172736	0.398672	−0.00664	−0.089046	−0.089046
1946	1	0.679288	0.179934	0.007219	−0.116798	−0.116798
1946	2	0.619311	−0.27687	−0.18738	−0.163579	−0.163579
1946	3	−0.49327	−0.31172	−0.26272	−0.200425	−0.200425
1946	4	−0.43763	−0.10456	−0.21163	−0.053182	−0.053182
1946	5	0.086843	−0.10788	−0.11169	0.02139	0.02139
1946	6	0.160356	−0.18243	−0.11207	0.139757	0.139757
1946	7	−0.4295	−0.28339	0.001459	−0.005313	−0.005313
1946	8	−0.09368	−0.06664	0.135309	−0.000365	−0.000365
1946	9	−0.07947	0.067976	0.370651	−0.111195	−0.111195
1946	10	0.25573	0.436041	0.188198	−0.26162	−0.26162
1946	11	0.238138	0.304368	0.047946	−0.278753	−0.278753
1946	12	0.888799	0.668838	−0.14723	−0.330201	−0.330201
1947	1	−0.1213	0.009152	−0.51295	−0.41186	−0.41186
1947	2	1.06041	−0.20088	−0.46619	−0.44915	−0.44915
1947	3	−0.83652	−0.67513	−0.61297	−0.50995	−0.50995
1947	4	−0.4586	−0.70995	−0.51334	−0.453648	−0.453648
1947	5	−0.56068	−0.58527	−0.52055	−0.396701	−0.396701
1947	6	−0.83616	−0.51824	−0.38214	−0.240711	−0.240711
1947	7	−0.19847	−0.25792	−0.21987	0.081471	0.081471
1947	8	−0.36537	−0.35724	−0.17985	−0.010096	−0.010096
1947	9	0.062218	−0.01617	0.047358	0.154337	0.154337
1947	10	−0.59191	−0.17404	0.310629	0.036947	0.036947
1947	11	0.58145	0.061936	0.233915	0.121562	0.121562

		1st Month	3rd Month	6th Month	9th Month	12th Month
1947	12	−0.1614	0.174064	0.26238	−0.07576	−0.07576
1948	1	0.165607	0.764418	0.11668	−0.10833	−0.10833
1948	2	0.758052	0.299897	0.089523	−0.088656	−0.088656
1948	3	1.021259	0.219452	−0.18168	−0.123734	−0.123734
1948	4	−0.70552	−0.33166	−0.39217	−0.136088	−0.136088
1948	5	0.423897	−0.03733	−0.22762	0.035125	0.035125
1948	6	−0.57808	−0.47098	−0.30024	−0.012825	−0.012825
1948	7	0.082686	−0.41374	0.025171	0.032382	0.032382
1948	8	−0.73733	−0.42442	0.097681	−0.041649	−0.041649
1948	9	−0.36456	0.116512	0.393852	0.080064	0.080064
1948	10	0.50596	0.687032	0.367852	−0.019776	−0.019776
1948	11	0.442661	0.689468	0.231786	−0.091882	−0.091882
1948	12	1.070792	0.667699	0.102124	−0.176136	−0.176136
1949	1	0.595307	0.07911	−0.31722	−0.360743	−0.360743
1949	2	0.362407	−0.24599	−0.38597	−0.448536	−0.448536
1949	3	−0.39472	−0.34798	−0.44669	−0.4035	−0.4035
1949	4	−0.26634	−0.49597	−0.47225	−0.377211	−0.377211
1949	5	−0.20357	−0.43969	−0.50471	−0.299613	−0.299613
1949	6	−0.77655	−0.48641	−0.41111	−0.297657	−0.297657
1949	7	−0.20949	−0.4074	−0.25917	−0.262124	−0.262124
1949	8	−0.32341	−0.53143	−0.1458	−0.266436	−0.266436
1949	9	−0.57602	−0.17156	−0.09697	−0.244075	−0.244075
1949	10	−0.5428	0.019975	−0.14374	−0.283542	−0.283542
1949	11	0.855807	0.386957	−0.04171	−0.247632	−0.247632

		1st Month	3rd Month	6th Month	9th Month	12th Month
1949	12	−0.06768	0.059452	−0.23042	−0.344323	−0.344323
1950	1	0.613103	−0.20449	−0.37401	−0.389141	−0.389141
1950	2	0.006892	−0.42429	−0.45739	−0.481672	−0.481672
1950	3	−0.74143	−0.40893	−0.45838	−0.427683	−0.427683
1950	4	−0.16201	−0.43437	−0.43546	−0.352397	−0.352397
1950	5	−0.22768	−0.44898	−0.48917	−0.185349	−0.185349
1950	6	−0.67456	−0.45793	−0.41338	−0.035087	−0.035087
1950	7	−0.31223	−0.3968	−0.26365	0.065379	0.065379
1950	8	−0.23825	−0.48088	0.078643	0.010727	0.010727
1950	9	−0.50184	−0.24183	0.343575	−0.186565	−0.186565
1950	10	−0.61996	−0.01325	0.405318	−0.078271	−0.078271
1950	11	0.699272	0.723494	0.348113	−0.22862	−0.22862
1950	12	0.135153	0.854542	−0.11982	−0.488541	−0.488541
1951	1	1.351345	0.804308	−0.09381	−0.516041	−0.516041
1951	2	1.001774	−0.07068	−0.55491	−0.586306	−0.586306
1951	3	−0.12475	−0.77489	−0.83414	−0.656859	−0.656859
1951	4	−0.64526	−0.63448	−0.8478	−0.566811	−0.566811
1951	5	−0.99724	−0.78116	−0.73759	−0.494222	−0.494222
1951	6	0.142972	−0.8394	−0.55721	−0.345514	−0.345514
1951	7	−1.05666	−1.00759	−0.48152	−0.308839	−0.308839
1951	8	−1.15997	−0.60022	−0.14948	0.028838	0.028838
1951	9	−0.25059	0.196209	0.243368	0.217008	0.217008
1951	10	0.643951	0.657713	0.417781	0.311938	0.311938
1951	11	0.350042	0.474596	0.452414	0.321122	0.321122

		1st Month	3rd Month	6th Month	9th Month	12th Month
1951	12	0.970199	0.328165	0.259723	0.299741	0.299741
1952	1	0.182765	0.223157	0.148631	0.210243	0.210243
1952	2	–0.17679	0.350021	0.203291	0.387259	0.387259
1952	3	0.557365	0.117616	0.231427	0.403334	0.403334
1952	4	0.488745	0.078636	0.185123	0.40644	0.40644
1952	5	–0.39828	0.112281	0.407786	0.46812	0.46812
1952	6	0.36189	0.352188	0.571811	0.612906	0.612906
1952	7	0.467661	0.307505	0.60789	0.67159	0.67159
1952	8	0.223505	0.776548	0.70914	0.734661	0.734661
1952	9	0.25745	0.842681	0.709042	0.676245	0.676245
1952	10	1.566362	0.886317	0.816386	0.527319	0.527319
1952	11	0.022732	0.422249	0.670261	0.326604	0.326604
1952	12	0.54736	0.539075	0.605234	0.241627	0.241627
1953	1	0.909171	0.768064	0.333164	0.266592	0.266592
1953	2	0.375615	0.768512	0.229501	0.154882	0.154882
1953	3	0.851609	0.520264	0.087235	0.145069	0.145069
1953	4	0.852031	–0.02316	0.05914	0.11043	0.11043
1953	5	–0.07091	–0.15938	–0.10217	0.026668	0.026668
1953	6	–0.6433	–0.25547	–0.05835	0.084735	0.084735
1953	7	0.307965	0.164358	0.212637	0.348154	0.348154
1953	8	–0.31142	0.024797	0.190598	0.386505	0.386505
1953	9	0.735965	0.342048	0.346224	0.383415	0.383415
1953	10	–0.06472	0.212228	0.405925	0.322369	0.322369
1953	11	0.262264	0.294216	0.582672	0.28772	0.28772

		1st Month	3rd Month	6th Month	9th Month	12th Month
1953	12	0.657269	0.373928	0.428657	0.251567	0.251567
1954	1	0.246321	0.640688	0.359401	0.32441	0.32441
1954	2	0.443142	0.711655	0.224397	0.394776	0.394776
1954	3	0.958087	0.348886	0.155619	0.38433	0.38433
1954	4	0.557669	0.115986	0.184792	0.282383	0.282383
1954	5	−0.29443	−0.12332	0.253738	0.284747	0.284747
1954	6	0.264855	0.00688	0.409462	0.390493	0.390493
1954	7	−0.18782	0.268821	0.39447	0.337593	0.337593
1954	8	0.070463	0.755925	0.596147	0.414884	0.414884
1954	9	1.179099	0.958876	0.617531	0.335037	0.335037
1954	10	1.043289	0.489963	0.306063	0.144587	0.144587
1954	11	0.308496	0.176359	0.160527	0.017737	0.017737
1954	12	0.10267	0.180358	0.030611	0.089426	0.089426
1955	1	0.535815	0.189476	−0.00143	0.234963	0.234963
1955	2	0.271847	0.092797	−0.07487	0.185863	0.185863
1955	3	−0.02479	−0.12408	0.014621	0.204088	0.204088
1955	4	0.202372	−0.12261	0.22806	0.249818	0.249818
1955	5	−0.29696	−0.16117	0.231206	0.293404	0.293404
1955	6	−0.05969	0.167637	0.408106	0.420043	0.420043
1955	7	−0.01374	0.604748	0.497227	0.48787	0.48787
1955	8	0.615141	0.758199	0.639199	0.507511	0.507511
1955	9	1.273647	0.748593	0.536748	0.602543	0.602543
1955	10	0.211728	0.239124	0.307572	0.614102	0.614102
1955	11	0.473983	0.279876	0.31592	0.576198	0.576198

		1st Month	3rd Month	6th Month	9th Month	12th Month
1955	12	0.317912	0.291582	0.547284	0.679159	0.679159
1956	1	0.420848	0.445206	0.750422	0.685519	0.685519
1956	2	0.460908	0.272739	0.608497	0.627421	0.627421
1956	3	0.46182	0.594518	0.746483	0.666153	0.666153
1956	4	0.044388	0.914422	0.749406	0.71668	0.71668
1956	5	1.167084	0.869377	0.785989	0.829505	0.829505
1956	6	1.260528	0.868804	0.699735	0.71731	0.71731
1956	7	0.034338	0.562451	0.591407	0.556001	0.556001
1956	8	1.20086	0.658224	0.771848	0.828718	0.828718
1956	9	0.387337	0.432675	0.504784	0.616056	0.616056
1956	10	0.094453	0.533549	0.45931	0.656964	0.656964
1956	11	0.79433	0.713169	0.876335	0.850398	0.850398
1956	12	0.787731	0.580379	0.71188	0.738901	0.738901
1957	1	0.713056	0.446152	0.692405	0.677981	0.677981
1957	2	0.41739	0.908791	0.825841	0.636966	0.636966
1957	3	0.280033	0.670342	0.727076	0.601661	0.601661
1957	4	1.610929	0.812283	0.738477	0.565459	0.565459
1957	5	−0.01088	0.727373	0.499655	0.359662	0.359662
1957	6	0.616386	0.75995	0.557697	0.438488	0.438488
1957	7	1.399675	0.641314	0.414962	0.491981	0.491981
1957	8	0.093723	0.225934	0.055744	0.20553	0.20553
1957	9	0.247074	0.239733	0.06712	0.283177	0.283177
1957	10	0.547632	−0.04767	0.28646	0.429581	0.429581
1957	11	−0.08492	−0.33093	0.179368	0.38755	0.38755

		1st Month	3rd Month	6th Month	9th Month	12th Month
1957	12	−0.37129	−0.0695	0.333376	0.338865	0.338865
1958	1	0.015824	0.64004	0.600795	0.386895	0.386895
1958	2	0.501843	0.472711	0.544514	0.385577	0.385577
1958	3	1.023722	0.471416	0.401048	0.40328	0.40328
1958	4	−0.13056	0.50733	0.266684	0.440357	0.440357
1958	5	0.503251	0.60045	0.35131	0.551622	0.551622
1958	6	1.097363	0.342026	0.367799	0.520458	0.520458
1958	7	0.15688	0.045091	0.400547	0.300151	0.300151
1958	8	−0.12119	0.061211	0.496383	0.500324	0.500324
1958	9	0.276637	0.409134	0.566196	0.742149	0.742149
1958	10	0.327682	0.800719	0.419406	0.695929	0.695929
1958	11	0.661879	0.842109	0.723848	0.659423	0.659423
1958	12	1.248425	0.712465	0.892127	0.576097	0.576097
1959	1	0.510071	0.037893	0.623009	0.437713	0.437713
1959	2	0.240807	0.5132	0.504159	0.43458	0.43458
1959	3	−0.30783	0.886252	0.453081	0.498108	0.498108
1959	4	1.292193	0.964256	0.546357	0.57754	0.57754
1959	5	1.271263	0.497685	0.402578	0.444215	0.444215
1959	6	−0.04313	0.058969	0.270626	0.45871	0.45871
1959	7	0.133051	0.135617	0.343444	0.513146	0.513146
1959	8	0.148204	0.296406	0.388087	0.560912	0.560912
1959	9	0.244593	0.607349	0.683704	0.498844	0.498844
1959	10	0.677406	0.564133	0.689472	0.364093	0.364093
1959	11	0.84675	0.34786	0.681882	0.272595	0.272595

		1st Month	3rd Month	6th Month	9th Month	12th Month
1959	12	0.268847	0.740516	0.468849	0.144795	0.144795
1960	1	0.171903	0.839047	0.266481	0.249323	0.249323
1960	2	1.438322	0.840098	0.186494	0.178429	0.178429
1960	3	0.391429	0.140719	−0.10436	0.195351	0.195351
1960	4	0.392285	−0.17828	0.007481	0.352275	0.352275
1960	5	−0.14769	−0.28085	−0.10436	0.349899	0.349899
1960	6	−0.551	−0.27739	0.238609	0.405761	0.405761
1960	7	−0.04548	0.224111	0.687094	0.730075	0.730075
1960	8	−0.11835	0.188103	0.815723	1.009002	1.009002
1960	9	1.102828	1.007049	0.860554	1.040393	1.040393
1960	10	−0.42543	1.116027	0.951181	1.070618	1.070618
1960	11	1.547827	1.24925	1.321945	1.012683	1.012683
1960	12	1.398001	0.656498	1.044074	0.93811	0.93811
1961	1	0.147207	0.777631	1.008587	0.725844	0.725844
1961	2	0.002107	1.287273	0.803915	0.81648	0.81648
1961	3	1.428601	1.204472	0.979291	1.064345	1.064345
1961	4	1.642167	1.11746	0.669508	0.890331	0.890331
1961	5	0.125376	0.311325	0.545309	0.660862	0.660862
1961	6	1.231279	0.705226	0.95615	0.746817	0.746817
1961	7	−0.29256	0.217585	0.743587	0.604295	0.604295
1961	8	1.170369	0.816201	0.868301	0.67291	0.67291
1961	9	−0.25551	1.164475	0.67635	0.416819	0.416819
1961	10	1.183886	1.214043	0.772729	0.536768	0.536768
1961	11	1.734812	0.731722	0.548908	0.443247	0.443247

		1st Month	3rd Month	6th Month	9th Month	12th Month
1961	12	0.034665	−0.01605	0.015349	0.120111	0.120111
1962	1	−0.60642	0.22641	0.140652	0.111777	0.111777
1962	2	0.678805	0.291855	0.260892	0.239175	0.239175
1962	3	0.376371	−0.03191	0.10804	0.162785	0.162785
1962	4	−0.00166	0.064606	0.057793	0.129178	0.129178
1962	5	−0.20204	0.250606	0.223751	0.135007	0.135007
1962	6	0.564881	0.258705	0.289507	0.516061	0.516061
1962	7	0.440719	0.072945	0.184702	0.645473	0.645473
1962	8	−0.16141	0.215977	0.032043	0.568018	0.568018
1962	9	0.029578	0.34784	0.622688	0.995751	0.995751
1962	10	1.031121	0.291511	0.922146	1.024923	1.024923
1962	11	−0.24401	−0.37421	0.735833	0.817159	0.817159
1962	12	−0.01288	0.85992	1.244229	1.076304	1.076304
1963	1	−0.36553	1.37515	1.29687	1.223197	1.223197
1963	2	1.786513	1.311304	1.123839	1.244826	1.244826
1963	3	1.366358	1.426786	1.095616	1.081583	1.081583
1963	4	−0.06926	1.041359	1.068673	0.96389	0.96389
1963	5	2.363264	0.868847	1.17626	1.086493	1.086493
1963	6	0.269346	0.692048	0.848565	0.716892	0.716892
1963	7	−0.33261	1.039653	0.895631	0.733768	0.733768
1963	8	1.97965	1.407954	1.162333	0.963763	0.963763
1963	9	1.263718	0.981713	0.635338	0.901381	0.901381
1963	10	0.440738	0.612263	0.425401	0.866457	0.866457
1963	11	0.968507	0.659204	0.645711	0.916091	0.916091

		1st Month	3rd Month	6th Month	9th Month	12th Month
1963	12	0.509869	0.187718	0.858342	0.859767	0.859767
1964	1	0.663366	0.292907	0.955939	0.770777	0.770777
1964	2	−0.43523	0.537022	0.940265	0.718895	0.718895
1964	3	0.496828	1.178662	1.031862	0.751032	0.751032
1964	4	1.050976	1.389649	0.925333	0.820035	0.820035
1964	5	1.647312	1.260723	0.801335	0.873932	0.873932
1964	6	0.933961	0.830091	0.485691	0.638555	0.638555
1964	7	0.800701	0.429321	0.477058	0.564825	0.564825
1964	8	0.61031	0.228206	0.584104	0.623248	0.623248
1964	9	−0.33854	−0.07013	0.391252	0.605338	0.605338
1964	10	0.405165	0.452268	0.569173	0.553447	0.553447
1964	11	−0.00923	0.823507	0.806913	0.548745	0.548745
1964	12	0.946419	0.807064	0.884403	0.602302	0.602302
1965	1	1.360521	0.724295	0.581526	0.727789	0.727789
1965	2	−0.06363	0.687709	0.361688	0.602427	0.602427
1965	3	0.555424	0.795659	0.448801	0.72456	0.72456
1965	4	1.171479	0.406531	0.695864	0.876546	0.876546
1965	5	0.535413	0.11604	0.561159	0.764802	0.764802
1965	6	−0.5178	0.13325	0.673643	0.67405	0.67405
1965	7	0.322017	0.959602	1.099875	0.941273	0.941273
1965	8	0.618121	1.072913	1.135952	1.026213	1.026213
1965	9	1.753871	1.269131	0.950686	1.041488	1.041488
1965	10	0.541463	1.133753	0.838938	0.688382	0.688382
1965	11	1.098236	1.009959	0.94938	0.768824	0.768824

		1st Month	3rd Month	6th Month	9th Month	12th Month
1965	12	1.440388	0.497686	0.90442	0.589514	0.589514
1966	1	−0.03349	0.510619	0.420223	0.452806	0.452806
1966	2	−0.69016	0.783562	0.577592	0.546776	0.546776
1966	3	1.300903	1.064866	0.55328	0.73295	0.73295
1966	4	0.850593	0.308932	0.404755	0.576752	0.576752
1966	5	0.75693	0.400471	0.434064	0.534934	0.534934
1966	6	−0.65287	0.070348	0.527211	0.512407	0.512407
1966	7	0.987242	0.500929	0.725324	0.704116	0.704116
1966	8	−0.08084	0.481375	0.611529	0.473307	0.473307
1966	9	0.602896	1.094461	0.757462	0.697055	0.697055
1966	10	1.087427	0.894659	0.744577	0.559257	0.559257
1966	11	1.264625	0.588606	0.440735	0.375132	0.375132
1966	12	0.074265	0.304979	0.492238	0.243955	0.243955
1967	1	0.379364	0.616431	0.374642	0.305578	0.305578
1967	2	0.725415	0.229173	0.229872	0.42485	0.42485
1967	3	0.636807	0.496468	0.168236	0.352586	0.352586
1967	4	−0.42263	0.158086	0.169903	0.367098	0.367098
1967	5	1.195924	0.243848	0.513263	0.570321	0.570321
1967	6	−0.33288	−0.09682	0.27509	0.511117	0.511117
1967	7	−0.15798	0.197542	0.499316	0.636115	0.636115
1967	8	0.298339	0.831311	0.775692	0.712466	0.712466
1967	9	0.662185	0.836463	0.876566	0.784384	0.784384
1967	10	1.380823	0.805747	0.832427	0.954462	0.954462
1967	11	−0.11535	0.504134	0.602665	0.809102	0.809102

		1st Month	3rd Month	6th Month	9th Month	12th Month
1967	12	0.74284	0.874768	0.763969	0.979463	0.979463
1968	1	0.971429	0.878549	0.996366	0.9353	0.9353
1968	2	0.986843	0.589676	0.848302	1.066663	1.066663
1968	3	0.558412	0.521645	0.946852	1.043073	1.043073
1968	4	0.218578	1.003783	0.922326	1.034671	1.034671
1968	5	0.759793	1.048996	1.287922	1.22426	1.22426
1968	6	1.766054	1.314407	1.297202	1.303932	1.303932
1968	7	0.417461	0.801389	1.018888	1.320984	1.320984
1968	8	1.451516	1.430521	1.26304	1.324897	1.324897
1968	9	0.283639	1.168582	1.192907	1.086239	1.086239
1968	10	1.756633	1.135868	1.4644	1.289161	1.289161
1968	11	0.744033	0.87986	1.210328	1.116654	1.116654
1968	12	0.331562	1.13936	1.02747	1.227699	1.227699
1969	1	1.492977	1.674118	1.328487	1.392725	1.392725
1969	2	1.343031	1.378055	1.139771	1.440364	1.440364
1969	3	1.763674	0.751625	1.186976	1.405318	1.405318
1969	4	0.34276	0.665051	1.104829	1.424599	1.424599
1969	5	−0.198	0.823353	1.437405	1.691336	1.691336
1969	6	1.842323	1.545575	1.695732	1.857065	1.857065
1969	7	0.753579	1.466956	1.692914	1.857725	1.857725
1969	8	1.614741	1.889679	1.938779	1.928783	1.928783
1969	9	1.573315	1.634768	1.847947	1.948987	1.948987
1969	10	1.741114	1.719713	1.900083	1.97577	1.97577
1969	11	1.066939	1.84132	1.857658	2.083161	2.083161

		1st Month	3rd Month	6th Month	9th Month	12th Month
1969	12	1.763819	1.887751	2.015838	2.120675	2.120675
1970	1	2.050875	1.949973	2.066386	2.009151	2.009151
1970	2	1.401607	1.801161	2.172248	1.732017	1.732017
1970	3	1.905396	2.011662	2.172122	1.741275	1.741275
1970	4	1.426154	1.861954	1.881935	1.812121	1.812121
1970	5	1.825468	2.42287	1.608988	1.786744	1.786744
1970	6	1.555676	2.116726	1.515347	1.761141	1.761141
1970	7	2.84834	1.750545	1.70172	1.764414	1.764414
1970	8	1.079571	0.572094	1.405851	1.504693	1.504693
1970	9	0.436671	0.722328	1.500149	1.57588	1.57588
1970	10	−0.09521	1.539904	1.687194	1.709915	1.709915
1970	11	1.391932	1.832595	1.766254	1.826202	1.826202
1970	12	2.050154	1.895553	1.852301	1.760478	1.760478
1971	1	1.314488	1.736464	1.751121	1.553208	1.553208
1971	2	1.668711	1.606282	1.728812	1.394242	1.394242
1971	3	1.722461	1.576847	1.5124	1.239511	1.239511
1971	4	0.626653	1.488683	1.309109	1.179169	1.179169
1971	5	1.694889	1.719902	1.212979	1.340877	1.340877
1971	6	1.587369	1.324216	1.002325	1.390221	1.390221
1971	7	1.318834	1.05324	0.980003	1.383504	1.383504
1971	8	0.734277	0.553932	1.074637	1.338211	1.338211
1971	9	0.826359	0.524518	1.316204	1.451949	1.451949
1971	10	−0.09566	0.790371	1.443549	1.452022	1.452022
1971	11	0.688189	1.36569	1.58155	1.569695	1.569695

		1st Month	3rd Month	6th Month	9th Month	12th Month
1971	12	1.398989	1.749049	1.755935	1.592615	1.592615
1972	1	1.656156	1.832893	1.719562	1.574783	1.574783
1972	2	1.794731	1.690481	1.600257	1.490252	1.490252
1972	3	1.558609	1.568808	1.357957	1.493391	1.493391
1972	4	0.923765	1.241499	1.253977	1.389567	1.389567
1972	5	1.648917	1.343524	1.303705	1.561668	1.561668
1972	6	0.670988	1.036399	1.398315	1.510158	1.510158
1972	7	1.229824	1.191478	1.401931	1.52037	1.52037
1972	8	0.923835	1.161033	1.581879	1.403995	1.403995
1972	9	1.107852	1.578972	1.604538	1.500648	1.500648
1972	10	1.189745	1.46013	1.569786	1.445805	1.445805
1972	11	1.806165	1.740552	1.443127	1.45706	1.45706
1972	12	0.888414	1.509695	1.406164	1.390993	1.390993
1973	1	1.933437	1.601549	1.381144	1.378414	1.378414
1973	2	1.354142	0.963545	1.174012	1.254147	1.254147
1973	3	1.057771	1.094668	1.214231	1.328515	1.328515
1973	4	0.157034	0.891034	1.14268	1.643897	1.643897
1973	5	1.719484	1.322608	1.386937	1.673332	1.673332
1973	6	0.434954	1.260072	1.414042	1.626162	1.626162
1973	7	1.305242	1.321689	1.870717	1.597828	1.597828
1973	8	1.596155	1.339004	1.732611	1.802382	1.802382
1973	9	0.568071	1.422157	1.663359	1.699249	1.699249
1973	10	1.357144	2.036174	1.622247	1.808644	1.808644
1973	11	1.680919	1.863835	1.88379	1.733398	1.733398

		1st Month	3rd Month	6th Month	9th Month	12th Month
1973	12	2.220019	1.742926	1.768142	1.804393	1.804393
1974	1	0.07383	0.893072	1.527468	1.448829	1.448829
1974	2	1.457369	1.833634	1.544132	1.627058	1.627058
1974	3	0.554266	1.607679	1.725621	1.526066	1.526066
1974	4	2.721005	2.044867	1.696155	1.731667	1.731667
1974	5	0.653805	1.027227	1.413547	1.530236	1.530236
1974	6	1.479529	1.701997	1.425412	1.816873	1.816873
1974	7	0.732124	1.229255	1.512949	1.765944	1.765944
1974	8	2.3023	1.669502	1.674985	1.684417	1.684417
1974	9	−0.00549	1.013345	1.742781	1.428325	1.428325
1974	10	1.707347	1.606631	1.862289	1.686632	1.686632
1974	11	0.63067	1.52268	1.614634	1.668816	1.668816
1974	12	1.754752	2.080063	1.570889	1.64632	1.64632
1975	1	1.585453	1.973954	1.671064	1.758994	1.758994
1975	2	2.18207	1.616161	1.671543	1.673513	1.673513
1975	3	1.335998	0.478637	1.119222	1.583358	1.583358
1975	4	−0.20581	0.845025	1.417253	1.500143	1.500143
1975	5	0.226618	1.588318	1.649205	1.628176	1.628176
1975	6	2.316299	1.677755	2.060904	1.849644	1.849644
1975	7	1.743276	1.861915	1.713621	1.582	1.582
1975	8	0.473012	1.553903	1.563162	1.661733	1.661733
1975	9	2.437187	2.042782	1.793454	1.738677	1.738677
1975	10	1.052563	1.475732	1.378291	1.462312	1.462312
1975	11	1.967065	1.405014	1.63039	1.35621	1.35621

		1st Month	3rd Month	6th Month	9th Month	12th Month
1975	12	0.696893	1.34623	1.460457	1.069391	1.069391
1976	1	0.786715	1.224312	1.397215	1.153638	1.153638
1976	2	1.87613	1.751813	1.239296	1.063805	1.063805
1976	3	0.083739	1.411846	0.827667	0.973701	0.973701
1976	4	2.233495	1.433568	1.059551	1.190948	1.190948
1976	5	1.211241	0.540905	0.589967	1.053212	1.053212
1976	6	−0.00164	0.217293	0.688703	1.090153	1.090153
1976	7	0.262644	0.637745	1.026406	1.150287	1.150287
1976	8	0.403312	0.637997	1.288969	1.233016	1.233016
1976	9	1.274356	1.207552	1.443279	1.245613	1.245613
1976	10	0.11408	1.295192	1.312232	1.054208	1.054208
1976	11	1.584071	1.63238	1.425631	1.082017	1.082017
1976	12	1.500721	1.543338	1.23641	1.09684	1.09684
1977	1	1.438426	1.29057	0.872706	1.135905	1.135905
1977	2	1.429218	1.078471	0.656902	1.150165	1.150165
1977	3	0.646298	0.679407	0.734533	1.121925	1.121925
1977	4	0.80196	0.340836	0.991464	1.219205	1.219205
1977	5	0.535308	0.268233	1.168003	1.137075	1.137075
1977	6	−0.31512	0.765212	1.32976	1.111725	1.111725
1977	7	0.522995	1.570451	1.576255	1.17127	1.17127
1977	8	1.817196	1.951035	1.527515	1.295659	1.295659
1977	9	1.815484	1.684186	1.200081	1.125968	1.125968
1977	10	1.548242	1.460975	0.854896	1.062933	1.062933
1977	11	1.263471	0.856863	0.84943	0.999062	0.999062

		1st Month	3rd Month	6th Month	9th Month	12th Month
1977	12	1.246124	0.401043	0.742794	0.866619	0.866619
1978	1	−0.67105	−0.04766	0.766946	0.741709	0.741709
1978	2	0.155373	0.73714	0.978156	1.035465	1.035465
1978	3	0.293842	0.849273	0.971486	1.244579	1.244579
1978	4	1.378636	1.265146	0.995655	1.181626	1.181626
1978	5	0.662294	1.161671	1.172487	1.112856	1.112856
1978	6	1.345614	1.045227	1.419608	1.149116	1.149116
1978	7	1.177373	0.683529	1.098777	1.40043	1.40043
1978	8	0.397312	1.101648	1.038281	1.387389	1.387389
1978	9	0.287314	1.603159	1.103602	1.355311	1.355311
1978	10	1.88421	1.372442	1.60797	1.381705	1.381705
1978	11	1.644295	0.763385	1.445002	1.203285	1.203285
1978	12	−1.01031	0.292485	1.170831	1.025542	1.025542
1979	1	0.778947	1.735966	1.335742	1.120522	1.120522
1979	2	0.956368	1.842899	1.314443	1.235303	1.235303
1979	3	2.394816	1.664405	1.236203	1.291842	1.291842
1979	4	1.027217	0.575239	0.630435	1.267622	1.267622
1979	5	0.189331	0.556047	0.780921	1.216318	1.216318
1979	6	0.445777	0.699401	1.02954	1.386628	1.386628
1979	7	0.919741	0.668541	1.535272	1.503932	1.503932
1979	8	0.586254	1.001126	1.491164	1.498087	1.498087
1979	9	0.342699	1.294293	1.587879	1.565294	1.565294
1979	10	1.624124	1.959597	1.734437	1.580143	1.580143
1979	11	1.287421	1.702738	1.623349	1.594104	1.594104

		1st Month	3rd Month	6th Month	9th Month	12th Month
1979	12	2.161093	1.71256	1.640993	1.44215	1.44215
1980	1	0.359797	1.345799	1.194168	1.240004	1.240004
1980	2	1.41122	1.440731	1.429932	1.55053	1.55053
1980	3	1.551703	1.346105	1.140399	1.498228	1.498228
1980	4	0.805873	0.873419	1.11423	1.305841	1.305841
1980	5	1.186968	1.314566	1.570386	1.511782	1.511782
1980	6	0.351191	0.861399	1.529323	1.613827	1.613827
1980	7	1.870717	1.286051	1.45448	1.554792	1.554792
1980	8	0.075124	1.673004	1.528944	1.483896	1.483896
1980	9	1.512028	1.874668	1.793426	1.571988	1.571988
1980	10	2.196764	1.474141	1.579077	1.464378	1.464378
1980	11	1.015054	1.213239	1.32582	1.174799	1.174799
1980	12	−0.07459	1.568855	1.320032	1.227299	1.227299
1981	1	1.986595	1.605523	1.401558	1.31877	1.31877
1981	2	1.787096	1.32769	1.067251	1.20199	1.20199
1981	3	0.116686	0.823273	0.913938	1.441894	1.441894
1981	4	1.264072	0.927687	1.045524	1.618482	1.618482
1981	5	0.854005	0.743989	1.100358	1.677478	1.677478
1981	6	0.362068	0.962407	1.712611	1.891227	1.891227
1981	7	0.806397	1.106447	1.826221	1.909993	1.909993
1981	8	1.431561	1.396097	1.951218	1.969092	1.969092
1981	9	0.765112	2.030591	2.065067	1.850213	1.850213
1981	10	1.480259	2.0835	2.060519	1.8675	1.8675
1981	11	2.388535	2.153278	2.06363	1.841584	1.841584

		1st Month	3rd Month	6th Month	9th Month	12th Month
1981	12	1.405056	1.936637	1.649071	1.607561	1.607561
1982	1	1.667892	1.893911	1.586162	1.661571	1.661571
1982	2	2.104167	1.892319	1.472998	1.509302	1.509302
1982	3	1.130571	0.98223	1.208423	1.097344	1.097344
1982	4	1.381679	0.817559	1.337209	1.127584	1.127584
1982	5	0.216725	0.795578	1.175496	1.141555	1.141555
1982	6	0.652526	1.361349	1.132678	1.290465	1.290465
1982	7	1.309096	1.744106	1.24157	1.408258	1.408258
1982	8	1.653554	1.478083	1.277795	1.449003	1.449003
1982	9	1.655615	0.780674	1.14711	1.487186	1.487186
1982	10	0.630193	0.538706	1.159462	1.457049	1.457049
1982	11	−0.99893	0.854935	1.368335	1.683211	1.683211
1982	12	1.265277	1.376122	1.725668	1.862121	1.862121
1983	1	1.422863	1.573171	1.811804	2.179766	2.179766
1983	2	1.284773	1.666071	2.002804	2.071936	2.071936
1983	3	1.624122	1.960774	2.087471	2.121697	2.121697
1983	4	1.563828	1.891195	2.54115	1.992428	1.992428
1983	5	1.979567	2.227993	2.261711	1.913632	1.913632
1983	6	1.284543	2.009802	2.112821	1.737458	1.737458
1983	7	2.483325	2.890083	1.924397	1.691624	1.691624
1983	8	1.462499	2.024989	1.678458	1.425791	1.425791
1983	9	2.928786	1.915474	1.510988	1.303393	1.303393
1983	10	0.410247	0.96837	1.040594	0.872404	0.872404
1983	11	1.483841	1.159687	1.029239	0.895362	0.895362

		1st Month	3rd Month	6th Month	9th Month	12th Month
1983	12	0.736113	0.822518	0.835862	0.686802	0.686802
1984	1	1.075968	1.10423	0.793598	0.784731	0.784731
1984	2	0.767233	0.788247	0.681208	0.624779	0.624779
1984	3	1.198852	0.69513	0.55671	0.782157	0.782157
1984	4	0.199987	0.411966	0.589767	0.818995	0.818995
1984	5	0.555733	0.57769	0.543895	1.104732	1.104732
1984	6	0.445684	0.420569	0.818769	1.198071	1.198071
1984	7	0.609078	0.752746	1.017251	1.343826	1.343826
1984	8	0.18497	0.498982	1.338351	1.561872	1.561872
1984	9	1.420836	1.215887	1.479207	1.78908	1.78908
1984	10	−0.32983	1.178353	1.512609	1.845744	1.845744
1984	11	1.647635	1.78332	1.84957	2.007252	2.007252
1984	12	1.366107	1.595253	1.966817	2.051393	2.051393
1985	1	1.889851	1.720146	2.140749	2.114076	2.114076
1985	2	1.230154	1.846094	2.078493	2.082999	2.082999
1985	3	1.603446	2.281221	2.282156	2.161749	2.161749
1985	4	2.157654	2.478165	2.295454	2.134908	2.134908
1985	5	2.227189	2.147296	2.151536	1.930481	1.930481
1985	6	1.746154	2.009653	1.995978	1.672929	1.672929
1985	7	1.672969	1.91094	1.884825	1.647763	1.647763
1985	8	1.907583	1.911844	1.734362	1.495623	1.495623
1985	9	1.436489	1.750044	1.413776	1.26827	1.26827
1985	10	1.677729	1.729882	1.456886	1.156288	1.156288
1985	11	1.59519	1.418303	1.216153	0.957525	0.957525

		1st Month	3rd Month	6th Month	9th Month	12th Month
1985	12	1.474779	0.866344	0.923172	0.752803	0.752803
1986	1	0.665202	1.045478	0.686158	0.532852	0.532852
1986	2	0.058853	0.883231	0.61915	0.600075	0.600075
1986	3	1.634789	0.815782	0.62833	0.581931	0.581931
1986	4	0.273866	0.288914	0.276207	0.443448	0.443448
1986	5	0.203462	0.384482	0.459838	0.384025	0.384025
1986	6	0.436009	0.438207	0.44225	0.396751	0.396751
1986	7	0.476144	0.273109	0.536653	0.495494	0.495494
1986	8	0.37348	0.554681	0.369973	0.630643	0.630643
1986	9	−0.08359	0.44274	0.271231	0.78048	0.78048
1986	10	1.269993	0.78795	0.566523	0.988536	0.988536
1986	11	−0.27601	−0.07906	0.639765	1.017134	1.017134
1986	12	0.931329	0.110711	0.930365	1.012984	1.012984
1987	1	−1.1582	0.37604	1.054067	1.066775	1.066775
1987	2	0.243331	1.029989	1.318525	1.310292	1.310292
1987	3	1.108734	1.34457	1.274678	1.253051	1.253051
1987	4	1.275182	1.513974	1.325745	1.180928	1.180928
1987	5	1.278963	1.535977	1.437242	1.106803	1.106803
1987	6	1.438262	1.117865	1.162852	0.904984	0.904984
1987	7	1.41371	1.059557	0.969289	0.931306	0.931306
1987	8	0.248354	1.214695	0.800562	0.753939	0.753939
1987	9	1.274503	1.11794	0.657178	0.893319	0.893319
1987	10	1.668879	0.758746	0.760708	0.868793	0.868793
1987	11	−1.01691	−0.03068	0.417828	0.741352	0.741352

		1st Month	3rd Month	6th Month	9th Month	12th Month
1987	12	0.469402	0.018069	0.772342	0.965531	0.965531
1988	1	0.546799	0.791422	0.894084	0.964036	0.964036
1988	2	−0.87707	0.658524	0.928396	0.910096	0.910096
1988	3	1.532683	1.128779	1.233627	0.907125	0.907125
1988	4	0.271467	0.894183	1.00519	0.819019	0.819019
1988	5	1.135216	1.137672	1.024643	0.812869	0.812869
1988	6	1.019661	1.262396	0.75882	0.605304	0.605304
1988	7	0.940695	1.063568	0.758954	0.5517	0.5517
1988	8	1.471939	0.837227	0.563634	0.601631	0.601631
1988	9	0.399567	−0.02892	−0.05009	0.370741	0.370741
1988	10	0.280644	0.222768	0.054875	0.417881	0.417881
1988	11	−1.11493	−0.03831	0.419288	0.525964	0.525964
1988	12	1.042136	−7E−06	0.562326	0.642486	0.642486
1989	1	−0.46998	−0.02241	0.48635	0.588481	0.588481
1989	2	−1.0236	0.665608	0.642238	0.826764	0.826764
1989	3	0.771596	0.794469	0.786295	0.852957	0.852957
1989	4	1.263391	0.769193	0.772552	0.959483	0.959483
1989	5	0.246419	0.614573	0.898665	0.808686	0.808686
1989	6	0.630805	0.751198	0.869894	0.825057	0.825057
1989	7	0.850246	0.747824	1.031049	0.689972	0.689972
1989	8	0.626593	1.161238	0.900079	0.594093	0.594093
1989	9	0.635996	0.956113	0.77404	0.547594	0.547594
1989	10	1.695993	1.206802	0.557223	0.563288	0.563288
1989	11	−0.60571	0.337859	0.173681	0.402403	0.402403

		1st Month	3rd Month	6th Month	9th Month	12th Month
1989	12	1.369569	0.536318	0.351942	0.635975	0.635975
1990	1	−0.47902	−0.32368	0.184808	0.471219	0.471219
1990	2	0.159157	−0.01681	0.358041	0.489321	0.489321
1990	3	−0.31825	0.114735	0.602106	0.451744	0.451744
1990	4	0.268166	0.446872	0.698496	0.554604	0.554604
1990	5	0.391944	0.641322	0.70785	0.549945	0.549945
1990	6	0.660825	1.054355	0.646921	0.509504	0.509504
1990	7	0.746167	0.924544	0.612482	0.354678	0.354678
1990	8	1.460833	0.76137	0.463835	0.38184	0.38184
1990	9	0.223318	0.003913	−0.05847	0.129244	0.129244
1990	10	0.232323	0.034421	−0.22299	0.418824	0.418824
1990	11	−0.45585	−0.17846	0.099505	0.424387	0.424387
1990	12	0.497351	−0.05263	0.219267	0.516223	0.516223
1991	1	−0.29538	−0.38303	0.557541	0.500177	0.500177
1991	2	−0.09945	0.250745	0.549646	0.454492	0.454492
1991	3	−0.35094	0.287635	0.630494	0.421108	0.421108
1991	4	0.967758	1.091651	0.754792	0.464667	0.464667
1991	5	0.189128	0.781596	0.546488	0.381109	0.381109
1991	6	1.865946	0.942876	0.501182	0.334439	0.334439
1991	7	0.22434	0.400673	0.08352	0.154466	0.154466
1991	8	0.626857	0.260567	0.052002	0.187221	0.187221
1991	9	0.362371	−0.22233	−0.32365	0.221461	0.221461
1991	10	−0.42466	−0.60344	−0.14355	0.302632	0.302632
1991	11	−0.72713	−0.40025	0.125835	0.574404	0.574404

		1st Month	3rd Month	6th Month	9th Month	12th Month
1991	12	−0.19755	−0.33166	0.434518	0.631096	0.631096
1992	1	0.183886	0.340881	0.606276	0.723858	0.723858
1992	2	−0.54975	0.432882	0.807783	0.8734	0.8734
1992	3	0.912125	0.767782	0.861775	0.866671	0.866671
1992	4	0.463471	0.73878	0.841739	0.727145	0.727145
1992	5	0.810462	1.102783	1.073985	0.73689	0.73689
1992	6	0.779158	0.918397	0.905604	0.717457	0.717457
1992	7	1.375946	0.908073	0.706866	0.60261	0.60261
1992	8	0.374738	0.973803	0.455054	0.377084	0.377084
1992	9	0.756502	0.83252	0.466896	0.304888	0.304888
1992	10	1.509451	0.322629	0.281659	0.290274	0.290274
1992	11	−0.83063	−0.59932	−0.07201	0.142803	0.142803
1992	12	−0.73687	0.017253	0.063584	0.240122	0.240122
1993	1	0.2216	0.318471	0.270785	0.281641	0.281641
1993	2	0.742026	0.253669	0.302986	0.275708	0.275708
1993	3	0.066813	0.018307	0.249592	0.199884	0.199884
1993	4	0.127797	0.205445	0.246679	0.228978	0.228978
1993	5	0.025015	0.349933	0.291156	0.388966	0.388966
1993	6	0.559653	0.47908	0.317343	0.412357	0.412357
1993	7	0.462938	0.299064	0.253442	0.317754	0.317754
1993	8	0.384181	0.238299	0.404902	0.204109	0.204109
1993	9	0.011269	0.060339	0.264731	0.122778	0.122778
1993	10	0.404798	0.093421	0.247073	0.163749	0.163749
1993	11	−0.12244	0.46051	0.168673	0.251087	0.251087

		1st Month	3rd Month	6th Month	9th Month	12th Month
1993	12	0.256599	0.49556	0.186463	0.320895	0.320895
1994	1	1.272674	0.463712	0.189592	0.33744	0.33744
1994	2	–0.01657	–0.13368	0.118812	0.160142	0.160142
1994	3	0.020121	–0.10427	0.202071	0.099721	0.099721
1994	4	–0.14095	–0.01482	0.268882	0.12523	0.12523
1994	5	–0.00327	0.309432	0.285291	0.118398	0.118398
1994	6	0.231007	0.505201	0.234462	0.052523	0.052523
1994	7	0.673557	0.56964	0.230813	–0.029291	–0.029291
1994	8	0.52909	0.275902	–0.0491	–0.075845	–0.075845
1994	9	0.431641	–0.21806	–0.48175	–0.056858	–0.056858
1994	10	–0.25879	–0.47332	–0.68367	0.043612	0.043612
1994	11	–1.1426	–0.69495	–0.33517	0.144615	0.144615
1994	12	0.195781	–0.67011	0.051638	0.252039	0.252039
1995	1	–0.94135	–0.7836	0.224551	0.26324	0.26324
1995	2	–1.12086	–0.08196	0.329924	0.270803	0.270803
1995	3	–0.21693	0.35454	0.441051	0.27192	0.27192
1995	4	0.583616	0.743004	0.535006	0.249016	0.249016
1995	5	0.616948	0.638554	0.421696	0.138745	0.138745
1995	6	0.886999	0.524105	0.232853	–0.046299	–0.046299
1995	7	0.321475	0.324367	–0.05152	–0.214845	–0.214845
1995	8	0.357352	0.16636	–0.29291	–0.194389	–0.194389
1995	9	0.334791	–0.25714	–0.71814	–0.164796	–0.164796
1995	10	–0.23575	–0.8901	–0.82296	–0.042638	–0.042638
1995	11	–1.12212	–1.15238	–0.45941	0.220839	0.220839

		1st Month	3rd Month	6th Month	9th Month	12th Month
1995	12	−0.95368	−1.14605	−0.08298	0.344129	0.344129
1996	1	−0.96799	−0.64873	0.236644	0.409984	0.409984
1996	2	−1.12226	0.011438	0.538722	0.415919	0.415919
1996	3	0.019589	0.372187	0.668375	0.421871	0.421871
1996	4	0.590118	0.696758	0.700983	0.390298	0.390298
1996	5	0.488686	0.939413	0.591058	0.290415	0.290415
1996	6	0.901504	0.933784	0.453708	0.15172	0.15172
1996	7	1.185022	0.683215	0.205988	0.063878	0.063878
1996	8	0.506384	0.166146	−0.25349	−0.092969	−0.092969
1996	9	0.12582	−0.34718	−0.71945	−0.097569	−0.097569
1996	10	−0.22889	−0.79093	−0.60051	−0.004646	−0.004646
1996	11	−1.04029	−1.04274	−0.28015	0.136762	0.136762
1996	12	−0.71485	−1.02714	0.046849	0.279914	0.279914
1997	1	−0.93731	−0.31055	0.257569	0.355805	0.355805
1997	2	−1.01939	0.206702	0.406242	0.358883	0.358883
1997	3	0.508131	0.508494	0.558189	0.419461	0.419461
1997	4	0.5457	0.556447	0.537296	0.311088	0.311088
1997	5	0.479001	0.562709	0.428698	0.200288	0.200288
1997	6	0.577684	0.598349	0.371246	0.048998	0.048998
1997	7	0.532768	0.509709	0.165091	−0.06215	−0.06215
1997	8	0.605609	0.273252	−0.10903	−0.09462	−0.09462
1997	9	0.32521	0.005669	−0.58851	−0.077824	−0.077824
1997	10	−0.26349	−0.56367	−0.69208	0.048484	0.048484
1997	11	0.019596	−0.85283	−0.36642	0.200484	0.200484

		1st Month	3rd Month	6th Month	9th Month	12th Month
1997	12	−0.99754	−1.22109	−0.07213	0.285907	0.285907
1998	1	−1.18104	−0.70658	0.259199	0.369008	0.369008
1998	2	−1.11479	−0.03062	0.440416	0.430088	0.430088
1998	3	0.010594	0.419395	0.60778	0.431363	0.431363
1998	4	0.510118	0.761884	0.659589	0.397069	0.397069
1998	5	0.680762	0.795933	0.628483	0.302548	0.302548
1998	6	0.941026	0.775184	0.441079	0.147624	0.147624
1998	7	0.605849	0.541222	0.176961	−0.051516	−0.051516
1998	8	0.662973	0.417023	−0.10887	−0.124132	−0.124132
1998	9	0.251384	−0.10198	−0.55948	−0.14802	−0.14802
1998	10	0.295304	−0.58976	−0.70186	−0.074695	−0.074695
1998	11	−1.14306	−1.13374	−0.53506	0.00339	0.00339
1998	12	−0.78283	−0.99764	−0.12456	0.262974	0.262974
1999	1	−1.08267	−0.69991	0.104856	0.325455	0.325455
1999	2	−0.68847	−0.10855	0.260708	0.337626	0.337626
1999	3	−0.19989	0.248968	0.529579	0.374886	0.374886
1999	4	0.375779	0.510096	0.598553	0.38472	0.38472
1999	5	0.539337	0.536608	0.527601	0.330521	0.330521
1999	6	0.556193	0.78957	0.452613	0.205389	0.205389
1999	7	0.425702	0.672095	0.311207	0.123252	0.123252
1999	8	1.217829	0.513877	0.15551	0.128463	0.128463
1999	9	0.173863	−0.0941	−0.4479	0.028878	0.028878
1999	10	−0.24255	−0.42003	−0.46408	0.106579	0.106579
1999	11	−0.08858	−0.57817	−0.15796	0.248397	0.248397

		1st Month	3rd Month	6th Month	9th Month	12th Month
1999	12	−0.42968	−0.75653	0.120259	0.342608	0.342608
2000	1	−0.69903	−0.39287	0.291116	0.40456	0.40456
2000	2	−0.68228	0.111103	0.431582	0.374711	0.374711
2000	3	0.149689	0.495416	0.579517	0.355901	0.355901
2000	4	0.546012	0.65461	0.626172	0.33504	0.33504
2000	5	0.729402	0.677508	0.491152	0.226331	0.226331
2000	6	0.558479	0.650439	0.280816	0.0368	0.0368
2000	7	0.591913	0.582561	0.142988	−0.066222	−0.066222
2000	8	0.696283	0.267841	−0.15604	−0.158611	−0.158611
2000	9	0.362531	−0.34662	−0.67142	−0.157409	−0.157409
2000	10	−0.54467	−0.79007	−0.77607	−0.135362	−0.135362
2000	11	−1.13337	−0.97188	−0.47657	−0.02988	−0.02988
2000	12	−0.28879	−0.92078	−0.03637	0.08237	0.08237
2001	1	−1.17143	−0.65115	0.086059	0.133408	0.133408
2001	2	−0.98662	−0.12018	0.180343	0.175275	0.175275
2001	3	0.047491	0.341187	0.279863	0.175566	0.175566
2001	4	0.258712	0.456532	0.328351	0.111583	0.111583
2001	5	0.691058	0.405659	0.301371	0.030406	0.030406
2001	6	0.35532	0.238189	0.091683	−0.181428	−0.181428
2001	7	0.121021	0.209412	−0.09643	−0.203956	−0.203956
2001	8	0.284776	0.19151	−0.30185	−0.162648	−0.162648
2001	9	0.327086	−0.14366	−0.69717	−0.158107	−0.158107
2001	10	0.020578	−0.78745	−0.67843	−0.066137	−0.066137
2001	11	−0.99336	−1.22768	−0.42285	0.046607	0.046607

		1st Month	3rd Month	6th Month	9th Month	12th Month
2001	12	−1.1123	−1.25791	−0.12098	0.186371	0.186371
2002	1	−1.18175	−0.46206	0.17589	0.257299	0.257299
2002	2	−1.11091	0.10672	0.336971	0.271708	0.271708
2002	3	0.39559	0.36507	0.485637	0.275899	0.275899
2002	4	0.467156	0.50483	0.445347	0.171418	0.171418
2002	5	0.299932	0.515613	0.346555		
2002	6	0.723929	0.597543	0.232899		
2002	7	0.465349	0.385045	−0.02903		
2002	8	0.543315	0.154206			
2002	9	0.083784	−0.39254			
2002	10	−0.29474	−0.94795			
2002	11	−1.04252				
2002	12	−1.12871				

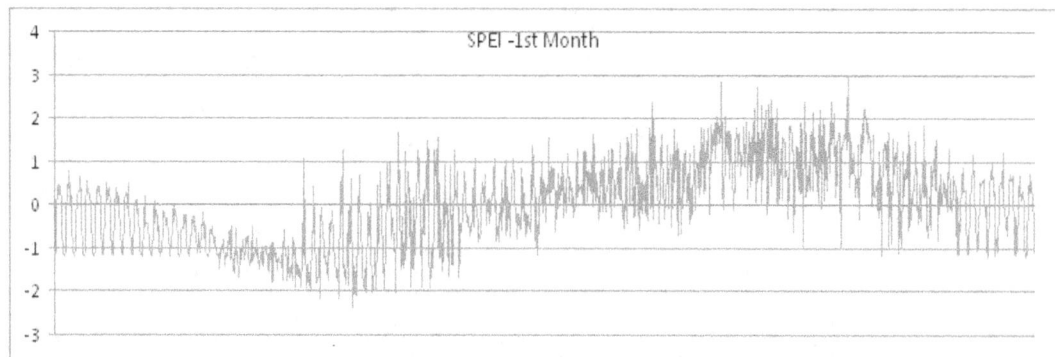

1901 to 2002

Description	Category	# of Times in 102 Years	Severity of Event
Extremely wet	2+	20	1 in 5 years
Very wet	1.5–1.99	74	1 in 1 years
Moderate wet	1.00–1.49	122	1 in 1 year

Near normal	(–)0.99 to 0.99	742	0
Moderate dry	(–)1 to (–)1.49	196	1 in 1 year
Serverly dry	(–)1.5 to (–) 1.99	52	1 in 2 years
Extremely dry	less than –2	8	1 in 13 years

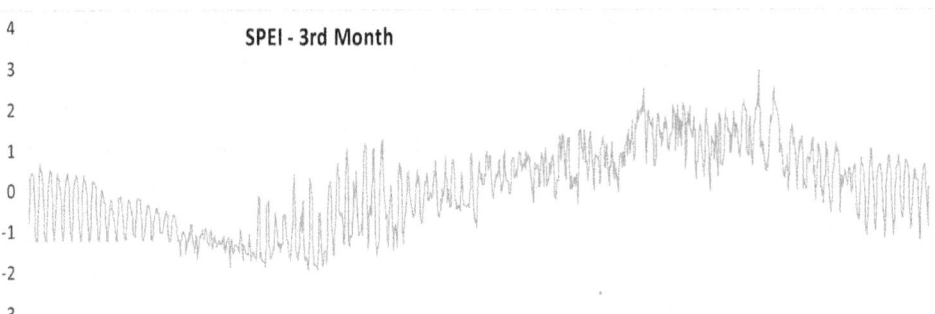

SPEI - 3rd Month

1901 to 2002

Description	Category	# of Times in 102 Years	Severity of Event
Extremely wet	2+	18	1 in 6 years
Very wet	1.5–1.99	76	1 in 1 years
Moderate wet	1.00–1.49	107	1 in 1 year
Near normal	(–)0.99 to 0.99	748	0
Moderate dry	(–)1 to (–)1.49	199	1 in 1 year
Serverly dry	(–)1.5 to (–) 1.99	71	1 in 1 years
Extremely dry	less than –2	0	

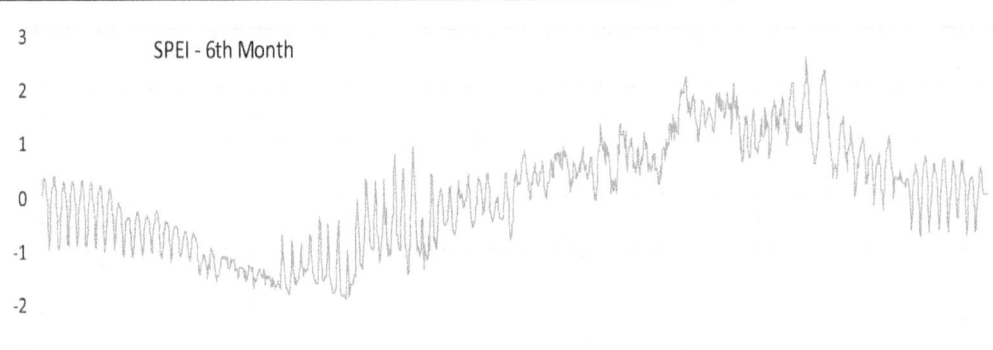

SPEI - 6th Month

1901–2002

Description	Category	# of times in 102 years	Severity of Event
Extremely wet	2+	18	1 in 6 years
Very wet	1.5–1.99	75	1 in 1 years
Moderate wet	1.00–1.49	115	1 in 1 year
Near normal	(–)0.99 to 0.99	754	0
Moderate dry	(–)1 to (–)1.49	158	1 in 1 year
Serverly dry	(–)1.5 to (–) 1.99	88	1 in 1 years
Extremely dry	less than –2	0	

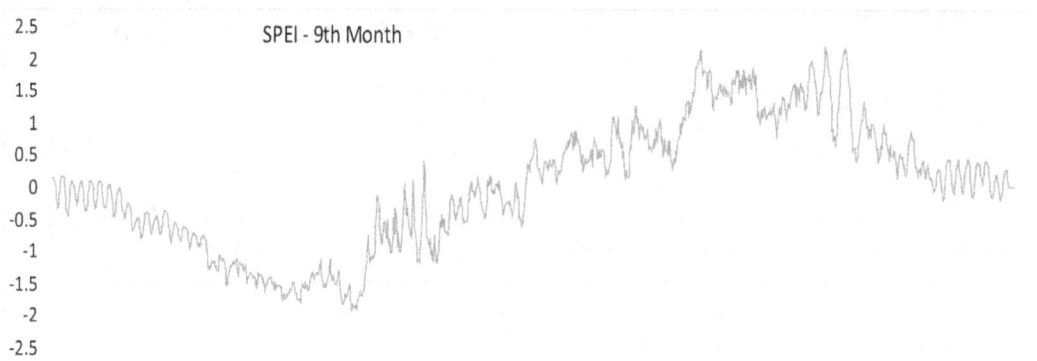

SPEI - 9th Month

1901–2002

Description	Category	# of times in 102 years	Severity of Event
Extremely wet	2+	12	1 in 9 years
Very wet	1.5–1.99	84	1 in 1 years
Moderate wet	1.00–1.49	120	1 in 1 year
Near normal	(–)0.99 to 0.99	758	0
Moderate dry	(–)1 to (–)1.49	0	1 in 1 year
Serverly dry	(–)1.5 to (–) 1.99	98	1 in 1 years
Extremely dry	less than –2	0	

SPEI – 12th Month

1901–2002

Description	Category	# of times in 102 years	Severity of Event
Extremely wet	2+	12	1 in 9 years
Very wet	1.5–1.99	84	1 in 1 years
Moderate wet	1.00–1.49	120	1 in 1 year
Near normal	(–)0.99 to 0.99	758	0
Moderate dry	(–)1 to (–)1.49	0	1 in 1 year
Serverly dry	(–)1.5 to (–) 1.99	98	1 in 1 years
Extremely dry	less than -2	0	

	Jan	feb	march	april	may	june	Jul	Aug	Sep	Oct	nov	dec
2000	-0.69903	-0.68228	0.14969	0.54601	0.72940	0.55848	0.59191	0.69628	0.36253	-0.54467	-1.13337	-0.28879
2001	-1.1714	-0.9866	0.0475	0.2587	0.6911	0.3553	0.121	0.2848	0.3271	0.02058	-0.9934	-1.1123

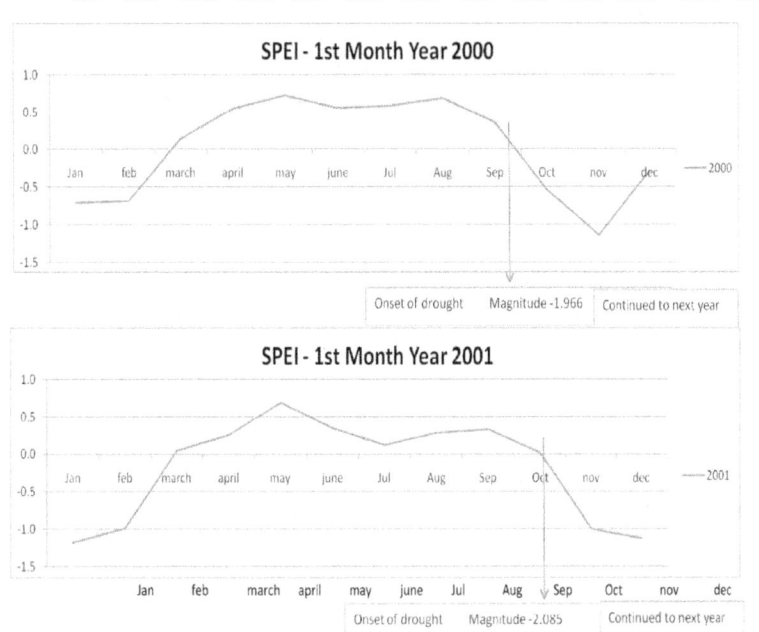

| 2000 | -0.3929 | 0.1111 | 0.4954 | 0.6546 | 0.6775 | 0.6504 | 0.5826 | 0.2678 | -0.3466 | -0.7901 | -0.9719 | -0.9208 |
| 2001 | -0.6512 | -0.1202 | 0.3412 | 0.4565 | 0.4057 | 0.2382 | 0.2094 | 0.1915 | -0.1437 | -0.7875 | -1.2277 | -1.2579 |

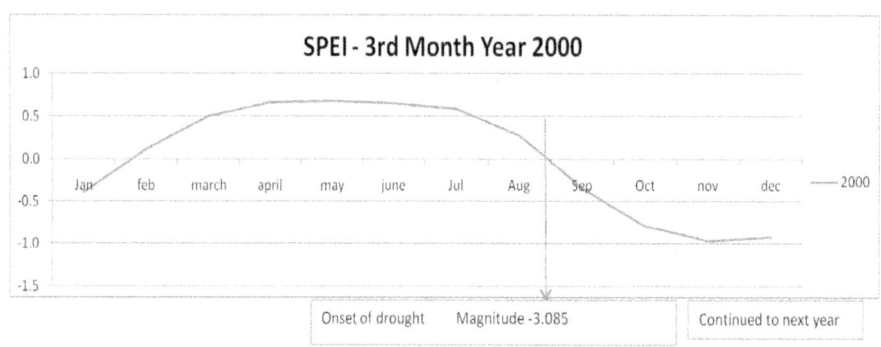

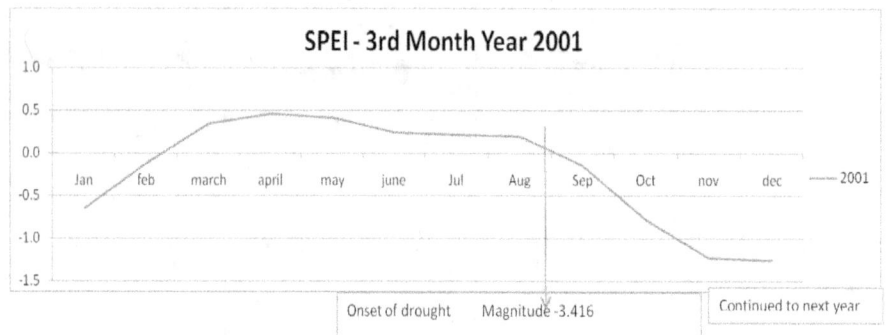

	Jan	feb	march	april	may	june	Jul	Aug	Sep	Oct	nov	dec
2000	0.29112	0.4316	0.5795	0.6262	0.4912	0.2808	0.143	-0.156	-0.6714	-0.7761	-0.4766	-0.03637
2001	0.08606	0.1803	0.2799	0.3284	0.3014	0.0917	-0.0964	-0.3019	-0.6972	-0.6784	-0.4229	-0.12098

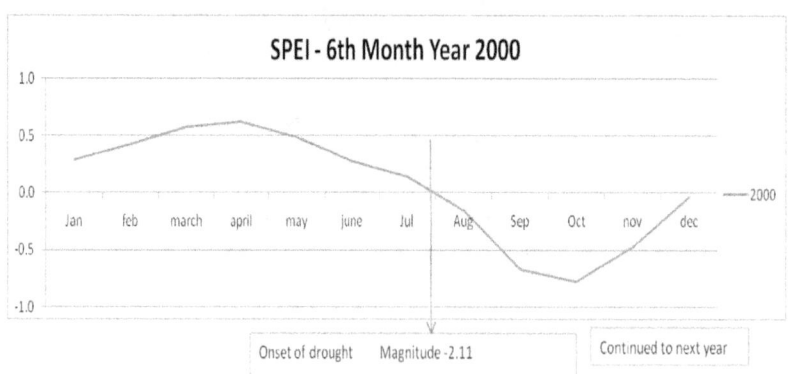

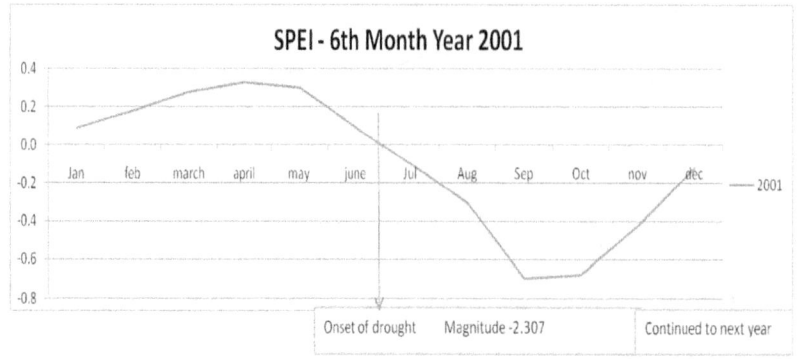

	Jan	feb	march	april	may	june	Jul	Aug	Sep	Oct	nov	dec
2000	0.40456	0.37471	0.3559	0.335	0.2263	0.0368	-0.0662	-0.1586	-0.1574	-0.1354	-0.0299	0.08237
2001	0.13341	0.17528	0.1756	0.1116	0.0304	-0.1814	-0.204	-0.1626	-0.1581	-0.0661	0.04661	0.18637

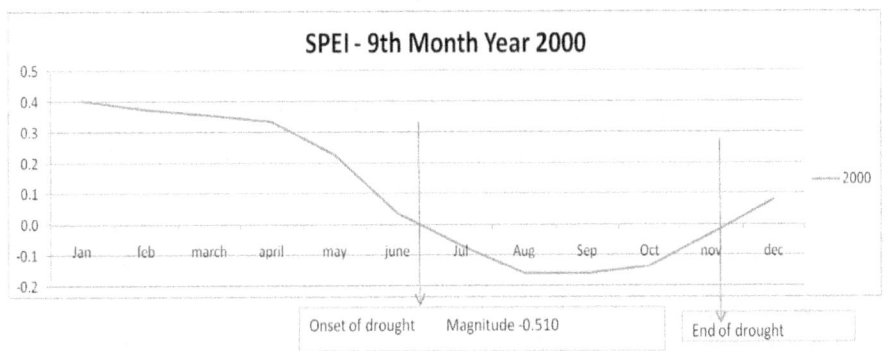

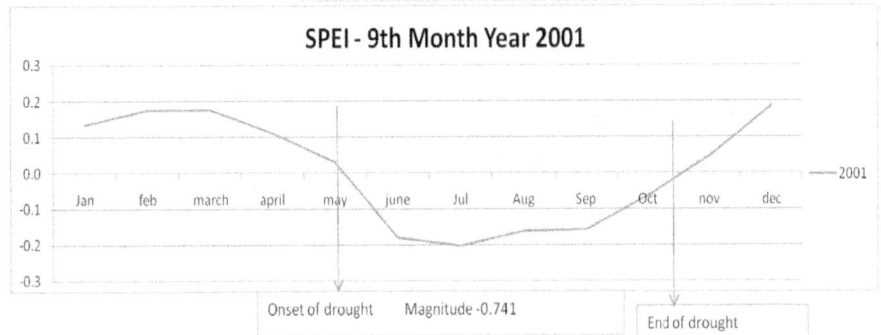

	Jan	feb	march	april	may	june	Jul	Aug	Sep	Oct	nov	dec
2000	0.40456	0.37471	0.3559	0.335	0.2263	0.0368	-0.0662	-0.1586	-0.1574	-0.1354	-0.0299	0.08237
2001	0.13341	0.17528	0.1756	0.1116	0.0304	-0.1814	-0.204	-0.1626	-0.1581	-0.0661	0.04661	0.18637

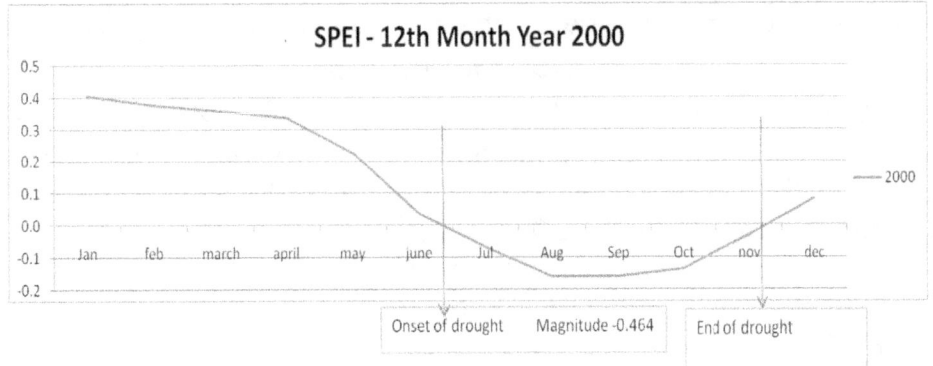

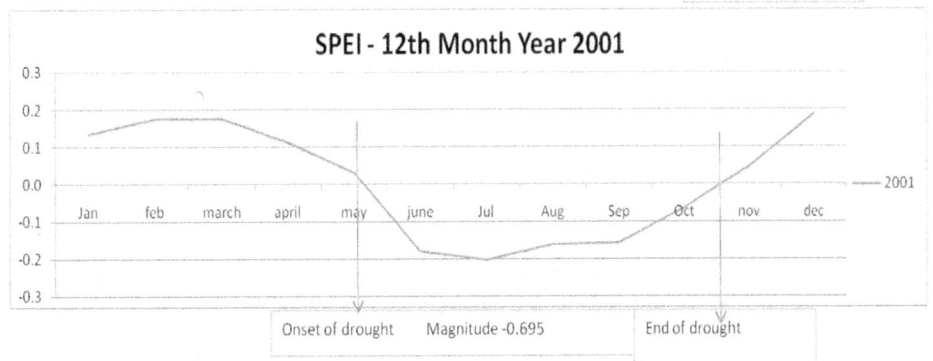

STANDARDIZED PRECIPITATION EVAPOTRASPIRATION INDEX (SPEI)

(spei_manual_en.pdf)

Description

The program calculates a time series of the spei at a given time interval from an input data file containing monthly time series of precipitation and mean temperature, plus the geographic coordinates of the observatory.

Usage

Spei (time interval) (input file) (output file)

Arguments

Time interval : A time interval, in months

Input file : The name of an input file with extension

Output file : The name of the output file to be produced with extension

Details

The spei index is a standardised monthly climatic balance computed as the difference between the cumulative precipitation and the potential evapotraspiration. Details on the index calculation and applications can be found in the reference below.

The spei can be calculated at the monthly scale with "time interval = 1 or accumulated at more than one month with "time interval" > 1. Typical values are 1, 3, 6, 12 & 24 months. If the accumulated index is calculated, the starting date of the resulting SPEI series will be lagged a number of months equal to "time interval"-1.

The input file (input file) can have any extension, but must be a plain text file (ASCII).

The file structure is a follows

Tampa

27.96

1900; 01

12

110.70; 14.30

...

The first line contains the name of the observatory and is only used for identification purpose. The second line is the latitude of the observatory in degrees.

The third line contains the year and month of the first record in the time series, separated by a semi-colon (;).

The fourth line contains the seasonality of the time series and must be set to 12.

Finally, from the fifth line the data series of monthly precipitation and mean temperature, separated by a semi-colon (;). The series must be continuous. Gaps and missing values are not allowed.

The output file (output file) can have any extension. It will be a plain text (ASCII) file with the following structure.

Tampa

27.959999

1900; 12

6

1.456516

…

The first three lines contain the name of the station, latitude and initial date of the spei series.

The fourth line contains a value indicating the cumulative parameter used, "time interval" (six months in the example). The spei time series is given from the fifth line on.

The program is run from the windows console. The easiest way is to locate the program and the input file(s) in the same directory. If you need to run the program from a different location, it might be necessary to modify the path system variable to include the path to the directory where the program was installed.

It is easy to create a batch script for automating the calculation of the spei over a large number of observatories or for several accumulated periods.

A hint on the usage of the program is obtained if the spei is involved with no arguments or with a wrong number of arguments.

See also

Spi program.

Examples

spi 1 tampa.txt tampa_spei_1.txt

spi 12 tampa.txt tampa_spei_12.txt

The above lines calculate the monthly SPEI and the 12 months cumulative SPEI time series for Tampa (Florida).

REFERENCES

Vicente – Serrano S.M. Lopez – Moreno J.I. Begueria S. "A multi-scalar drought index sensitive to global warming. The standardized precipitation evapotranspiration